Proceedings of the 33rd International MATADOR Conference

Springer
London
Berlin
Heidelberg
New York
Barcelona
Hong Kong
Milan
Paris
Singapore
Tokyo

David R. Hayhurst (Ed)

Proceedings of the 33rd International MATADOR Conference

Formerly The International Machine Tool Design and Research Conference

With 338 Figures

Springer

David R. Hayhurst, BSc, PhD, ScD, CEng, FIM, FIMechE
Department of Mechanical Engineering, University of Manchester Institute of Science and Technology, PO Box 88, Sackville Street, Manchester, M60 1QD, UK

ISBN-13: 978-1-4471-1200-6 e-ISBN-13: 978-1-4471-0777-4
DOI: 10.1007/ 978-1-4471-0777-4

British Library Cataloguing in Publication Data
Proceedings of the 33rd International MATADOR Conference
1.Manufacturing processes - Congresses 2.Production engineering - Congresses 3.Computer integrated manufacturing systems - Congresses 4.Manufacturing processes - Automation - Congresses 5.Machine-tools - Congresses
I.Hayhurst, D.R. II.International MATADOR Conference (33rd : 1999 : Manchester, England)
621.9'02
ISBN 1852333235

Library of Congress Cataloging-in-Publication Data
A catalog record for this book is available from the Library of Congress

Softcover reprint of the hardcover 1st edition 2000

Typesetting: Camera ready by contributors

34/3830-543210 Printed on acid-free paper SPIN 10763553

Proceedings Of The

Thirty-Third International

MATADOR

Conference

Associate Editors
S. Hinduja, J. Atkinson, M. Burdekin,
R. G. Hannam, L Li, A W Labib

Organising Committee
Professor D. R. Hayhurst *(Chairman)*
Professor S. Hinduja
Dr J. Atkinson
Dr M. Burdekin
Professor L Li
Dr R. G. Hannam
Dr A. W. Labib

Organising Secretary
Mrs C. Collins

UMIST

Manufacturing Division
Department of Mechanical Engineering
University of Manchester Institute of Science and Technology

Foreword

by

Conference Chairman

It is my pleasure to introduce this volume of Proceedings for the 33rd MATADOR Conference. The Proceedings include 83 refereed papers submitted from 19 countries on 4 continents.

The spread of papers in this volume reflects four developments since the 32nd MATADOR Conference in 1997: (i) the power of information technology to integrate the management and control of manufacturing systems; (ii) international manufacturing enterprises; (iii) the use of computers to integrate different aspects of manufacturing technology; and, (iv) new manufacturing technologies. New developments in the manufacturing systems area are globalisation and the use of the Web to achieve virtual enterprises. In manufacturing technology the potential of the following processes is being realised: rapid prototyping, laser processing, high-speed machining, and high-speed machine tool design. And, at the same time in the area of controls and automation, the flexibility and integration ability of open architecture computer controllers are creating a wide range of opportunities for novel solutions. Up-to-date research results in these and other areas are presented in this volume.

The Proceedings reflect the truly international nature of this Conference and the way in which original research results are both collected and disseminated. The volume does not, however, record the rich debate and extensive scientific discussion which took place during the Conference.

I trust that you will find this volume to be a permanent record of some of the research carried out in the last two years; and, to provide an excellent reference source for both established and young researchers in the field of Manufacturing Technology and Systems Management.

Professor David R Hayhurst
Manchester, 2000

Contents

MANUFACTURING SYSTEMS I

Agent-based Distributed Manufacturing System Modelling for Agile Manufacturing System Control

WT Goh[1] and Z Zhang[2]

[1]Dept. of Eng., Liverpool University, Liverpool L69 3GH, mgwt@liv.ac.uk
[2]Dept. of Eng., Liverpool University, Liverpool L69 3GH, zhengwen.zhang@liv.ac.uk

Abstract: As product life cycle is becoming much shorter and world market is becoming more customer driven, market changes become increasingly turbulent. In order for a manufacturing enterprise to survive, it must be agile to cope with turbulent market changes. As none of the existing manufacturing modelling tools are capable of supporting the rapid remodelling and reconfiguring of manufacturing system in response to market changes, an agent-based modelling approach termed as Autonomous Agent Network (AAN) is proposed. AAN is constructed by clusters of autonomous distributed agents that have the ability to respond to stimuli, make decisions, solve problems and act upon decisions made.

1. Introduction

Today, manufacturing enterprises are subjected to changing market environment rife with uncertainties. Market demand that changes unpredictably manifest itself not only locally, but also globally. In order for manufacturing enterprises to survive and to preeminence, they must move in tandem with current market strategy, which requires rapid response to market changes i.e. agility. The main aim of this research is to present a dynamic control strategy that enable manufacturing enterprise to attain rapid responsiveness in an agile manufacturing environment. This control strategy is focusing on manufacturing system modelling using a novel agent-based approach termed as Autonomous Agent Network (AAN). Three well-developed and widely used modelling tools (existing modelling tools) have been reviewed. They are Petri-Net (PN), Integrated Computer-aided Manufacturing Definition (IDEF) and Hierarchical Object Oriented Manufacturing System Analysis and Definition (HOOMA). As these modelling tools are found to be incapable of supporting the rapid remodelling and reconfiguring of manufacturing system in response to market changes (Section 2), AAN is proposed.

2. Literature Review on Existing Modelling Tools

2.1 Petri-net

PN is used to describe and represent information of any processing system (e.g. synchronous, asynchronous, etc.). PN models consist of places and transitions represented by circles and rectangle bars respectively. Arcs link places and transitions together. Places represent resources (e.g. machines, raw materials, etc.). Transitions represent instantaneous events (e.g. machines finish machining, etc.). PN has the ability to govern process flows by the use of tokens. Graphically, tokens are represented as black dots within places. Tokens indicate the availability of the resources. The firing of tokens in places, activates the places' corresponding transitions. During the firing process, tokens are removed from input places and added to the output places. The action of transferring tokens completes the given tasks. Unfortunately, PN suffers from the following pitfalls:

- It is difficult to include details. As PN is limited to event-based modelling, it losses the ability to model other relevant information apart from activities that happen within the modelled system. Relevant information are tool change data, availability, capacity, information exchanged between production model (PN) and other functional sections such as Production Scheduler.
- High level PN model, such as H-EPN (Hierarchical Time-Extended Petri-Nets model [1]) tends to mix events, manufacturing objects (e.g. buffers), supplier and customer within a single PN model. Such model is confusing.
- There is no clear distinction between information flow and process flow. The reason is that both the information and process flow are modelled using similar type of arrow. This will cause problems like confusing users, causing difficulties in information retrieval (which arrows to look at) as well as difficulties in distinguishing between modelled objects and modelled processes.
- The difficulty with model reuse. As PN is an event-based modelling tool, changes in the modelled system require the rebuild of the overall model.

2.2 Integrated computer-aided manufacturing definition (IDEF)

IDEF [2] consists of many modules. Two modules are discussed in this paper as they are directly related to manufacturing system modelling. Firstly IDEF0, which is the Function Modelling Method. It is developed to describe the functional relationship of any manufacturing environment by modelling corresponding decisions, actions and activities. A function block is the basic element of IDEF0. Each function block consists of input, output, control and mechanism. Input represents raw material, semi-finished part, etc. Mechanism represents the required resource during the input to output transformation process (e.g. operator, tool, etc.). Control only exerts influences on the transformation process, but it is not involved in the process itself. Output represents the outcome(s) of the transformation process. Unfortunately, IDEF0 suffers from the following pitfalls:

- It tends to be interpreted wrongly as a sequence of activities (process flow). In fact IDEF3 is developed for process flow modelling.
- It tends to become complex and not easily read when modelling complex system with large data. This is due to the graphically modelled arrows (input, control, output and mechanism arrows) that exist within the same layer.
- It excludes raw materials, transportation devices (i.e. robots, etc.) and other possible "mechanisms" (i.e. operators, etc.) that exist in a manufacturing system. As each mechanism has its own corresponding events (i.e. operators loading machines), neglecting these mechanisms means neglecting some events that are happening within the manufacturing systems.
- There is no clear distinction between information flow and process flow.
- The difficulty with model reuse since it is an event-based modelling tool.

Secondly IDEF3 which is the Process Description Capture Method. It has the ability to collect and document processes by capturing the precedence and causality relations between situations and events. IDEF3 consists of two description modes: (1) Process Flow Description, which describes processes and the relationships between processes and (2) Object State Transition Network (OSTN), which describes the transitions an object may go through during a specified process. Unfortunately, IDEF3 suffers from the following pitfalls:

- There is no clear distinction between information flow and process flow.
- The difficulty with model reuse since it is an event-based modelling tool.

2.3 Hierarchical object oriented manufacturing system analysis and definition (HOOMA)

HOOMA is an object-oriented (O-O) system analysis method that utilises the features of O-O analysis. It is developed to support the requirements of system analysis and definition within manufacturing context [3]. HOOMA has the ability to represent manufacturing system in both the dynamic and the static structure. Dynamic structure represents the state of manufacturing processes (i.e. whether a particular process is still in processing, etc.) and the interaction among processes (i.e. the flow of processes). Static structure represents the information (i.e. quality, price, etc.) of the modelled objects. Manufacturing processes are regarded as objects in HOOMA. The functionality of these objects is attained by defining the set of messages to which these objects respond. Unfortunately, HOOMA still suffers from a number of pitfalls that other modelling tools share (section 2.4).

2.4 Overall pitfalls of the existing modelling tools

- These modelling tools generate "passive" rather than "active" models. Passive models are models that do not have the ability to respond to stimuli. Decisions are accomplished by a centralised problem solver. Active models are models that have the ability to respond to stimuli and make decisions.

- These modelling tools model merely the given system configuration(s) rather than incorporating the generation of alternative system configurations within manufacturing system modelling.

3. Autonomous Agent Network (AAN)

Autonomous Agent Network (AAN) is an agent-based model constructed by clusters of autonomous distributed "active entites" termed as agents. One definition of agent is: "... a piece of software code that acts upon receiving data from sources and reasons using its residing intelligence, which comes from its internal knowledge" [4]. All manufacturing resources (i.e. machines, robots, etc.) within a given manufacturing system are represented as agents. Each agent consists of separate classes of objects implemented in an object-oriented manner, which enhance model reuse. AAN is capable of modelling and representing all detail information of the modelled manufacturing system by having each agent encapsulating its corresponding data. These data are classified as either static or dynamic data. Static data are data that remain unchanged with respect to time (e.g. maximum capacity of a machine, etc.) and are modelled in Static Layer. Dynamic Data are data that will change with respect to time (e.g. operations performed by a machine, etc.) and are modelled in Dynamic Layers (one system configuration per Dynamic Layer). Figure 1 depicts both the Static Layer and a Dynamic Layer. A clear distinction between process and information flows is achieved by modelling processes in solid arrows and information in dotted arrows.

(a)

(b)

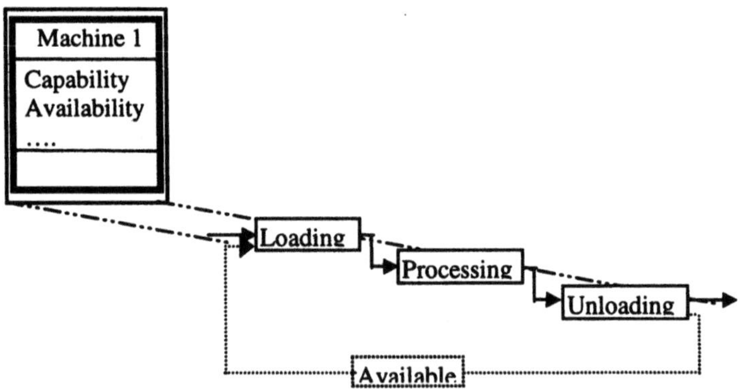

Figure 1: (a) Static layer with static data, (b) Dynamic layer with dynamic data

All agents residing within AAN have the ability to respond to stimuli, assess received stimuli, make decisions, solve problems, act upon decisions made and communicate with one another. With these abilities a distributed problem-solving environment with multiple problem solvers (i.e. agents) rather than a centralised problem solver (practice of traditional manufacturing system control approach) is created. These distributed problems solvers are capable of breaking complex operational tasks into small and specific tasks to be managed by individual problem solvers. Figure 2 depicts the general architecture of AAN agents. ***Event Monitor*** is concerned with event surveillance. Agents detect the occurrence of events via Event Monitor. ***Event Database*** consists of all responded events by agents. Within AAN, events are considered as problems. ***Inference Engine*** is concerned with problem examination and decision making. Agents evaluate the arose problems using their information residing within Knowledge-base and Database. ***Knowledge-base*** consists of sets of rules governing the production operations (e.g. precedence rules, etc.). ***Database*** consists of information regarding agents (e.g. processing speed, capability, etc.). ***Action Edition*** is concerned with action generation. Actions are generated based upon decisions made. ***Action Database*** consists of all possible actions (functions) to be carried out by agents in response to certain problems.

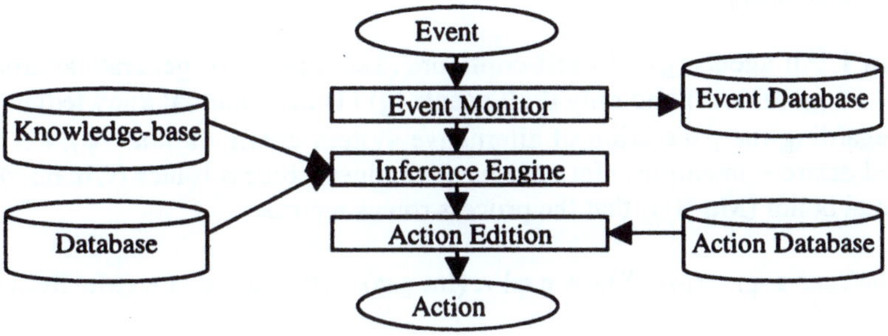

Figure 2: General architecture of AAN agents.

Furthermore, agents residing within AAN are also viewed as Belief-Desire-Intention (BDI) systems (one of the most successful agent-based architecture). Beliefs represent the information that an agent possessed regarding the world. Desires represent the allocated tasks upon an agent. Intentions represent the committed desires. Researchers in Artificial Intelligent use Temporal Logic to represent knowledge and belief of an agent. KL_n and BL_n are the Temporal Logic of knowledge and belief respectively. A formula $[j]\varphi$ means agent j knows (belief) φ. If an agent knows φ, then φ must be true. In contrast, even though agent belief φ, φ could be false. Detail discussion on KL_n and BL_n can be found in [5].

Once market changes (an event/problem), a new and common goal (i.e. meet market requirements rapidly) as well as desire of all agents (i.e. generate an optimal system configuration rapidly) will arise. In order to achieve the goal, agents will evaluate the newly changed market requirements and decide their roles in the production of different product ranges (i.e. whether to take part in the production of one product, all products, etc.) based on their capability (resided within database). Pursuant to the process routes obtained from Production Scheduler, agents will self-configure to generate alternative system configurations via negotiation. AAN has considered the generation of alternative system configurations in order to evaluate all possible ways and propose an optimal way of configuring the given manufacturing system. All generated system configurations will be represented in Dynamic Layers for simulation purpose. Dynamic simulation and bidding (bid as a group of system configuration) will be executed concurrently via continuous supply of up-dated production schedules (i.e. processing time, idle time, etc.) from

Production Scheduler to select an optimal system configuration. The selected optimal system configuration will be used to remodel, reprogram and reconfigure the given manufacturing system. Figure 3 depicts all committed desire/tasks (i.e. intentions) of agents in order to achieve their common goal.

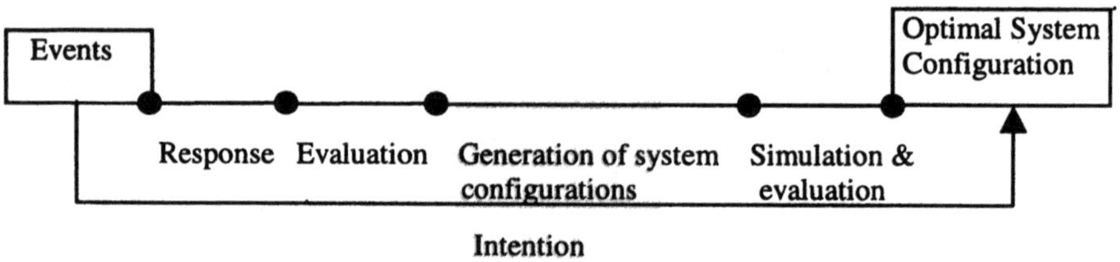

Figure 3: Agents' intentions.

Definition 1. All agents (g) will self-configure (Self-Conf) and generate alternative systems configurations if and only if all agents (g) (1) have mutual knowledge (Mu-Know) regarding the generation of alternative system configurations (φ); (2) have committed desire – intentions (Int); (3) have obtained process routes (Obtain *r*) and have mutual belief (Mu-Bel) that the process routes are true.

$$(\text{Self-Conf } g\ \varphi) = (\text{Mu-Know } g\ \varphi) \wedge (\text{Int } g\ \varphi) \wedge ((\text{Obtain } g\ r) \wedge (\text{Mu-Bel } g\ r))$$

4. Conclusion

This paper has presented a new modelling methodology for manufacturing systems. This methodology has the ability of helping any manufacturing organisation to respond rapidly to market changes through rapid remodelling and reconfiguring of existing manufacturing systems using a novel agent-based modelling approach. This approach does not convey the pitfalls found in existing modelling tools, rather it excels those tools by possessing the ability mentioned previously. Besides, it is also an easily used and understood modelling approach.

References

[1] Ramaswamy S., Valavanis, K.P. and Barber, S., 1997, Petri Net Extensions for the Development of MINO Net Models of Automated Manufacturing Systems. *Journal of Manufacturing Systems*, Vol. 16(3), 175-191.

[2] Wu, B., 1995, *Manufacturing System Design and Analysis*, Chapman & Hall London.

[3] Wu, B., 1995, Object-oriented System Analysis and Definition of Manufacturing Operations. *International Journal of Production Research*, Vol. 33(4), 955-974.

[4] Moore M.L., Reyns R.B., Kumara S.R., et al., 1997, Distributed Intelligent Agents for Logistics (DIAL). *IEEE International Conference on Systems, Man, and Cybernetics*, **3**, 2782-2787.

[5] Fisher, M., Wooldridge, M. and Dixon, C., 1996, A Resolution-Based Proof Method for Temporal Logics of Knowledge and Belief. *Proceedings of the International Conference on Formal and Applied Practical Reasoning*, Springer-Verlag London.

Integrated Manufacturing Systems Control Using a Multi-Agent System

M K Lim[1] and Z Zhang[2]

[1] Dept of Eng., Liverpool University, Liverpool L69 3GH, mklim@liv.ac.uk
[2] Dept of Eng., Liverpool University, Liverpool L69 3GH, zhengwen.zhang@liv.ac.uk

Abstract: With conventional manufacturing control strategies, enterprises face difficulties in dealing with dynamic customers requirements in a global and dynamic market. In order to preeminence and vie with other competitors, the agility of enterprises in responding to rapid changes in market needs should be the ultimate competitive strategy. This paper introduces a new agent-based manufacturing systems control framework to integrate dynamic process planning and scheduling to maximise the responsiveness of manufacturing system towards changes, as a means to achieve agile manufacturing. The objectives of this framework are to cope with dynamic changes in customer requirements, to integrate process planning and production scheduling, to optimise the utilisation of manufacturing resources, and to assess alternative configurations of manufacturing systems.

1. Introduction

As customer requirements are changing dramatically with respect to time, manufacturing enterprises begin to realise that flexibility and agility of manufacturing systems for dynamic market adaptation, has now become the competitive strategy to vie with business competitors. Competitive pressures are moving manufacturers towards shorter product cycles, lower inventories, higher equipment utilisation, and shorter lead times [1]. *Agile manufacturing* is a new concept of manufacturing developed recently which provides a competitive edge for enterprises. In this concept, agile enterprises integrate design, engineering, and manufacturing with marketing and sales in such a way that products are customised to the exact needs of customers [2].

In manufacturing, conventional approach of separation between process planning and production scheduling has become an obstacle to achieving agile manufacturing (i.e. to obtain high responsiveness of manufacturing systems). This is because the optimum process plan does not take scheduling constraints such as machine capacity and availability into account and hence such plan becomes infeasible in real production. Many problems arise within the manufacturing systems due to this approach [3][4].

Real-time manufacturing control and scheduling is an increasingly popular research domain in a number of academic fields, including engineering, artificial intelligence, and multi-agent systems [5]. An agent is a real physical or a virtual entity which is capable of acting on itself and on the surrounding world populated by other agents [6]. A *Multi-Agent System* (MAS) is a *Distributed Artificial Intelligence* (DAI) system composed of a number of autonomous agents that collaborate closely to reach common goals. In our study, an integrated dynamic process planning and dynamic scheduling framework is investigated and presented in this paper. The concept of MAS constitutes the backbone of this study. All tasks related to process planning, scheduling and manufacturing resources optimisation are performed by autonomous agents which are capable of interacting and negotiating with each other to bid for jobs and of making decisions by themselves.

2. An Agent-Based Integrated Dynamic Process Planning and Dynamic Scheduling Framework

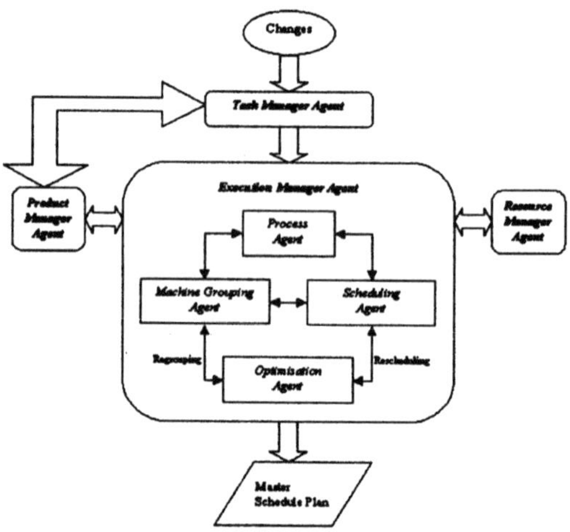

Figure 1: An agent-based integrated dynamic process planning and scheduling framework.

Figure 1 depicts the architecture of an agent-based integrated dynamic process planning and dynamic scheduling framework developed in this study. This figure also reveals the relationships between agents. This framework consists of four manager agents as shown in figure 1: *Task Manager Agent* (TMA), *Product Manager Agent* (PMA), *Resource Manager Agent* (RMA), and *Execution Manager Agent* (EMA). These four manager agents collaborate closely and share data/information with each other to achieve common objectives: 1) to integrate dynamic process planning and production scheduling; 2) to dynamically generate a *Master Schedule Plan* for all the products (including the introduction of new products or existing products with changing demand levels) to be manufactured in response to the changes; 3) to optimise the utilisation of manufacturing resources dynamically; and 4) to assess alternative configurations of manufacturing systems.

3. Descriptions of Agents

3.1 Product Manager Agent (PMA)

PMA consists of product agents which represent detailed information of physical products that are currently and previously manufactured in the shop floor. The information includes product identification, due dates, quantity, quality requirements, CAD drawing on dimensions, manufacturing processes involved, etc. PMA serves as a data agent to provide other agents with the products information. It monitors and updates all the product agents from time to time and creates new product agent if there is new product information brought into the framework.

3.2 Resource Manager Agent (RMA)

All the resources in the shop floor are represented by agents, for instance *Machine_Milling1 Agent*, *ConveyorBelt Agent*, *Tool Agent*, etc. Each resource agent stores updated and detailed information of its representative resource. For example, *Machine_Milling1 Agent* contains the information on machine type and identity, machine capability and reliability, current jobs, maintenance records, etc. of Machine Milling 1. All the machine agents are capable of bidding for jobs. The bidding process is briefly discussed in section 3.4.2. On the other hand, resource agents such as *ConveyorBelt Agent* and *Tool Agent* are responsible for providing other agents with updated information when requested.

3.3 Task Manager Agent (TMA)

When there are market changes (either changes in product mix or demand patterns), TMA will be informed and provided with detailed information of the changes. The detailed information includes geometric features, quantity, due date, CAD drawing, quality requirements, etc. in the case of changes in product mix. In the case of changes in demand patterns of an existing product mix, the information is on new demand levels. If there are changes in product mix (i.e. a new product introduction), TMA will send a request to PMA to create a new product agent to store new product information. However, if the changes affect only demand patterns (i.e. existing products with new demand levels), PMA will update the particular product agents with the new demand levels. Eventually, TMA will announce this product information to EMA in order to generate an updated Master Schedule Plan for all the products (including the new product or new demand levels of existing products) to be manufactured in the shop floor.

3.4 Execution Manager Agent (EMA)

EMA is responsible for initiating and monitoring its sub-agents producing an updated Master Schedule Plan for all the products to be manufactured whenever there are external changes (market changes) or internal changes (e.g. machine breakdowns). Internal changes will lead to rescheduling to take place which is discussed in section 3.4.3. As shown in figure 1, the elements constituting EMA are *Process Agent* (PA), *Machine Grouping Agent* (GA), *Scheduling Agent* (SA) and *Optimisation Agent* (OA).

3.4.1 Process Agent (PA)

When TMA announces to EMA that a new Master Schedule Plan needs to be generated due to changes, EMA will request PA to carry out its task to ensure the manufacturing processes for the products are available and working in good conditions in the shop floor. PA first acquires PMA to provide a list of processes recommended for each feature of the products. There is usually more than one process that is able to produce an identical feature. For instance, milling process and drilling process are capable of producing a hole. Based on the processes recommended, PA produces a list of possible process sequences.

3.4.2 Machine Grouping Agent (GA)

GA is responsible for generating alternative machine groupings based on the list of process sequences provided by PA. A machine may provide more than one process and there may be more than one machine providing the same process. Therefore, there will be a number of alternative machine groupings generated. GA announces the processes required for the process sequences to all the corresponding machine agents for bidding purposes. All the corresponding agents enter into a bidding process based on their own capability and availability. Each machine works out its possible shortest processing time for a particular job and reply to GA. Total processing time can be obtained as $\{\mathbf{T}_{tc} + \mathbf{T}_{setup} + \mathbf{T}_{mc}\}$, where $\mathbf{T}_{tc}$ is tool change time, $\mathbf{T}_{setup}$ is setup time, and $\mathbf{T}_{mc}$ is machining time.

3.4.3 Scheduling Agent (SA)

SA schedules all the alternative machine groupings into the existing Master Schedule Plan to generate alternative scheduling options. In each scheduling option, machines involved in a particular machine grouping are scheduled with respect to the current loading of the machines. From this scheduling option, bottlenecks incurred within each machine can be identified. The alternative machine groupings, scheduling options and bidding results (i.e. the shortest processing times) will then be sent to the OA to generate a new optimum Master Schedule Plan.

Master Schedule Plan is the plan that has to be carried out accordingly in the shop floor to manufacture all the required products. SA is responsible for monitoring this plan continuously and updating it whenever there is an optimum scheduling option being determined for a product. If there are any changes in the shop floor (e.g. machine breakdowns or any production problems such as tool breakage and urgent jobs introduction) that Master Schedule Plan could not be carried out accordingly or no optimum scheduling option could be chosen, Master Schedule Plan has to be altered. SA will send a request to EMA to trigger rescheduling process. In the rescheduling process, all the products to be manufactured will be given a priority ranking. The most important product will have the priority to be rescheduled in a new Master Schedule Plan. The four agents in EMA will be requested to work out the optimum machine grouping/scheduling option for each product, to be rescheduled in the plan.

3.4.4 Optimisation Agent (OA)

The function of OA is to evaluate the alternative machine groupings/scheduling options beyond the only criterion (i.e. total processing time) taken into account by bidding process. The evaluation/optimisation technique is based on total production cost, total production time, bottlenecks of each machine grouping and last but not least, the reliability of individual machine in each machine grouping.

The total production cost involved in every machine grouping is worked out as $\{\Sigma \mathbf{M}_i(\mathbf{C}_{setup} + \mathbf{C}_{mc}) + \mathbf{C}_{hand} + \mathbf{C}_{con}\}$, where $\mathbf{M}_i$ is machine i involved in a particular machine grouping, $\mathbf{C}_{setup}$ is setup cost, $\mathbf{C}_{mc}$ is machining cost, $\mathbf{C}_{hand}$ is handling cost, and $\mathbf{C}_{con}$ is configuration cost. The structure of the manufacturing system may have to be reconfigured if the new process sequence required differs from the existing. Therefore, there must be a configuration cost involved in the reconfiguration of manufacturing system structure with respect to the process sequence of existing products.

OA calculates the total production time taken in each machine grouping to manufacture a product as $\{\Sigma \mathbf{M}_i(\mathbf{T}_{setup} + \mathbf{T}_{tc} + \mathbf{T}_{mc} + \mathbf{T}_{hand})\}$, where $\mathbf{M}_i$ machine i involved in a particular machine grouping, $\mathbf{T}_{setup}$ is setup time, $\mathbf{T}_{tc}$ is tool change time, $\mathbf{T}_{mc}$ is machining time, and $\mathbf{T}_{hand}$ is handling time (flowing time) of a part from $\mathbf{M}_{i-1}$ to $\mathbf{M}_i$ including bottlenecks at $\mathbf{M}_i$.

OA is also responsible for identifying bottlenecks based upon data on the time availability of machines and handling times, machining times, setup times and tool change time of machines from machine agents. Alternative scheduling options generated by SA are also helpful in tracing bottlenecks in each machine grouping. Finally, the reliability of each machine grouping is determined by multiplying the reliability of each machine in the particular machine grouping.

Many researchers select an optimum solution for production problems with single criterion. However, the framework developed within this study foresees the importance of various criteria which should be taken into considerations. The first attempt of the optimisation technique is to discard those machine groupings/ scheduling options with tardiness. For example, there are six machine groupings (G1, G2, G3, G4, G5, and G6) generated by GA. They are all scheduled into Master Schedule Plan and resulted in six scheduling options produced. Assume that machine groupings G1 and G3 are unable to meet the delivery due date. These two machine groupings are discarded. The remaining machine groupings, G2, G4, G5 and G6 will be prioritised in terms of least production cost, least production time, most reliability and least bottlenecks. A ranking order number is assigned to each machine grouping that has been prioritised in each criterion as shown in Table 1. A ranking index is developed to sum up the ranking order numbers assigned to each machine grouping, i.e. $\mathbf{RI}(\mathbf{G}_i) = \mathbf{RO}_c + \mathbf{RO}_t + \mathbf{RO}_b + \mathbf{RO}_r$, where $\mathbf{RI}(\mathbf{G}_i)$ is ranking index of machine grouping i, and $\mathbf{RO}_c$, $\mathbf{RO}_t$, $\mathbf{RO}_b$, $\mathbf{RO}_r$ are ranking order numbers assigned to machine grouping i in cost criterion, time criterion, bottlenecks criterion, and reliability criterion respectively.

Table 1: Ranking order based on four criteria.

Ranking Order	Least Production Cost	Least Production Time	Least Bottlenecks	Most Reliability
1	G4	G6	G4	G4
2	G6	G4	G6	G2
3	G5	G2	G5	G6
4	G2	G5	G2	G5

The lowest ranking index will be selected as the optimum machine grouping with the best scheduling option, to be the new Master Schedule Plan because it contains the best combination of the four criteria mentioned earlier. In this example, G4 has the lowest ranking index, $\mathbf{RI}(\mathbf{G}_4) = 5$.

4. Conclusions

Conventional approach of separation between process planning and scheduling has caused many problems in production and in addition, due to dynamic changes of customers requirements in a global market, enterprises begin to realise that agility and responsiveness of enterprise towards dynamic market changes, is an competitive strategy to be emphasised. This paper introduces an integrated dynamic process planning and dynamic scheduling framework developed based on the concept of multi-agent systems, as a means to achieve agile manufacturing. This framework enables alternative machine groupings/scheduling options to be evaluated dynamically in response to market changes. It also facilitates dynamically optimised utilisation of manufacturing resources and contributes a platform where alternative reconfiguration options of manufacturing systems can be assessed.

References

[1] Parunak H. V., 1994, *Autonomous Agent Architectures: A Non Technical Introduction.* Industrial Technology Institute.

[2] Kidd P. T., 1996, *Agile Manufacturing: A Strategy for the 21st Century.* IEE Colloquium (Digest), n.179.

[3] Huang H., Zhang H., and Smith M., 1995, *A Progressive Approach for the Integration of Process Planning and Scheduling.* IIE Transactions, v.27, n.4, 456-464.

[4] Kempenaers J., Pinte J., Detand J., et al, 1996, *A Collaborative Process Planning and Scheduling System.* Advances in Engineering Software, v.25, n.1, 3-8.

[5] Ouelhadj D., Hanachi C., and Bouzouia B., 1998, *Multi-Agent System for Dynamic Scheduling and Control in Manufacturing Cells,* Proceedings of the 1998 IEEE International Conference on Robotics & Automation, v.3, 2128-2133.

[6] Ferber J., 1999, *Multi-Agent Systems: An Introduction to Distributed Artificial Intelligence.* Addison-Wesley.

Capturing the Process of Manufacturing Strategy Formulation

Neil Darlow[1] and Tim Baines[2]

Cranfield University, Manufacturing Systems Department, Building 30, School of Industrial and Manufacturing Science, Cranfield University, Cranfield, Bedfordshire, MK43 0AL, [1]n.r.darlow@cranfield.ac.uk
[2]t.s.baines@cranfield.ac.uk

Abstract: A 'process' is a sequence of actions that occur in the formation of a decision. This paper presents and illustrates a methodology for identifying the process by which a company has arrived at its present manufacturing strategy. The intention is to bring increased understanding of the process of manufacturing strategy, to make implicit actions explicit, and to enable companies to evaluate and learn from past actions.

1. Introduction

After three decades of academic endeavours, the concepts of manufacturing strategy have not been widely disseminated into industry, and are found confusing by many [1,2]. In particular, understanding of the process of formulating and implementing manufacturing strategy is weak [3,4]. This situation is concerning, since the literature suggests that a formally developed strategy can be beneficial [5,6,7]. This paper addresses such issues. From an investigation of the existence within companies of processes such as those recommended in the literature, we found evidence of a partial theory-practice gap. We argue that this gap is partly attributable to lack of understanding within companies of the steps of manufacturing strategy process that they already use. The intention of the process mapping approach developed in this paper is to enable companies to evaluate what steps are already used in manufacturing strategy formulation, so that the appropriate steps can be adopted from another process. In this way, the approach can assist in increasing industrial understanding of manufacturing strategy process concepts.

The paper first presents a summary of how the literature recommends manufacturing strategy should be formulated. The role of a process mapping approach for increasing understanding and adoption of formal processes is discussed. A description is given of how the approach was developed, and how it is used in practice.

2. Background

Manufacturing strategy is often viewed in terms of content and process. Content can be considered in terms of competitive edge criteria and structural and infrastructural decision areas, and there is general agreement on what constitutes these [8]. Following on from the work of Skinner [9], manufacturing strategy process has often been conceptualised in the literature as a sequence of steps that a company could follow in developing a strategy. Table 1 collates and summarises the views of several authors in the field. Authors such as Swink and Way [11] and Skinner [1] have voiced concern as to whether such processes as prescribed in the literature are actually used in practice.

Table 1: Steps in the formulation of manufacturing strategy [10]

Define corporate objectives	A top-down approach to manufacturing strategy requires manufacturing strategy to be derived from corporate strategy. Corporate objectives may include, for example, a proliferation of new products, low costs, or a high level of product customisation
Select product families	Products made at the plant may be grouped in terms of their competitive requirements, followed by the formulation of a manufacturing strategy for each group
Audit of the external conditions	This audit addresses market requirements and competition in order to determine what is required from the firm's products to meet each of the competitive criteria. This stage uses input from the marketing function of the company in order to view the competitive environment. Benchmarking may assist in understanding who the competitors are and what threat they represent
Audit of internal capabilities	The internal audit considers the manufacturing capability of the firm in the context of the existing manufacturing strategy. This stage assesses the state of the current manufacturing facilities, technology and infrastructure with respect to the intended manufacturing strategy
Gap analysis between actual and desired performance	A comparison is carried out between the internal and external audits in order to assess how the manufacturing capabilities of the firm need to change to meet the competitive criteria, whilst remaining congruent with the corporate objectives and business strategy of the firm
Prioritise key issues and define objectives of manufacturing	After gap analysis, the manufacturing task is defined. The shortcomings of the manufacturing system (relative to the competitive strategy) are translated into a set of tangible priorities and objectives
Choose manufacturing strategies	This step develops action plans for manufacturing strategies that could be employed. Evaluation of alternatives may be done subjectively, or by using computer models

An earlier part of the research reported in this paper [10] gained an insight into how strategies are formulated in practice. This enabled comparisons to be made between practice and theory, and provided a basis for establishing that within a sample of UK companies, wholesale adoption of full formalised approaches to manufacturing strategy formulation had not taken place. The finding that processes have been at least partially adopted suggests that subsequent work should assist companies in identifying those steps of the process that they do not carry out explicitly at present, rather than advocating that they adopt any particular manufacturing strategy formulation process in full. Our intention is that a 'process mapping' approach would be used to identify which steps in a formal planning process are currently followed by the company in formulating their manufacturing strategy. This exercise would make explicit the steps that the company follows, and would utilise knowledge already in the company. This would avoid heavy dependence on consultants or the wholesale adoption of a time-consuming formal planning process. Selective adoption of steps from a published process could then take place to complete the manufacturing strategy formulation process for the next strategy formulation activity.

3. Mapping the Process of Manufacturing Strategy

A 'process mapping' approach has been developed and evaluated by the authors. First, a specification for an approach to mapping manufacturing strategy formulation was developed. Second, existing methods of recording information were compared against the specification. These included manufacturing strategy content charting [12], project management tools such as Critical Path Analysis and Gantt charts, mind mapping, cognitive mapping, Rich Pictures, and Ishikawa diagrams. Based upon this comparison, elements of the various approaches were formed into the principles of a manufacturing strategy process mapping approach. The resulting process mapping approach (Figure 1) uses a horizontal axis to represent time. The vertical axis splits the map into portions for what: strategic events and decisions; how: the procedural steps for reaching the decision; why: the reasoning and analysis behind the decision; and who: the individuals or teams responsible for the strategy. The procedure for completing the map consists of three parts: data collection, process map construction, and analysis of the map.

The strategy process mapping starts from a major strategic event at the company. Data are gathered about the event to establish the inputs and outputs leading to that strategic event, by asking who, when, why, and how questions. Representative questions are:

1. Why? - What was the reasoning behind this decision? What was the trigger for this event? What event did this decision give rise to?
2. Who? - Who initiated this process? Who was involved in this project team? Who took responsibility for data collection? Whose idea was this?
3. When? - What was the timescale for this project? How long did this stage of the analysis take? How often do you review the manufacturing strategy document?
4. How? - What procedural steps comprise the process for formulating this manufacturing strategy?

From the responses to the inquiry, the process map is constructed. First, the key event triggering manufacturing strategy formulation or review is recorded in a box in the 'What?' row of the process map. The year in which the event occurred is recorded below the event on the 'When?' axis. As causes and effects are probed, they are recorded on the map in the appropriate rows and columns. The 'When?' axis assists in structuring the map into a logical chronological sequence. Upon completing the map to a level of detail that interviewee and interviewer consider adequate, the map contains a compact representation of the company's manufacturing strategy formulation process. The map can be analysed by row or by column. The 'How?' row will contain the way in which the company followed the steps of the manufacturing strategy formulation process, and the 'What?' row will summarise the content of the manufacturing strategy. The 'Why?' row will provide explanations for the decisions, and the 'Who?' row will show which company personnel managed the various steps of manufacturing strategy formulation.

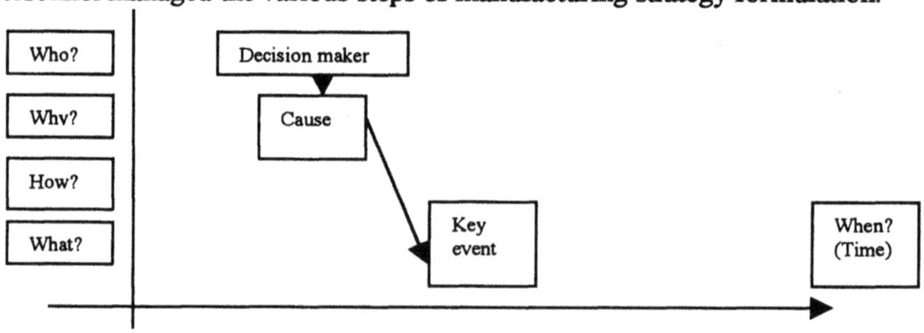

Figure 1: Simplified representation of the process mapping approach

4. An Illustration of the Process Mapping Approach

The process mapping approach was used in a sample of four manufacturing companies in the USA. Figure 2 shows a manufacturing strategy process map for one of the companies. From the map, the following information can be interpreted.

From the 'What?' row, the recent key event triggering the manufacturing strategy was the decision to expand the plant. From the 'Who?' row, it can be seen that audits were done by the heads of manufacturing and marketing (Cell 1) against targets set by the company president (Cell 9). Manufacturing engineers at the company headquarters planned the expansion (Cell 2). After the proposed investment had been financially justified, responsibility for the consequences of the expansion rested with the Plant Manager. The 'Who?' row demonstrates that the plant level decisions are controlled financially by the parent corporation, but the nature of the individual decisions is the responsibility of the management of the plant. The 'Why?' row shows a focus strategy over time, the continuation of which into the late 1990's was enabled by the plant expansion in 1993. However, the specific reason for the decision to expand the plant was the lack of necessary capacity (Cell 6). The 'How? row shows the manufacturing strategy formulation process. The company president, the general manager, and marketing managers formed a team to set goals for projected sales growth and profit. These goals, together with marketing analysis, manufacturing costs, and other financial

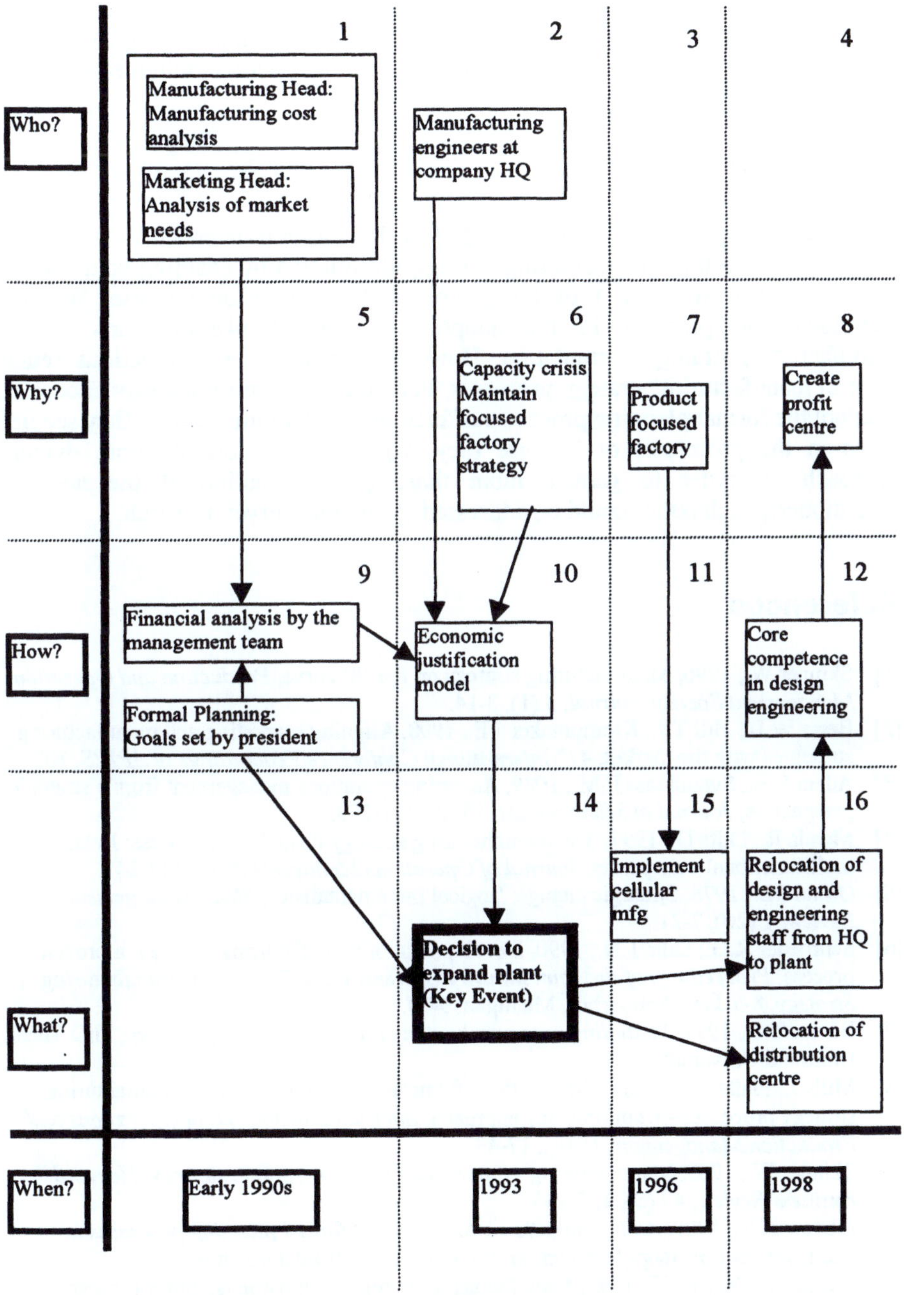

Figure 2: Manufacturing strategy process map for an American manufacturer

information, formed the inputs to an economic justification (Cell 10) to evaluate the expansion. Detailed system design was handled by the Manufacturing Engineering function, while the manufacturing head evaluated the capacity requirements.

5. Conclusions

The process mapping approach developed in this paper is intended to support the application of formal manufacturing strategy formulation by enabling companies to visualise which steps of a planning process they do or do not use. We have addressed one problem in the adoption of formal planning processes in manufacturing strategy formulation. However, many research directions remain open in manufacturing strategy processes. For example, companies may choose not to adopt the formal planning processes offered in the literature because they see little value in the processes, or because they require a less formal, more dynamic approach. In order to gain a more thorough appreciation of the needs of practitioners, such issues could be addressed by broader survey research.

References

[1] Skinner W., 1996, Manufacturing strategy on the 'S' curve, *Production and Operations Management Society Journal*, 5 (1), 3-14.
[2] Berry W.L., Hill T.J., Klompmaker J.E., 1998, Aligning marketing and manufacturing strategies with the market, *41st International Conference Proceedings of APICS*, 101.
[3] Adam E.E., Swamidass P.M., 1989, Assessing operations management from a strategic perspective, *Journal of Management*, 15 (2), 181-203.
[4] Menda R., Dilts D., 1997, The manufacturing strategy formulation process: linking multifunctional viewpoints, *Journal of Operations Management*, 15, 223-241.
[5] Quinn J.B., 1978, Strategic change: 'logical incrementalism', *Sloan Management Review*, 1 (20), 7-21.
[6] Schroeder R.G., Lahr T.N., 1990, Development of manufacturing strategy: a proven process, *Proceedings of the Joint Industry University Conference on Manufacturing Strategy*, 8-9 Jan., Ann Arbor, Michigan, 3-14.
[7] Baines T.S., 1994, *Modelling in the evaluation of a manufacturing strategy*, PhD Thesis, Cranfield University.
[8] Mills J., Platts K., Gregory M., 1995, A framework for the design of manufacturing strategy processes - a contingency approach, *International Journal of Operations and Production Management*, 15 (4), 17-49.
[9] Skinner W., 1969, Manufacturing - the missing link in corporate strategy, *Harvard Business Review*, 47 (3), 136-145.
[10] Darlow N.R., 1999, *An investigation into the role of formal planning processes in manufacturing strategy formulation*, PhD Thesis, Staffordshire University.
[11] Swink M., Way M.H., 1995, Manufacturing strategy: propositions, current research, renewed directions, *International Journal of Operations and Production Management*, 15 (7), 4-26.
[12] Mills J., Neely A., Platts K., et al, 1998, Manufacturing strategy: a pictorial representation, *International Journal of Operations and Production Management*, 18 (11), 1067-1085.

Mapping the Manufacturing Technology Acquisition Process

Tim Baines[1] and Neil Darlow[2]

Cranfield University, Manufacturing Systems Department, Building 30, School of Industrial and Manufacturing Science, Cranfield University, Cranfield, Bedfordshire, MK43 0AL,
[1]t.s.baines@cranfield.ac.uk
[2]n.r.darlow@cranfield.ac.uk

Abstract: Manufacturing technology acquisition may be thought of as a 'process' – a sequence of activities that lead to a decision. Every company has a technology acquisition process, though it is not necessarily structured and formalised. Before a technology acquisition process can be analysed and refined, it needs to be properly understood. This paper presents a method for identifying and mapping the processes that a company actually follows in forming its technology acquisition decisions.

1. Introduction

Some companies are heavily reliant on the capabilities of their manufacturing process technology for product competitiveness. Likewise, the capabilities of a manufacturing technology are dependent on how the technology was acquired by the host company. The challenge to companies that are reliant on manufacturing capabilities for product differentiation is to ensure that they always have access to the 'world's best' manufacturing technologies. The process of manufacturing technology acquisition is concerned with identifying new technologies and applying them within the company.

Manufacturing technology acquisition may be thought of as a 'process' – a sequence of activities that lead to a decision. Every company that has an operating function has a technology acquisition process, though it is not necessarily structured or formalised. Where formalised procedures do exist, in reality the practising engineer may not follow them. However, before a technology acquisition process can be analysed and refined, it needs to be properly understood. Therefore, this paper presents a method for identifying and mapping the processes that a company actually follows in forming its technology acquisition decisions.

The paper presents the development and application of this mapping approach. The first sections discuss the need to map technology acquisition processes. The following sections describe the development of the methodology, and illustrate the basic components of the mapping approach. A case study is then given, which illustrates the mapping approach in action. Finally, the paper draws some conclusions about the approach.

2. Exploring processes in action

In recent years, researchers have shown increased interest in the 'process' aspects of decision making in an operations management context (for example, [1]). Every organisation uses decision processes, although these are not necessarily highly structured or formalised [2]. The challenge is to tease out the decision processes that operations managers actually use. Understanding how organisations arrived at their present situation is fundamental to improvement [3].

A decision process can be defined as the mechanism through which policies are formed in an organisational setting [4]. Within an organisation, some decision making may be highly formalised and explicitly documented. In practice, this is not always the case [5]. While a formal planning framework may exist within an organisation, individuals may form decisions intuitively in a less obviously structured manner. However, in hindsight a logical structure may be apparent in the process. The challenge is to identify the whole process in action that formed a decision.

Since the decision process does not entirely comprise formal planning, it is not feasible to thoroughly research processes by consulting organisational planning documents. Some method of identifying the actual processes followed is needed. Mintzberg et al [2] suggested that most decision processes follow a basic structure, with various amounts of emphasis on the different steps of the process. However, an important challenge is to capture not just what the steps of a process are, but how they are operationalised within an organisation. As managers are more concerned with results than means, a method of capturing the processes by focusing on decisions made is required.

3. Research aim and method

In order to address the need for a method for capturing technology acquisition processes, the aim of the research reported in this paper has been to develop and evaluate a methodology for identifying the actual processes used to form decisions in practice.

The first phase of the research created the methodology by considering previous research that has attempted to capture decision processes, and augmenting this with a mapping approach.

Secondly, the methodology was evaluated in the context of manufacturing technology acquisition processes in a large company. This case study formed the basis for the evaluation of the methodology.

4. Creating a methodology for investigating technology acquisition processes

The first phase of the research was concerned with forming the methodology. To achieve this, our approach was to first review existing methods of collecting and representing data about informalised processes. Existing techniques that perform this task can be considered as either highly structured or loosely structured methods. This categorisation is related to the complexity of the graphical syntax used within the technique [6]. Highly structured methods include systems analysis techniques such as $IDEF_0$, Structured Analysis and Design Technique (SADT), Integrated Enterprise Modelling, associated with Business Process Reengineering (BPR), and Structured Systems Analysis and Design Methodology (SSADM). More loosely structured methods, characterised by a more relaxed graphical syntax, include Rich Pictures, part of Soft Systems Methodology, Inference Diagrams in Qualitative System Dynamics, and Mind Mapping [7]. In a Mind Map, from a central idea, branches representing themes are made to evolve, and individual trains of thought can be branched from the themes. On review, structured techniques provide a mechanistic approach to process mapping. This favours development of highly consistent maps that are easy to share among personnel who are familiar with the associated syntax. Their limitation is in data collection, as the real decision making processes tend not to be naturally expressed in such regimented forms. Figure 1 illustrates our expectations of decision process forms.

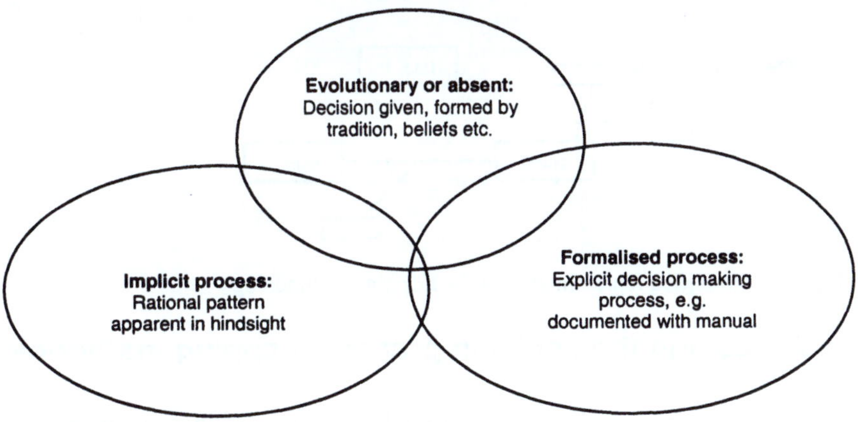

Figure 1: How decisions can be formed (adapted from [4])

The more loosely structured techniques appear to be more suited to capturing decision processes in action. However, paradoxically they lack sufficient structure to aid the practitioner attempting to map their technology acquisition processes. A resolution therefore, may be to tailor a semi-structured technique to this specific task. Mintzberg et al [2] investigated several 'unstructured' decision processes and, on the basis of empirical work, identified an underlying structure in the form of a flow diagram that illustrates the steps which may take place in the formation of decisions. Mintzberg et al's [2] decision process structure can form the backbone of the methodology by providing a sequence of possible steps in a decision process. These steps guide the data collection through semi-structured interviews with decision-makers in an organisation. Next, means of recording data were considered. A method was required that would allow technology acquisition decisions to be mapped out, using the Mintzberg et al [2] decision process steps as a starting point. The technique known as 'mind mapping' described in Section 2 was considered to be appropriate, since it starts from a central thought, which could be one of the Mintzberg et al [2] steps.

The resulting methodology works as follows. The data collection is done by means of a semi-structured interview. The interviewees are asked to describe how they went about their most recent technology acquisition decisions, and the processes are explored by identifying links between actions and marking them on the mind maps (Figure 2). Examining the maps then allows a view of the dominant processes within the organisation to be developed. After representing the process as a flow diagram, for example, it is shown to the company personnel for discussion. The goal is to share a common understanding of the process in action at the present time, as a basis for future refinement.

In summary, the methodology consists of four phases. First, the identification of a possible structure for the process being studied. Second, the selection of 'experts' in the activities that make up the process to be identified. Third, mapping the processes via semi-structured interviews and mind mapping methods. Finally, collating the findings into a structured process and discussing its credibility and implications with the process owners.

Figure 2: Simplified representation of the mapping methodology

5. Assessing the decision process mapping methodology

The organisation studied is a good example of a manufacturing company that needs to maintain a competitive edge through excellence in manufacturing technology. In order for its products to compete, innovative manufacturing processes must be developed and implemented. The company has a dedicated division to perform this role. They wish to understand and implement 'best practice' in manufacturing technology acquisition. To achieve this, the first step was to identify and understand the processes used within the company to form manufacturing technology acquisition decisions. This problem was an opportunity to apply the decision process identification methodology.

To assess the methodology, three criteria suggested by Platts [8] were adopted: feasibility, usability and utility. Feasibility asks if the methodology could be followed, usability refers to how easily it could be followed, and utility relates to the output – did the output meet expectations and was it useful? In the case of identifying a manufacturing technology acquisition process, for a high utility using the methodology would need to lead to an output that would be useful in further refining the process of manufacturing technology acquisition.

In accordance with a case study method [9], an outline of the possible form of the process of manufacturing technology acquisition was attempted. The literature gives some guidance as to the steps that could occur in this process. For example, Gregory et al [10] propose that the process of technology management includes consideration of issues relating to technology identification, selection, acquisition, exploitation and protection. Such issues were used as a guide in the subsequent interviews and mind mapping.

Next, interviewees were identified. In this large organisation, there is not simply one Manufacturing Director or Manufacturing Manager, as could be the case in a smaller organisation. Expertise in manufacturing technology acquisition is distributed around the organisation. Discussions with company personnel identified fifteen of the key managers and engineers involved in developing manufacturing technologies in the various divisions of the main UK sites. Each interview lasted about two hours. Each interviewee was asked to describe how they had gone about some recent technology acquisitions. By exploring the issues identified from the literature, maps of the processes apparent in the actions of the company were built up. Following the round of interviews, the maps were rationalised into a diagram showing the common threads in the process of technology acquisition practised in the company, together with notes on where different processes were apparent in the actions of different interviewees. The process was then discussed with company personnel. This process comprised a flowchart illustrating six main process steps, sub-divided into actions taken to execute each step, together with an indication of who took responsibility for carrying out each step of the process, and tools and techniques that support decision making. Throughout the research, observations were made on where the process identification methodology contributed to the information gathered, and an assessment was made of the feasibility and usability. Utility was assessed based on the discussion with staff about the usefulness of the findings.

6. Findings on the methodology

From the assessment of the methodology, several findings on its feasibility, usability and utility emerged. These are presented as discussion points below.

Finding 1: The methodology is feasible and usable

Platts [8] states that if the phases of a methodology were followed, then the methodology is proved feasible. Each of the four phases was successfully executed, so as a tool for researchers the methodology is feasible. Within the context of this research, the methodology was found to be easy to apply (usable), within the limitation that it was applied and assessed by the same researchers who developed it.

Finding 2: The methodology produced a useful output

Discussion with company personnel found that the output captured the flavour of their role accurately, and also identified some instances of where in the organisation expertise could be found in certain aspects of the process, for example benchmarking. Therefore, it was found that the methodology is capable of producing a useful, credible output, and also helps in directing future work.

Finding 3: The methodology captures the nature of the process as well as its details

Earlier sections of the paper identified that the process a company actually follows is a combination of formal intentions and implicit actions. In the case of this company, various formal frameworks and documents were found to exist and tie in to the actions of the people within the divisions. The methodology therefore identifies the role of formal procedures within the overall process, as well as the steps of the process.

7. Conclusions

This paper has described a methodology for identifying the technology acquisition processes in action in a manufacturing company. In organisations with highly institutionalised formal planning procedures, the methodology could provide a check on how closely such procedures are actually being followed. In other organisations, the methodology makes processes explicit and can assist in disseminating good practice. The output of the methodology is an understanding of the processes currently followed, and is a credible starting point for future process improvement efforts. Further work could help to assess and develop the methodology further. For example, using company facilitators, with or without researcher intervention, would give a fuller assessment of feasibility and usability.

Acknowledgement

The research reported in this paper was supported by EPSRC grant GR/M65748.

References

[1] Platts K.W., 1993, A process approach to researching manufacturing strategy, *International Journal of Operations and Production Management*, 13, 8, 4-17.

[2] Mintzberg H., Raisinghani D., Theoret A., 1976, The structure of unstructured decision processes, *Administrative Science Quarterly*, 21, June, 246-275.

[3] Mills J.F., Neely A.D., Platts K.W., Gregory M.J., 1998, Manufacturing strategy: a pictorial representation, *International Journal of Operations and Production Management*, 18, 11, 1067-1085.

[4] Baines T.S., Whitney D.E., Fine C., 1999, Manufacturing technology sourcing practices in the USA, *International Journal of Production Research*, 37, 4, 939-956.

[5] Mintzberg H., 1978, Patterns in strategy formation, *Management Science*, 24, 9, 934-948.

[6] Baines T.S., 1994, *Modelling in the evaluation of a manufacturing strategy*, PhD Thesis, Cranfield University.

[7] Buzan T., 1989, *Use your head (revised edition)*, BBC Books, London.

[8] Platts K.W., 1990, *Manufacturing audit in the process of strategy formulation*, PhD Thesis, University of Cambridge.

[9] Yin R.K., 1994, *Case study research: design and methods*, Sage, Thousand Oaks, CA.

[10] Gregory M.J., Probert D.R., Cowell D.R. 1996, Auditing technology management processes, *International Journal of Technology Management*, 12, 3, 306-319.

Re-engineering SMEs for Information System Implementation

P. J. Armstrong[1] and P. J. Coyle[2]

[1]School of Mechanical and Manufacturing Engineering, Queen's University Belfast
Ashby Building, Stranmillis Road, Belfast, N. Ireland BT9 5AH
pj.armstrong@qub.ac.uk
[2]School of Mechanical and Manufacturing Engineering, Queen's University Belfast
Ashby Building, Stranmillis Road, Belfast, N. Ireland BT9 5AH
pj.coyle@qub.ac.uk

Abstract: The lack of implementation of modern information systems in manufacturing SMEs is discussed. The problem is investigated by consulting relevant publications, reviewing the commercial software available to SMEs and by carrying out a series of interviews with local companies. On the basis of the results, a comprehensive programme is outlined which aims to assist SMEs with the specification and implementation of information systems appropriate to their needs.

1. Introduction

It is widely recognised that manufacturing SMEs are important, in terms of the flexibility they contribute to the manufacturing sector and the potential they hold for employment growth. However, it is also evident that manufacturing SMEs are not taking full advantage of modern information systems (where the term is used here to cover commercial MRPII and ERP systems, CAPM systems in general and bespoke systems for decision support). SMEs, for example, are much less likely to have MRPII systems in place than larger companies [1] and, as a general indicator of IT uptake in SMEs, a recent survey suggests that only 10% of SMEs with between 10 and 100 employees have adopted e-mail [2].

Instead of formal information systems, small companies tend to rely on informal communication and ad hoc arrangements [3]. However, as companies grow they reach a stage when their operations need to be systematised and streamlined, in order to provide a sound foundation for expansion [4]. As part of this process it is particularly important that they introduce formal systems for production and inventory planning and control. Failure to take this step or to achieve a successful outcome is likely to be a factor in the low output per employee associated with SMEs [5] and the fact that less than 10% of the manufacturing lead time in a typical SME is devoted to productive work [6]. It is also becoming strategically important that SMEs implement formal information systems as a basis for inclusion in future supply chain systems and virtual enterprises. The current paper concerns initial work on a comprehensive programme to assist manufacturing SMEs with the successful implementation of formal information systems.

2. An Analysis of Relevant Literature

Ostensibly an information system (IS) implementation should be less demanding in an SME compared with a larger company, because internal relationships are less complex [7] and compatibility with legacy systems is less likely to be a problem. However, an SME will often face the dilemma that it does not have the necessary expertise, but also lacks the financial resources to employ consultants. In addition, it is widely recognised that the costs and benefits of an IS implementation are difficult to quantify [8] and, if estimates are possible, the payback period may be 3 to 5 years [9]. This possibility and the fact that implementations in SMEs tend to take at least one year [10] are at variance with the tendency of SMEs to focus on the short term [11]. The specification and selection of software for an IS are also time consuming tasks and, in the absence of a formal method [12], SMEs tend to use an ad hoc approach involving excessive reliance on software suppliers [13]. This can lead to the purchase of an unsuitable product [14]. Matters are not helped by reports of dissatisfaction from those who have completed IS implementations [12, 6].

An SME planning to implement an IS may underestimate the extent to which its existing practices and procedures will have to change. With this in mind, some authors have advocated a BPR exercise before an IS implementation [8]. This is logical in the sense that BPR involves taking a holistic systems approach which focuses on the business processes which the IS will support [15]. However, BPR has its roots in the problems of large corporations [11] and limited guidance is available as to how an SME should carry out a BPR exercise [16]. Apart from the more obvious differences it has been noted that SMEs are likely to consider BPR during a period of expansion, whereas large companies turn to BPR when they are underperforming [17]. SMEs will therefore be reluctant to risk radical change and a more modest restructuring of the type generally associated with continuous process improvement may be more palatable.

In addition to the limited information available on BPR methodologies for SMEs, there is also a lack of suitable support tools for business process re-engineering or analysis. Some authors advocate the development of an enterprise model as a precursor to a BPR exercise or an IS implementation [18]. Systematic approaches have been proposed which aim to make it easier for SMEs to create enterprise models, using either the IDEF [19] or CIMOSA [15] architectures. However, in practice it has been noted that standard flow diagrams are the only support tool used in the great majority of BPR projects [20]. Such diagrams, as is the case with current enterprise models, can only represent the static behaviour of an enterprise and it has been claimed that failure to take the enterprise's dynamic behaviour into account has adversely affected many BPR projects [21]. If dynamic simulation models of the enterprise were used instead then a range of benefits would result, including the ability to compare alternatives quantitatively, the opportunity to assess the effects of variability and randomness on system performance and the availability of animated graphical displays showing systems in operation. While simulation software has traditionally been regarded as expensive and difficult to use, business process simulation packages have recently become available which are both user-friendly and relatively inexpensive [22, 23].

Although it may be logical to model an enterprise as a basis for specifying and selecting an IS, it must be recognised that ERP software, for example, already assumes that the enterprise conforms to a particular model. In fact it has been suggested that business processes should be designed to fit such in-built models [24]. Clearly this pre-empts the choice of software and an approach whereby an independent model is created for analysis purposes and then matched against the assumptions incorporated in a range of packages would be more logical [8].

The above comments presuppose that the IS implementation will employ commercial software. Some authors have argued that because SMEs lack the resources needed for software development they will invariably rely on commercial packages [13, 7] and successful applications have been reported [10]. Others argue that commercial software is generally inappropriate for SMEs [1] and that implementations are more likely to be unsuccessful [25]. This may mean that in the final analysis bespoke software may have to be produced [8]. At a more fundamental level, it has been argued that the MRP technique which is at the core of many software packages is not suited to the environment in which SMEs operate, particularly make-to-order companies [9, 14]. In response to such arguments, a number of research groups have developed original software specifically for SMEs [9]. Furthermore, a recent trend has been to demonstrate that suitable software can be created using the tools available in the Microsoft Office suite [26]. This approach has the advantages that development time is reduced and low cost software with a familiar user interface can be produced. For the most part, work to produce software specifically for SMEs has been based on empirical studies of SME requirements. However, a theoretical basis for SME-oriented information system design may emerge from current research on holonic manufacturing systems [18]. The premise is that holonic concepts, which stress the need to retain autonomy and co-operative behaviour within a system, better reflect the human-centred organisation of many SMEs, which arguably provides them with their characteristic flexibility. In effect, the structured nature of commercial information systems may deprive SMEs of one of their most important advantages over larger companies.

To summarise, the literature suggests that various factors are preventing SMEs from taking advantage of modern information systems, including the fact that:

- SMEs generally lack the necessary resources, both financially and in terms of time and expertise.
- Investments in information systems are difficult to justify on a cost-benefit basis and long term strategic thinking is required.
- There is a lack of assistance and independent advice available on the specification and selection of software.
- There is insufficient guidance available to help SMEs re-engineer their business processes and an absence of suitable tools to carry out the task.
- It is often unclear as to whether existing commercial software will prove to be suitable for a particular SME.
- If SMEs require unique software, there is no consensus at present as to the form that this should take.

The tendency to regard SMEs as a homogeneous group is an underlying problem which often produces empirical results and leads to conclusions which appear to be contradictory. In practice, SMEs exhibit wide variation in terms of the stage of development they have reached, the business environment they operate in and the production systems they employ. Although some research has been reported which seeks to classify SMEs [27, 28], more work is required to identify and characterise subgroups for which generic results can be produced and generic systems derived.

3. A Preliminary Research Study

In order to explore further some of the issues raised in the literature review, two surveys were carried out. In the first survey, MRPII and ERP software available commercially was reviewed. Initially over 200 suppliers with offices in Ireland or the UK were identified. This total was then reduced to 74 unique products or product families by eliminating specialist software and multiple resellers of the same product. Information was obtained on the 74 products and analysed in detail. One result of relevance to this paper was the high proportion of packages which were claimed to be suitable for small companies. This applied to almost 80% of the packages marketed as MRPII software and around 15% of the products in the ERP category. Since many of the simpler packages are modestly priced it is clear that the main problem faced by an SME is one of software selection rather than a lack of low cost packages. This assumes, of course, that commercial software is suitable.

The second survey carried out involved visiting 40 manufacturing SMEs in Northern Ireland with between 10 and 100 employees. Northern Ireland is particularly dependent on its SMEs with over 95% of its manufacturing companies employing less than 100 people. A structured questionnaire was used to interview the owner/manager of each of the 40 SMEs visited. Results of relevance included the fact that, while around 95% of SMEs had computer-based accounting systems, less than 8% had implemented MRP. Even fewer companies (2.5%) had appointed an individual to be responsible for IT, thus underlining the SME's resource problem in terms of expertise. The vast majority of owner/managers appreciated the need to introduce some form of computer-based IS, but almost all admitted that they required help with system selection and implementation. Of the options offered, some type of advice centre was regarded as the most suitable form of assistance and the University and its students were seen as potential sources of expertise. Over 50% of the owner/managers also favoured networking with similar SMEs.

4. A Programme to Assist SMEs

A programme has been devised to address some of the problems highlighted in the literature review, along with those identified as a result of the preliminary research study. The programme has three main elements, one of which involves setting up a training and advice centre at the University. This will feature a software demonstration facility and will also promote networking by forming an SME club.

Work is in hand on this initiative and the objective is to assist SMEs with software selection and implementation.

The second element of the programme aims to provide SMEs with access to expertise through linkages with the University. To date this has involved setting up three Teaching Company Scheme (TCS) programmes with different SMEs and placing a number of students in companies to carry out project work. In each case the students were trained in the use of a particular MRPII package before being asked to assess its implementation in a collaborating SME. Both TCS programmes and, to a lesser extent, student placements have the potential to alleviate the knowledge and time resource problem in SMEs [8, 25, 29].

The third element of the programme consists of a series of research projects. The objectives are to develop practical tools to enable SMEs to re-engineer their businesses and specify their IS requirements and also to develop systems and software for SME environments not catered for by existing commercial packages. Current work involves the assessment of business process simulation as the basis of a re-engineering and IS specification toolset. Alternatives to MRP are also being developed and assessed, including systems which exploit the potential of simulation techniques. The possibility of embedding simulation in Microsoft Office products to provide both decision support and transactional processing is also being examined. In addition, as contacts with SMEs expand, it is hoped that a contribution can be made to the development of a classification system for manufacturing SMEs which creates generic groups with characteristics, problems and requirements in common, and hence holds out the possibility that generic solutions can be proposed.

References

[1] Little D., Peck M., Rollins R., et al, 1998, Responsive planning and scheduling architectures. IEE Workshop on Responsiveness in Manufacturing, London, UK, 11/1-4.

[2] Sillince J.A.A., Macdonald S., Lefang B., et al, 1998, Email adoption, diffusion, use and impact within small firms: a survey of UK companies. Int. J. of Information Management, 18(4), 231-242.

[3] Denton P.D., Hodgson A., 1997, Implementing strategy-led BPR in a small manufacturing company. Proc. 5th Int. Conf. on Factory 2000, Cambridge, UK, 2-4 Apr 1997, 1-8.

[4] Whitfield R.C., Szeto I.C., 1997, Business process reengineering for small building products manufacturer. J. Management in Engineering, 13 (6), 84-93.

[5] Wittenberg G., 1994, Small and medium-sized manufacturing enterprises in the UK. Purchasing & Supply Management, June 1994, 20-22.

[6] Turner E., Saunders D., 1994, Surveying the manufacturing planning and control environment. Proc. 4th Int. Conf. on Factory 2000, York, UK, 3-5 Oct 1994, 613-619.

[7] Wastell D., Sewards A., 1995, An information system profile of the UK manufacturing sector. J. Information Technology, 10(3), 179-189.

[8] Irani Z., 1999, Difficulties of IT/IS investment justification: an interpretivist empirical case study. Proc. 32nd Hawaii Int. Conf. on System Sciences, Maui, HI, USA, 5-8 Jan 1999, 255.

[9] Halsall D.N., Muhlemann A.P., Price D.H.R., 1993, Production planning and scheduling framework for smaller manufacturing enterprises. Proc. IEEE Int. Conf. on Systems, Man and Cybernetics, Pt. 4, Le Touquet, France, 17-20 Oct 1993, 33-38.

[10] Tinham B., 1999, The MRP/ERP user satisfaction survey 1999. Manufacturing Computer Solutions, 5(7), 19-22, 25, 26, 29.
[11] Cagliano R., Spina G., 1998, Designing BPR support services for small firms. Int. J. Operations and Production Management, 18(9/10), 865-885.
[12] McGarrie B., 1998, Case study: production planning and control - selection, improvement and implementation. Logistics Information Management, 11(1), 44-52.
[13] Little D., 1998, Development of a package selection method and tool for small and medium sized enterprises (SMEs). Computer Integrated Manufacturing Systems, 11(4), 351-355.
[14] O'Toole B., Etheridge R., Warrington A., 1993, Assessing the suitability of manufacturing control software for small make-to-order manufacturing companies. BPICS Control, 19(6), 29-33.
[15] Smart P.A., Maull R.S., Childe S.J., 1997, Integration in small and medium enterprises: specification of a business process re-engineering methodology. Proc. Int. Conf. on Enterprise Integration and Modeling Technology, Torino, Italy, 28-30 Oct 1997, 449-58.
[16] Childe S.J., Maull R.S., Bennett J., 1994, Frameworks for understanding business process re-engineering. Int. J. Operations and Production Management, 14(12), 22-34.
[17] Chang L-J., Powell P., 1998, Towards a framework for business process re-engineering in small and medium-sized enterprises. Information Systems J., 8(3), 199-215.
[18] Toh K.T.K., Newman S.T., Bell R., 1998, Information systems architecture for small metal-working companies. Proc. Institution of Mechanical Engineers, 212(B2), 87-103.
[19] Cantamessa M., Paolucci E., 1998, Using organizational analysis and IDEF0 for enterprise modelling in SMEs. Int. J. of Computer Integrated Manufacturing, 11(5), 416-429.
[20] Hlupic V., Robinson S., 1998, Business process modelling and analysis using discrete-event simulation. Proc. 1998 Winter Simulation Conf., Washington, USA, 13-16 Dec 1998, Pt. 2, 1363-1369.
[21] Tumay K., 1995, Business process simulation. Proc. 1995 Winter Simulation Conf., Arlington, VA, USA, 3-6 Dec 1995, 55-60.
[22] Benjamin P.C., Delen D., Erraguntla M., 1998, Introduction to using Prosim for business process simulation and analysis. Proc. 1998 Winter Simulation Conf., Washington, DC, USA, 13-16 Dec 1998, Pt. 1, 315-321.
[23] Sadowski D., Bapat V., Drake G., 1998, Arena product family: enterprise modeling solutions Proc. 1998 Winter Simulation Conf., Washington, USA, 13-16 Dec 1998, Pt. 1, 205-212.
[24] MacDonald G., 1996, Business transformation with SAP R/3. Proc. IEE Colloquium on IT Support for Business Process Re-Engineering, London, UK, 11 Jun 1996, 2/1-2/5.
[25] Saxena K.B.C., Tam M.M.C., Chung W.W.C., et al, 1994, Institutionalization of decision support technologies in small manufacturing enterprises of Hong Kong. IFIP Trans., A-45, 139-158.
[26] Braun M., Mohle S., 1998, Developing complex information systems from standard software components: a case study within manufacturing resource planning. Proc. 31st Hawaii Int. Conf. on System Sciences, Kohala Coast, HI, USA. 6-9 Jan 1998, Pt. 6, 143-52.
[27] Carrie A.S., Macintosh R., Scott A., et al, 1994, Linking strategy to production management structures and systems. Int. J. of Production Economics, 34(3), 293-304.
[28] Raymond L., 1987, Information systems evolution in small business. Proc. 20th Hawaii Int. Conf. on System Sciences, Kailua-Kona, HI, USA, 6-9 Jan 1987, Pt. 3, 329-335.
[29] Drake P.R., John E.G., Petheram C.L., et al, 1998, Slipping systems into SMEs. Manufacturing Engineer, 77(5), 217-220.

MANUFACTURING SYSTEMS II

Environmental Sustainability in Supply Chain Management

Dr. Richard Y. K. Fung[1], Ms Barbara Morton[2] and Mr. Steven Chong[3]

[1] Department of Manufacturing Engineering and Engineering Management
City University of Hong Kong, 83 Tat Chee Ave., Kowloon, Hong Kong
E-mail: richard.fung@cityu.edu.hk
[2] Centre for Research on Organisations, Management and Technical Change
Manchester School of Management, UMIST, Manchester M60 1QD, UK
E-mail: barbara.morton@umist.ac.uk
[3] Department of Manufacturing Engineering and Engineering Management
City University of Hong Kong, 83 Tat Chee Ave., Kowloon, Hong Kong
E-mail: mestevec@cityu.edu.hk

Abstract: The journey of environmental conservation and protection in supply chain management begins with the evaluation and selection of materials and their appropriate suppliers based on their environmental performance. Increasing market influence is being imposed by proactive customers world-wide on their vendors to operate environmentally consciously and responsibly. This paper outlines the work carried out by a project collaboration between City University of Hong Kong and UMIST of UK in identifying, formulating and disseminating good practices in environmentally sustainable supply chain management. The background and long-term significance of the work as well as the relevant advances in various industrial and business sectors in the subject area are discussed.

1. Introduction

Increasingly stringent environmental regulations, legislation as well as social pressure on organisations to become more environmentally conscious and responsible positively stimulate and drive environmental improvements with the partners throughout their supply chains [1]. The market influence of large corporate customers on their suppliers is apparent. Organisations in both the private and public sectors increasingly recognise the potential environmental impact of the goods and services supplied to them. As customers become more environmentally conscious, good environmental practice becomes a crucial factor when selecting a material supplier [2]. Recent studies argue that conflicts do not necessarily exist between satisfying the needs for environmental performance and improving profitability simultaneously [3] [4]. However, many organisations have found it difficult to address the complex issues associated with communicating and implementing environmental policies and practices through their supply network, mainly owing to the diversity of the trading partners involved in their supply chain, and the complexity in integrating 'environment' with existing management mind-set, culture and systems [5].

This paper discusses the findings, knowledge and experience acquired through the work of a joint research project between the Department of Manufacturing Engineering and Engineering Management (MEEM), City University of Hong

Kong and the Centre for Research on Organisations, Management and Technical Change (CROMTEC), UMIST, UK. This work is financially supported by the Hong Kong Environmental and Conservation Fund and the UK ESRC respectively. This collaborative project addresses on those crucial issues in environmentally sustainable supply chain management (SCM) as identified in previous studies [6] [7] and [8] [9]. The project aims to identify, formulate and disseminate good practices in fostering environmental responsibility throughout global supply networks. Furthermore, it also examines the roles that can be played by academic / professional institutions and business enterprises in this meaningful movement.

2. Background and the Long-Term Significance of the Project

This project was conceived on the belief that the movement of environmental conservation and protection in industry should begin with the sourcing and selection of materials and their respective suppliers throughout the entire supply network . It is because the properties of materials used on a product or service often pre-determine or dictate the methods by which they are processed, hence the pollution and wastes associated with the processes. In other words, the materials used on a product have a profound impact on the entire supply chain right from the acquisition, storage, processing, through to the assembly, distribution and the ultimate disposal or recycling of the used items. Many environmentally proactive companies are already using their market powers to foster and shape the development of a green market, and systematically strengthen the movement towards the goal of sustainable improvement. The emergence of supplier environmental appraisal and assessment systems is an evidence of the very real commercial threat now confronted by those companies which have been overlooking their environmental impacts and responsibilities [2]. On the other hand, proactive vendors also consider the supply of green products as an opportunity for establishing a healthier market profile to competently respond to increasingly stringent customer environmental demands.

3. The Global Advances in Environmentally Sustainable Supply Chain Management

This section examines the steps being taken by major industrial sectors in their pursuit of environmental sustainability.

3.1 Environmental supply chain practices in the automotive industry

Pressures on companies to demonstrate their commitment to environmental improvement in the supply chain is well illustrated by recent announcements from the automotive sector. Ford and General Motors have "... told their manufacturing suppliers – well over 10,000 businesses in all – to implement environmental management systems within the next 2-3 years" [10]. Having made progress in implementing the ISO 14001 standard for their own sites, these companies have now committed themselves to extending environmental management practices throughout their supply chains. For instance, Rover and Jaguar (a Ford subsidiary) having already taken steps to promote ISO 14001 in the supply chain [11].

3.2 Environmental supply chain practices in the retail and service sectors

Under scrutiny from environmental pressure groups, the do-it-yourself retailer, B&Q in the UK examined its sourcing policies on timber and has since developed the QUEST system into one of the most comprehensive supplier assessment approaches. International chain stores, such as Marks & Spencer, Body Shop, Boots etc., also have established and implemented their policies in environmental purchasing and supply in their outlets world-wide including Hong Kong.

A major driver for supply chain improvement in the UK has been the influence of the Business in the Environment Index of Corporate Environmental Engagement [12] [13]. Using a survey technique BiE aimed to determine the progress that FTSE 100 companies consider they have made against ten key attributes of environmental management. One of the survey questions specifically refers to the existence of an environment-focused supplier programme in the company.

In the first year (1996) this was the weakest parameter and many companies were prompted to address the issue of supply chain as a matter of some urgency. The Index has since been extended to the FTSE top 350 companies and, not surprisingly, early results have shown similar deficiencies.

3.3 Environmental supply chain practices in the electronic industry

The electronics industry is also very active in their environmental supply network, particularly from the point of view of product stewardship. Many companies in this industry have adopted life cycle assessment techniques in their efforts to conduct environmental management, taking a 'cradle to grave' approach. Being a global industry, the electronics sector has dictated the pattern of environmental activity in many countries. The pattern of adoption of ISO 14001 in this industry illustrates this point well. For instance, the number of ISO 14001 certificates awarded has been remarkably high in Malaysia, in response to the perceived demand for the standard from their international clients. It appears that accreditation to ISO 14001 is regarded in some parts of the world at least as a necessary part of the 'licence to operate' package for international businesses.

The electronics industry is particularly active in this field largely under the influence of European environmental regulation on waste electronic and electrical equipment. This is part of a suite of directives concerned with producer responsibility. Other areas subject to existing or forthcoming regulation include packaging waste and end-of-life vehicles. Operating in global markets, companies in the affected sectors are therefore driven to tackle the environmental impacts of their suppliers world-wide. Thus companies such as Nortel Networks, Motorola,

Kodak and Xerox are developing their approaches to environmentally sustainable supply chain management. The situation facing companies operating in international markets is far from simple, however. Despite globalisation, international firms face highly differentiated pressures from different regional markets.

3.4 Environmental supply chain practices in the utilities sector

Companies offering consumer goods are not alone in their concerns for environmental management in the supply chain. In the UK, BT was among the first companies to address the issue and its supplier environmental assessment questionnaire GS13 is well known to prospective suppliers. Other sectors prompted to address environmental improvement through the supply chain include the utility companies, whose score against this parameter more than doubled from 1996 to 1997 [13]. Water, gas and electricity supply companies have begun to implement programmes of improvement with their suppliers. The extent to which they have been assisted in this process is debatable, however. A proposed scheme to assist suppliers to implement environmental management systems in a staged approach was eagerly anticipated by large organisations seeking to make progress. The approach ran into difficulties before it could deliver significant achievements and in recent months, both the UK Department of Trade and Industry and the Hong Kong SAR Department of Environment and Food have announced their commitments along similar lines, aimed at assisting smaller companies to improve their environmental performance.

3.5 Environmental supply chain practices in the public sector

Reference to the above market sectors leads to consideration of another important aspect in the discussion of environmentally sustainable supply chain management. This concerns the influence, actual or potential, of public sector purchasing policies and practices. A recent report for the European Commission on 'Integrated Product Policy' highlighted the potentially crucial role to be played by public sector purchasers in stimulating demand for sustainable goods and services [14].

The US – Asia Environmental Partnership engages US and Asian multinational companies to promote corporate environmental policies and practices among their suppliers. A survey of 30 global companies revealed that many firms have environmental criteria for their suppliers, particularly in the textiles and apparel, automobile and microelectronics industries.

The Canadian Government, through Environment Canada, has been amongst the most active in producing guidance for buyers seeking to source environmentally sustainable products and services. The US Environmental Protection Agency has also addressed the issue by making guidance available.

Since 1995, the Danish Government has required that all government institutions and departments include environmental criteria on equal terms with price, quality, delivery, etc in tendering processes [15]. Guidelines have been produced in Denmark using a streamlined approach to LCA [16].

The European Green Purchasing Network has, as one of its aims, the promotion of green purchasing as an essential component of a sustainability policy. The International Council for Local Environmental Initiatives has established ' Eco-Procura' which runs a number of projects aimed at assisting 'greener' public sector purchasing. Governments in many of Europe are implementing 'green house-keeping' programmes to ensure that their own purchasing practices are in line with commitments to encouraging sustainable development.

In the UK, in addition to what have been discussed in the earlier sections, the Department of Environment, Transport and the Regions and the many others have produced a 'Green Guide for Buyers' in an attempt to deliver on its commitments to sustainable development.

In Hong Kong, the Department of Environment and Food categorically stipulated in their Waste Reduction Framework Plan the Government's commitment to Environmentally Responsible Purchasing (Green Purchasing). They have been encouraging both the public and private sectors to buy products with higher recycling contents or with less packaging material through the Waste Reduction Task Force for Government. As a major public buyer, they have introduced a number of products with recycled contents into the Government Supplies Department's stores e.g. recycled photocopying paper, paper towel, etc. Environmental specification and identification of new "green" products require careful consideration on market supply and procurement cost. A project is being commissioned to develop green specifications for a list of environmentally sensitive items procured by the government.

5. Conclusion

Realising the importance of tackling the environmental conservation and protection issue in the industry at source, i.e. from the careful selection of materials and suppliers, a joint project has been carried out by the City University of Hong Kong and UMIST of UK to promote the concepts of environmental SCM and assist the industry to put the ideas into practice. In this paper, the background and the long term significance of the work are explained, and the current environmental supply chain practices in a number of industrial and business sectors are reported. Some encouraging progress has been made in many instances, however they only mark the beginning of a very long journey of sustaining a cleaner living surrounding for mankind through more environmentally conscious and responsible SCM.

Acknowledgements

This research discussed in this paper is supported by the HKSAR Environment and Conservation Fund (Project No: 17/99) and the UK ESRC Fund respectively

References

[1] Knight, A., 1995, *B&Q, its suppliers and the environment*, UNEP Industry and Environment, April-September, pp. 32-34.

[2] Noci, G., 1997, *Designing 'green' vendor rating systems for the assessment of a supplier's environmental performance*, European Journal of Purchasing and Supply Management, Vol. 3, No. 2, pp. 103-114.

[3] Sroufe, R., 1996, *The environmental reporting logistics system*, Logistics Spectrum, Vol. 30, Issue 2, pp. 13.

[4] Verma, R., 1998, *An analysis of the supplier selection process*, Omega, International Journal of Management Science. Vol. 26, No. 6, pp. 739-750.

[5] Enarsson, L., 1998, *Evaluation of suppliers: how to consider the environment*, International Journal of Physical Distribution and Logistics Management, Vol. 28, No. 1, pp. 5-17.

[6] Fung, R.Y.K., Wong, C.F.Y. and Pun, K.F., 1999a, *Environmentally Conscious Purchasing: An Overview of International Practices*, Proceedings of the 4th International Conference on ISO9000 and TQM (4th ICIT), Hong Kong, April 7-9, pp. 201-206.

[7] Fung, R.Y.K. and Wong, C.F.Y., 1999b, *Analysing the Criteria For Environmentally Conscious Supplier Selection / Evaluation*, Proceedings of the 5th International Conference on Manufacturing Technology, Beijing, China, October 31 – November 3.

[8] Green, K., Morton, B. and New, S., 1996, *Purchasing and Environmental Management: Interactions, Policies and Opportunities*, Business Strategy and the Environment, Vol. 5, No. 3, pp. 188-197.

[9] Green, K., Morton, B. and New, S., 1998, *Green purchasing and supply policies: do they improve companies' environmental performance?*, Supply Chain Management, Vol. 3, No. 2, pp. 89-95.

[10] ENDS Report, 1999a, *Ford, GM push ISO 14001 down their supply chains*, ENDS Report 297, October, p. 39.

[11] ENDS Report, 1998, *Rover and Jaguar raise the stakes in greening of corporate supply chains*, ENDS Report 279, April, pp 23-24.

[12] BiE, 1996, *Index of Corporate Environmental Engagement*, Business in the Environment, London.

[13] BiE, 1997, *Index of Corporate Environmental Engagement*, Business in the Environment, London.

[14] Ernst & Young / SPRU, 1999, *Integrated Product Policy – report to European Commission*, Ernst & Young / SPRU.

[15] Sorensen, S. Y., 1999, *Environmental criteria in tendering procedures*, Presentation to the Waste & Water / Eco-Procura Conference, Copenhagen, June 1999

[16] Schmidt, A., 1999, *Using the life cycle approach in green procurement*, Presentation to the Waste & Water / Eco-Procura Conference, Copenhagen, June 1999.

Achieving Competitiveness through the Manufacturing Technology Supply Chain

T S Baines[1] and D K Harrison[2]

[1]Cranfield University, Bedfordshire, MK43 0AL, t.s.baines@cranfield.ac.uk
[2]Glasgow Caledonian University, Cowcaddens Road, Glasgow G4 0BA, dha2@gcal.ac.uk

Abstract. This paper argues that integration into the manufacturing technology supply chain can be a valuable weapon in defending a company's position in the product supply chain.

The paper approaches this argument by first examining how manufacturing supply chains in the west have restructured in recent years, and how through this restructuring some companies have become vulnerable. The paper then suggests that some of this vulnerability can be overcome by the company integrating into manufacturing technology source for a few choice technologies. Benefits of selective integration are then explored. Finally the paper presents a methodology that is intended to help a company decide the level of integration for these technologies.

1. Introduction

The design of the product supply chain for an organisation is a complicated task, and is far to complex to be determined by the policy of 'downsize and outsource'. Rather, an organisation should carefully and systematically consider and monitor its position in the product supply chain, frequently questioning the boundaries between what to make and what to buy. However by one means or another, there has been massive and ongoing restructuring of supply chains in western companies in recent years [1, 2, 3]. It is quite possible for organisations to go too far in outsourcing products. If the contribution that an organisation makes to the supply chain becomes to small, then either suppliers or customers can marginalise and by-pass the now redundant organisation. So what other strategies can be used to lever competitiveness through supply chain design?

There is a second supply chain that is often overlooked, or quickly dismissed by companies in the west, namely the infrastructure supply chain. This broad term embraces the supply of manufacturing technologies (processes). This paper will argue that integration into the manufacturing technology supply chain can be a valuable weapon in defending a company's position in the product supply chain. It is important to emphasise that the author's are advocates for a major reversal in product sourcing policies; the theme here is not 'integrate and grow' as such a policy would be as unrealistic as the one it replaces. Neither does this paper diminish the value of a well thought through product sourcing strategy, the intention here is to simply improve the awareness of the opportunities offered by integration into a frequently neglected supply chain.

2. Vulnerability and the Product Supply Chain

Contrasting western and Japanese manufacturing practice has been a rich source for research in recent years and one area that has been given particular attention is product sourcing and the associated relationship between a manufacturing company and component supplier. Such authoritative texts as 'The machine that changed the world' [4], and 'Japanese manufacturing techniques' [5] highlight the differences between the product sourcing structure of companies such as Toyota [6] and their traditional western equivalent. Authors such as Fine and Whitney [7] argue that Toyota helps to maintain a Lean manufacturing system by outsourcing the manufacture of many components and systems of their product. However, a difference also exist in another sourcing practice that is often overlooked, namely procurement of manufacturing technology and general infrastructure.

While Japanese companies buy much of their product components, they make much of their infrastructure [7]. This difference in sourcing structure is illustrated in Figure 1. The vertical axis of this graph represents integration in the product supply chain, while the horizontal axis represents integration in the manufacturing technology (process) supply chain. In a recent study of manufacturing companies in the USA and UK Baines et al [8] observed that manufactures are advanced in their adoption of Japanese style sourcing practises for products, but have tended to ignore process integration. Indeed, the response of most manufacturers, when questioned about process supply chain integration, is captured well by the quotation '*our business is cars not robots*' [7]. Referring back to Figure 1, this subsequent trend in sourcing practices can be thought of as a migration from 'make' to 'buy' on the vertical axis of the graph.

The consequential structure of a typical western company, that has heeded the call to 'down size and outsource', is also represented in Figure 1. This supply chain structure has a significant implication on the host organisation. Retraction in the product supply chain, while maintaining position in the process supply chain, reduces the number of activities for which an organisation is directly responsible. The more you buy the less you build. In Figure 2, this reduction is represented by the smaller area beneath the boundary line, and this area also symbolises other changes to the host organisation.

Organisations only have direct control over the activities they own, fewer activities means lower control, and increased exposure to the inabilities and fallibilities of the supply base. The opportunity to add value is also reduced as more components are purchased, and their subsequent combined value creeps towards the product selling price. More significantly, the knowledge base of the organisation is eroded as evermore products are outsourced. Suggesting integration into the manufacturing technology supply chain is contentious. It is easy to dismiss this suggestion, seeing it as a näive proposal for a company to make such established equipment as CNC machining centres. Clearly, there is little benefit from reproducing technologies that are commonly available. If however, an organisation can build on an existing technology, tailoring the equipment to produce better and cheaper products faster, then a distinct contribution to competitiveness starts to emerge. The organisations knowledge, control and value adding capabilities all benefit. The key to success is the selection of a few manufacturing technologies that are core to defending and differentiating a company's contribution to the product supply chain.

3. Determining the Manufacturing Technology Source

Higgins [9] shows that there are right times to make and right times to buy. Researchers such as Bruck [10] capture the thoughts of many in their argument that no amount of observation of prevailing practice in different industries will reveal whether high or low vertical integration creates competitive advantage. Every company must find its own optimum level of integration, and as the authors previous work has shown, this may differ for technologies across an organisation.

The key is to define the technical interface, this is the point at which a supplier's contribution fits into the manufacturers own technologies. The extent of integration for all technologies can be established by assessing the role of the manufacturing technology in supporting the competitive strategy of the organisation, contrasted against the characteristics of the company supply base. A company that has a weak supplier base may be forced into the design and construction of a technology not considered to be core.

Alternatively, a company based in a supplier rich location may develop strong supplier relationships that enable the outsourcing of core technologies. To aid the practitioner in this decision, the authors have formed a methodology that is shown in overview in Figure 2. This methodology illustrates the linkage between the manufacturing technology sourcing decision and the saleability of products in the market. The methodology commences by recognising that products sell because of distinctive capabilities. Some of these are dependent on the manufacturing activity, such as product features or lead-time. Others are dependent on marketing or after-sales support.

The competitive strategy of the business should support the distinct product capabilities identified above. This means that where such a distinction has some dependency on manufacturing, the manufacturing strategy has to ensure that complementary production capabilities are developed. For most companies, not all components and sub-assemblies provide an equal contribution to product competitiveness. Those with a high contribution may be referred to as core or key components. The remainder are often referred to as commodities. For example, consider a manufacturer of Formula One racing cars whose success is dependent on chassis stiffness and design. The manufacturing strategy may well evolve around developing capabilities in composite fibre production. Composite fibre components are core, while simple nuts and bolts are treated as commodities as they offer no distinction to product competitiveness. Only if a decision has been made to manufacture a component in-house, does this imply that the associated manufacturing technologies could be considered core. These core technologies are those that have the greatest sustainable impact on product cost or differentiation. This is closely associated with the concept of core competencies.

Quinn [11] argues that a company should concentrate its own resources on core competencies where it can achieve definable pre-eminence and provide unique value for customers. However, integration into the manufacturing technology supply chain should again be based upon the relationship between the strategic importance of the capabilities of the technology, contrasted against the supply base characteristics. The outcome of this decision will be a policy governing the extent of integration into the supply chain for the specific technology being considered. This final decision process is considered and discussed in detail in Baines et al [8].

4. Concluding Remarks

Organisations can go too far in outsourcing products. There is a second supply chain that is often overlooked, or quickly dismissed by companies in the west, namely the infrastructure supply chain which incorporates the supply of manufacturing technologies. This paper has argued that integration into the manufacturing technology supply chain can be a valuable weapon in defending the product supply chain, and has outlined a methodology to help companies manage this situation. This paper has not advocated 'integrate and grow', neither has it dismissed the benefits of a well thought through product sourcing strategy. The intention here has simply been to improve the awareness of the opportunities offered by integration into a supply chain that is often overlooked by many companies.

References

1. Porter, M.E., (1985), *Competitive Advantage - Creating and Sustaining Superior Performance*, The Free Press, New York.

2. Skinner, W. (1985). *Manufacturing - The Formidable Competitive Weapon,* John Wiley & Sons, New York.

3. Hayes, R.H. and Wheelwright, S.C., (1984), *Restoring Our Competitive Edge: Competing Through Manufacturing*, John Wiley & Sons, New York.

4. Womack, J.P., Jones, T.J., and Roos, D., (1990) *The Machine that Changed the World,* Rawson Associates, Macmillian

5. Schonberger, R.J., (1982) *Japanese Manufacturing Techniques*, The Free Press, Macmillan Publishing.

6. Shingo, S., (1990), *A study of the Toyota production system from an industrial engineering viewpoint*, Productivity Press.

7. Fine, C.H. and Whitney, D.E., (1999), *Is the make-buy decision process a core competence?* Presented at the International Symposium on Logistics, Florence, Italy, 11-14 July.

8. Baines T.S., Whitney D.E. and Fine C.H. (1998), *Manufacturing technology sourcing practices in the USA*, International Journal of Production. Vol 37, no 4, 939-956.

9. Higgins, C.C.H., (1955), *Make-or-buy re-examined,* Harvard Business Review March-April, Vol. 33, 2, 109-113.

10. Bruck, F., (1995), *Make versus buy: The wrong decision costs,* McKinsey Quarterly, 1, 28-47.

11. Quinn, B.Q. , (1995), *Strategic outsourcing.* McKinsey Quarterly, 1, 47-69.

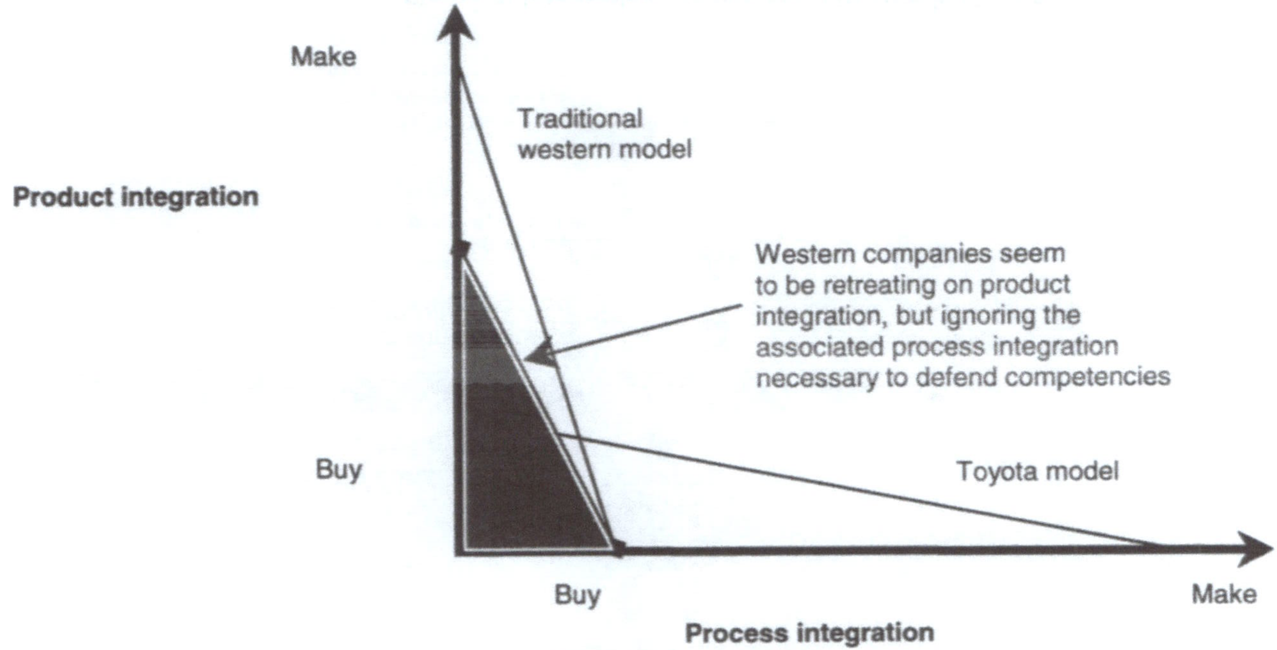

<u>Figure 1: Becoming a vulnerable organisation - a contrast between a traditional western model and the Toyota model</u>

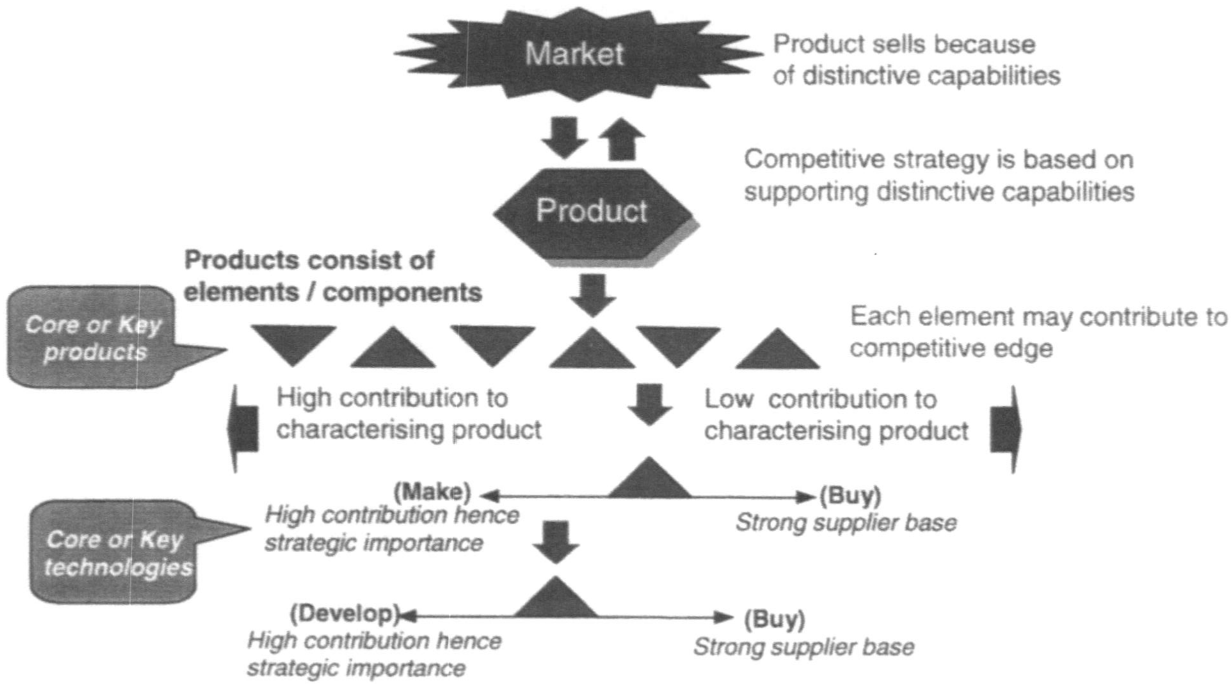

Figure 2: Integration management in supply chains

A Study of Technology Uptake Processes in Manufacturing SMEs

Rainer Seidel[1], Barbara Simpson[1], David Tweed[2], Darl Kolb[1], Judy McGregor[2], John Henley-King[2]

[1] The University of Auckland, Private Bag 92019, Auckland, New Zealand,
Email rha.seidel@gmx.net
[2] Massey University, Private Bag 11222, Palmerston North, New Zealand
Email D.M.Tweed@massey.ac.nz

Abstract: This paper presents the findings of a three-year study of the ability of manufacturing SMEs to take up new or improved technologies and use them effectively in their efforts to innovate. Based on empirical fieldwork, an integrated model that covers a broad range of factors that influence technology uptake in manufacturing SMEs has been developed. The model has been tested, refined and extended to provide a list of competencies necessary for the successful uptake and management of technology in SMEs. Manufacturers are able to use the model and competency list as an audit tool to identify their areas of strength and opportunities for improvement by comparing their competencies with those needed to successfully manage technology uptake and innovation.

1 Introduction

Continuous innovation in all areas of business and the strategic uptake of new technologies are key determinants for the ability of organisations to stay competitive. As there are many factors that affect the progress and the results of innovation, and to due to the large impact of these factors on business competitiveness, innovation and technology uptake have attracted much research from different academic disciplines. Some of the areas involved are manufacturing engineering, product design and development, information systems, project management, organisation and management theory, entrepreneurship, and economics. Due to this diversity and in particular the complexity of the issues involved there have been few attempts so far in analysing innovation and technology uptake processes from a viewpoint that integrates the large amount of knowledge and expertise from the various disciplines [1]. This is particularly true in the domain of Small and Medium sized Enterprises (SMEs), as most studies have concentrated on innovation in larger organisations, e.g. [2]. However, SMEs play an essential and universally recognised role in world economics [3, 4].

The Technology for Success study examines technology uptake and innovation in New Zealand manufacturing SMEs from a broad multidisciplinary basis. The main objectives of the study are to identify the managerial competencies that are essential for successful technology uptake in these organisations, to develop a model of the technology uptake process and to develop action learning tools to improve the management skills and knowledge.

2 The Manufacturing Environment in New Zealand

New Zealand is a relatively small country with around 3.7 million inhabitants. Historically the country's economy has largely depended on exports of agricultural, horticultural and forestry products and other natural resource-based industries. The manufacturing industry in New Zealand that developed in this environment has traditionally concentrated on the provision of engineering support to the primary sector and on the processing of the country's natural resources. A long period of almost total protection for the industry through tariffs and import restrictions ended in the mid-1980s when these barriers were gradually removed and a 'level playing field' free market economy was established.

An important feature of the New Zealand manufacturing environment is the lack of big organisations and the large number of small and medium sized businesses. The vast majority - 98% - of the New Zealand manufacturing organisations are categorised as SMEs (which in New Zealand are defined as having less than 100 staff [5]), whereas for example only 43% of manufacturers in the European Union employ fewer than 100 people [6].

In respect to technology and innovation much of the New Zealand manufacturing industry is still at a fairly early stage of using technology as a tool for creating sustainable competitive advantage [7]. As has been noted in a number of studies of the industry environment [8, 9], the use of technology in most New Zealand businesses is relatively low, and in particular the smaller companies have relied on 'Kiwi ingenuity' for their innovation needs. While this ability to improvise has served the industry well when the country was relatively isolated from the main world markets, the opening of the New Zealand economy and the globalisation of international trade have clearly shown the limits and dangers of the resulting industry culture. Examples for this culture are the unwillingness or inability of many organisations to use outside providers for their technology development [10] and the relatively common view amongst smaller manufacturers that R&D staff are not relevant to either product or process innovations in their firm [8]. Recent studies have indicated that many New Zealand manufacturers still do not take a strategic approach to technological development. The majority of firms do not appear to manage innovation and technology in an integrated way by aligning the development of new processes and technologies with strategic and human resource practices [10]. Generally large deficiencies in generic management skills and knowledge in particular in manufacturing SMEs have been identified, and there have been repeated calls for more emphasis on removing these impediments to strategic innovation and technology uptake by industry.

3 The Technology for Success Programme

3.1 Programme set-up and objectives

The Technology for Success (T4S) programme was established in 1998 as a response to the call of industry bodies and academic researchers to investigate the particular problems of New Zealand manufacturing SMEs in using new technology to improve their business competitiveness. T4S is a 3-year study funded by the New Zealand Government aimed at 'Removing Impediments to the Uptake of New and Improved Technology by Small and Medium Sized Manufacturing Enterprises'. The multidisciplinary research team from 2 different New Zealand universities has backgrounds in the areas of small business management, manufacturing engineering, information technology, management theory, organisational learning, and technology management and innovation, which provide an almost complete cover of the multi-facetted and diverse research field.

As the its title suggests, the main goal of the study is to contribute to the removal of obstacles for technology uptake of SMEs in New Zealand. Particular emphasis in the research is put on managerial competencies that affect the ability of New Zealand manufacturing SMEs to innovate, and on organisational learning which is the basis for improvement of their innovation behaviour [5, 11].

3.2 Field research

In order to achieve this the research programme was structured into a number of objectives and logical steps. In the initial phase a knowledge base of innovation, technology management and small business management in the New Zealand manufacturing sector was established by identifying relevant research as well as existing support mechanisms and structures and institutions involved in technology transfer and innovation. The main objective of this work was to achieve a sound understanding of innovation and SMEs as a basis for the subsequent field study.

The field study approach was chosen to capture all relevant aspects and to make best use of the broad and complementary research skills of the project team. Another important factor was the intention to perform the research inductively on the basis of grounded theory [12] to develop an unbiased perspective of the unique situation of New Zealand manufacturing SMEs.

The initial phase of the field involved twelve SMEs and was carried out over a period of several months by conducting interviews with top management, owners, and technical and other key staff, by observing company processes, operations and technologies applied, and by collecting documents and any other relevant information. The result was an extensive data base of qualitative and quantitative information on company history, structure, management, and operations, and in particular on innovation behaviour and mechanisms. Following the grounded theory approach the information in the data base was analysed [13], and in a subsequent study phase solidified and extended through further studies performed in a nation-wide series of focus groups and seminars.

3.3 Model for technological learning

The first main outcome of the field study was a framework of recurring themes that appear in each of the case studies [14]. As a first steps of theory building a process model for technological learning was derived from the framework [5]. Learning is at the centre of the process which is shown in Figure 1. The four knowledge components Know Why, Know How, Know Whom and Know Yourself are closely interrelated, indicated by the circular arrows. Learning about new technology happens when all four components are involved, otherwise there are impediments that affect the innovation performance of the business.

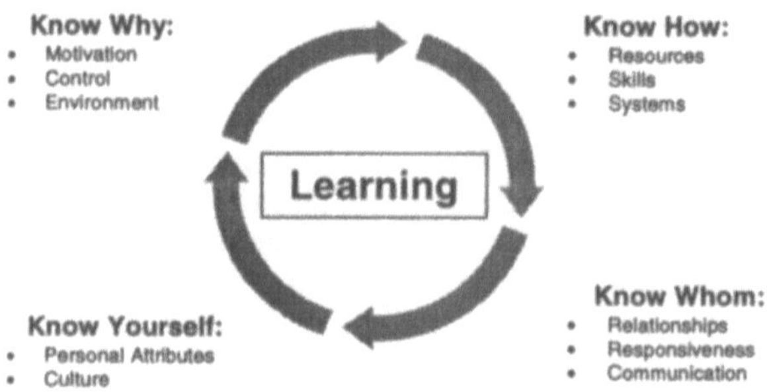

Figure 1: Process model for technological learning [14]

3.4 Integrated model of technology uptake

The above model provides a sound conceptual basis to the understanding of technological learning from an academic viewpoint, in particular when its four learning stages are not perceived as a two-dimensional cycle, but more as a three-dimensional, spiralling process where each revolution leads to a higher degree of technological maturity. However like similar concepts and approaches published in the literature [11, 14] this model does not provide much direct support for managers to improve their organisation's innovation performance. In order to achieve this tools have to be provided that allow managers to identify their personal and their organisation's areas of weakness [15].This is the concept behind the integrated model of technology uptake developed in the next step of this research [16]. The diverse range of factors affecting innovation is captured in the model by the three dimensions *Process*, *Style* and *Skills* shown in Figure 2. The three-dimensional nature of the model reflects the close interdependencies between the three aspects: Managers require and/or are more likely to need certain *Skills* depending on the *Style* of their company and their current stage or stages of the technology uptake *Process* [16].

Most innovation models in the engineering management literature are based on the product development *Process* and are basically (development) stage models [17, 18]. They are characterised by a sequential view of the process of technology uptake and use process stages that are separated by milestones or events. Stage models appeal to engineers as they emphasise the quantitative aspects of innovation; they also have direct applications in the project management of technology uptake. In the model the Process dimension has five stages: Motivation for Change, Source of technology solutions and ideas, Development, Implementation, and Review and decision making.

The dimension of *Style* captures many of the more qualitative aspects of an organisation's innovation environment. At this stage 13 characteristics of innovating SMEs have been defined as dichotomous features [16]. Typical features include Team vs. individualistic culture, Strategy vs. opportunity- driven approaches to planning, Change vs. stability orientation, and Systematic vs. informal approach.

The third dimension of the model, *Skills*, relates to 54 management skills and knowledge areas that have been identified in the field studies to have an impact on an organisation's ability to innovate. This dimension addresses the issue of managerial competencies, whose improvement is a key objective of the T4S programme.

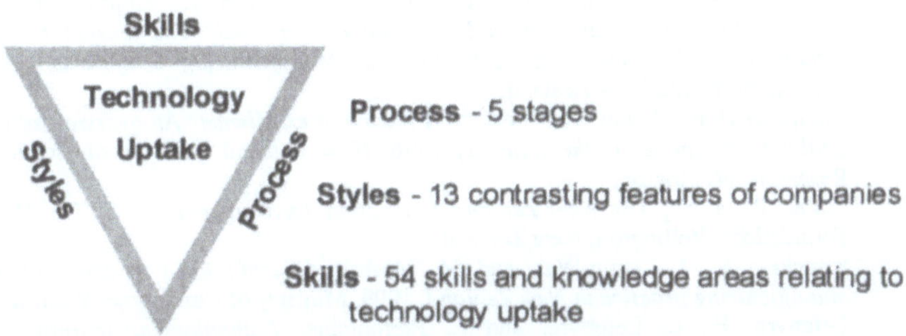

Figure 2: Integrated Model of Technology Uptake

A first prototype of the integrated model has been tested in a nation-wide survey [19] and at a professional development seminar. Comments on the usefulness of the survey and the model by participating managers have been very encouraging. Current work in the project concentrates on the refinement of the model parameters, and in particular on the development of a management training programme on the basis of the survey results.

4 Conclusions

The work in the T4S research programme has provided new insights in the nature of innovation and the factors that affect technology uptake in manufacturing SMEs. The grounded theory approach applied in the research has proven to be very successful in a scenario that is characterised by a large range of diverse factors, and has resulted in prototypes of a new model of the technological learning process and an integrated model of innovation in SMEs. Current work in the programme continues on the refinement of the models and on the development of a training programme for SME managers to become better innovators.

References

1. Tang, H.K., *An integrative model of innovation in organizations.* Technovation, 1998. 18(5): p. 297-309.
2. Dougherty, D. and C. Hardy, *Sustained product innovation in large, mature organizations: Overcoming innovation-to-organization problems.* Academy of Management Journal, 1996. 39(5): p. 1120-1153.
3. Lefebvre, E., L.A. Lefebvre, and L. Prefontaine. *The role of technological capabilities in the internationalisation of R&D intensive SMEs.* in *32nd Hawaii International Conference on Systems Sciences.* 1999. Honolulu: IEEE.
4. Blackburn, R. and P. Jennings, eds. *Small firms: Contributions to economic regeneration.* . 1996, Paul Chapman Publishing: London.
5. Simpson, B., *et al.*, *Learning in the Manufacturing Sector.* University of Auckland Business Review, 2000. 2(1): p. 38-50.
6. OECD, *Globalisation and small and medium enterprises - Volume 1 Synthesis Report*, . 1997, Organisation for Economic Co-operation and Development: Paris.
7. Hettema, H. and R. Seidel. *Can innovation be managed? The tension between technology and business.* in *Technology for Manufacturing Conference.* 1997. Palmerston North, New Zealand.
8. Campbell-Hunt, C. and L.M. Corbett, *A season of excellence? An overview of New Zealand enterprise in the nineties*, 1996, New Zealand Institute of Economic Research: Wellington.
9. Frater, P., *et al.*, *The New Zealand Innovation Environment*, 1995, The BERL Foundation: Wellington, New Zealand.
10. Knuckey, S., J. Leung-Wai, and M. Meskill, *Gearing up - A study of best manufacturing practice in New Zealand*, 1999, Ministry of Commerce: Wellington.
11. Lefebvre, E., L. Lefebvre, and L. Prefontaine. *Technological learning and organizational context: Fit and performance in SMEs.* in *Thirtieth Hawaii Conference on Systems Sciences.* 1997. Honolulu: IEEE.
12. Glaser, B. and A. Strauss, *The discovery of grounded theory: Strategies for qualitative research.* 1967, New York: Aldine Publishing Company.
13. Strauss, A. and J. Corbin, *Basics of qualitative research: Grounded theory procedures and techniques.* 1990, Newbury Park: Sage.
14. Simpson, B., *et al. Towards a new model of technological learning: Experiences from manufacturing SMEs in New Zealand.* in *British Academy of Management Conference.* 1999. Manchester.
15. Farrukh, C., R. Phall, and D. Probert. *Tools for technology management: Dimensions and issues.* in *PICMET 99 - Portland International Conference on Management of Engineering and Technology.* 1999. Portland, OR.
16. Kolb, D., *et al.*, *Technology Uptake in New Zealand Manufacturing: Report to New Zealand Manufacturers' Association*, 1999, The University of Auckland and Massey University: Auckland and Palmerston North.
17. Cooper, R., *Winning at new products: Accelerating the process from idea to launch.* 1993, Reading, MA: Addison-Wesley.
18. Crawford, C., *New products management.* 4th edition ed. 1994, Burr Ridge, Ill.: Irwin.
19. Tweed, D., *et al.*, *Technology for Success survey: Helping SMEs to prosper*, 1999, The University of Auckland and Massey University: Auckland and Palmerston North.

A Feature-Based Design System for Concurrent Engineering

Rainer Seidel[1] and Michael Norton[2]

[1]Department of Mechanical Engineering, The University of Auckland, PO Box 91092, Auckland, New Zealand, Email rha.seidel@gmx.net
[2]Criterion Manufacturing Ltd, PO Box 58646, Greenmount, Auckland, New Zealand, Email michael@criterionfurniture.com

Abstract: This paper presents the development of a feature-based design framework that facilitates communication between the different functions in the manufacturing cycle. Particulár emphasis is placed on the integration of sales and marketing considerations in the design process through the extension of feature-based design principles upstream from design into the sales and marketing functions. The framework is based on Viewpoint Relationship Diagrams that formalise the specific information requirements of the various functions involved in the product design and manufacturing cycle and encourage a greater understanding of other functions involved in the design process.

1 Introduction

The functional roles in the manufacturing cycle can be generally defined as some combination of consumer, retailer, marketing, concept design, detail design, planning, production, sales, retailer, and consumer [1,2,3]. A great deal of attention is currently paid to enhancing customer/supplier relationships and encouraging 'customer focus' [4,5] between both internal and external functions within the manufacturing cycle. Using the feature-based design and concurrent engineering concepts it is common for design engineers to work with production engineers on the joint development of a new product and its production process with this approach often leading to significant savings in development time and cost [6,7].

One of the largely unexplored opportunities of feature-based design utilisation is the integration of design with its upstream functions such as marketing and sales. Traditionally the marketing function has acted as a filter in the information flow between the consumer/retailer and design, and it is still common practice that product briefs are 'thrown over the fence' by marketing towards design. Conversely, when design is undertaken as an internal function of a manufacturing company, the designer's mindset often leans towards the produceability of the design based on the designer's subjective understanding of the customer's parameters. The design function can then find itself in the middle of personality or stereotype barriers between marketing and production. Such barriers may be the most difficult of all communication barriers to reduce or eliminate [8]. Because of this an opportunity exists to seek mechanisms that enhance understanding and help to build trust between functions [9].

Besides the often differing interests and viewpoints of members of a cross functional group [10], a further hindrance to working more closely together is the often technical and sometimes hidden nature of manufacturability and design constraints. This area of design can be perceived by some as a mysterious 'black box' [11] controlled by experts who think and talk in complicated terminologies that particularly marketing and sales oriented people are neither willing, able, nor should be expected to understand.

In this paper we present a summary of our approach to addressing this black box communication issue by constructing a feature-based design framework around the multiple viewpoints of the different manufacturing functions mentioned above. Our work has been directed towards the panel furniture industry where we have worked with a large Australasian manufacturer to identify the design viewpoints and communication issues that are most relevant to this application.

2 Application Area

The panel furniture industry produces panel based hard furniture such as office furniture (desks, storage cabinets, etc.), home furniture excluding soft furnishings (tables, wall units, entertainment centres, etc.) and kitchens. The products may be made from solid wood or composite board materials, and may be sold fully built up or ready to assemble.

Customer driven design is particularly relevant in panel furniture production where changing fashions and styles are an important consideration and where the ratio of cosmetic to functional features is often high. Therefore this industry offers particularly good opportunities for a feature-based design system that allows the design customer (either marketing, sales, or ultimately the consumer) to make their own decisions regarding feature modifications to the product's design.

In the mass produced panel furniture market cost is a key sales driver and the efficient utilisation of materials and mass production machinery is very important. In such an environment design is not totally flexible because of cost requirements, and good communication and design understanding between the sales, marketing, design and production functions is very important. This is the target area of our multiple viewpoint feature-based design framework [8], which consists of an optimised design structure including a set of feature-based design and communication tools .

3 Objectives and Development Issues

The primary goal of the design framework is to establish a basis for improved cross-functional communication in the design/manufacturing cycle of panel furniture products from sales and marketing through to production. Other important objectives are the utilisation of feature-based design concepts beyond the technical detail design function and the support of the specific interests and viewpoints of the different functions. This requires in particular to provide marketing with some form of feature-based design tools that allows them to participate more actively in the technical design process.

There are a number of issues that are important for the successful application of the design framework; they are in particular manufacturability, sustainability, and methodology. Features as means of design communication have the significant potential of forcing the consideration of manufacturability from an early stage of a design. Generally features that directly indicate manufacturability can be very beneficial to the design process, and if structured in an appropriate way can also open doors for non-design people to undertake limited product design.

Implementing any design system that attempts to cover a broader spectrum of manufacturing functions than just design itself is a very complex task and requires significant investments in development cost and in staff. It is thus very important that the framework for the design system is sustainable and does not have to be reinvented each time a new opportunity for system is identified and implemented. Analysis method and framework should facilitate as much 'foresight' into such opportunities as can reasonably be achieved.

The methodology for creating a feature-based design framework should preferably not be completely new and should already have some acceptance inside normal design practice. If this can be achieved then it will aid the understanding of design process and procedures by all parties involved. Both process and procedures should facilitate obtaining the maximum possible contribution (and resultant ownership) from each party involved in the design of new products.

4 Function Viewpoint Analysis

To build a framework for feature utilisation and design communication that extends beyond the design function itself, it is necessary to find or create a consistent means to analyse the viewpoints and information relationships of the design function, as well as those functions that precede and follow design in the product development cycle. It is important for the method to successfully address the issues outlined above. The new method is entitled the Viewpoint Relationship Diagram (VRD) and is based on the House of Quality analysis technique which is a well known analysis method that is generally used to help develop a new product [12].

While Viewpoint Relationship Diagrams have a similar structure as the House of Quality, the contents of the diagrams differ significantly. Instead of mapping customer attributes against engineering characteristics, then engineering characteristics against parts characteristics etc., VRDs map the viewpoints of the various manufacturing functions against each other [13]. Typical choices for viewpoints are *Marketing* (perhaps combined with *Retailing*), *Conceptual Design, Detail Design, and Planning* (perhaps combined with *Production*). Figure 1 shows an example VRD for *Marketing* and *Conceptual Design.*

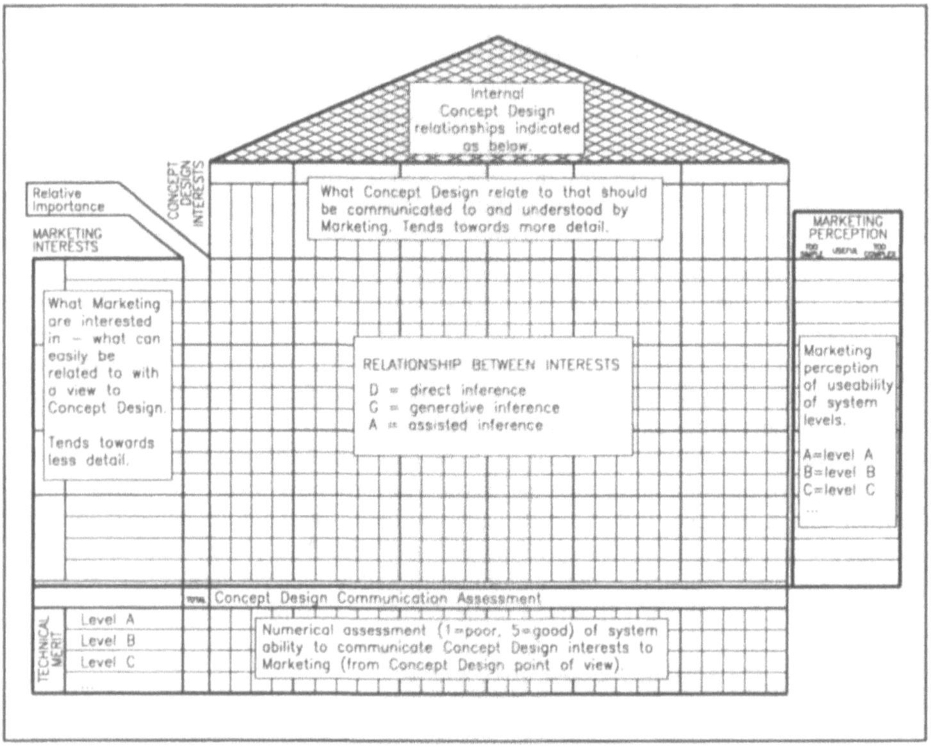

Figure 1: Viewpoint Relationship Diagram for *Marketing* and *Conceptual Design* [8]

For each viewpoint appropriate primary and secondary attributes, or interests, have to be defined. Typical primary interests for a *Conceptual Design* viewpoint in the panel furniture industry are *construction*, *connections*, and *surfaces*. Example secondary attributes of the *surface* interest are *type*, *finish*, *material*, etc. As the relationships between interests are mapped against each other, it is beneficial if the interests are defined jointly between the functions involved.

The central matrix of the diagram is used to chart the different complexities of relationships between the primary and secondary attributes (or interests) belonging to each viewpoint. The relationship types are *direct* (where a simple logical one to one relationship exists), *generative* (where a complex many to one relationship exists) and *assisted* (where the relationship cannot reasonably be expressed using logic). Particularly the direct and generative relationships of such an analysis highlight opportunities for streamlining or automating communication of design parameters and development of design. This analysis is used to support the conceptualisation of feature levels within a design framework proposed for whatever situation the diagram is created for.

After a framework has been proposed, the VRD is extended to include an analysis of the value of each proposal in terms of supporting the defined interests of each viewpoint [8].

5 Application of the Design Framework

As the scope and amount of design information in the framework is so large, it is appropriate to represent this information in a number of relating levels where each level is a progression of design/manufacturing detail from the previous level. The data types chosen should be multidisciplinary, that is all users of the information need to be able to relate to or at least comprehend the structure and reasons for each data type.

It is easiest to first determine the data type for the highest level. Generally this is the *product*, representing the finished, packaged, saleable item, for example a computer desk or an entertainment centre. Other levels are determined by considering the stages of manufacturing process a typical *product* goes through. To gain acceptance from the various functions involved however, care must be taken that the data type definitions are not too manufacturing oriented. For ease of use and communication it is also important that each data type can be represented graphically.

The data type for the lowest level is selected by considering what the minimum information is that will describe the maximum necessary design detail of a component for manufacture. The lowest design data level in many applications is likely to be a *feature* of some kind, for example *form feature*, *functional feature*, *assembly feature*, etc. In the panel furniture framework one *design feature* may be made up of one or several *manufacturing features* and may also include items of purchased hardware.

Having determined the data types at each end of the framework, it remains to create the intermediate level data types. For the development of the design framework for panel furniture, two more levels were defined between the product level at the top and the feature level. The first level down from the product level was defined as the *component* level. This level covers finished manufactured components, but also items of purchased assembly hardware such as a screws or hinges that are not fitted directly to a component as part of its manufacture.

The level between the already defined feature and component levels was determined as that of the *strip*. In panel furniture manufacturing a majority of panel components can be viewed as being made in a long *strip*, finished length-wise on two edges, then cut to length and finished on the remaining two edges, then perhaps drilled, then becoming a completed *component*. Having a *strip* level in the description of the makeup of a *component* provides a direct linkage to the actual process the component undergoes in order to be manufactured.

Thus four levels or layers of information have been chosen to represent the transition from for instance, a single edge shaping tool to a finished product. Each level is a progression in detail from the previous with different attributes ascending through the levels to define the design of the finished product.

Based on the set of viewpoints and data levels for the panel furniture design framework, the next step was the creation of data instances throughout the four levels. The primary considerations in this exercise were how the data are used from within each viewpoint and how the data can be used to link one viewpoint to the next so that the progression of data through the levels and viewpoints is as automated as possible.

6 Conclusions

The above approach to multiple viewpoint feature-based design has provided many different opportunities to improve processes in panel furniture production. In the pilot company the framework has provided a pathway of learning into the design process. Depending on an individual manufacturer's approach and requirements, this pathway may be structured in such a way that a design toolbox can be created and passed over to marketing and sales functions. These functions can then undertake design work and specification in their own right. Alternatively, the pathway may work in reverse to more easily allow a designer to communicate directly with one of marketing or sales own end customers.

References

[1] Pahl, G., Beitz, W. (1988): Engineering Design; A Systematic Approach, Edited by Wallace, K., The Design Council, Great Britain, pp. 40-44.

[2] Pugh, S. (1993): Total Design; Integrated Methods for Successful Product Engineering, Addison-Wesley, Great Britain, pp. 5-11.

[3] Hales, C. (1993): Managing Engineering Design, Longman, Great Britain, pp. 5-9.

[4] Price, R.C. (1995): TQM in the R&D Function, Quality Progress, Vol. 28, No. 7, pp. 109-111.

[5] Sommer, E. (1993): Fundamentals of Quality Work in Companies, Chemie Ingenieur Technik, Vol. 65, No. 12, pp. 1457-1464.

[6] Turino, J. (1994): Accelerating the Engineering to Manufacturing Transition, Proceedings Westcon Conference Los Angeles, pp.368-371.

[7] Youssef, M.A. (1994): Design for Manufacturability and Time to Market (Part 1), International Journal of Operations and Production Management, Vol. 14, No. 12, pp. 6-21.

[8] Norton, M.H. (1998): Multiple Viewpoint Feature-Based Design, PhD Thesis, The University of Auckland

[9] Griffen, A., Hauser, J.R. (1996): Integrating R&D and Marketing: A Review and Analysis of the Literature, Journal of Product Innovation Management, Vol. 13, No. 3, pp. 191-215.

[10] Shooshtarian, F.D., et al. (1993): Design and Implementation of a Relational Database for Automated Process Planning, Computers and Industrial Engineering, Vol. 25, No. 1, pp. 309-312.

[11] Jones, J.C. (1992): Design Methods, Van Nostrand Reinhold, New York, [12] Hauser, J.R., Clausing, D. (1994): The House of Quality; in Clark, K.B., Wheelwright, S.C: The Product Development Challenge, A Harvard Business Review Book, Harvard Business School, USA, pp. 299-317.

[13] Seidel, R.H.A, Norton, M.H., Tedford, J.D (1997): A Feature-Based Framework for Concurrent Engineering, Proceedings Technology for Manufacturing Conference, Feb. 1997, Massey University, Palmerston North, New Zealand, pp. 427-435.

INTELLIGENT MANUFACTURING SYSTEMS I

Multiple Criteria Approach to the Stochastic Assembly Line Balancing Problem

A. J. R. Smith[1] and A. Bhandari[2]

[1]Department of Mechanical and Manufacturing Engineering, The University of Melbourne, Victoria 3010, Australia, ajrs@mame.mu.oz.au
[2]PricewaterhouseCoopers Management Consulting Services, Melbourne, Victoria 3000, Australia, amit.bhandari@au.pwcglobal.com

Abstract: The line-balancing practitioner frequently encounters multiple objectives and uncertainty in the available data as well as precedence. Whilst the research literature has examples of these two issues being considered independently it is difficult to find reference to their simultaneous consideration. This paper presents a chance-constrained, binary integer, goal programming approach to the mixed model stochastic assembly line balancing problem. The proposed method is illustrated by a numerical example, and a diverse range of objectives that a decision-maker may face while balancing lines have been considered.

1. Introduction

The common problem associated with the design of assembly lines is the best allocation of tasks to each workstation. Traditionally the assembly line balancing problem has been formulated and solved as a single objective problem with the objective being either to minimize the cycle time given a number of stations or minimize the number of stations utilized given the required cycle time. However, the assembly line balancing literature reveals a number of varied, and potentially conflicting, objectives to be faced when balancing assembly lines.

Some researchers have started acknowledging the multiple objective nature of the assembly line balancing situation in practice. The approaches have been varied, such as using mixed integer goal programming, maximizing profit by lowering the cycle time in steps and measuring the balance efficiency, and forming a single composite weighted function composed of different objectives that may be optimized. Whilst this last approach, as demonstrated in Malakooti [1], is initially appealing it does suffer from the requirement to precisely allocate weights and does not ensure that any one goal will be fully satisfied.

Most of the multiple criteria literature is concerned with single models and very little work from a multiple objective perspective has been reported in the mixed model category. One of the few exceptions is the work of Gokcen and Erel [2] who have presented an integer goal-programming model for the deterministic mixed-model assembly line balancing problem. Researchers have also long recognized the variable nature of task times and some of the earlier works have approached the problem by increasing each task time before assigning work to stations. Whilst this method allows the stochastic problem to be attacked with a deterministic solution procedure it often leads to very poor utilization of resources.

Both stochastic behaviour and multiple objectives have been considered in assembly line balancing separately but due to the complexity involved in stochastic multiple objective models the combined problem has been virtually ignored in the literature. This paper attempts to begin redressing that situation.

2. Goal programming formulation

Goal programming is a multiple criteria decision-making technique that is based on the concept of '*satisficing*'. This implies that it seeks a solution that fully satisfies as many goals (or targets) as possible rather than optimizes a single goal. A pre-emptive approach to goal programming is utilized in which the higher priority goals are satisfied first and only then are the lower priority goals considered.

The model presented here assumes that the goals can be specified and ranked in the order of their importance. Further, similar models are made on the same production line and the precedence for the various models is known precisely. A few other assumptions are also made:

- Task splitting is not allowed. If it is unavoidable each split task becomes a separate work element.
- Task times are normally distributed with known means and variances.
- The tasks are independent of each other, except through precedence.
- Number of stations is the same for all the models. i.e. each station has at least one task from each model.

Typically a mixed-model assembly line will have several tasks common to the various models manufactured on the line with similar precedence relations amongst these common tasks, otherwise separate lines may be more cost effective.

Thomopoulous [3] has presented the concept of combining two or more precedence diagrams, one for each model, to represent a mixed model system. Essentially, the combined precedence diagram transforms the mixed model system into an equivalent single model system so that single model line balancing techniques may be used. The mixed-model nature of the problem is preserved through the setting of constraints on the allocation of tasks to stations.

2.1 Earliest and latest task assignments

In the balancing problem presented here the variables used by Patterson and Albracht [4] are modified to accommodate mixed models and task time, t_i, variation. This is done by using the upper bound on the cycle time for a particular model 'm', C_{ubm}, in the calculation of the earliest, E_i, and latest stations, L_i, to which a particular task may be assigned.

$$E_i = \begin{cases} 1 & \text{if} \quad (E(t_{im}) + \sum_{j \in PR_i} E(t_{jm})) = 0 \\ [(E(t_{im}) + \sum_{j \in PR_i} E(t_{jm})) / C_{ubm}]^{+} & \text{otherwise} \end{cases} \tag{1}$$

$$L_i = \begin{cases} K & \text{if} \quad (E(t_{im}) + \sum_{j \in S_i} E(t_{jm}) = 0 \\ K + 1 - [(E(t_{im}) + \sum_{j \in S_i} E(t_{jm})) / C_{ubm}]^{+} & \text{otherwise} \end{cases} \tag{2}$$

where $[z]^+$ is the smallest integer greater than or equal to z, PR_i is the set of all tasks preceding task i and S_i is the set of all tasks following task i.

The earliest and latest stations that tasks for combined precedence can be assigned to are $\text{Max}_{m=1,\dots P}\{E_{im}\}$ and $\text{Min}_{m=1,\dots P}\{L_{im}\}$ respectively.

2.2 Assignment and precedence constraints and goals

Assignment constraints assure that each task is assigned to only one workstation. Assignment constraints can be found in most scheduling models [4] and are normally not suitable for consideration as goals. However, if the exclusion of a non-essential task allowed the satisfaction of some higher level goal, an assignment goal could be made by adding a deviational variable, d_i^-, to the constraint equation:

$$\sum_{k=Ei}^{Li} V_{ik} + d_i^- = 1 \tag{3}$$

where V_{ik} represents the assignment of task i to station k ,
and $d_i^- = 1$ would allow the exclusion of task i.

The precedence relation between task 'a' and task 'b', where 'b' is an immediate follower of 'a', can be assured by suitable constraints [4]. As for assignment constraints, precedence constraints are not appropriate for consideration as goals since it is unlikely that one would desire a change in the task precedence. However, a possible goal involving precedence could be if it is desirable to have task 'a' completed within υ workstations of task 'b' and this can be formulated as in equation (4). Here $d_p^+ = 0$ if task 'a' precedes 'b'; $d_p^+ > 0$ if 'a' does not precede 'b'.

$$\sum_{j=Ea}^{La} j.V_{ak} - \sum_{k=Eb}^{Lb} k.V_{bk} - \upsilon - d_p^+ = 0 \tag{4}$$

2.3 Chance constraints and goals

Chance constraints are generally expressed in the form shown below :

$$\text{Prob} \left(\sum_j a_{ij} X_j \leq b_i \right) \geq \beta_i \tag{5}$$

where the terms in parentheses represent desirable constraints/goals and β_i represents the desired probability of goal achievement. In the line balancing problem addressed here only variation in task times, represented above by a_{ij}, will be considered. For this situation, the conversion of the chance constraint to its deterministic equivalent takes a non-linear form [5]. This presents a difficult situation that normal linear programming techniques cannot solve. A method [6] of converting the non-linear portion of the chance constraint to its deterministic linear equivalent is presented in this paper and is based on Naslund [5]. In an assembly line with stochastic task times, the decision-maker will often desire that the probability of the sum of task times at a particular station exceeding the cycle time should be less than a specified probability. This constraint can be expressed as

$$\text{Prob}\left(\sum_{i \in W_{km}} t_{im} V_{ik} \leq C_m\right) \geq \beta_i \qquad k = 1,\ldots,K,\ m = 1,\ldots,P \qquad (6)$$

This chance constraint after converting to deterministic equivalents and subsequently using Naslund's approximation [5] can be expressed as in equation (7). The deviational variable, d^+_{km}, denotes over-achievement of the cycle time for each station for model m and d^-_{km} denotes the idle time in station k for model m.

$$E\left(\sum_{i \in W_{km}} t_{im} V_{ik}\right) + F^{-1}(\beta)\sqrt{\left[V\left(\sum_{i \in W_{km}} t_{im} V_{ik}\right)\right]} + d^-_{km} - d^+_{km} = E(C_m) \qquad (7)$$

2.4 Workstation constraints and goals

To implement the assumption that each model utilizes the same number of stations, station constraints developed by Deckro [6] are employed. Let $X_{km} = 1$ if workstation k is needed for model m and $X_{km} = 0$ otherwise. $||W_{km}||$ represents the cardinality of the set of all tasks that can be assigned to station k for model m, W_{km}. Then,

$$\sum_{i \in W_{km}} V_{ik} - ||W_{km}|| X_{km} \leq 0 \qquad k=1,\ldots K;\ m=1,\ldots P \qquad (8)$$

$$\sum_{m=1}^{P} X_{km} - P.A_k = 0 \qquad k = 1,\ldots K. \qquad (9)$$

where A_k is equal to 1 if station k is utilised for all models; otherwise it is 0.

By adding a desired number of stations, Φ, and the appropriate deviational variables, the workstation goal can be written as in equation (10). The classic objective of minimizing the number of stations along the line can be achieved by maximizing d^+_{st}.

$$\sum_{k=1}^{K} A_k + d^-_{st} - d^+_{st} = \Phi_k \qquad (10)$$

2.5 Compatible and Incompatible zoning constraints and goals

In compatible zoning, it is desirable or essential to perform a set of tasks in the vicinity of each other (e.g. sharing tooling) or it might be desired that task 'a' should directly succeed task 'b' (e.g. cleaning and painting operations). The compatible zoning constraint can be expressed as a goal as follows [6]:

$$\sum_{j=Ea}^{La} j.V_{ak} - \sum_{k=Eb}^{Lb} k.V_{bk} + d^-_{comp} - d^+_{comp} = 0 \qquad (11)$$

If d^-_{comp} and d^+_{comp} are minimized to zero, the compatible tasks 'a' and 'b' will be assigned to the same workstation.

An incompatible zoning constraint indicates that a work element might interfere with another and should therefore not be located in close proximity. This can be achieved by setting the right hand side of equation (11) to 'ω', where ω represents the number of stations by which it is desired that tasks 'a' and 'b' should remain apart. The deviational variables could be renamed d^-_{inc} and d^+_{inc}

3. Illustrative Example

A problem based on Gocken and Erel [2] is used to illustrate the approach outlined above. It consists of two models, M1 and M2. The generality of this problem can be seen from table 1 where the same tasks for different models have different mean times, E(t) and variances, V(t). The generality of this example is further captured by the choice of goals, one is a deterministic goal, another a stochastic goal and the third is an optimizing goal. The methodology is not limited to three goals, but these should be adequate to display the characteristics. The goals are, in priority order,

Goal A: Task 2 and 8 should not be assigned to the same station.

Goal B: The probability of the summation of task times at each station being less than the cycle time, 24, for model 1 and 26 for model 2 should be greater than 0.96.

Goal C: Minimize the number of stations.

Table 1. Elementary operations, their mean unit times, variances and dependencies

Task	1	2	3	4	5	6	7	8	9	10
$E(t_{M1})$	10	4	7	7	4	-	-	12	7	11
$V(t_{M1})$	1	0.16	0.49	0.49	0.16	-	-	1.44	0.49	1.21
$E(t_{M2})$	10	-	9	-	5	12	6	13	-	10
$V(t_{M2})$	1	-	0.81	-	0.25	1.44	0.36	1.69	-	1
Depends on	-	1	1	1	4	3	1	3, 5	2	6,7,8,9

The problem was solved using a commercial integer-programming package on a standard personal computer. The results indicate that the number of stations has been minimized to three (see table 2), after satisfying goals A and B.

Table 2. Task assignments and corresponding mean station times

Station	Tasks	Model 1		Model 2	
		Tasks	Station time	Tasks	Station time
1	1,2,3,7	1,2,3	21	1,3,7	24
2	4,5,8	4,5,8	23	5,8	18
3	6,9,10	9,10	18	6,10	22

The results of a sensitivity analysis are shown in table 3. The last two sequences attribute goal C (the minimizing goal) with the highest priority. All the tasks in these two sequences have been assigned to one station thereby satisfying the precedence relationships, but this has also resulted in goal A and B not being satisfied. This clearly shows that the placement of optimizing goals is very important in the priority order. Careless placement will result in most (if not all) following goals not being satisfied. This can also occur if a very restrictive goal is placed first, or an unreasonable upper bound for the number of stations is chosen.

Table 3. Sensitivity analysis by rotation of priorities

Priority 1	Priority 2	Priority 3	Unsatis-fied goal	Value of non-zero deviational variable	No. of stations
Goal A	Goal B	Goal C	-	A1+A2+A3+A4+A5= 3	3
Goal A	Goal C	Goal B	Goal B	CYC1+CYC2 = 14.649	2
Goal B	Goal C	Goal A	-	A1+A2+A3+A4+A5= 3	3
Goal B	Goal A	Goal C	-	A1+A2+A3+A4+A5= 3	3
Goal C	Goal B	Goal A	Goal A Goal B	ZNG =1 CYC1 + CYC2 = 85.63	1
Goal C	Goal A	Goal B	Goal A Goal B	ZNG = 1 CYC1 + CYC2 = 85.63	1

4. Conclusions

The proposed method has provided sensible results to a more complex line balancing problem than has previously been reported. That such problems can be solved on a PC has shown a potential for addressing real problems in industry.

The sensitivity analysis has also enabled a basic guideline to be established, namely that when multiple goals incorporate an optimizing function (such as minimizing the number of stations), the optimising goals should be placed last for sequential goal solvers. The sensitivity analysis conducted emphasizes the central importance of multiple goals being carefully formulated and that the order of priority be given due consideration.

Utilizing a combined precedence diagram has considerably reduced the size of the model for the mixed model problem. The use of variables such as the earliest and latest stations to which a task can be assigned to has also limited the number of decision variables and constraints. Problems with 20 tasks and 3 models and several goals have been successfully solved using this method.

The model presented in this paper is believed to be the first goal programming model for the stochastic mixed model line balancing problem.

References

[1] Malakooti, B., 1990, A multiple criteria decision making approach for assembly line balancing problem, Int. J. Prod. Res., Vol. 29, No.10, 1979-2001.

[2] Gokcen, H., and Erel, E., 1997, A goal programming approach to mixed-model assembly line balancing problem, Int. J. Prod. Econ.,Vol.48, 177-185.

[3] Thomopolous, N.T., 1970, Mixed-model line balancing with smoothed station assignments, Mgmt Science, Vol. 16, No.9, 593-603.

[4] Patterson, J.H. and Albracht, J.J., 1975, Assembly line balancing zero-one programming with Fibonacci search, Operations Research, Vol. 23, 166-172.

[5] Naslund, B., 1966, A model of capital budgeting under risk, J. Bus., Vol. 39, 257-271.

[6] Bhandari, A., 1999, Chance constrained goal programming model for the assembly line balancing problem, unpublished M. Eng. Sci. thesis, The University of Melbourne.

[7] Deckro, R.F, 1989, Balancing cycle time and workstations, Trans IIE, Vol. 21,106-111.

An Adaptive Multi-Agent Acquisition System for Co-operative Design in Early Product Development.

Evan Govender[1], Joe Y Shiau[2], Monica Lagazio[3], George Valtchanov[4]

[1] University of Nottingham, MEOM, University Park, Nottingham, NG7 2RD, epxeg@unix.ccc.nottingham.ac.uk
[2] University of Nottingham, MEOM, University Park, Nottingham, NG7 2RD, epxys@unix.ccc.nottingham.ac.uk
[3] University of Nottingham, Dept. of Politics, University Park, Nottingham, NG7 2RD, ldxml@nottingham.ac.uk
[4] University of Nottingham, MEOM, University Park, Nottingham, NG7 2RD, epxgvx@nottingham.ac.uk

Abstract: The paper reports on a new theoretical platform and pilot environment for rapid distributed product design in extended enterprises using collaborative, autonomous agents. The approach is based on exchange of requests and information between design, manufacturing, engineering and marketing agents. The resulting decisions are based on design criteria which is then optimised by an intelligent adaptive design moderator which interacts with the different agents to generate new solutions. This research aims to provide designers with an adaptive assessment tool to be used at conceptual design stage in complex virtual enterprises.

1. Introduction

In the effort to ensure that products are sufficiently profitable, the ability of companies to efficiently utilise available time for product development becomes a question of survival. In order to dramatically reduce the cost at the early stages of product development and produce a competitive product for the market place, the planners require methodologies and tools that would allow them to judge the implications of product design specification on the required manufacturing processes, facilities and marketing requirements.

In choosing a suitable methodology, the following factors should be considered: the increasing complexity of the interrelation across the main variables affecting the design process as well as the maintenance and up-dating of the chosen tool.

The increased globalisation of competition, more demanding customer requirements, and rapid technological developments have recently resulted in previously unknown levels of complexity and dynamism in the process of product development. Furthermore, the design tasks become even more complicated when the new product design process is carried out in a virtual enterprise environment where the capabilities of different geographically distributed business entities have to be taken into account. This means that more and more interrelated domain specific areas can affect the likelihood of success in the design development. As a consequence of this increasing number of variables, knowledge management requires new methodologies that are able to cope with these changing requirements that affect design performance.

Maintenance issues are also affected by the complexity characterising the design scenario. Because the nature of the design process changes over time, a decision making tool is needed that is easily adaptable and has low maintenance cost associated with it.

The system presented below attempts to address some of these issues by considering a multi-agent framework where the responses from the various distributed agents are combined using artificial intelligence techniques to adapt the current responses with past design scenarios.

2. Multi-Agent Approaches to Collaborative Design

A key characteristic of a concurrent engineering environment is that design is performed in a co-ordinated and collaborative manner which allows the consideration of the multiple factors involved in the process. Co-ordination is the key to maintain the consistency and adaptation of concurrent decision-making and the integrity of information in the product development lifecycle [4]. However, co-ordination in collaborative design takes time and requires knowledge, but if the proper tools are available it could reduce the lead-time and let designers focus on the essential problems of design [5].

A concurrent engineering framework based on a multi-agent planning environment is reported by Balasubramanian et al [1] and is further developed by Gu et al [2], where the proposed methodology assumes an agile manufacturing environment based on intelligent machines capable of making autonomous planning decisions.

Another multi-agent framework is proposed by Tan et al [3]. The framework draws on design-team behaviour in different domains with the goal of providing the infrastructure to enable designers, planners, and managers from different functional areas to collaborate using conflict resolution techniques in improving the initial product design.

3. System Architecture

In agent-supported collaborative design, design teams can co-ordinate and communicate with the manufacturing teams through computer software agents associated with the teams. In this context the human user acts as a complementary external expert who interacts with the software agent via a user interface.

The multi-agent approach presented here introduces a new concept: the intelligent adaptive design moderator. This moderator co-ordinates the process of exchange of requests and information between design, manufacturing, engineering and marketing agents, and evaluates the results of requests at each iteration. Figure 1 illustrates the architecture for the multi-agent system.

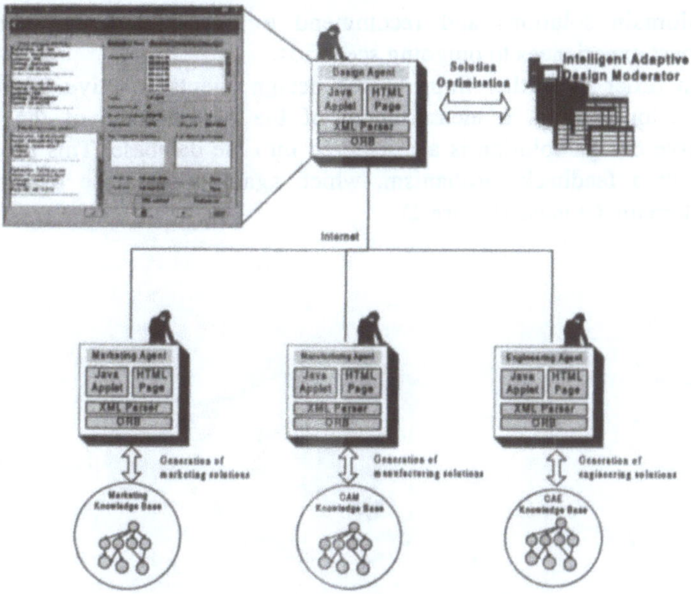

Figure1. Multi-agent system architecture

In constructing the design moderator, neural network techniques are used as a key component in the structure. This offers an alternative approach for efficiently integrating the domain-specific solutions generated by the individual agents and produces a collective solution from the responses it receives. Indeed, the design agent's knowledge is developed following the assumption that reasoning and searching for solutions on presently complex design tasks is an inductive process directed by examining past experiences, drawing analogies, reshaping patterns in the data corpus and adapting past solutions to newly experienced problems [6].

To achieve the necessary flexibility, the knowledge of the design agent is split into two components: a problem domain database and an adaptive decision tool based on the neural network algorithms.

The problem domain database stores information on the design specifications, designer queries, individual agent solutions, and design outcome of all previously encountered design cases as well as the new generated design scenarios. As new data from new design scenario is entered in the database, the problem domain is updated.

Once the responses generated by the individual agents using parameters from the initial design problem are entered in the problem domain database, the adaptive neural network tool is used to generate the collaborative solution. This co-operative process is aided by the ability of the adaptive tool to learn from past data as well as new design scenarios provided by the autonomous agents. By finding the general pattern characterising previously successful design projects and then adapting it to the new design scenario, the design moderator is able to optimise the individual solution generated by the single agents. In a single computational effort the neural decision tool is able to integrate the individual specific domain solutions and recommend a successful design strategy by adapting past experiences to outgoing scenarios.

In order to further improve the design agents adaptive capability for future decision-making, a measurement of the performance of the endorsed collaborative design solution is also entered into the database. This provides the system with a feedback mechanism, which again updates the design agent's problem domain database (Figure 2).

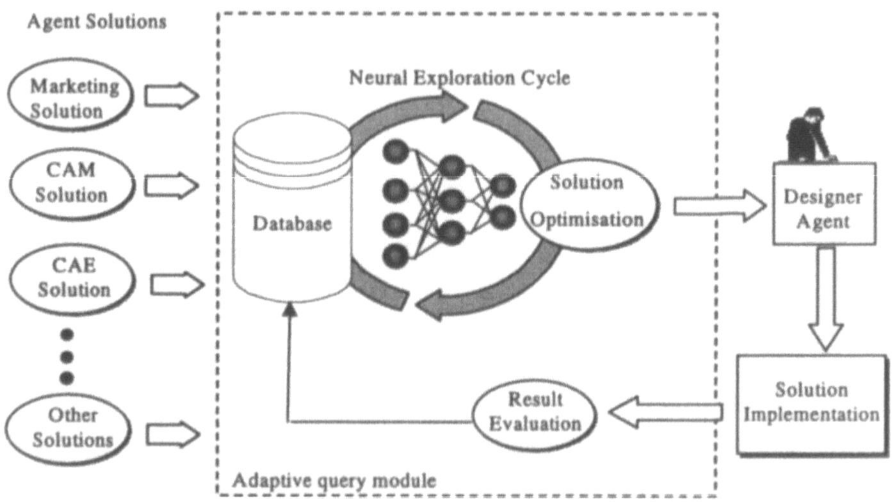

Figure2. The intelligent adaptive design moderator

The advantages of using a neural network implementation in contrast to other approaches, such as case based reasoning, is that in the situation of an incompletely described design query, or when no similar past cases exist, the system has the ability to internally manipulate the existing data to produce an adaptive solution. The neural network tool so provides an effective possibility to perform an adaptive task in a Case Based Reasoning environment.

4. Agent Communication

This section provides a detailed description of the languages and protocols utilised by the multi-agent systems in communicating within the suggested framework.

The agent abstraction is utilised as a mechanism for encapsulating and exchanging the distributed knowledge. The agent knowledge can be acquired implicitly by having the agents make assumptions independently or by jointly utilising support from other agents for co-ordinated activities by explicit communication.

Recently, researchers in the ARPA Knowledge Sharing Effort [7] have defined the components for communication, called the Agent Communication Language (ACL), which satisfies these needs. However, in many cases XML (eXtensible Markup Language) could be used to substitute ACL as a more broadly understood - albeit less expressive form of representation [8].

The Internet is increasingly becoming a popular, distributed, and open environment and hence attracts more people to investigate agent related knowledge and techniques on it. Web pages are essentially constructed using Hypertext Markup Language (HTML). However, by considering XML to encode the necessary information and services using content specific structures and semantics, it could easily be understood by agents as well. Other researchers have created XOL (XML-based Ontology exchange Language) [9] which is designed to provide a format for exchange ontology definitions that is easily published on the WWW for exchange among a set of application developers. Ontobroker [10] provides a framework which combine ontologies and XML documents.

In the single agent communication, a method is proposed which involves isolating the design domain, manufacturing domain, engineering domain and marketing domain, and modelling each domain using an object-relational data model. The structured data provided by the XML meta tags are then stored and linked to the domain knowledge. The communication scenario between the agents using XML is as follows:

A design agent sends a query to the autonomous agents for a particular manufacturing solution. This message is delivered to the XML parser/generator where the message is then parsed by the design application module to generate the code which includes details such as the location of other agents and the requirement string. This request is sent to the receiving agent where the needed operations are executed according to the content of the message from the sending agent, using its domain knowledge base. A reply message using similar steps detailed above, will be sent to the design agent. The result is a response from the domain specific agents to a query from the design agent within a distributed environment.

5. Conclusion

In this paper, a framework for intelligent multi-agent decision making is presented. The individual autonomous agents with their domain specific knowledge are interrogated by the design agent for a domain specific solution. The response from these individual distributed agents are collectively filtered by the intelligent adaptive design moderator where comparisons are made with outcomes of previous design scenarios and a feasibility decision is suggested by the system to the design agent. The collaborative and adaptive nature of the decisions put forward by the design agents match the evolving nature of the design problem. Indeed, the knowledge of the design moderator automatically adapts with the inclusion of new scenarios. This self adaptation based on new cases provides this system with a low maintenance, low cost, and flexible approach for decision recommendation in early product design that could provide a competitive advantage to its users.

References

[1] Balasubramanian, S. Maturana F. P. Norrie D. H., 1996, Multi-Agent Planning and Coordination for Distributed Concurrent Engineering. International Journal of Co-operative Information System. 5:153-179.

[2] Erdmann, M. Studer R., 1999; Ontologies as Conceptual Models for XML Documents. Twelfth Workshop on Knowledge Acquisition, Modeling and Management, KAW'99, Voyager Inn, Banff, Alberta, Canada.

[3] Genesereth, M. R. Ketchpel S. P., 1994, Software Agents. Communications of the ACM; 37(7):49-53.

[4] Gu, P. Balasubramanian S. Norrie D. H., 1997; Bidding-based Process Planning and Scheduling in a Multi-Agent System. Computers and Industrial Engineering. 32(1):477-496.

[5] Karp, P. Pangea Systems Inc., 1999, XOL: An XML-Based Ontology Language for Bioinformatics. Objects in Bioinformatics'99; Silicon Valley, San Jose.

[7] Lassila, O., 1998, RDF Metadata and Agent Architectures. Workshop on Compositional Software Architectures; Monterey, California.

[8] Lottaz, C. and Smith I., 1997, Collaborative Design Using Constraint Solving. Lausanne, Switzerland: Swiss Federal Institute of Technology (EPFL).

[9] Murthy, S. Akkiraiu R. Rachin J. and Wu F., 1998, Agent-based Co-operative Scheduling. Yorktown Heights, NY USA: IBM T.J. Wrson Research Center.

[10] Tan, G. W. Hayes C. C. Shaw M., 1996, An Intelligent Agent Framework for Concurrent Product Design and Planning. IEEE Transactions on Engineering Management. 43:297-306.

Using the Analytical Hierarchy Process (AHP) to formulate a Continuous Improvement Process (CIP)

Jinesh Shah and Ashraf Labib

University of Manchester Institute of Science and Technology (UMIST), Manufacturing Division, P.O. Box 88, Manchester, M60 1QD, U.K. Ashraf.Labib@umist.ac.uk

Abstract: The growing importance of setting up continuous improvement strategies in manufacturing organisations is proving to be a difficult process. Managers seem to have difficulties in choosing strategies that will improve productivity, availability and quality. A Multiple Criteria Decision-Making (MCDM), based on the Analytical Hierarchy Process (AHP), is used as a means of systematising the decision process. A generic framework was developed to formulate an ideal strategy for a Continuous Improvement Process (CIP) using AHP. In addition, the issues and implications regarding consistency and sensitivity analysis are also discussed.

1. Introduction

The globalisation of the economy and the rise in competitive challenges has created a problem for managers in the manufacturing environment to take decisive actions, and be responsive in implementing a strategy that gives better quality, availability and process efficiency. External environmental conditions such as demand, puts pressure on manufacturing personnel to improve their processes to fuel this demand before competitors begin entering the market. Internal environmental conditions perceived by the company would be supply in the form of production capacity that can be improved by achieving the objectives of the company.

MCDM methods are used to support managers in choosing a strategy for a CIP. AHP is one such method that breaks down a complex situation into a hierarchic order to facilitate pair-wise comparisons to determine the relative importance of each alternative in terms of each criterion. In addition, for AHP to be effective it needs to be coupled with a sensitivity analysis phase plus a feedback measure on consistency analysis discussed later.

The following sections illustrate the strategies available to set up continuous improvement, the structure of the hierarchy developed followed by a detailed analysis including sensitivity analysis and consistency measures.

2. Continuous Improvement Strategies

The strategies presented in Table 1 are those key ones that were identified to have an influential effect on achieving one, several or all the objectives. These objectives are *minimising changeover (set-up) times*, *zero defects*, *zero waste*, *zero delays* and *zero breakdowns*. Further details of P-M Analysis can be found in [1], 5S's in [2], SMED in [3], and Eliminating Minor Stoppages in [4]. Achieving any of the above objectives would indirectly contribute to a high overall equipment effectiveness

(OEE) value. OEE is a performance measure expressed as a percentage and is the product of availability, performance and quality [5].

Table 1: Continuous Improvement Strategies and their features.

Strategy	Features
1. P-M (Phenomenon Mechanism) Analysis	Physically analyses chronic defects according to inherent principles in order to reduce them.
2. 5S's and Autonomous Quality Maintenance	A discipline approach in standardising operating procedures, good house keeping and maintaining equipment through vigorous cleaning by operators to identify causes for machine breakdowns/failures.
3. SMED (Single Minute Exchange of Die) and Defect Free Changeovers	Reducing changeover times and also making sure first batch of products made are defect free through elimination of adjustments and readjustments.
4. Eliminating Minor Stoppages & Speed Losses	Aims to reduce intermittent stops during normal production and increase actual cycle time to design specification.

3. Structure of the Hierarchy Model.

Using the goal as obtaining a strategy for a Continuous Improvement Process (CIP), a hierarchy is logically constructed and shown in Figure 1. It is important that elements at each level are important to elements in the level below. The different levels are described in more detail below.

3.1 Scenarios (Level 1)

The first level deals with the perceived external and internal environmental scenarios relative to the company. The external factors would be economic e.g. high/low demand according to Arbel and Orgler [6]. A high demand scenario corresponds to an expanding economy. The internal factor would therefore be high/low supply that is affected by the objectives to be achieved. Therefore coupling the two factors together helps creating the operating scenarios that can be classified as follows:

Scenario 1: *High Demand and Low Supply (HD, LS)* – where existing production capacity cannot meet increasing orders due to lengthy set-ups or high breakdowns.

Scenario 2: *High Demand and High Supply (HD, HS)* – existing production capacity is able to meet the high demand.

Scenario 3: *Low Demand and High Supply (LD, HS)* – a stable economy with sufficient capacity to meet orders/sales.

Scenario 4: *Low Demand and Low Supply (LD, LS)* – a stable economy with low production capacity.

Other scenarios can also be added to the above ones depending upon the specific circumstances of the company.

3.2 Actors (Level 2)

According to [6], the 'actors' are members of the shopfloor management team who make decisions. An actor is an individual or a group which plays a significant role in responding to forces that shape current events, and therefore future outcomes [7]. The actors identified are the *Production Manager (PM)*, *Maintenance Manager (MM)* and *Quality Managers (QM)*.

3.3 Objectives (Level 3)

The objectives identifies were mentioned in Section 2. In essence, achieving zero waste relates to eliminating time used in searching and moving items i.e. tools, parts and work-in-progress. Zero delays relates to eliminating minor stoppages.

3.4 Options (Level 4)

The last level involves the specific options strategies that are relevant in implementing a continuous improvement strategy. These were described in Table 1.

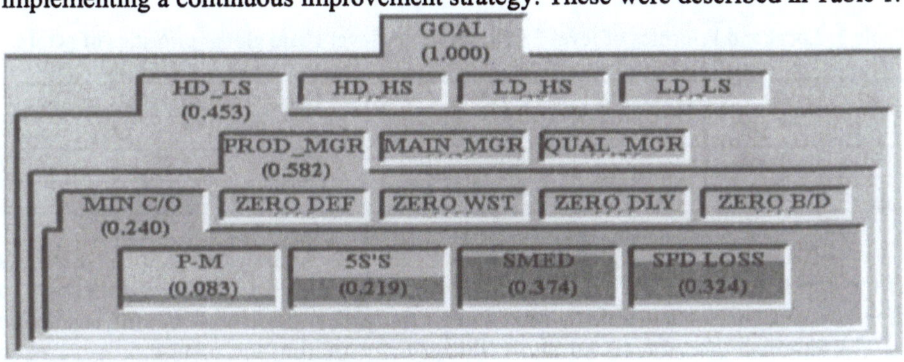

Figure 1: The hierarchy for formulating a CIP.

4. An Analysis of AHP using the developed Model

Once the hierarchy is formulated, the quantitative evaluation through pair-wise comparisons can be carried out for all the levels. Although the judgements are made using qualitative information available on the options, AHP allows one to convert this into quantitative data. This highlights the powerful nature of AHP. Pair-wise comparisons are undertaken with members of one level with respect to a member of the next higher level. To start the evaluation, the relative likelihood of the four scenarios with respect to the goal is assessed to determine which one is most favourable in creating an environment where CIP thrives.

From Table 2, it is evident that high demand situations favour the CIP with low supply being the most likely. This is because the company can increase its production capacity by achieving the objectives outlined. Therefore the priority of a high demand and low supply situation is 0.447. The computation of the relative likelihood is as follows; the elements of each column are divided by the column sum and then row averages are calculated to obtain local priorities.

The next stage assesses the priorities of each actor as considered with respect to one of the scenarios. In the case of high demand situations, the importance of production and maintenance managers are more significant due to the need to increase productivity while during low demand the quality and maintenance managers are more significant. Maintenance managers are important in all scenarios as good maintenance provides more production capacity and better quality. The results shown in Table 3 demonstrate the importance of each manager under different scenarios. Note that column values add to unity.

Table 2: Relative likelihood of scenarios.

Goal	*HD, LS*	*HD, HS*	*LD, HS*	*LD, LS*	Relative Likelihood
HD, LS	1	2	3	4	0.453
HD, HS	1/2	1	5	3	0.331
LD, HS	1/3	1/5	1	1/2	0.089
LD, LS	1/4	1/3	2	1	0.127
					CR = 0.06

Table 3: Local and Priorities of level 2 with respect to level 1 and global priorities of actors.

	HD, LS	*HD, HS*	*LD, HS*	*LD, LS*	**Global Priority of actors**
PM	0.582	0.443	0.109	0.124	0.436
MM	0.258	0.387	0.309	0.517	0.338
QM	0.105	0.169	0.582	0.359	0.201

To determine the global priorities of the actors, the relative local priorities of the scenarios in Table 2 are weighted with the priorities of actors with respect to the various scenarios in Table 3. This reveals the global priorities exhibited in Table 3. The rationale for this is to show the priority of the actors not only to each scenario but to the overall goal as well.

The prioritisation of the first two levels is now complete. The next step is to make pair-wise comparisons of all the objectives with respect to each actor and scenario. As these would take too long to describe, the assessment of different actors' objectives in the scenario high demand and low supply are exhibited in Table 4. Note that rows add to unity.

Table 4: Judgements of different actors under a high demand and low supply scenario.

HD, LS	*Minimising c/o times*	*Zero defects*	*Zero waste*	*Zero delays*	*Zero breakdowns*
PM	0.240	0.075	0.063	0.252	0.370
MM	0.109	0.226	0.065	0.129	0.472
QM	0.058	0.356	0.133	0.076	0.378

Table 4 suggests that during high demand and low supply situations, all three actors prefer to achieve zero breakdowns. In addition, the maintenance and production managers, to increase productivity also aim for zero delays and minimising changeover times. As always, the quality managers' priority lies towards reducing waste and defects. Finally for the last level, pairwise comparisons of the options are made against each objective. The information exhibited in Table 1 is used to make the judgements when finding the local priorities of the options with respect to each of the objectives. For the high demand and low supply scenario, a summary of the local priorities of the options is shown in Table 5. To be concise, comparisons with respect to all scenarios will not be shown. Table 5 shows that under the stated scenario, a 5S's strategy is able improve productivity by reducing breakdowns, delays and changeover times. Elements under minimising changeover times' column of Table 5 relate to those exhibited on the last level in Figure 1.

Table 5: Production manager's option priorities for the scenario '*HD, LS.*'

HD, LS	*Minimising c/o times*	*Zero defects*	*Zero waste*	*Zero delays*	*Zero breakdowns*
P-M	0.083	0.101	0.199	0.078	0.206
5S's	0.219	0.267	0.599	0.226	0.572
SMED	0.374	0.229	0.135	0.111	0.080
Speed Losses	0.324	0.402	0.067	0.585	0.141

Table 6: Global priorities of Strategies.

Strategies	Priorities
P-M	0.146
5S's	0.391
SMED	0.178
Speed Losses	0.292

To determine the global priorities of the options, the local priorities of the objectives are weighted with the global priorities of the actors. The result of this is weighted with the local priorities of the options to give the global priorities of the various strategies. This is exhibited in Table 6. Although there are several factors that need to be considered e.g. economic criteria, organisational culture and so on which is beyond the scope of this paper, from the ones considered above, a 5S's strategy would be an ideal one for implementing a CIP. Saaty [8] describes the mathematics behind the calculation of the global priorities. Consistency of making comparisons is another critical aspect of AHP that will be briefly discussed below.

5. Sensitivity Analysis and Inconsistency Ratios

Once judgements for all levels are made, Expert Choice Software [9] allows one to visualise the 'total picture' of the problem by performing sensitivity analysis which helps to examine different scenarios and re-addresses the importance of criteria in a 'what-if' approach [10]. The variation of options can be examined with respect to one objective as displayed in Figure 2. The x-axis gives the relative priority of minimising changeover times while the y-axis shows the importance of the strategies. The intersection of the vertical line with other lines indicates the priority of each of those strategies. If the priority of minimising changeover times is below 0.45, the 5S's strategy is preferred. Moving the vertical line between 0.45 and 0.75, the importance shifts to Speed Losses. Similarly, a priority greater than 0.75 shifts the importance of minimising changeover times to a SMED strategy.

While making judgements, perfect consistency cannot be expected when working with AHP. Therefore, the issue is how much inconsistency can be accepted which Saaty (1994) [8] says a CR less than 10% (0.1) is appropriate.

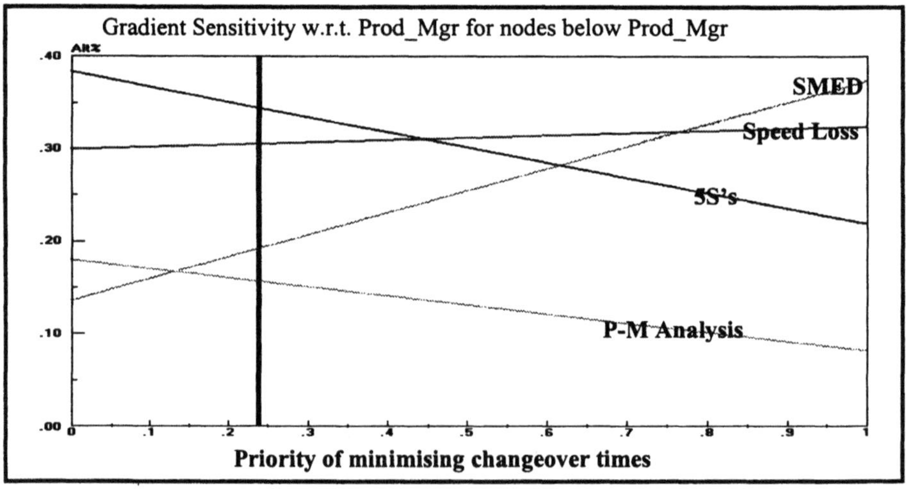

Figure 2: Sensitivity Analysis Graph.

6. Summary and Conclusion

The model offers a methodology to ensure sustainability of a CIP. It also offers a responsive and adaptive framework that is influenced by the external environment. Figure 2 shows the complexity of AHP when making judgements.

References

[1] Shirose, K., 1990, *P-M Analysis*, Productivity Press.
[2] Hirano, H., and Talbot, B., 1995, 5 *Pillars of the Visual Workplace*, Productivity Press.
[3] Shingo, S., 1985, A revolution in manufacturing: The SMED System, Productivity Press, Portland, Oregon.
[4] Suehiro, K., 1992, *Eliminating Minor Stoppages*, Productivity Press.
[5] Nakajima, S., 1988, *Introduction to TPM*, Productivity Press.
[6] Arbel, A. and Y. E. Orgler, 1990, *An application of AHP to banking strategic planning: the mergers and acquisition process,* European Journal of Operations Research, v48 no. 4.
[7] Labib, A., Williams, G. and O'Connor, R., (1996), *Formulation of an appropriate productive maintenance strategy using multiple criteria decision making*, Maintenance Journal, Vol. 11 No. 11, April.
[8] Saaty, T. L., 1994, *Fundamentals of decision making and priority theory with the analytical hierarchy process*, RWS Publications, Pittsburgh, USA.
[9] Expert Choice Software, 1996, Expert Choice Inc., Pittsburgh, PA.
[10] Labib, A., *A logistics approach to managing the millennium information systems problem,* Logistics Information Management, v 11 no. 5, 1998, 285-294.

INTELLIGENT MANUFACTURING SYSTEMS II

Neural-Based Grasping for Industrial Grippers

Franco Failli[1], Gino Dini[1]
[1]Department of Mechanical, Nuclear and Production Engineering, University of Pisa
Via Bonanno Pisano, 25/b - 56126 Pisa (Italy)
franco.failli@ing.unipi.it, dini@ing.unipi.it

Abstract: The problem of automatic grasp planning for industrial applications involving solid bodies is approached by using Artificial Neural Networks (ANNs). Three kinds of industrial grippers are considered: two-jaw grippers, magnetic grippers and vacuum grippers. To allow the automatic solution of a wide set of cases, the ANNs are trained using different sets of practical situations where geometrical and technological features of grasped objects are considered. This paper, in particular, is mainly focused to the discussion of the results obtained in different examples of application aiming at showing the performance of the proposed approach.

Keywords: Gripper, Grasp planning, Neural networks.

1. Introduction

The problem of grasp planning is very popular in assembly plants, where a large number of different and complex objects has to be handled using mechanical grippers. The problems to be solved mainly concern three aspects of grasping: stability, accessibility and accuracy. A grasp has to be stable to avoid undesired movements of the handled object during approaching and placing phases [1,2]. The accessibility of the surfaces to be grasped has to be verified to allow the grasp [3]. A grasp has to guarantee the accurate execution of the programmed task, especially in the field of assembly where a high precision is required. In a large number of grasping problems, different combinations of surfaces can be selected to fulfil these requirements, thus it is often necessary to reach a compromise. The choice of the best compromise represents the main objective of a human planner or a computer-aided system. Different grasp planning criteria for industrial cases can be found in [4,5]. The application of *Artificial Intelligence* software tools is also an important topic [2,6,7]. Moreover, some authors have been taken into consideration *Artificial Neural Networks* (ANNs) related to some aspects of robotic grasping [8,9].
In this paper the ANNs are used to combine the traditional experiences in the field of grasp planning, based on the recognition of the object features, with the most recent and effective Artificial Intelligence technique, to obtain a high speed software tool to be used in robot task planning systems.

2. Overall description of the proposed system

Three kinds of gripper are considered in the proposed system: *two-jaw grippers*, with pivoting or parallel motion of fingers; *magnetic grippers*, with a single circular magnetic plate; *vacuum grippers*, with a single circular vacuum cup. The system can be extended to other kinds of industrial grippers (such as *three-jaw* or *expandable grippers)* simply by adding other ANNs trained and validated for the grasps performed by these devices.

Fig.1 gives an overview of the system where the following aspects are highlighted:

- *inputs:* represented by morphological and technological aspects of the object and information about the task to be performed (object orientation, accessible surfaces, direction of insertion, etc.);
- *pre-processing of inputs:* the inputs are pre-processed in order to select the potential grasps which satisfy the minimum requirements from a geometrical point of view (e.g.: correct orientation and position of surfaces);
- *evaluation of grasps:* the previously selected grasps are evaluated by ANNs. Each network produces an evaluation on the basis of the knowledge acquired in training;
- *final results:* the result is a list including: kind of gripper to be used, surfaces to be grasped (one or more combinations for each gripper), a "judgement" for each proposed solution.

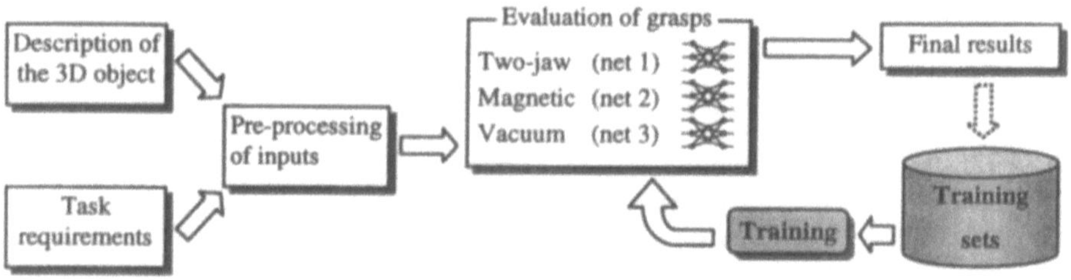

Fig.1 – Flow-chart of the system

3. Neural networks

The ANN architecture used in the system is the *Radial Basis Function* (RBF) formed by an input layer, a hidden layer and an output layer.

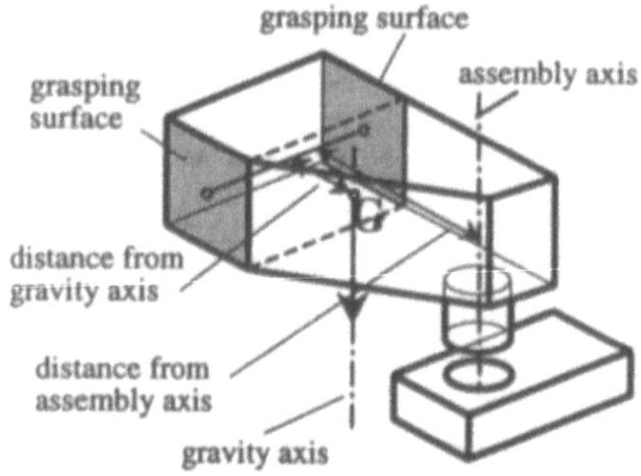

Fig.2 – ANN input parameters

The inputs of the ANNs are represented by a set of parameters taking into account the different features of a generic grasp. For example, for a two-jaw gripper, they are (Fig.2): 1) shape of grasping surfaces, 2) available area for grasping, 3) slenderness of available area, 4) distance between surfaces, 5) mutual orientation of surfaces, 6) position of grasp (external or internal), 7) distance from gravity axis, 8) orientation with respect to the gravity axis, 9) distance from assembly axis, 10) orientation with respect to the assembly axis, 11) surface roughness, 12) surface regularity (presence of holes and/or protruding parts on the grasped surfaces). The outputs of the ANNs are the grasp judgement represented by one of the following linguistic variables: *optimum, good, quite good, acceptable, bad.* The activation of two contiguous outputs can also be accepted (e.g.: *good – quite good*)

to mean an intermediate result.
ANN training is performed through a recursive procedure which allows a minimisation of the sum squared errors of the outputs and the optimisation of the training set size. Sets of families of elementary objects such as blocks and hollow cylinders are used as training and validation examples. This approach gives the chance to acquire the knowledge without using complicated heuristic rules, usually difficult to be formulated and managed, but presenting to the system a set of simple grasping examples together with their evaluation given by a human expert.
More detail about the input parameters and the training procedure can be found in [10]. The main scope of this contribution is to demonstrate the capability of the system to approach and solve typical grasp planning problems which can occur in industrial practice. For this purpose, three case studies are described and discussed in the following sections.

4. Results

The three ANNs used for the different kinds of grippers have been trained and validated by the following classes of examples:

- *class A*: consisting of a family of boxes obtained varying the three dimensions, the centroid position, the assembly axis position and the surface regularity;
- *class B*: consisting of a family of hollow cylinders obtained varying the height, the inner and the outer diameter, the centroid position, the assembly axis position and the surface regularity;
- *class C*: consisting of a family of prisms obtained varying the height, the shape of the base surface, the centroid position, the assembly axis position and the surface regularity.

Table 1 summarises the parameters and the examples used in training and validation of ANNs.

Table 1 - Parameters and examples used in training and validation of ANNs.

	ANN		
	Two-jaw gripper	**Magnetic gripper**	**Vacuum gripper**
Classes of examples	A, B and C	A	A
No. of examples in training set	326 (198 of A, 92 of B, 36 of C)	80	100
Max. sum squared error	4.0	1.0	1.0
No. of neurons[1]	191	64	85
Computational time[2]	1 h 33'	4'	10'
1: in hidden layer; 2: obtained using a 486 PC (66 MHz)			

4.1 First case study

The component to be grasped is shown in Fig.3a. It is a cover obtained by drawing a metal sheet. The peculiarities and the difficulties of this component related to the selection of grasping surfaces are the simultaneous presence of curved and flat surfaces having different reciprocal orientations, as well as the presence of thin surfaces difficult to be grasped.

The part is considered placed on the workbench by the surface 17. Therefore, this surface and all the internal ones are not accessible for grasping.

The selected grasps and their evaluation are reported in Table 2. Observing the results, the following considerations can be made:

- the best evaluations have been associated to the grasps performed on the surfaces 2 and 4, by a two-jaw gripper, and on the surface 1, by a magnetic or a vacuum gripper. This fact is mainly due to the great available area of these grasps, as well as the short distance from the centre of gravity of the object;
- the grasps on the curved surfaces 7, 8, 9 and 10 have been evaluated as *quite good* since their shape guarantees a correct stability of the grasp, even if the available area for gripping is quite small;
- the possibility to grip the surfaces 13, 14, 15 and 16 have been discarded in the pre-processing of inputs due to their very low value of slenderness.

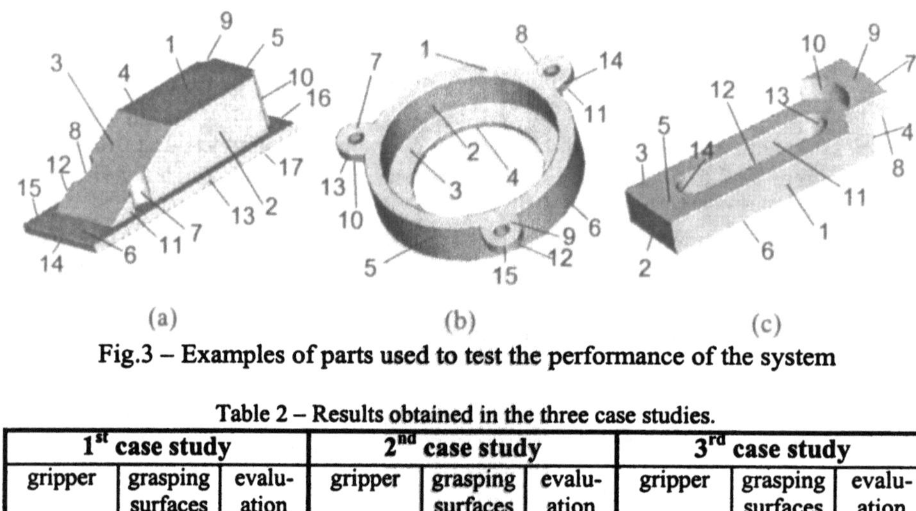

(a) (b) (c)

Fig.3 – Examples of parts used to test the performance of the system

Table 2 – Results obtained in the three case studies.

1st case study			2nd case study			3rd case study		
gripper	grasping surfaces	evalu-ation	gripper	grasping surfaces	evalu-ation	gripper	grasping surfaces	evalu-ation
two-jaw	2 – 4	5	two-jaw	2 – 2	5	two-jaw	1 – 3	5
magnetic	1	4 ÷ 3	two-jaw	5 – 5	4	two-jaw	8 – 9	4
vacuum	1	4 ÷ 3	two-jaw	4 – 4	4 ÷ 3	two-jaw	4 – 13	3
two-jaw	7 – 8	3	two-jaw	2 – 5	4 ÷ 3	two-jaw	7 – 10	3 ÷ 2
two-jaw	7 – 9	3	magnetic	1	4 ÷ 3	magnetic	5	3 ÷ 2
two-jaw	8 – 10	3	two-jaw	2 – 13	3	two-jaw	1 – 12	2
two-jaw	11 – 12	1	two-jaw	2 – 14	3	two-jaw	2 – 4	2
magnetic	6	1	two-jaw	2 – 15	3	two-jaw	3 – 11	2
vacuum	6	1	two-jaw	4 – 5	3	two-jaw	11 – 12	1
			two-jaw	8 – 11	2	two-jaw	13 – 14	1
Evaluations: 5 = Optimum			two-jaw	9 – 12	2	two-jaw	2 – 14	1
4 = Good			two-jaw	7 – 10	2	vacuum	5	1
3 = Quite Good			magnetic	3	1	magnetic	9	1
2 = Acceptable 1 = Bad			vacuum	1	1	vacuum	9	1

4.2 Second case study

The second case study concerns a metallic ring (Fig.3b) which offers different solutions of external and internal grasps. Furthermore, the surface 5 presents three protruding parts influencing the choices of the system. The object is placed on the workbench by the surface 6.
Table 2 shows the 14 best results generated by the system (the other grasps have not been reported for brevity). The following considerations can be made:

- the best evaluation has been given to the grasp on the surface 2 performed by a two-jaw gripper. This grasp has been preferred to that one acting on the surface 5 since, although it has a smaller available area and it is an internal grasp, it does not have any protruding parts which can obstacle a correct approach of fingers;
- a lower evaluation has been assigned to the grasps on the surface 4 by a two-jaw gripper and on the surface 1 by a magnetic gripper. This solutions, in fact, represent good grasps due to their central positions but they have small and thin available areas for gripping;
- the grasp on the surfaces 2 and 5, although it does not have a central position, has received an evaluation equal to the previous solutions. This fact can be explained considering three positive aspects considered by the system: a) a greater area of the surfaces; b) a lower slenderness; c) a favourable orientation which prevents slippages due to the weight of the object.

4.3 Third case study

The object to be grasped is shown in Fig.3c and presents different alternatives of grasping. As the previous cases, it is placed on the surface 6 which therefore is not accessible.
The results are listed in Table 2. Relevant comments can be expressed on the following aspects:

- despite to its small available area, the grasp performed on the surfaces 8 and 9 has obtained a good evaluation. This result can be explained by the orientation of the surfaces with respect to the gravity axis and by the small distance between surfaces;
- the very low evaluation given to the internal grasp on the surface 11 and 12 cannot be justify by geometrical or technological reasons. The grasp appears as a quite good solution, especially for its central position and its great available area, and the behaviour of the system in this case can be explained only by an incomplete or incorrect training of the ANN for two-jaw grippers.

5. Discussion and Concluding Remarks

The satisfactory results obtained in the three previous cases demonstrate the feasibility of a grasp planning system with a knowledge base structured by ANNs. Particular emphasis has to be given to the capability to acquire new knowledge by very simple examples of grasps performed on elementary objects such as boxes, cylinders or prisms.

A critical point is undoubtely represented by the selection of proper training examples and their distribution on the technological domain which the system has to cover. For example, the error made by the system in evaluating the grasp on the surfaces 11 and 12 of the third case study could be generated by an improper selection and distribution of examples in the domain of internal grasps having geometrical and technological features similar to that one generating the error. However, the great advantage of this approach is the possibility to use the incorrect responses for enhancing the knowledge: the situations which give incorrect solutions can be easily added to the set of examples and the training procedure can be repeated, improving the performances of the system during its use (Fig.1).
Another positive aspect is represented by the very low computational time typical of ANNs. The responses are practically instantaneous in all situations. Also the training and validation phases are not time-consuming (Table 1) considering that these operations have to be performed a limited number of times.
Some open issues still remain to complete the system. One of them is the verification of the grasping surface accessibility. At the current status, the system does not perform any collision check among the gripper, the object and the external environment. This control, in fact, can be accomplished only if the geometrical aspects of the gripper, the fixtures and the devices used in the surroundings and the trajectory of the robot are completely known. This control could be implemented into a successive simulation module, where the alternatives proposed by the grasp planning system can be analyzed, under this point of view, and finally selected.

References

[1] Montana DJ.; The conditions for contact grasp stability. Proceedings of IEEE International Conference on Robotics and Automation, 1991. p. 412-417.
[2] Dini G.; A module for the automated planning of grasps in robotized assembly operations. Annals of the CIRP, 1993. 42(1): 1-4.
[3] Jones JL., Lozano-Perez T.; Planning two-fingered grasps for pick-and-place operations on polyhedra. Proceeding of IEEE International Conference, 1990. p. 683-688.
[4] Dubey A., Nnaji BO.; Task-oriented and feature-based grasp planning. Robotics and Computer Integrated Manufacturing, 1992. 9(6): 471-484.
[5] Boubekri N., Ramanujam G.; Design of grasping methodologies for rotational assembly components. International Journal of Flexible Manufacturing Systems, 1995. 7: 373-388.
[6] Canzi U., Cugini U., Rigillo F., Kali; a knowledge-based system for grasp planning. Proceedings of 5th International Conference on Advanced Robotics, 1991.
[7] Pham DT., Tacgin E.; Grippex: a hybrid expert system for selecting robot gripper types. International Journal of Machine Tools and Manufacturing, 1992. 32(3): 349-360.
[8] Wang C., Chang E.; Neural network and striped lighting pattern-based autograsping technology for flexible robotic assembly, Journal of Robotic Systems v 14 n 7 Jul 1997. p 559-569
[9] Lee Y.H., Moon S., Lee H.; New approach for automated parts recognition using time series analysis and neural networks, Journal of Intelligent Manufacturing v 8 n 3 Jun 1997. p 167-175
[10] Dini G., Failli F.; Planning optimal grasps for industrial robotized applications using neural networks (to be published)

A Knowledge-Based Layout Selection Methodology for Flexible Assembly Lines

Asar Khan[1] and Andrew J. Day[2]

[1]University of Bradford, Department of Mechanical & Medical Engineering, University of Bradford, England (UK), BD7 1DP. A.Khan19@scm.brad.ac.uk
[2]University of Bradford, Department of Mechanical & Medical Engineering, University of Bradford, England (UK), BD7 1DP. A.J.Day@bradford.ac.uk

Abstract: In this paper a knowledge-based methodology is presented for the selection of an appropriate layout for a Flexible Assembly Line. The selection is made from straight-line, U-shaped, zigzag and odd-angled line layouts, based on the overall dimensions of available space and the dimensions of each individual workstation on the line. The information is analysed systematically in a knowledge-based approach which considers the strategic and technical factors in order to reach a consistent decision. A straight-line layout solution is considered first, followed by the simultaneous consideration of U-shaped and zigzag line layouts. Finally the odd-angled line layout is considered. The end result is advice on the best line layout to utilise the available space and to justify the line layout requirements for the system under consideration economically. The methodology is developed in AM which is a rule-based Expert System Shell operating under Microsoft Windows.

1. Introduction

The design of Flexible Assembly Lines must consider the layout of the associated workstations and facilities. Although, in principle, layout design is relatively straightforward, many industrial Flexible Assembly Lines have not fully considered all the design parameters prior to their installation. As a result the layout can often be sub-optimal and the performance of the assembly line can be affected.

A manufacturing system essentially includes production activities and the effective arrangement of these activities is an important factor in achieving the goal of economical and profitable production. One basic objective of any manufacturing organisation must therefore be to facilitate an efficient flow of materials, parts and people from the beginning of the manufacturing process to the end. The flow of work must be planned from the bottom-up as well as the top-down, to ensure that individual operations and sub-assemblies points have an appropriate relationship with the complete plant layout.

This paper addresses the question of manufacturing layout design methodology from a knowledge-based approach. A method is developed, presented and illustrated by example, based on the overall dimensions of available space and the dimensions of each individual station on the line, together with some initial basic manufacturing information. Such information (e.g. demand volume, product variety, operational requirements, part requirements and the appropriate product-operation-part relationship) is part of the requirements of a broader manufacturing assembly line system design process including an appropriate balancing [1, 2, & 3]. These data are then analysed systematically to recommend the number of appropriately configured workstations in a line layout selected from a straight-line, U-shaped, zigzag, or odd-angled configuration [4]. The analysis is then concluded by consideration of any further strategic or technical factors to "fine-tune" the line layout. The research does not consider facility layout design for jobshop and batch production.

In this paper a brief background to the subject is given, followed by the theoretical basis of facilities layout. The flexible assembly line layout analysis methodology developed is described and an example of its application shown. The paper concludes with discussion and conclusions.

2. Background

The first analysis of assembly line manufacturing appeared in the mid 1950's [5]. Since then, the topic of assembly line design has been of continuing interest to researchers. Assembly line manufacturing system selection, analysis, design, and balancing have been tackled in a number of research papers and textbooks, and line design with deterministic task times has received most attention. Assembly line layout design is considered to be relatively straightforward and therefore less work is seen on this topic.

Many industrial manufacturing companies do not fully consider the design of their assembly line layout which then can result in sub-optimal processes and production. In practice, layout design is a complicated process which has to consider many factors including available space, capacity, demand, process machinery design, material handling craftsmanship, quality and scheduling flexibility, especially when space is restricted, the products are varied, or the line is extensive [4, 6]. Once the line layout is decided upon and installed, any change in the design will incur considerable time and cost penalties. For this reason future expansion or the introduction of new models should be planned for, early in the manufacturing planning stage [7].

From these studies, it is clear that, where used, assembly line analysis has been mainly directed towards the comparison and modification of the existing layouts. Typically these start from a simple consideration and move towards a more efficient layout so as to fulfil the desired gaols of production rates and cost effectiveness for that particular organisation. Detailed station design and the interrelationships between them have been analysed only for optimal production. However the analysis and selection of the assembly line layout has not been reported fully. Line layout has a strong impact on the future of any enterprise involving assembly processes and is the focus of this paper.

3. Notation

L =Length dimension of the available space (m)
LS_i =Length of station i (m)

N	=Number of station on the line
S_i	=Station i on the line
U_{max}	=Maximum length for a U-shaped line (m)
W	=Width of the available space (m)
WS_{max}	=Maximum station width on the line (m)
WS_i	=Width of station i (m) on the line

4. Layout Design in Manufacturing Systems

Economical facility layout design for any type of manufacturing system demands that the required physical facilities are arranged in such a way to ensure the efficient, safe, and effective flow of materials. It should start with the analysis of the required production when the individual work areas are planned, and their interrelationships are analysed. Any related activities from raw material to finished product (e.g. milling, grinding, and assembly) can be considered in the final stages of facilities layout design.

Optimal material flow, efficient use of manpower, and effective utilisation of the available space in the system provides the basis for the layout design. Many flow planning techniques (e.g. Relationship Diagrams, From-To Charts and Critical Path Network method [4]) are in common use. These have found applications in many types of facility layouts and material handling systems related to process, product, or group methods of layouts [1]. The cost:quantity:distance equation is commonly taken as the basis of layout planning and material handling [8].

Material flow has particular importance in product-oriented layouts where its effective management plays a key role in decreasing product unit cost and increasing production rate. This is because of standardised production; the products possess similar manufacturing characteristics and quality standards. Certainly assembly lines are one of the most popular manufacturing plants based on product-oriented facilities layouts where the physical facilities are usually organised in a straight-line, U-shaped, zigzag or odd-angled layouts. The straight-line layout is one in which the workstations are arranged one after the other in a linear fashion, and requires adequate space and length. The U-shaped line layout is one in which the line can take several 90° turns to finish next to the start. This layout is recommended where the same facilities are required for loading and unloading purposes. The zigzag line layout also has a number of 90° turns but does not necessarily finish next to the start. The odd-angled line layout is increasingly popular and is recommended when the production space is enough for the system but due to space limitation, non of the other line layouts are possible. The basic guidelines of the minimum dimension requirements for a straight-line and a U-shaped or zigzag line layout is shown in Fig.1.

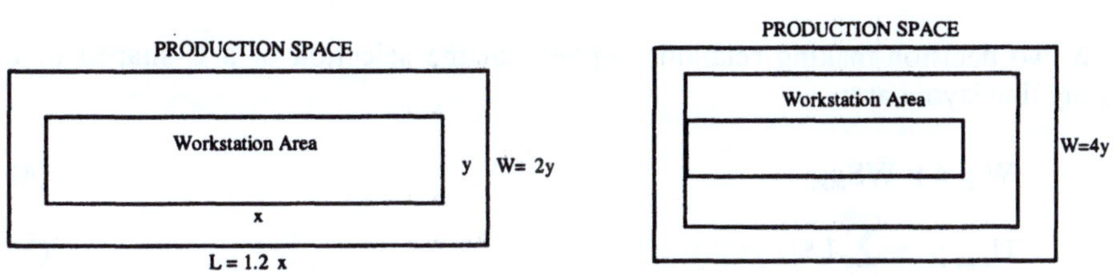

Fig.1 Basic guidelines for a straight-line and a U-shaped layout design

5. The Flexible Assembly Line Layout Analysis Method

The flexible assembly line analysis method presented here can be applied to a single, multi or mixed-product assembly line. The production rules have been developed through the use of an Expert System Shell, called AM (Application Manager) [9] which is Microsoft Windows based. The analysis starts with input information in terms of the number of operations allocated to each workstation, the tooling needed at each workstation, the type and number of auxiliary parts if required at the workstation, and the appropriate supply condition for the parts to the workstation. The operations, combined or separate, and the workstation and line configuration must already have been decided [3]. The total number of workers, the number of workers on each side of each workstation on the line must also be provided as a pre-requisite [10].

The overall dimensions of the available production space are defined as length of the available space, and width of the available space. The workstation dimensions include the space for machines or equipment, space for the worker, space for buffer storage (if required) and space for materials and tools defined as LS_i (length for workstation i) and WS_i (width for workstation i). Each dimension is measured in metres. The total required space for the workstations on the line is calculated and compared with the total available production area. If the total available production area is less than 1½ times the area required for all workstations on the line, the user is warned about a possible space limitation and appropriate guidance is given for the minimum requirement of production space. The method considers the four types of line layouts described earlier, viz. straight-line layout, U-shaped line layout, zigzag line layout, odd-angled line layout.

After qualifying the available production area, the first analysis is applied for the straight-line layout. This is the most desirable layout for ease of design and application [6], but does require suitable space. The method verifies a straight-line layout by the relationship:

$$L \geq 1.2 \times \sum_{i=1}^{i=N} LS_i \qquad (1)$$

If Eq. (1) is not true, then the U-shaped and zigzag line layouts are considered next by analysing the receiving and shipping activities with specific location requirements. The maximum workstation width on the line and the maximum length for U-shaped and zigzag lines are defined as:

$$WS_{max} \geq WS_i \quad \text{such that } 1 \leq i \leq N \qquad (2)$$

$$U_{max} = \{2(L - WS_{max}) + (W - 3WS_{max})\} \qquad (3)$$

The two decision-making relations applied for the selection of a U-shaped or a zigzag line layout are:

$$W \geq 4 \times WS_{max} \qquad (4)$$

and $$U_{max} \geq \sum_{i=1}^{i=N} LS_i \qquad (5)$$

The selection between a U-shaped and zigzag line layout is then made by a knowledge-based rule. In the case when the available production area is sufficient, but none of the equations (1), (4) and (5) are true, then the odd-angled line layout is recommended. As soon as the analysis is completed it decides on the best line layout design, displays the decision arrived and provides an appropriate explanation for the line layout selection to the system user.

6. Illustrated Example

The method is illustrated by the following example. A suitable line layout is required for a mixed-product assembly line where the total working load of 65.28 hours is distributed among 6 workstations operated by 9 workers in an 8-hour single shift per day. The maximum available time at Workstations 2, 3 and 5 is 16 hours and therefore 2 workers each share these stations. The overall dimensions of available space and each individual workstation are presented in Table 1.

Table 1: Input Information

Space Available: (Length = 5m, Width=3.8m)		
Stn. No	Stn. Length	Stn. Width
1	2m	1m
2	2m	1m
3	2m	1m
4	2m	1m
5	2m	1m
6	2m	1m

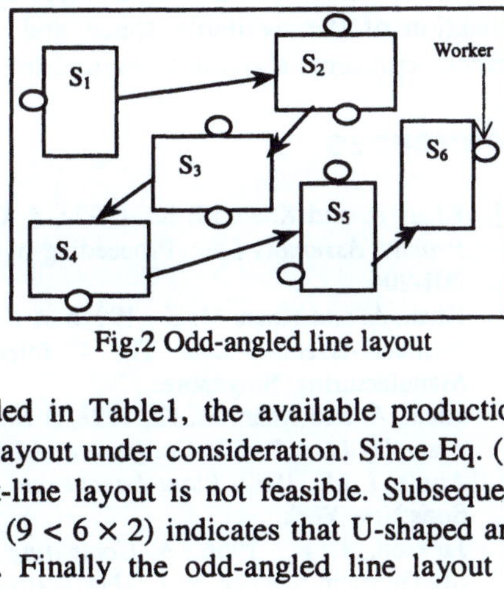

Fig.2 Odd-angled line layout

Based on the information provided in Table1, the available production space ($5 \times 3.8 = 19$) is enough for the line layout under consideration. Since Eq. (1) is not true (as $5 < 1.2 \times 12$), the straight-line layout is not feasible. Subsequent analysis, Eq. (4) ($3.8 < 4 \times 1$) and Eq. (5) ($9 < 6 \times 2$) indicates that U-shaped and zigzag line layouts are also not feasible. Finally the odd-angled line layout is recommended as the most desirable design, see Fig. 2.

7. Discussions and Conclusions

Although the complete flow analysis for an actual Flexible Assembly System can be very complicated, the method presented here provides a useful quantitative insight into the initial layout design for flexible assembly lines in the start-up stages. The analysis seems comparatively straightforward but in spite of this it is often ignored in facility layout design. However such analysis is key to the design of an efficient production process. Any non-optimal decision in any type of facilities layout design always results in production loss and therefore increased production costs. It is essential that layouts are properly designed with appropriate consideration of future changes if required.

The method presented here works well for flexible assembly lines where the design of the operations and workstations has been established. Thus it is best used in conjunction with a full manufacturing design analysis, such as the flexible assembly line design and balancing methodology presented elsewhere by the authors. Most importantly, though, this methodology enables the layout to be fully considered at the design stage so that the layout design is analysed and not just left to develop ad hoc.

This methodology advances the subject knowledge because, based on the 4 categories of assembly line layout, the rules for selection have been derived. It is important in the design of manufacturing systems and has been incorporated into a comprehensive knowledge-based line balancing and design methodology [1, 2, &3].

The methodology presented successfully analyses the strategic and technical factors associated with flexible assembly line layout in a step by step knowledge-based approach. Basing the line layout on 4 specific forms, viz. straight-line, U-shaped, zigzag, and odd-angled line layouts depending on the available space information and the designed workstation requirements is sufficient to cover most requirements. The production rules are able to analyse the available information and recommend the most appropriate line layout design to satisfy the utilisation of the available space and economically justify the line layout. The method represents a viable technique for such applications.

References

[1] Khan, A. and Khan, M. K., 1997a, *A Knowledge-Based Planning System for the Multi Product Assembly Line.* Proceeding of the 32nd International MATADOR Conference, 201-206.

[2] Khan, A. and Khan, M. K., 1997b, *A Knowledge-Based Selection System for the Multi-Product Assembly Line.* The 4th International Conference on Computer Integrated Manufacturing, Singapore.

[3] Khan, A. and Khan, M. K., 1998, *A Knowledge Based Balancing for a Mixed-Product Assembly Line.* Pacific Congress on Manufacturing Management, Australia.

[4] Apple, J. M., 1977, *Plant Layout and Material Handling, 3rd Edition.* John Wiley & Sons New York.

[5] Jackson, J. R., 1956, *A Computing Procedure for a Line Balancing Problem.* Management Science, Vol.2, No.3, 261-271.

[6] Bhattacharyya, S. K., Roy, R. and Low, M. J., 1993, *A Computer Simulation System for the Evaluation of Man Assignments on Car Assembly Tracks.* Technical Article, Simulation, Vol.61, No.2, 124-133.

[7] Ruehl, G. and Ing, 1974, *Work Structuring, Part II.* Industrial Engineering, Vol.6, February, 52-56.

[8] Besant, C. B. and Lui, C. W. K., 1986, *Computer-Added Design and Manufacture,* John Wiley & Sons New York.

[9] AM for Windows, 1995, *Reference Guide, Version 8.2.* Intelligent Environment.

[10] Khan, A., 1998, *A Knowledge-Based System for the Design and Analysis of Assembly Lines.* Ph.D. Thesis, University of Bradford.

Modelling Knowledge used in the Design of Hosiery Machines

K. OLDHAM

Knowledge Engineering and Management Centre, Room Q102, Coventry University, Priory Street, Coventry CV1 5FB. e-mail: Keith.Oldham@kbe.coventry.ac.uk

Abstract: This paper looks at some of the developments within ESPRIT project 25418, 'Methodology and tools Oriented to Knowledge-based engineering Applications'. The methodology covers the entire lifecycle of a knowledge-based engineering application, from identifying the opportunity to using the system to gain business benefits. Maintenance is regarded as a subset of these activities. The techniques are illustrated with the knowledge required to design part of an industrial half-hose knitting machine; namely the mechanism for driving the needle-carrying cylinders when knitting the heel and toe of a sock. This demonstrates how much knowledge can be involved and how important it is to define clearly the scope of any knowledge-collection exercise. There must also be a clear end-use in mind so that the knowledge is appropriate and manageable.

1. Introduction

Knowledge is being recognised increasingly as a valuable asset to a company. This is reflected in the phrase 'knowledge capitalisation'. When it comes to design, how can this knowledge be captured and of what use is it? One possibility is to build a knowledge-based engineering (KBE) application. The usual way of doing this is to collect the necessary knowledge and code it directly into a KBE package. However, this approach causes problems when the system requires maintenance, particularly if the development team has dispersed and the system needs to be maintained by a different group. Although some packages claim to be self-documenting, this is inadequate for maintenance of the knowledge and a better approach is required. Similar considerations apply if the knowledge is to be reused in another application. In this case, any limitations on the validity of the knowledge must be scrutinised to ensure that they are not violated in the new applications.

2. MOKA Methodology

This paper looks at some of the developments within ESPRIT project 25418, 'Methodology and tools Oriented to Knowledge-based engineering Applications' [1]. The methodology covers the entire lifecycle of a knowledge-based engineering application, from identifying the opportunity to using the system to gain business benefits. Maintenance is regarded as a subset of these activities. The associated software tool supports the structuring and modelling of the knowledge that has been collected in order to build a KBE application. The collected knowledge is structured into forms with predefined fields. There are separate forms for Illustrations, Constraints, Activities, Rules and Entities, known collectively as ICARE forms. For example, the Activity form contains fields for objective (goal), input requirements, description (method), associated entities, rules etc. The forms rapidly grow in number as they are filled in, so graphical charts are used to show how the forms relate to one another. The forms and charts are designed to be easily understood so that the relevant experts can check the content. This structuring and validation identifies any gaps, inconsistencies or ambiguities in the collected knowledge and so may prompt the collection of additional knowledge. This approach ensures that the structured knowledge is of a consistently high quality. The set of completed forms and charts is known as the 'Informal Model'.

The structured knowledge in the Informal Model can be coded directly by hand into the desired KBE package. This requires a good understanding of the capabilities of the KBE package and the way in which it holds the knowledge. This may be satisfactory if only one package is being used for all applications and this has an assured future. A different approach has been adopted here.

The Informal Model is converted into a Formal Model, which is a graphical, object-oriented representation of engineering knowledge at one level of abstraction above application code. The Formal Model consists of two models, namely Product and Design Process. This separation again facilitates the reuse of knowledge. To build a formal model, developers require a modelling language. MOKA has based its on UML (Unified Modelling Language), the leading industrial object-oriented modelling language for software engineering, supplemented by a constraint language, such as OCL (Object Constraint Language). UML is a general-purpose language and it was decided that a more dedicated language could be developed that would support KBE application development more readily. This decision is in the spirit of UML where users can extend the standard definition to target a particular domain. The resulting extended subset is the MOKA Modelling Language (MML), a graphical object-oriented modelling language to represent engineering design knowledge at a user level for deployment in KBE applications. MML inherits all the modelling capabilities of UML and therefore users can resort to standard UML when MML is found lacking.

The aim of MML is to provide a framework to build models. The framework, or meta-model, acts in an advisory role, where pre-defined components show what is commonly required to model engineering knowledge. Users select the pre-defined components and use them as stereotypes to populate the models as required. Figure 1 shows the pre-defined meta-classes for the Structure View of the Product Model. However, there is provision for the developer to add new meta-classes or attributes to the meta-models should this prove necessary. As a result, these model instances can be processed by computer and, once suitable translators are available, translated into the desired KBE package with relative ease. This means that users are not restricted to one package and so knowledge from one application can be reused easily in another even if it is based on a different KBE package. This ability is valuable if the preferred KBE package becomes obsolete.

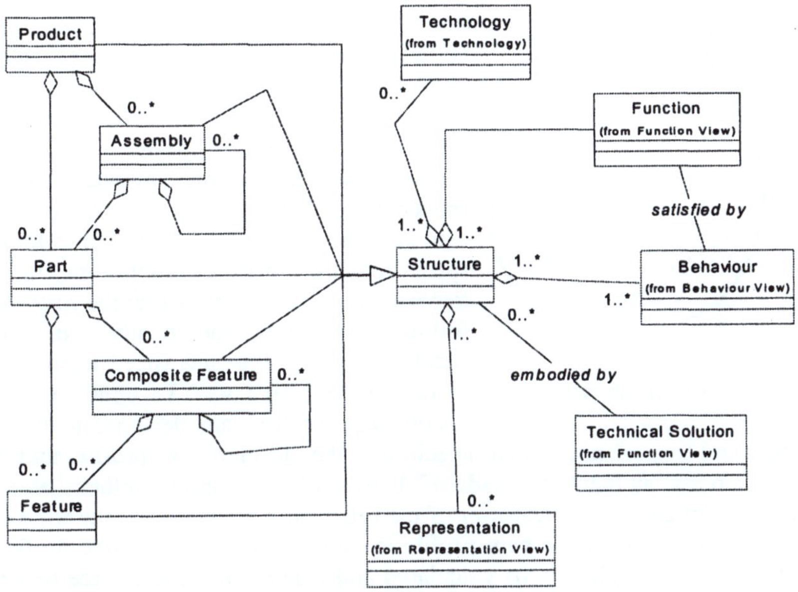

Figure 1: MOKA meta-model structure view

3. Application

The techniques just described are illustrated with the knowledge required to design part of an industrial half-hose knitting machine. These machines have two coaxial cylinders carrying double-headed latch needles in longitudinal slots. The needles can transfer between the cylinders so enabling the machine to knit plain and purl (rib) stitches. The focus of the application is the mechanism required to drive the cylinders when knitting the heel and toe of a sock. The heel and toe each form a pouch inserted into the tubular part of the leg and foot. To achieve this effect, the cylinders have to rotate backwards and forwards about their common axis through an angle of either 376 or 380 degrees depending on whether the machine leaves the socks connected together to form a string or automatically separates them.

Various mechanisms can be used to drive the cylinders. Rotating cams with swinging followers provide good motion control but require a gear train to achieve the desired amplitude, which increases the effective inertia of the cylinders at the follower. Spring return of a follower increases loading on the cam, while a closed track cam causes skidding of the roller when it changes sides of the track. Hence a pair of conjugate cams sharing a common axis of rotation and with followers linked together was adopted, as shown in Figure 2. The designer is free to specify one profile completely but the conjugate profile is governed by the need to keep the two rollers in contact at all times. Other constraints apply if the two profiles are to be identical [2]. Obviously, the cams are overconstrained and so must be manufactured accurately if jamming is not to occur.

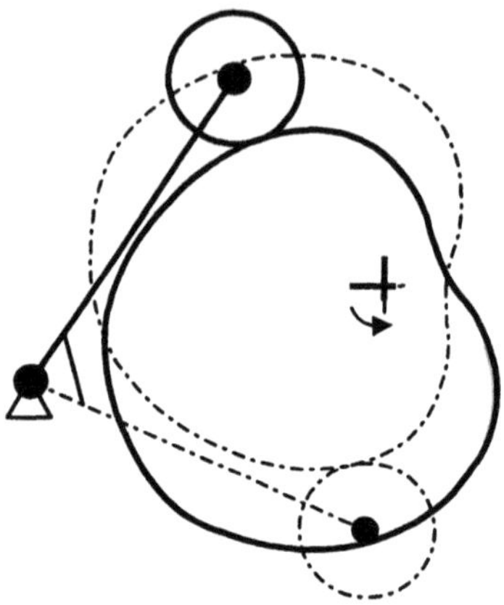

Figure 2: Conjugate cam

The follower motion can be specified in terms of a parameterised trigonometric cam law. This avoids step changes in acceleration (infinite jerk) while giving scope for a considerable range of cam laws, such as cycloidal, simple harmonic and modified trapezoidal. The cam designer is free to specify the limits of follower angular movement and the angular movement of the cam in which this must occur. In addition, specifying the relative proportions of the trapezoidal acceleration profile, the maximum follower velocities, the relative maximum accelerations and the cam speed is sufficient to define the cam law. To obtain the best results requires the cam designer to use his knowledge and experience when specifying the acceleration profile and the maximum velocity, as there is a trade-off between the maximum velocity and the maximum acceleration. The diameter of the rollers and the length of the followers must also be specified for the cam profile to be determined. Once the profile is known, the pressure angle can be calculated and hence the contact force between the roller and the profile. Knowing the maximum force, the designer can assess whether the roller is likely to deflect under it and so decide whether a barrelled roller is required rather than a cylindrical roller. The radius of curvature of the profile can then be calculated and hence the Hertzian contact stress. In light of this, a material can be selected for the cam that will give a satisfactory life.

This relatively small application illustrates how much knowledge can be involved in engineering design. The types of knowledge include theoretical, empirical and heuristic and cover knowledge related to conceptual design, manufacturing constraints, geometric dependencies, stress analysis and material selection. If such a small example contains so much knowledge, it is obvious that the scope of any knowledge-collection exercise must be defined clearly so that the collected knowledge is both appropriate and manageable.

4. Informal Model

In structuring the design knowledge, it has been found useful to have diagrams or charts to provide an overview of the domain, both of the product and the associated design process. Any type of diagram that is helpful can be used. For example, the structure chart in Figure 3 shows the various cam mechanism components and how they are related. The codes in each box are those used to identify the associated ICARE form. The key is PA - Physical Assembly, PP - Physical Part and LP - Logical Part or Feature. The linked follower levers are labelled 'quadrant' as they form part of the segment of a spur gear. Only one possible option has been shown for attaching the rollers to the quadrant. The camshaft and follower axle have been omitted for clarity, as have most of the logical parts. IDEF0 diagrams and/or state-transition diagrams are useful for depicting activities in the design process.

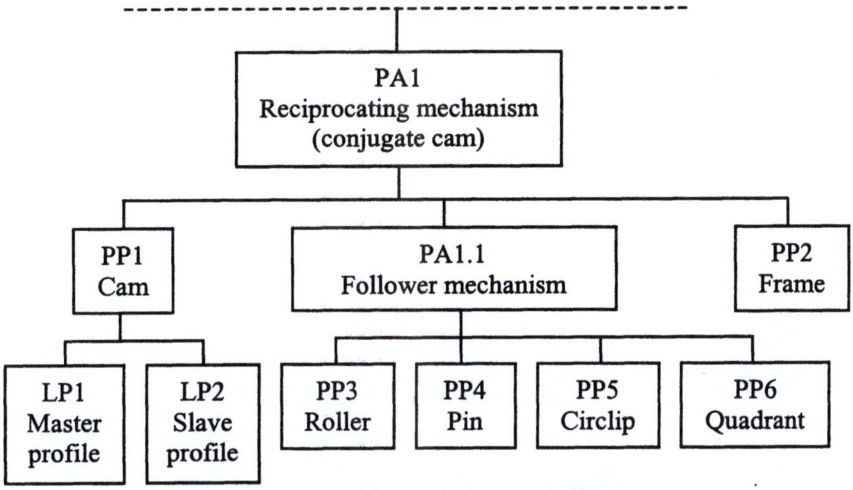

Figure 3: Structure chart for cam mechanism

The details of the assemblies, components etc are entered on the ICARE forms. The Entity form contains fields for Name, Reference, Functions, Behaviour, Context and Validity, Description and Related Activities, Entities, Constraints, Rules and Illustrations as well as Information Origin, Author, Date, Version number and Status. For example, the Function of the reciprocating mechanism, PA1, is 'to convert rotary motion to oscillatory motion'. The Description of the conjugate cam, PP1, includes its structure, i.e. it consists of two profiles linked together, together with facts such as the designer is free to specify the master profile, LP1, but that the slave profile, LP2, is derived from it. It rotates about a fixed axis (the camshaft) located in the frame, PP2, with a constant rotational speed.

Having a structured form with predefined fields encourages the collector (knowledge engineer) to ask questions to complete the form that otherwise might go unasked. This results in a more complete knowledge model. Limiting the length of an ICARE form to one side of A4 has the advantage of concentrating attention on one small aspect of the design or process but this can fragment the knowledge and result in a large number of forms. The prototype MOKA tool helps to manage these. Hyperlinks may be an alternative means of navigation in some software.

5. Formal Model

Using the Structure meta-model in Figure 1, a partial product model has been developed for the conjugate cam, PP1, showing the 'has-part' relationships, see Figure 4. This indicates that the conjugate cam has attributes such as centre of rotation and rotational speed. The representation of the cam has been refined into three parts, namely two disc cams (master and slave) spaced apart and connected together by a cylinder. Each disc cam has a feature with a superclass of profile and each inherits all its attributes and associations, such as being defined in terms of a set of points. The dashed arrow indicates that the slave profile depends on the master profile due to the constraint that both rollers should be in contact with their respective profile at all times. The numbers attached to the relationships indicate cardinality. For example, a conjugate cam has only one master cam, one slave cam and one cylinder and vice versa.

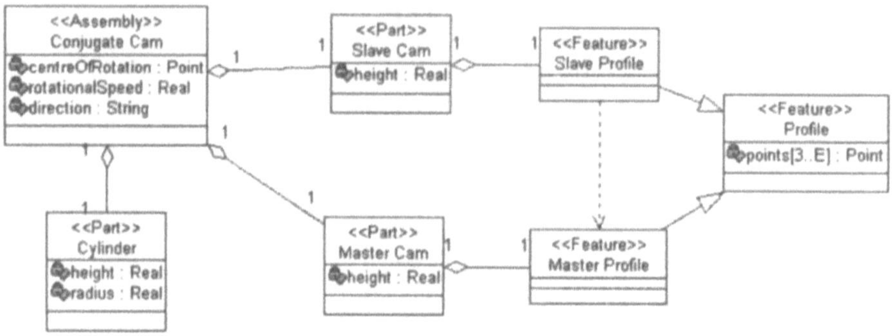

Figure 4: Part of product model for conjugate cam

6. Conclusion

The MOKA project is satisfying a user-driven requirement for a methodology in a growing market for knowledge-based engineering applications. Its use will significantly improve business competitiveness by encouraging the adoption of such systems where it is vital to capture strategically important design and manufacturing knowledge, in all its richness, and transfer it to other people, who can make good use of it. In this way, Knowledge-Based Engineering will be able to realise its full potential and contribute to the efficient management of engineering knowledge. This paper has concentrated on developing KBE applications. Maintenance of the models and applications raises a whole series of issues, which deserve a paper in their own right to do them justice. Further information can be found on the web-site at http://www.kbe.coventry.ac.uk/moka .

Acknowledgements

The author acknowledges the contribution of the MOKA project partners to the ideas presented in this paper. The partners are Aerospatiale Matra (leader), BAE SYSTEMS, Coventry University KEM Centre, DaimlerChrysler, Decan, Knowledge Technologies International and PSA Peugeot Citroën.

The financial contribution of the CEC under the ESPRIT programme is also acknowledged.

References

[1] Oldham, K, Kneebone, S, Callot, M, Murton A and Brimble, R, MOKA – A Methodology and tools Oriented to Knowledge-based engineering Applications, in N. Mårtensson, R. Mackay and S. Björgvinsson, *Changing the Ways We Work*, IOS Press, Amsterdam, 1998, pp. 198-207.

[2] Oldham, K, Discussion on 'Contributions to the Geometry of Cam Mechanisms with Oscillating Followers' by W. Wunderlich, *Mechanism and Machine Theory*, 7 (1972), 3, pp.371-372.

COMPUTER-AIDED ENGINEERING AND PROCESS PLANNING

An Incremental Interacting Feature Recognition Approach Based on Design-by-Feature

WD Li, SK Ong, and AYC Nee
Mechanical & Production Engineering Department
National University of Singapore, 10 Kent Ridge Crescent, Singapore 119260
Email: mpeongsk@nus.edu.sg

Abstract: Manufacturing applications typically require that a part created in a design-by-feature system be described through STEP-based manufacturing features. This paper presents a manufacturing feature recognition processor that utilizes the high-level design feature information retrieved from a design-by-feature system to incrementally generate a STEP-based manufacturing feature model efficiently by systematically representing and handling the relationships between interacting features.
Keywords: Feature recognition, design-by-feature, STEP

1. Introduction

The traditional feature recognition approach meets with much difficulty in recognizing interacting features from the solid models of design parts. As the design-by-feature approach becomes mature and popular in product design, it is imperative to integrate this with the feature recognition to enhance the ability of the latter by utilizing the high-level design feature information [1-3].

In this paper, an incremental interacting feature recognition approach based on design-by-feature and STEP AP 224 [4] is proposed to generate a manufacturing feature model. The design feature models of a part are created through a design-by-feature system. Based on the analysis of the relationships between interacting features, from a design feature CSG tree, a feature recognition processor incrementally recognizes a manufacturing feature tree in which the manufacturing features are defined according to STEP AP 224.

2. Design Feature Models and Manufacturing Features

A design feature CSG tree of a part is maintained to organize design features in a hierarchical structure, and reflects the evolving design process. The root of a CSG tree is the initial design stock of the part. Each intermediate node in the CSG tree is a Boolean union or difference operator. Its corresponding leaf node is an additive design feature volumetrically added on, or a subtractive design feature decreased from the part. A transition design feature is attached to the root, a leaf node, or between two nodes. A design feature CSG tree can be converted to a B-Rep after Boolean computing. The face set of a feature in volumetric representation of the CSG tree remaining on the B-Rep is the *trace* of the feature.

STEP AP 224 defines four groups of manufacturing features: machining features, transition features, replicate features and compound features [4]. Manufacturing features in this research are defined based on STEP AP 224.

3. The Relationships between Interacting Features

The interacting relationships between a design feature DF that is added to a part P (P' is the updated part of P) with a feature FE in P are summarized in Table 1, where V and ∂ are the volume and face boundary of a feature. In the adjacency, cross and nesting relationships, $V(DF)\cap^* V(FE) \neq \phi$ and $V(DF)$ can be divided into k partitions by FE. Each partition belongs to one of the two sets: $S = \{s_{i(1 \le i \le m)} \mid s_i \in (V(DF) -^* V(FE))\}$ and $S' = \{s'_{i(1 \le i \le n)} \mid s'_i \in (V(DF)\cap^* V(FE))\}$, where $m + n = k$, and if $m = 0$, $S = \phi$.

Table 1 The relationships between interacting features

Relationships	Conditions	Representations
Abutment	$V(DF)\cap^* V(FE) = \phi$ and $\partial(DF)\cap^* \partial(FE) \neq \phi$	$(DF) - abut - (FE)$
Adjacency	(1) If both DF and FE are subtractive, $m = 1$ and $V(s_1)\cap^* V(P) \neq \phi$ $(s_1 \in S)$; or (2) if both DF and FE are additive, $m = 1$.	$(DF) - adja - (FE)$
Cross	(1) If both DF and FE are subtractive, $m > 1$ and for each $s_i (s_i \in S)$, $V(s_i)\cap^* V(P) \neq \phi$; (2) if DF is additive and FE is subtractive, $m \ge 1$; or (3) if both features are additive, $m > 1$.	$(DF) - cross - (FE)$
Nesting	(1) When $m = 0$, if DF is subtractive and FE is additive, or if DF is additive and FE is subtractive, $V(DF) \subset V(FE)$ and $V(DF)\cap^* V(P) \neq \phi$ for a subtractive DF, or $V(DF) \subset V(FE)$ and $V(DF)\cap^* V(P') \neq \phi$ for an additive DF; or (2) when $m > 0$, if DF is subtractive and FE is additive, for each $s_i (s_i \in S)$, $V(s_i)\cap^* V(P) = \phi$.	$(DF) - nest - (FE)$
Co-trace	The traces of DF and FE can be combined into a single trace extended by a 2-D profile along a path	$DF - co\_trace - FE$
Co-axis	Both DF and FE are holes and their central axes have co-centre tolerance.	$DF - co\_axis - FE$
Replicate-Pattern	DF is copied and constrained to satisfy certain arrangement pattern.	$DF - repl - (pattern)$

A few examples of above defined relationships are shown in Fig. 1.

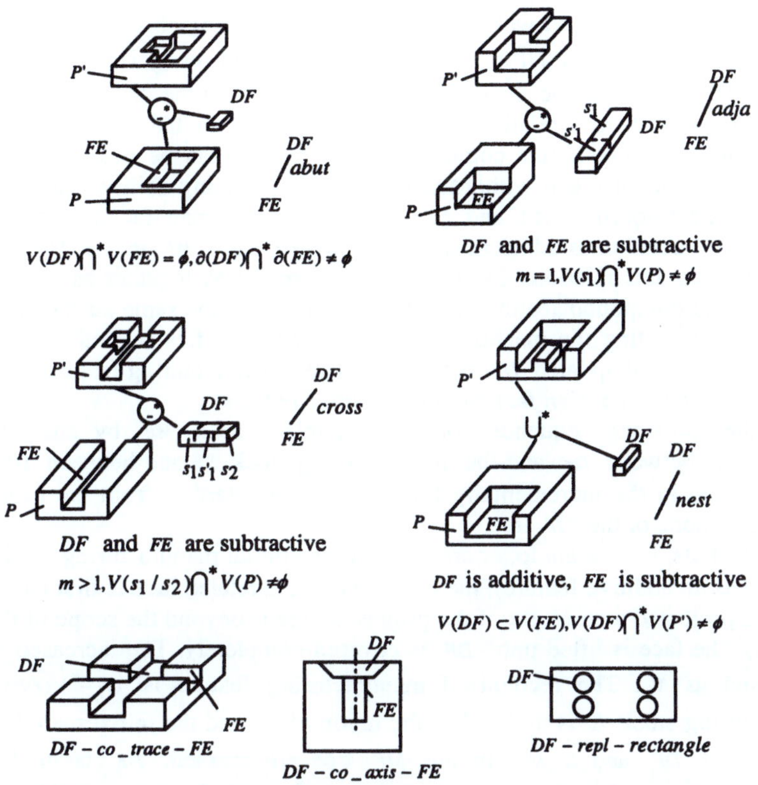

Fig.1 A few examples of interacting relationships between features

4. Incremental Interacting Feature Recognition Processor

The proposed feature recognition processor incrementally converts a design feature in a design feature CSG tree to a manufacturing feature model by traversing the CSG tree from its root to the highest leaf according to the following steps:

(1) Recognizing the initial manufacturing stock of a part

The root of the CSG is recognized as the initial manufacturing stock S. The traversal process in the CSG tree continues and the design feature DF being recognized is handled according to the following steps:

(2) If the recognized design feature is a subtractive feature

In the following sequence, DF is recognized as a manufacturing feature MF by analyzing the relationships between DF and the manufacturing stock S, and between DF and a MF traversed in the incrementally formed MF tree starting from the root to the highest leaf node of the MF tree.

(a) If DF nests on the stock S (S is treated as an additive feature), DF is modified as $V(DF) \cap^* V(S)$; (b) If DF abuts or is adjacent to a MF in the MF tree, and they have a co-trace relationship, DF and MF are united together as a single manufacturing feature if this combined single feature can be machined with the same surface tolerance and without violating the machining ability. Otherwise, DF is modified as $(V(DF) -^* V(MF))$, the updated DF is the recognized manufacturing feature, and MF is kept the same; (c) If DF and a MF in the MF tree are crossed, and they have a co-trace relationship, DF and MF are united together as a single manufacturing feature if this combined single feature can be machined with the same surface tolerance and without violating the machining ability. Otherwise, the two features are not modified, and DF is the recognized manufacturing feature; (d) If a hole DF abuts or is adjacent to another hole feature MF in the MF tree, and they have a co-axis relationship, DF and MF are united together as a compound feature if this compound feature can be machined with the same surface tolerance and without violating the machining ability. If DF satisfies several of the above conditions, the final updated DF is the recognized manufacturing feature.

(3) If the recognized design feature is an additive feature

In the following sequence, DF is recognized as a MF by analyzing the relationships between DF and the manufacturing stock S, and between DF and a MF traversed in the incrementally formed MF tree starting from the root to the highest leaf node of the MF tree.

(a) If DF abuts, is adjacent to, or crossed with the initial manufacturing stock S (S is treated as an additive feature), the boundary of the related face is first enlarged to cover the projected area of DF if the projected area is beyond the scope of the face. Secondly, the face is lifted until DF is covered completely. The increased volume is denoted as V_i. The recognized manufacturing feature is $(V_i -^* V(DF))$. The manufacturing stock S is updated as the union of V_i and the initial manufacturing stock; (b) If DF and a MF in the MF tree are crossed, DF is modified as $(V(DF) -^* V(MF))$, and MF is updated as $(V(MF) -^* (V(DF) \cap^* V(MF)))$; (c) If DF nests on a MF in the MF tree, DF joins MF as a single manufacturing feature, $(V(MF) -^* V(DF))$.

(4) If the recognized design feature has a transition feature

A transition feature that consists of a single planar face connecting two faces is recognized as a chamfer. A transition feature that is a concave surface with C' or C'', smoothly connecting two faces is recognized as a fillet. Lastly, a transition feature that is a convex surface with C' or C'', smoothly connecting two faces is recognized as a round edge feature.

(5) If several features have replicate-pattern relationships

In the finally recognized MF tree, several manufacturing features are grouped as a replicate feature in a certain mode if they have replicate-pattern relationships.

A prototype system based on the above approach has been implemented using C++ language and Parasolid V10 kernel functions, in a Visual C++ 6.0 environment. A test part – CAM-I ANC 101, its design feature CSG tree, and the finally recognized manufacturing feature tree are shown in Fig. 2.

In Fig. 3 – Fig. 5, some explanations of the recognition process are given. In Fig. 3 (a), $DF1$ is an additive feature added to the initial manufacturing stock S. By

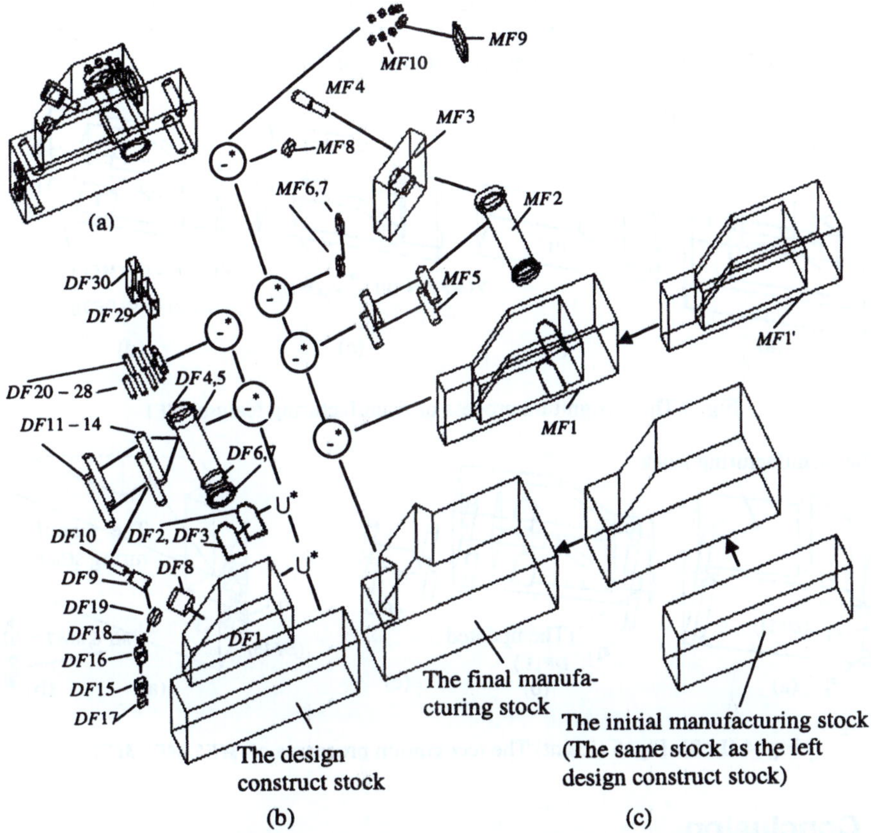

Fig. 2 (a) A test part; (b) The design feature CSG tree; (c) The manufacturing feature tree

lifting and projecting operations, V_1 is obtained in Fig. 3 (b). In Fig. 3 (c), $(V_1 -^* V(DF))$ is the recognized manufacturing feature $MF1'$ from $DF1$. Since $DF2$ and $DF3$ nest on $MF1'$, the manufacturing feature is updated as $V(MF) = (V(MF1') -^* V(DF2) -^* V(DF3))$ in Fig. 3 (d). In Fig. 4 (a), $DF11$ is crossed with the stock S, $DF11$ is updated as $V(DF11) \cap^* (S)$. In Fig. 4 (b), the updated $DF11$ is adjacent to a recognized feature $MF1$ in the MF tree, so it is further modified as $(V(DF11) -^* V(MF1))$. $DF12$ - $DF14$ are handled on the same steps, and they are arranged in a rectangular way as a replicate feature $MF5$ in Fig. 4 (c). In Fig. 5 (a), $DF15$ nests on S and is modified as $(V(DF15) \cap^* V(S))$. The updated $(V(DF15) \cap^* V(S))$ and $(V(DF17) \cap^* V(S))$, $(V(DF16) \cap^* V(S))$ and $(V(DF18) \cap^* V(S))$ have a co-base relationship respectively, so they are combined as two features in Fig. 5 (b), $MF6$ and $MF7$. Other features can be handled similarly. $MF3$ is recognized from the additive $DF8$. $DF29$ and $DF30$ are crossed and have a co-base relationship. They are combined as a single manufacturing feature $MF9$. $DF4 - DF7$, and $DF9 - DF10$, are combined as two compound hole features, $MF2$ and $MF4$. Parts of $(V(DF20 - 28)\cap^* V(S))$ form another replicate feature $MF10$.

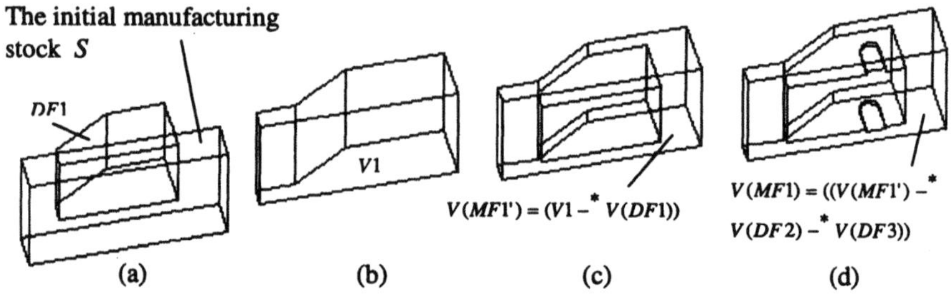

Fig. 3 The recognition process of manufacturing feature $MF1$

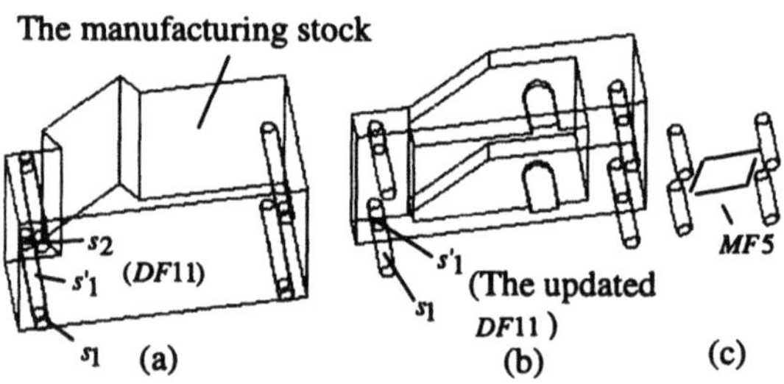

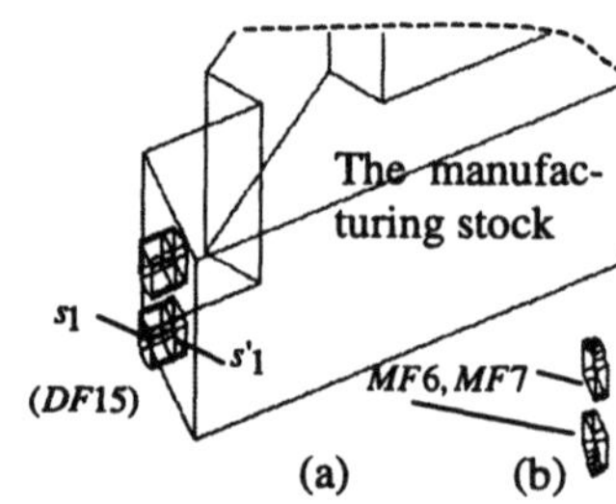

Fig. 4 (Left), Fig. 5 (Right) The recognition processes of $MF5, MF6, MF7$

5. Conclusion

Using a systematic analysis and process, the relationships between interacting features are represented. Based on these representations, an incremental recognition processor recognizes the manufacturing feature tree from a design feature CSG tree according to the evolving design process. In the future work, a method will be investigated to select an optimum machining sequence from the recognized manufacturing feature tree to serve the manufacturing applications.

References

[1] Laakko, T. and Mantyla, M. Feature modelling by incremental recognition. Computer Aided Design, 1993, 25(8), pp. 479-492.

[2] Martino, T.D. Falcidieno, B. and Habinger, S. Integration of design-by-feature and feature recognition approaches through a unified model. In Modeling in Computer Graphics Methods and Applications, Springer-Verlag, 1993, pp. 423-437.

[3] Lee, J.Y. and Kim, K. A feature-based approach to extracting machining features. Computer Aided Design, 1998, 30(13), pp. 1019-1035.

[4] ISO 10303-224, Mechanical product definition for process planning using form features. Document ISO TC184/WG3.

Vertical Integration of Tooling Functions from Design to Process Planning.

P.G. Maropoulos[1], R.P. Baker[2] and K.Y.G. Paramor[3]

[1,3]Design and Manufacturing Research Group, The School of Engineering, University of Durham, Durham. DH1 3LE. p.g.maropoulos@dur.ac.uk, k.y.g.paramor@dur.ac.uk
[2] Unigraphics Solutions Ltd. Parker's House, 46 Regent Street, Cambridge. bakerrob@ugsolutions.com

Abstract: To meet the competitive demands of modern manufacturing, it is necessary to reduce design times and enrich decision making by integrating process planning into the design activity using Concurrent Engineering principles. Cutting tool selection is an essential thread linking feature-based design of machined parts to process planning. Consequently, the implementation of tooling considerations into design is an important requirement for an integrated CAD/CAPP system. This paper defines an architecture to enable the vertical integration of tooling considerations from early design to process planning and scheduling.

1. Introduction

The organisation of a machining facility of any kind is a complex task involving a myriad of different skills. Examples of the different tasks involved include routing, tool selection, set-up analysis, cutter path optimisation, machinability assessment and time and cost estimation. The subject areas and methods that encompass these functions are also wide-ranging and include feature technology, artificial intelligence, mathematical modelling and manufacturing systems philosophies (cellular manufacturing, just in time and agility) amongst others [1, 2]. To provide enough time for the decision making process associated with these tasks, it is desirable to integrate process planning within all the stages of product development from conceptual design through to the introduction of the finished design on to the shop floor. Since many activities within design and process planning have been implemented as computer software in the form of computer aided design (CAD) and computer aided process planning (CAPP), it is now becoming feasible to automate process planning tasks as a design takes place and information about the component being designed becomes available. This paper describes the characteristics of an intelligent design system with embedded tool selection and process planning functionality.

2. Structuring Manufacturing Information

The framework used within this research for structuring manufacturing information is that of the time-phased Aggregate, Management and Detailed (AMD) architecture for the classification of process planning methods according to input data requirements and the required interfaces with design [3]. The objectives of each stage of AMD are as follows. Aggregate process planning, corresponding to conceptual and embodiment design, involves identifying processing options for a component and creating production routings. Management process planning assesses and rationalises the processing options identified at Aggregate level and takes place during the final, detailed stages of design. Detailed process planning, which takes place once the design is complete, optimises the remaining options to provide approved process plans.

The most important stages of design are the conceptual and embodiment stages, since the greatest scope for cost reductions exist during the early stages of product development [4, 5, 6]. Any proposed architecture for the implementation of process planning activities within CAD has to consider the fact that during the early stages of design, there is insufficient data available to construct a detailed product model, without which it is impossible to prepare detailed process plans [7]. Aggregate methods use the limited data available during the early stages of design to formulate processing options. The methods invoked during Management process planning carry out multi-criteria assessment of the options identified during the Aggregate stage in order to create a short list of options that are most likely to be used. These methods are different from those required for Detailed process planning, which use data analysis and optimisation methods along with shop floor expertise to determine accurate data necessary for machine programming, scheduling and shop floor control.

3. Mapping Form Features to Tooling Considerations

The tasks carried out by various CAPP systems are highly varied. However, the process planning function that can most easily be integrated into design, is the task of tool selection [8]. The three main reasons for this are as follows.

Firstly, the building blocks for component design are features that represent the technology required to manufacture the component being developed [9]. Tooling issues can be mapped to both geometrical feature representations and manufacturing requirements and so act as a bridge between the geometry of design and the technology of manufacture. Since tool selection is one of the first steps of process planning, tooling issues have the potential of influencing early design decisions [10]. Secondly, the selection of cutting tools influences the cutting conditions used and hence the performance of machining operations employed. Thus, careful consideration of tooling issues throughout the various stages of design will help to optimise machine selection. Finally, the selection of tools constrains feature sequencing and the selection of appropriate tool approach directions [11]. This in turn affects batch set-up times and machine utilisations. Consequently, by performing tool selection during design, it is possible to start planning for resolving possible scheduling problems. This is important, because by migrating some of the short and medium term scheduling problems into the long range domain [12], it is possible to eliminate some of the uncertainties in manufacturing [13].

From the above discussion, it can be deduced that tooling issues represent a 'thread' that links feature-based design with a variety of process planning activities. The ability to link tooling functions with process planning in a time-phased manner has been demonstrated through the concept of the five levels of tool selection [10]. The objective of the five levels of tool selection is to provide a framework within which to develop a tool resource structure (TRS). TRS represents a set of tool options for each feature and is made up of ordered lists of tools that are capable of carrying out the individual operations required to manufacture the component, and also includes the machines on which the tools are mounted.

4. An Architecture for the Integration of Tool Selection with Design

It is necessary to provide a formal categorisation of tooling issues and their effect and contribution to various process planning activities. This categorisation is provided in the form of the five levels of tool selection [10]. The time-phased, vertical integration of the various design stages, with the AMD architecture and the five levels of tool selection is shown in Figure 1.

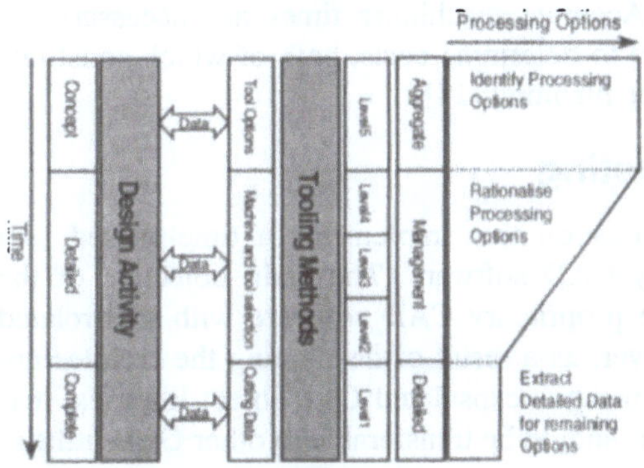

Figure 1. Vertical integration of tooling methods with design and the AMD process planning architecture

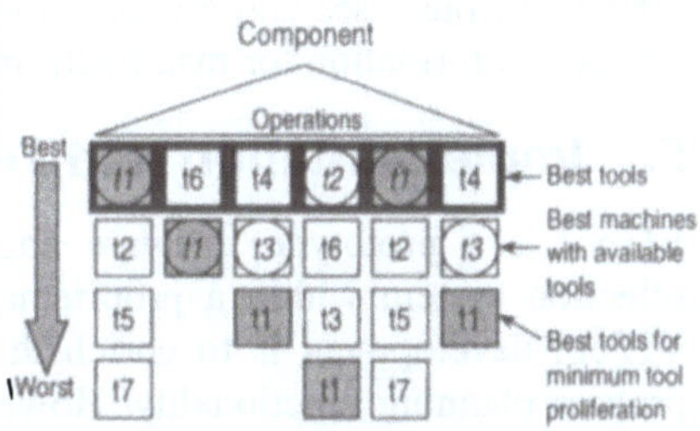

Figure 2. Using the TRS for machine and tool selection

The main objective of level 5 tool selection is to identify all possible machines containing tools that can manufacture all or part of the component. Matching tools to each feature of the component is performed by using the dimensions of the features as the basis for defining ideal tool dimensions along with the identification of 'alternative features' which correspond to alternative tool approach directions and create additional tool selection options. The identification of alternative features creates more options from which to select the final machining strategy.

The tasks carried out in levels four, three and two are grouped together, as the activities of these levels cannot be performed in isolation. During these stages enough information about a design is available to enable rationalisation of the processing options identified at level 5. The first task is to identify set-ups. A distinct set-up is a group of features requiring a specific tool approach direction. Consequently, the task of set-up identification is to decide which feature from a set of feature alternatives should be used. The next task in this level is to create a TRS such as the one shown in Figure 2. The TRS as an ordered set of tools that are able to carry out each operation required to manufacture the component. The ordering of tools is performed based on machining time, suitability of tool dimensions and the availability and location of tools. It is then possible to select machines and tools according to three possible strategies outlined in Figure 2. The first of the three strategies is the selection of the best machine or machines to manufacture the component. As part of this strategy, it is necessary to select the best machines that are able to completely manufacture each set-up and select the best tools in the TRS that are attached to those machines. The next strategy is the selection of the best tools and the final strategy is the reduction of tool proliferation by selecting common tools. Although three separate strategies are shown, perhaps the most important task is to assess these strategies simultaneously. By weighting the importance of each strategy, machines and tools can be selected such that a compromise is reached between the results of the strategies.

Level 1 tool selection is invoked after the design has been completed and when machine and tool selection and rationalisation has taken place. The responsibility of this level is to produce accurate cutting conditions from which machining times can be extracted. Accurate machining times are necessary for scheduling purposes and are also used to determine costs, both of which constitute essential information for manufacturing planning [14].

5. Implementation and Testing

VITool is a prototype software application that implements a time-phased tool selection system within a proprietary CAD software. The main objective of the VITool development is to enrich the proprietary CAD software with tool-related process planning functionality. However, as a result of developing the architecture described thus far and programming using encapsulated C++ object libraries, it is also intended that the VITool functions should be transferable to other CAD utilities as an open system.

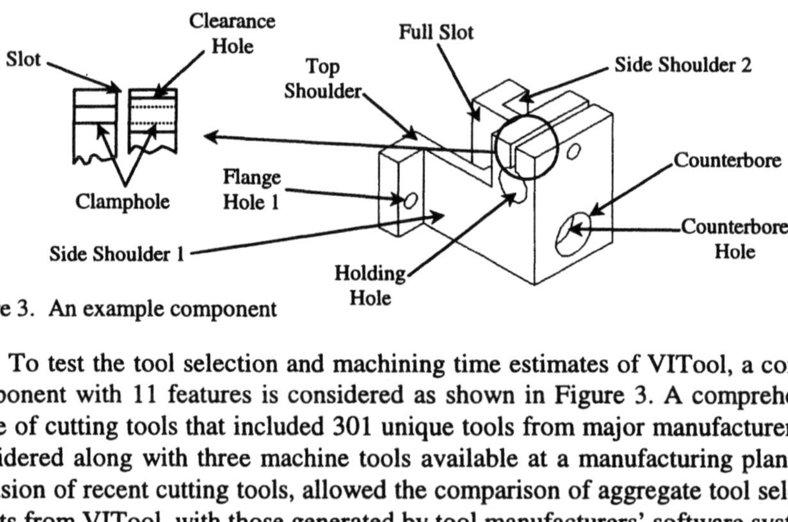

Figure 3. An example component

To test the tool selection and machining time estimates of VITool, a complex component with 11 features is considered as shown in Figure 3. A comprehensive range of cutting tools that included 301 unique tools from major manufacturers was considered along with three machine tools available at a manufacturing plant. The inclusion of recent cutting tools, allowed the comparison of aggregate tool selection results from VITool, with those generated by tool manufacturers' software systems.

The summary results for the component shown in Figure 3 are shown in Table 1. In particular, for each feature VITool identified the cutting operations required and suitable tools were selected for each operation. From the range of possible tools, optimised tools were selected on the basis of minimum machining time. For the optimal tools selected, the machining times estimated are shown in the last column of Table 1.

Table 1: Summary results of VITool.

Feature	Operation	No. of tools selected	Optimised tools	Machining time (sec.)
Full Slot	Roughing	3	Seco Solid EndMill 25	161.4
	Profiling	3	Seco Solid EndMill 25	23.22
Counterbore	Drilling	9	Coromat U Drill Short 20	0.96
Counterbore Hole	Drilling	1	Coromant_Delta_C_Drill_Long_dia_5	3.24
Clamphole	Drilling	1	Coromant_Delta_C_Drill_Long_dia_5	1.32
Clearance Hole	Drilling	2	Coromant_Delta_C_Drill_Short_dia_6.9	2.16
Flange Hole 1	Drilling	2	Coromant_Delta_C_Drill_Short_dia_6.9	1.68
Flange Hole 2	Drilling	2	Coromant_Delta_C_Drill_Short_dia_6.9	1.68
Holding Hole	Drilling	2	Coromant_Delta_C_Drill_Short_dia_15.51-16	12.84
Side Shoulder 1	Roughing	6	Secodex-C_Sq_Sh_80	55.14
	Profiling	6	Secodex-C_Sq_Sh_63	5.1
Side Shoulder 2	Roughing	6	Secodex-C_Sq_Sh_80	55.14
	Profiling	6	Secodex-C_Sq_Sh_63	5.1
Top Shoulder	Roughing	7	Secodex-C_Sq_Sh_50	84.18
	Profiling	7	Secodex-C_Sq_Sh_50	8.4
Total tools selected		63	15	
Total unique tools		27	8	
Total mach. time				394.14

Rationalising selected tools reduces the final tool count from 27 unique tools to 8, a decrease of 70%. The eleven features were clustered into five set-ups according to their tool approach direction and a single machine tool was selected to perform all operations. It is seen from the results produced by VITool that correct types of tools were selected for each type of feature, for example square shoulder mills to produce shoulders, slot and drilling endmills to produce slots and twist and u-drills to produce hole features.

Further testing, not detailed within this paper, proves that the rationalisation of tooling is highly dependent upon the strategy employed. Reducing tool proliferation by selecting common tools is an efficient method that minimises tool count, however, the remaining tools may not be optimal. To the other extreme, selection of best tools or machines can leave a high tool count if further rationalisation is not performed. It was therefore found that the best compromise was achieved using a combination of all methods. It was also found that tool count optimisation is highly dependent upon the quantity of tools in the initial TRS and the percentage of drilled hole features as these have less scope for tool rationalisation, as multi-purpose tools cannot be used.

6. Conclusions

Using tooling issues as the link between design and process planning, through the integration of the five levels of tool selection within the AMD architecture, a framework for the vertical, time-phased CAD/CAPP integration has been developed in the form of VITool. It is envisaged that the result of this research will provide Industry with a tool to assess the potential impact that designs might have on the shop floor. This will lead to more efficient product and process development since likely problems will be highlighted during the early stages of design, rather than being discovered prior to, or even during production.

To demonstrate the validity of the tool selection functions and machining time calculations, a complex component was planned using VITool and a range of tools was selected with cutting data and accurate machining time estimates.

Acknowledgements

The support of the Engineering and Physical Sciences Research Council (EPSRC) UK is gratefully acknowledged (Grant GR/L07017). The authors also wish to thank the collaborating companies and acknowledge the help of Dr. H.D. Bradley.

References

[1] Alting L, Zhang HC. *Computer-aided process planning - the state-of-the-art survey.* International Journal of Production Research 1989; 31(9): 2139-2160

[2] Leung HC. *Annotated bibliography on computer-aided process planning.* International Journal of Advanced Manufacturing Technology 1996; 12(5): 309-329

[3] Maropoulos PG. *A novel process planning architecture for product-based manufacture.* Proceedings of the Institution of Mechanical Engineers Part B 1995; 209: 267-276

[4] Goldense B, Vermette D. *Driving forces for product development speed.* Rapid News 1997; 5(6): 12-17

[5] CAMI: Computer-Aided Manufacturing International Inc. Cited in: A smarter way to manufacture. Business Week; April 30 1990: 110

[6] Li F, Chow JG. *Development of a CAD-based intelligent manufacturing system.* Computers and Industrial Engineering 1994; 27(1-4): 135-138

[7] Maropoulos PG, Bradley HD and Yao Z. CAPABLE: *An Aggregate process planning system for integrated product development.* Journal of Material Processing Technology, Vol.76/1-3, 1998, 16-22.

[8] Chen CL, LeClair SR. *Integration of design and manufacturing: solving set-up generation and feature sequencing using an unsupervised-learning approach.* Computer Aided Design 1994; 26(1): 59-75

[9] Gao JX, Huang XX. *Product and manufacturing capability modelling in an integrated CAD/process planning environment.* International Journal of Advanced Manufacturing Technology 1996; 11: 43-51

[10] Maropoulos PG. *Cutting tool selection: an intelligent methodology and its interfaces with technical and planning functions.* Proceedings of the Institution of Mechanical Engineers Part B 1992; 206(1): 49-60

[11] Kamhawi HN, LeClair SR, Chen CL. *Feature sequencing in the rapid design system using a genetic algorithm.* Journal of Intelligent Manufacturing 1996; 7: 55-67

[12] McKay KN, Safayeni FR, Buzacott JA. *Job-shop scheduling theory: what is relevant?* Interfaces 1988; 18(4): 84-90

[13] Bermejo J, Calinescu A, Efstathiou J, Schirn J. *Dealing with uncertainty in manufacturing: the impact on scheduling.* Proceedings of the 32nd Matador Conference, UMIST, Manchester, July 10-11 1997: 149-154

[14] Dewhurst P, Boothroyd G. *Early cost estimating in product design.* Journal of Manufacturing Systems 1988; 7: 183-191

Neural network based application and data management for cutting values

Prof. Dr.-Ing. habil. Dieter Fichtner[1], Dr.-Ing. Andreas Nestler[1]
Dr.-Ing. Andrew Kaldos[2], Dipl.-Ing. Christiane Rehm[1]
[1]Institute of Production Engineering, Dresden University of Technology, 01062 Dresden, Germany, nestler@mciron.mw.tu-dresden.de
[2] School of Engineering, Liverpool John Moores University, Byrom Street, Liverpool, L3 3AF, UK, a.kaldos@livjm.ac.uk

Abstract: The manufacturing processes have to guarantee a flexible, high quality production at low cost. In order to fulfil the demands, intelligent information systems are required with special respect to the technology data type. A central point of technological data is the cutting values, i.e. feed and speed. The high speed milling process is one of the newly established processes with the need to set the optimum cutting parameters. The rapid development of advanced materials, tools etc. the results of a recent project* to solve cutting parameter generation problems by means of neural network, mathematical model and technology database system.

1. Introduction

A very promising creative approach to generate technology data is the application neural networks. Neural networks offer a possibility to solve these problems by training networks using known cutting values and its ability to generalise and translate them for a forthcoming application phase [1].

A cutting tool application catalogue or applications of mathematical models for technological data determination give useful initial cutting parameter values for the application in different technological software systems. A mathematical process model for the simulation of high speed milling was successfully developed and used for generating cutting parameters, for cutting process energy analysis considerations and for process model verification. Nevertheless it is necessary to verify the cutting values after practical tests, e.g. by verifying a new CNC-program on the machine tool. The use of know-how within a technology database is necessary for the determination of cutting values. Primary cutting value

---

* The funding provided by the British Council and the Deutscher Akademischer Austauschdienst (DAAD) through the Academic Research Co-operation (ARC) programme in support of this project is greatly appreciated and acknowledged.

is necessary for the determination of cutting values. Primary cutting value parameters, which were obtained from industrial experience or from detailed cutting experiments are more valuable and accurate than calculated cutting values. Also, for technological problems, which have too many input and output variables with non-linear interactions, therefore it would be very difficult to describe them by mathematical equations, it is better to use suitable knowledge based methods for process modelling.

The neural networks are able to integrate known cutting values from different sources and to condense this set of data step by step. A suitable cutting process parameter database has to be designed in support of the selected neural network configuration. The selection of the architecture of the neurones and the connections in the hidden layers largely depend on the specific application. In the applications of neural networks it has been demonstrated how a neural network can be used for decision support before or during manufacturing. This project also demonstrates the results obtained when using neural network, mathematical model and technology database system for milling and high speed cutting.

2. Neural networks for cutting values determination

The traditional way for the determination of cutting values is to use tables, handbooks and database information systems. Unfortunately the complexity of machining conditions and the interactions between cutting parameters cannot be fully described with only one of these methods. In addition, intuition and personal experience are often considered as extremely useful and important support in setting-up a neural network based process modelling system

The basic calculation methods contain mathematical models and knowledge based methods, such as neural network, fuzzy logic or genetic algorithm in connection with a database [2]. The present and future trends are moving towards the hybrid, intelligent structures.

Application of mathematical models for technology data determination can give useful initial parameters for using in different technology software systems. Computer models, which allow for the prediction of the process output parameters, complete with supporting experimental data, are essential in determining cutting parameter values [3].

Neural networks are applied for knowledge acquisition and processing in intelligent machining and process modelling. Unknown correlation between input and output parameters can be learned and reproduced by neural networks. It is necessary to get useful parameters to describe a technology situation and to select and provide a data set for a learning database.

The provision of a learning database method offers a possibility to calculate cutting values by training networks using known cutting values. It is also feasible to apply different data sources for cutting process parameter selection [1].

3. Guidelines for establishing a neural network application

The procedure for establishing a suitable network for application includes the main stages as follow [1]:

- description of the problem;
- data selection;
- design of neural network;
- training and test of neural network;
- practical performance test;
- integration of network in a framework of an application.

The first step includes considerations about the present target and the purpose of the later application. In preparation phase the available data sources have to be analysed.

4. Data sources

The aim is to verify results of a process simulation model for high-speed machining and fine-tuning a current database for cutting value prediction [4]. Firstly, the two basic types of data sources have to be checked and verified:

- Data from tool manufacturer catalogue [4,5] and
- Data from results of calculation module for HSC- simulation [6].

Allocation of typical cutting parameters for high-speed cutting (HSC) obtained from tool catalogues is different, as shown in Table 1. The application of the simulation model makes it possible to show the

Table 1: Examples for HSC-input (I) and -output (O) parameters based on tool catalogues

Parameter	Unit	Hoffmann-Catalogue	Fette-Catalogue
Workpiece material, Code	-	I	I
Workpiece material, name	-	I	I
Workpiece material, tensile strength. R_m	N/mm^2	-	I
Cutting material, Code	-	O	-
Cutting material, name	-	O	-
Cutting speed v_c	m/min	O	O
Feed per tooth f_z	mm	O	O
Number of tooth z	-	I	-
Tool diameter D	mm	I	I
Depth of cut, axial	mm	I	I
Width of cut, radial	mm	I	-
Spindle speed n	min^{-1}	O	O

Correlation between the cutting parameters and their effects on the simulated results, as illustrated in Figure 1.

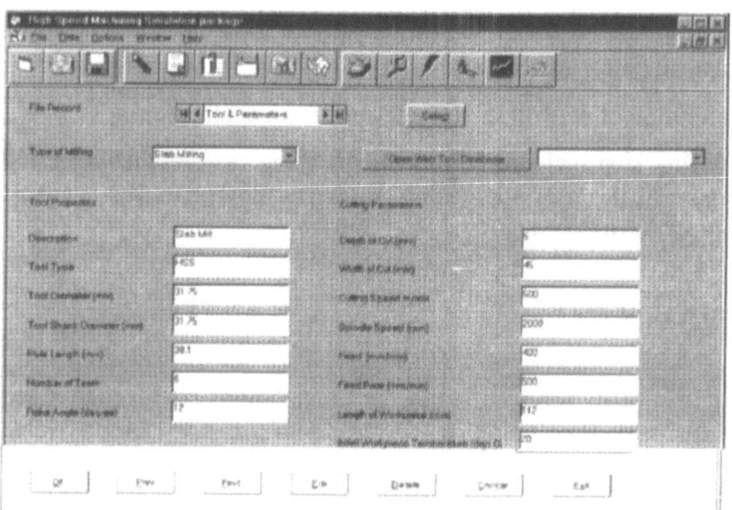

Figure 1: Input- and output parameter values of the HSC-simulation model [6]

It can be clearly seen that the input parameters selected by the user greatly influence the high speed cutting process. The developed process simulation module is supported by an integrated technology database for the data acquired for workpiece and tooling, mechanical and thermal properties, cutting conditions, workpiece geometry, tooling geometry and machine tool parameters. This process model also included the provision of easy communication with other databases and user sites via the Internet connection.

5. Data management to establish a neural network

The creation of the necessary neural networks, the preparation of the network training data and the selection, training and testing of the neural network requires an extended data management system. Figure 2. shows the integration of trained networks into the framework of a technology database system. Figure 3. Illustrates the data use from a tool catalogue.

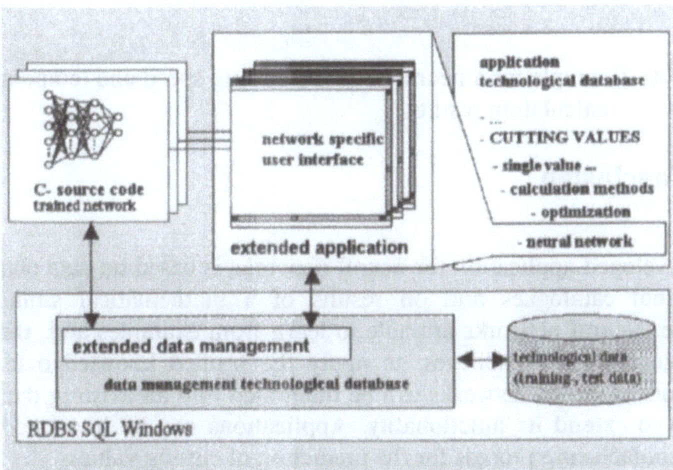

Figure 2: Extended data management for neural networks

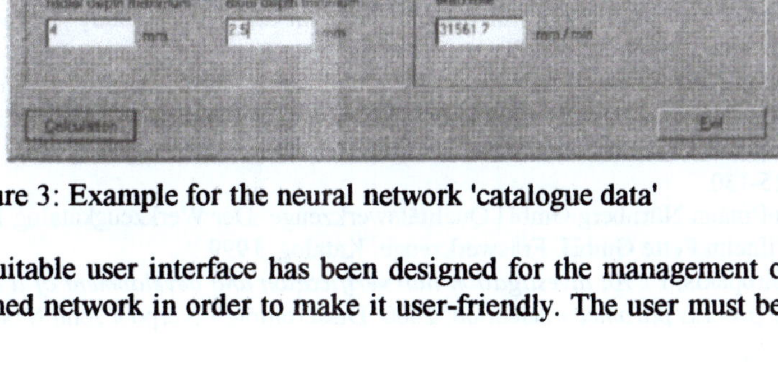

Figure 3: Example for the neural network 'catalogue data'

A suitable user interface has been designed for the management of the trained network in order to make it user-friendly. The user must be able to work with the network based modelling system without in-depth for the use of neural network based modelling to establish cutting temperature for selected conditions.

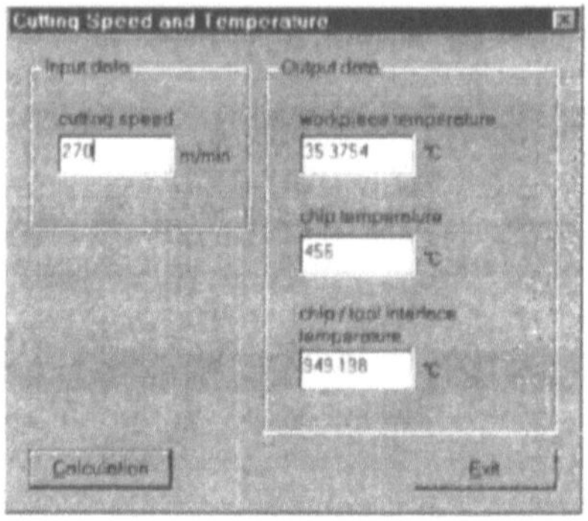

Figure 4: Example for a neural network 'cutting speed and temperature' based on calculation results

6. Conclusion

The developed application for neural networks is based on data obtained from tool catalogues and on results of a mathematical simulation module. Neural networks are able to learn from examples and, through good generalisation abilities, to apply the trained knowledge to new situations. Trained networks can be integrated into an existing database system to extend its functionality. Applications can be integrated into the manufacturing process for the prediction of cutting values.

References

[1] Fichtner, D.; Nestler, A.; Schulz, G.1999, *Wissensakquisition für Schnittwerte beim Fräsen unter Nutzung von neuronalen Netzen*. Fortschr.-Berichte VDI Reihe 20, Nr. 304, Düsseldorf: VDI-Verlag, ISBN 3-18-330420-1

[2] Balic, J., Nestler, A.; Schulz, G 1999, *Prediction and Optimization of Cutting Values Using Neural Networks and Genetic Algorithm*, Proceedings of the 4th International Conference on Design for Manufacture in Modern Industry DMMI'99. Maribor, Slovenia,, ISSN-1580-1330, pp. 192-203

[3] Kaldos, A.; Boyle, A.; Fichtner, D.; Dagiloke, I., 1998, *Development of an integrated process parameter selection model for high-speed machining*, Proceedings of the 14th National Conference on Manufacturing Research, University of Derby, Derby, UK, pp. 125-130.

[4] Hoffmann Nürnberg GmbH Qualitätswerkzeuge. Der Werkzeugkatalog 1997/98.

[5] Wilhelm Fette GmbH. Fräswerkzeuge. Katalog, 1999

[6] Maropakis, P.: *An investigation into verification and development of a computer model for process parameter selection,* 2000, Dissertation, Liverpool John Moores University

A Knowledge-Based System for the Selection of Cutting Tools and Data for High Speed Machining of Moulds / Dies

W.M. Sim [1], R.C. Dewes [1] and D.K Aspinwall [1]

[1] School of Manufacturing & Mechanical Engineering, The University of Birmingham, w.m.sim@bham.ac.uk, r.c.dewes@bham.ac.uk, d.k.aspinwall@bham.ac.uk

Abstract: The paper details the development of a PC-based, knowledge-based system (KBS), for assisting users to select appropriate cutting tools and machinability data, for the high speed machining of moulds and dies. The system has been designed to incorporate fuzzy logic and genetic algorithm functions. Selection criteria used include workpiece specification (tool steel designation and hardness), machine tool specification (spindle taper, spindle speed and feed rate), nature of the operation (face milling and ball nose end milling) and type of cut (roughing and finishing). Outputs include recommended cutting tools and operating parameters.

1. Introduction

High speed machining (HSM) is a technology that requires machine tools capable of high rotational speeds (10,000-100,000 rpm), high feedrates (10,000-80,000 mm/min) and high accelerations / decelerations (10-20 m/s^2). Small diameter (<10 mm) ball nose end mills are generally employed. The emergence of new generation cutting tool materials / coatings and CAD/CAM systems, has accelerated the uptake of this technology. Currently, the aerospace and mould / die industries have seen the largest growth of HSM, with benefits including reduced costs and shorter lead times. With moulds and dies, this is largely due to the elimination / reduction of electrical discharge machining (EDM) and manual polishing operations. Other advantages include longer cutting tool life [1] and up to 25% longer die life due to the generation of compressive residual stresses in the machined surface [2]. HSM allows the cutting of a wide range of tool steels in the fully hardened state (30-65 HRC) using coated carbide tools without the need for cutting fluid.

2. Background to Machinability Database Systems

The correct selection of cutting tools and machinability data leads to reduced cost, increased productivity and improved component quality (lower workpiece surface roughness and higher dimensional accuracy) [3]. This selection process is complex and requires expertise and experience. In contrast with established machining processes such as turning and drilling, structured comprehensive machinability data for HSM is not widely accessible despite widespread information in academic journals. Adding to this is the fact that, end milling and in particular, ball nose end milling, is a more complex operation than for example, turning, as it involves many additional factors. These include up & down milling, axial & radial depths of cut, a range of workpiece inclination angles and cutting directions (β angles). HSM also involves a wide variety of tool types and formats, e.g. solid / indexable tools, ball nose / corner radius end mills and parallel / taper cutters. Another problem that relates to moulds / dies is their complexity in terms of draft angles, corner geometries, etc. All of these factors mean that the optimisation of HSM for such components is extremely difficult if not impossible.

Over the past decade, cutting tool developments have resulted in a large array of commercial products. As a consequence, it is difficult for end users to identify the most appropriate tools (type / grade / format / etc.) and machinability data for their application. In addition to standard reference books and manuals which offer generic machinability data, many of the leading cutting tool manufacturers / suppliers have developed database systems using CD-ROMs [4-6], via the Internet or in more traditional paper-based catalogues. Table 1 gives examples of computerised machinability database systems available on the market. These are intended for a cross-section of industry sectors and are, in general, not specific to a certain application area, eg. mould and die. Not surprisingly, many systems are developed solely for commercial / marketing purposes, and are therefore biased towards the sale of a specific manufacturer's cutting tools.

Table 1: Examples of commercial machinability database systems

Supplier, system name and reference	System inputs	System outputs
Sandvik SSEM 98 [4]	Workpiece material / hardness, type of tool, tool diameter, a_p, a_e, R_{zmax}, N_{max}, milling procedure, cutting fluid	Cutting tool specification, v_c, N, f_z, v_f, P, L, Q, T
Seco Wintool Compact [5]	Workpiece material, type of tool, type of operation, tool diameter, insert, grade	Cutting tool specification, v_c N, a_e, a_p, h_m, f_z, v_f, P
Gühring Guroguide 2.0 [6]	Workpiece material, type of tool, type of operation, tool diameter, a_p, a_e, shank type, tool material, cutting fluid,	Cutting tool specification, v_c, N, f_z, v_f, P, F, T

a_p:axial depth cut; a_e:radial depth cut; R_{zmax}:workpiece surface roughness; N_{max}:max. spindle speed; v_c:cutting speed; N:rotational speed; f_z:feed per tooth; v_f:feedrate; P:power consumption; L:length cut; Q:metal removal rate; T:tool life; h_m:mean chip thickness.

2. Knowledge-Based System (KBS) Development

A KBS represents human expertise on a specialist topic within a computer program in order to perform functions similar to those normally performed only by a human expert [7]. A KBS is useful when specialised knowledge / expertise is not easily accessible. The previously detailed machinability databases are not true knowledge systems in that they cover a wide range of different / general machining operations and lack any built-in intelligence to guide a user to select appropriate tools and data for a specific application. True KBS examples for machining operations have been published [8-14]. These are primarily based on academic project work and are rule-based. Most focus on the selection of cutting tools and machinability data for turning prismatic parts at conventional cutting speeds using a limited range of workpiece materials.

The current work involves the development of a prototype KBS containing the facts and rules that govern the selection of cutting tools and machinability data for mould and die machining. The prototype system is standalone, however, future research will involve the development of an online system to enable CAD/CAM and CNC programmers to select appropriate tools and data at the CAD workstation or on the machine tool. The system includes commonly used materials (AISI H13 hot work tool steel & AISI P20 cold work tool steel) and incorporates face milling & end milling with coated indexable & solid carbide tools. The KBS is able to handle core / cavity machining, machine tool aspects, cutter path strategies and tool wear. A Pentium II PC with Windows 98 was used to run a rule-based expert system incorporating fuzzy logic (FL) and genetic algorithms (GA), i.e. Attar XpertRule [15]. The rule-based aspect of the system was used for structuring the knowledge, enabling FL to deal with vagueness in decision making and GA for searching optimal solutions. The software had two types of inference engine (forward & backward chaining), to deduce the knowledge decision tree. It interfaced with other Windows software to allow data transfer between applications.

The knowledge base within the prototype system comprised: selection of cutting tools / machinability data, algorithms, cutter path strategies and tool wear. The knowledge was obtained from experts (CAM programmers, machine operators & a mould / die specialist from a cutting tool supplier), academic literature, handbooks, tooling catalogues and visits to toolmakers. The system included databases of tooling (toolholders & inserts) and machinability information, see Figure 1. The system was designed as a number of main modules: machining operations (face milling – roughing and finishing & end milling – roughing, semi-finishing and finishing), user interface, machine tool capacity check, cutter path strategies and tool wear. Figure 2 shows a sample flow chart for rough machining.

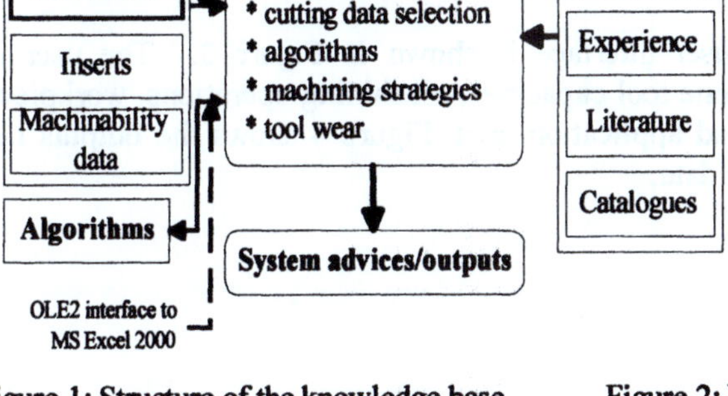

Figure 1: Structure of the knowledge base

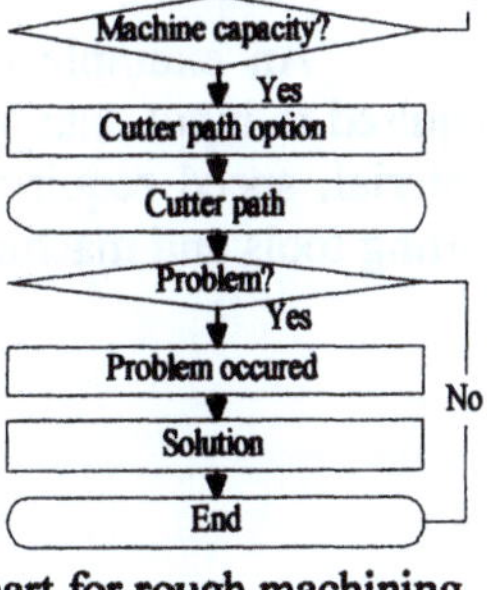

Figure 2: Flow chart for rough machining

The criteria and factors for the selection of cutting tools and machinability data are too extensive to report here, however, as an example, rules are given in Table 2 [16] for the selection of axial and radial (stepover / pick feed) depth of cut (rough machining), based on cutter type and diameter. Similarly, in respect of workpiece surface roughness for finishing operations, the 'Makino concept' was used which employs the same values for radial depth of cut and feed per tooth, in order to obtain uniform surface topography [17].

Table 2: Rules for axial and radial depth of cut for rough machining [16].

Cutter type	Axial depth of cut, a_e	Radial depth of cut, a_p
Round insert end mill	< 0.66 Dc	< 0.3 Ic
Indexable ball nose end mill	< 0.3 Dc	< 0.25 Dc
Solid corner radius end mill	< 0.3 Dc	< 0.1 Dc
Solid ball nose end mill	< 0.1 Dc	< 0.05 Dc

Dc : cutter diameter; Ic : insert diameter

Curve fitting / regression analysis, was used to derive equations for the recommended data i.e. feed per tooth, ramping angle, workpiece hardness compensation factor, etc. Algorithms were created in XpertRule and MS Excel 2000, via an Object Linking Embedded (OLE2) interface, to calculate the spindle rotational speed and feedrate for different types of cutting tools. MS Access 2000 was used to store tooling / machinability data and was accessed via an Open Database Connectivity (ODBC) interface. The database was grouped as follows: cutter diameter, cutting tools, inserts, feed per tooth and cutting speed.

Cutter path strategies are an important issue in end milling. Two commonly used strategies for rough machining are ramping and circular interpolation. The ramping angle used is affected by the type of cutter and cutter diameter. Table 3 details maximum ramping angle and axial depths of cut [16]. A tool wear module was also incorporated into the KBS to provide solutions to any problems that may occur when using the recommended machinability data [18].

Table 3: Rules for maximum ramping angles and axial depths of cut [16].

Cutter type	Max. ramping angle (°)	Max. axial depth of cut, a_p
Round indexable end mill	13	< 0.3 Ic
Indexable ball nose end mill	30	< 0.3 Dc
Solid corner radius end mill	10	< 0.1 Dc
Solid ball nose end mill	10	< 0.05 Dc

An example of the user interface is shown in Figure 3. The user is required to input data on machine tool capacities, machining operations, workpiece material, speed requirement and application area. Figure 4 shows the outputs i.e. cutting tools and machinability data.

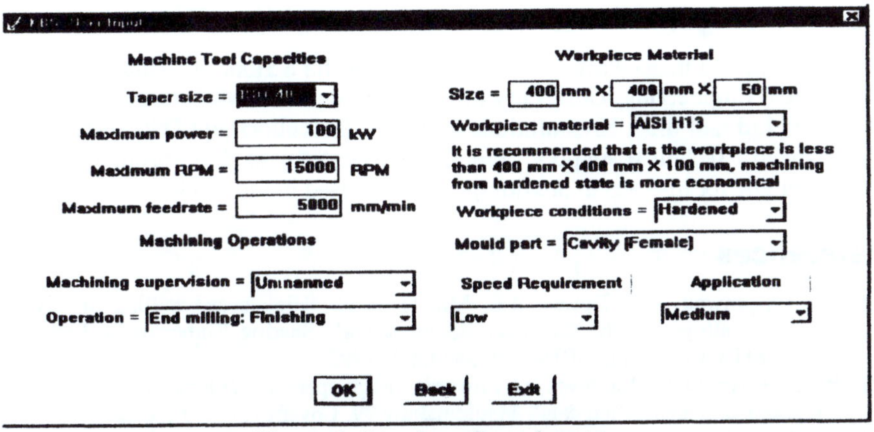

Figure 3: User interface - inputs

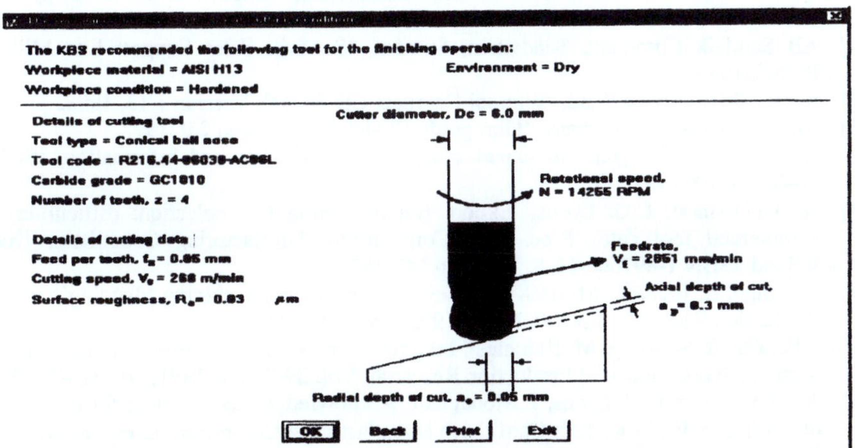

Figure 4: User interface - outputs

3. Conclusions

- A major requirement of the work and KBS programming in general, is elicitation of expert knowledge, which is very time consuming (~3 months for the present prototype system).
- The commercial software package employed (cost ~£2000) proved appropriate for the knowledge task.
- The prototype system is suitable for CAM programmers and machine operators involved in the HSM of moulds and dies.
- The system will benefit from future data enhancement with additional cutting tools and incorporation into machine tool controllers (PC-based) and CAD / CAM systems.

Acknowledgements

We would like to thank Prof. A. A. Ball, Head of the School of Manufacturing and Mechanical Engineering, University of Birmingham, John McTernan of Bridgeport Machines Ltd and the UK Committee of Vice-Chancellors and Principals (CVCP) for financial / technical support. We are also grateful to Bob Hirons of Sandvik Coromant UK for his help in providing expertise used in the KBS.

References

[1] D.K. Aspinwall, R.C. Dewes, A.L. Mantle, et al, 'High speed milling of advanced aerospace alloys and hardened mould/die steels', Sharing Tomorrow's Technology Today, MTTA and DTI, ISBN-0-907348-04-1, 1998.

[2] R. C. Dewes, D. K. Aspinwall, D. Edwards, 'High speed machining of moulds and dies – research and application data', Mouldmaking 99, Coventry, UK, February, 1999.

[3] Y. Chen, A. Hui and R. Du, 'A fuzzy expert system for the design of machining operations', Int. J. Machine Tools and Manufacture, Vol.35, No. 12, 1995, pp. 1605-1621.

[4] AB Sandvik Coromant, Sandviken, Sweden, 'Sandvik Sales Support End Mill CD-ROM', 1998.

[5] Seco, Zürich, Switzerland, 'Wintool Compact CD-ROM', Version 1.13, 1998.

[6] Gühring, Albstadt, Germany, 'Guroguide CD-ROM', Version 2.0, 1998.

[7] A. Goodall, 'The guide to expert systems', Oxford: Learned Information, ISBN 0-904933-49-0, 1985.

[8] A. McDermott, C.G. Lyons, 'Expert system cutting tool selection: difficulties with commercial tool data', Proc. 14th Conf. Irish Manufacturing Committee, Dublin, Ireland, ISBN 1-897606-16-8, 1997, pp.183-192.

[9] R. Singh, S. Raman, 'METEX – An expert system for machining planning', Int. J. of Production Research, Vol. 30, No. 7, 1992, pp. 1501-1516.

[10] S.H. Yeo, Y.S. Wong, M. Rahman, ' Integrated knowledge-based machining system for rotational parts', Int. J. of Production Research, Vol. 29, No. 7, 1991, pp.1325-1337.

[11] A. McDermott, C.G. Lyons, J. Monaghan, 'A knowledge based system for the selection of cutting tools', Proc. 12th Conf. Irish Manufacturing Committee, Cork, Ireland, ISBN 0-907134-335, 1995, pp.1001-1008.

[12] B. Arezoo, K. Ridgway, 'The applications of expert systems to the selection of cutting tools and conditions for machining operations', Proc. 1st Conf. Artificial Intelligence and Expert Systems in Manufacturing, ISBN 1-85423-075-1, 1990, pp. 113-123.

[13] R. Razfar and K. Ridgway, 'EX-CATSMILL - An expert system for selecting cutting tools and conditions for milling', Factory 2000 - Advanced Factory Automation, 1994, IEE Conf. Publication, No.398, 1994, pp. 203-208.

[14] I.A. Choudhury, M.A. El-Baradie, 'Computer aided selection of optimum machining parameters in turning operations', Proc. 12th Conf. Irish Manufacturing Committee, Cork, Ireland, ISBN 0-907134-335, 1995, pp.205-212.

[15] Attar Software, 'XpertRule Reference Manual', Release 3.63, 1997.

[16] Sandvik Coromant, 'Rotating Tools' catalogue, C-1100:6-ENG, 1999.

[17] M. Sato, 'Advanced machining technology for die and mould – concept and examples of flush fine cutting', Proc. Int. Conf. and Exhibition on Design and Production of Dies and Molds, Istanbul, Turkey, 1997, pp. 111- 116.

[18] AB Sandvik Coromant, Sandviken, Sweden, 'CoroKey – The productivity guide catalogue', 5th Edition, 1999, C-2903:5 ENG, p.178.

CUTTING TOOL SELECTION BY USING "MICROSOFT ACCESS" DATABASE SOFTWARE

Erhan GÜNEL and Can ÇOĞUN

Mechanical Engineering Department,
Gazi University, Maltepe, Ankara, TURKEY
cogun@mikasa.mmf.gazi.edu.tr

Abstract: This paper summarises the algorithm of selection of cutting tool/toolholder system for turning operations by using 'Cutting Tools' database with 'Microsoft Access Professional 97' software. A database, called Cutting Tools, is created for easy data entrance, access and for the selection of turning cutting parameters. Access Basic Modules are developed as interface programs between the database and the prepared forms. The cutting tool/toolholder systems and grades given in Sandvik Coromant Turning Tools Catalogue are used in the database. The database only includes the information about General Turning Operations, i.e. external and internal turning.

1. Introduction

Success in metal cutting depends on selection of the proper cutting tool (material and geometry) for a work material. Work material characteristics, part characteristics (geometry, accuracy, finish, and surface-integrity requirements), machine tool characteristics including the workholders (adequate rigidity with high power, and wide speed and feed ranges) and support systems (sensors, controls, and method of lubrication and chip removal) are the effective parameters in selection of cutting tools. Cutting tool selection includes the selection of indexable inserts and toolholders/boring bars with the insert clamping system. In order to specify toolholders and inserts, their designations should be given which are standardised by the ISO (standard 5608-1989 for external machining, ISO 6261-1984 for boring bars and ISO 1832-1985 for inserts). Nowadays, machinability database systems are designed to access the available data for the selection of the specified or the optimum cutting conditions and cutting tools/toolholders.

The aim of the study is to develop a software in selection of cutting tools/toolholders by using a database system created in Microsoft Access 97 software package. In this paper, the structure of the created database which is formed by the information given in Sandvik Coromant Turning Tools Catalogue [1] and the Access Basic Modules that are developed as interface programs between the database and the prepared forms are given.

2. Cutting Tools Database

The Cutting Tools database is created to select cutting tools/toolholders within the structure of Microsoft Access 97 [3-6]. This database includes all necessary information about inserts and toolholders for cylindrical turning [1, 7]. The data in the database are composed of geometrical properties of inserts/toolholders (size, shape, tolerances and angles), tooling and clamping systems and grades. The designations and details of the turning tooling system used in this study can be found from turning tools catalogue of Sandvik Coromant [1]. The Cutting Tools Database also includes some other information in form of tables useful in selection of tools such as work material, hardness and working area. Figure 1 summarises the menu layout of the created database.

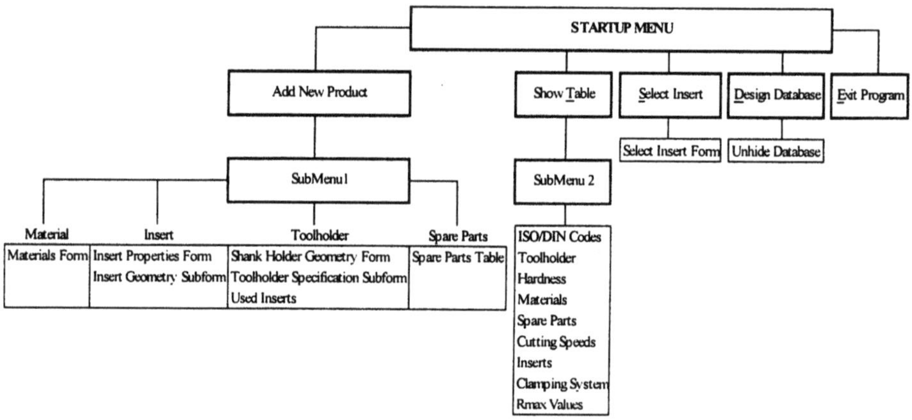

Figure 1 Menu layout of the developed database

2.1 Tables

In the study 19 tables are created to store all the data. The names of the created tables in the study are as follows: Cutting Edge Length, The Geometry Recommendation Tables, Hardness Comparison Table, Insert Geometry and Insert Properties Tables, Material Tables, Nominal Cutting Speed/Feed Table, Nose Radius/Feed/R Table, Rmax Values Table, Selecting Cutting Speed Table, Spare Parts Table, Tool Life Table, Toolholder Table, Toolholder Clamping System Tables, Insert Table, Working Area Table

2.2 Forms

Cutting Tools Database is supported with 11 forms. Four of them are the main forms, namely, *Geometry Recommendation, Insert Properties, Materials and Toolholder Geometry forms*. All of these forms are used for editing or entering the database information.

2.3 Queries

In the Cutting Tools database, four queries are used for filtering data [2,8,9]. The first query, *Insert Selection Query* initiates the insert selection procedure. The result of this operation not only includes insert codes, but also shows cutting parameters (speeds, feeds, power and machining time). *Toolholder Selection Query* determines toolholder codes, entering angles and some limiting factors. The final query runs the same filtering operation to eliminate repeated records. Text boxes on the Insert Selection Form control the functions of these queries.

2.4 Reports

Most of the information in a report comes from an underlying table, query, or SQL (structural query language) statement. Some other information, which is not appeared in the tables, is stored in the report's design.

2.5 Macros and modules

Cutting Tools database includes totally 27 macros. The macros are used to open the related tables and activate some function buttons such as toolholder, filter insert, print, areas, material codes, material property, hardness difference, report and close, as shown in Figure 2. Cutting Tools database modules are created for management of data entry. In this study, the form modules are designed to perform string operations. When a new record is entered in a field, the other related fields are automatically updated and shown on the screen. The characters of the ordering codes of the inserts or toolholders are transformed to their descriptions as standardised in the DIN norms and stored into the related fields.

2. Operation Principles of the Developed Software

3.1 Flow of the developed software

The menu layout of the developed software is given in Figure 1. When the program is initiated, a 'Startup Menu' appears on the screen. When 'Add New Product' button is clicked, a macro runs and opens a new form called 'Submenu 1'. This submenu opens forms for entering new data. 'Material' button opens 'Materials Form'. 'Insert' button runs a macro to open 'Insert Properties Form' with *Insert Geometry* subform. 'Toolholder' button opens 'Shank holder Geometry Form' together with *Toolholder Specification* and *Used Insert* subforms. When 'Spare Parts' button is clicked, 'Spare Parts Table' is opened for entering a new record. The 'Back' button is used to return the program pointer to 'Startup Menu'. 'Show Table' command button on the 'Startup Menu' form opens 'SubMenu2' form that has 10 new command buttons. Each of these 10 buttons opens related table, which can not be edited at this stage.

'Select Insert' button runs a macro to open 'Select Insert Form' (Figure 2). In this form, there are 15 fields for choosing inserts and toolholders. First, the material (to be machined) code is entered into the 'DIN Code of Workpiece Material' field. Then, the tab moves to the 'Selection Degree'. This field has a value of 5 for the best choice by default. Next step is the decision of 'Working Area'. If material is machined in extreme finishing conditions, a string character "A" is assigned for this field. For continuous cut conditions, the 'Intermittent' field has a value of 1, and for non-continuous cut conditions, this value must be 5 for the best choice. 'Ra (μm)' field is a combo box that contains surface finish value in terms of average profile depth for the calculation of minimum nose radius of the tool. 'Coromant Grade' is the most critical area in data entrance. These grades are divided into 12 groups, which the values between 1 and 7 are assigned for basic grades, and the values between 8 and 12 are assigned for supplementary grades. The material selected brings different grouping for the Coromant grade. The value of 1 defines the higher resistance to tool wear. With the increasing value (up to 7), the toughness increases (wear resistance decreases). The value of 8 indicates the higher resistance to wear, and a value of 12 is used for indicating tougher tools for supplementary grades. The first selection degree in the field should be done according to the recommendations given in the manufacturer's catalogues. For the tougher grade, this value can be increased. This value can be reduced for the higher wear resistance of the tool. After determining of the Coromant grade, 'Tool Life', and 'Hardness Difference' are entered to calculate the corrected cutting speed. The next step is the entrance of 'Cutting Depth' in mm in the combo box. If the external surface will be machined, the "External Machining" option in the 'Operation' combo box should be selected. On the other hand, internal machining can also be selected from the box depends on the required operation. The available selections for the 'Machining Types' in the box are finishing, roughing, finishing for copying and roughing for copying. For determining cutting parameters, 'Machining Diameter' and 'Machining Distance' (length) are entered in mm.

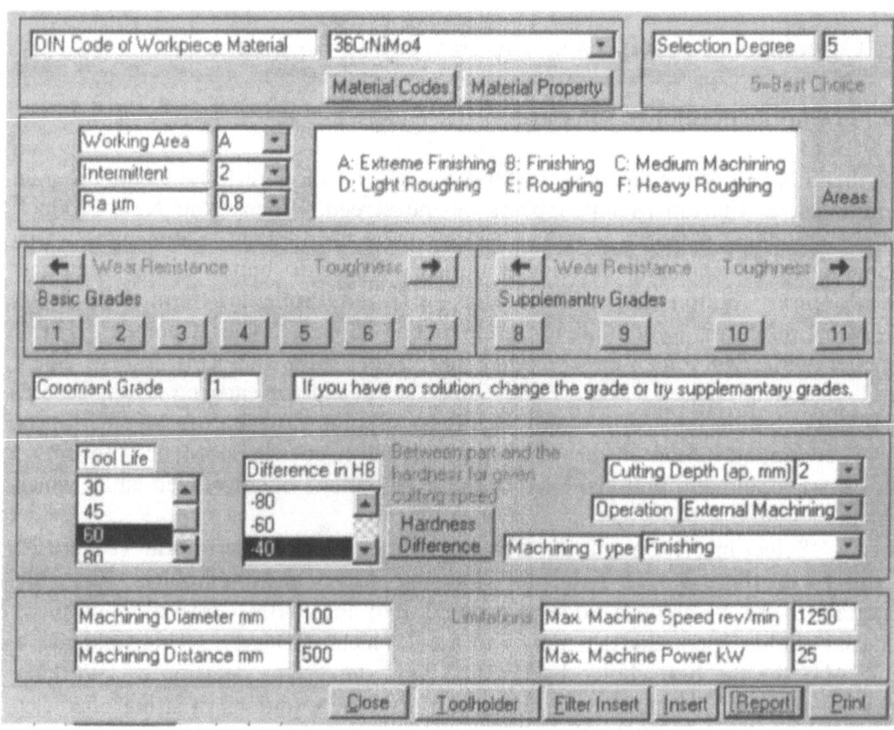

Figure 2 Main selection form

There are two limitations to be entered about the existing machine tool, namely, the 'Maximum Machine Speed' and the 'Maximum Machine Power'. These parameters are used for reducing the number of insert type selections. 'Toolholder' button runs a macro, which has a duty to run the 'Types of Holders' query. 'Filter Insert' button applies a filtering operation and shows the ordering code of the selected inserts. The 'Insert' button is clicked to see the selected cutting parameters. The last two command buttons are related to the 'Reports' and 'Printing' of the results.

'Design Database' button on the 'Startup Menu' runs an event called *Unhide* (Figure1). This selection provides the ability of changing and managing database, adding new tables, form and reports, setting new relationships, queries and other properties. New features such as profiling, drilling, threading, and milling operations can be added to the database by this function.

3.2 Sample Run

The Figure 3 shows two reports, which are obtained as a result of the input data given in Figure 2. The sample run illustrates the selected tools to machine a low alloy steel part (DIN 36CrNiMo4) with 390 HB hardness, 100mm diameter, 500mm length and 2mm depth of cut. The desired surface roughness (Ra) is 0.8 μm. The 'Areas' button opens a table on the screen and 2mm depth of cut in the table corresponds to a working 'Area' of extreme finishing. It can also be seen from the 'Hardness Difference' button that all cutting data of this material is given in reference to 350 HB. So, the difference between desired hardness and the 350HB hardness is +40 HB. If high resistance to wear is desired, Coromant grade must have a value of 1. A value of 5 can be entered for intermittent cut, although no solution can be generated for this value. If this value is decreased to 2, the solution can be generated by the query. To eliminate some of the cutting tool selections, the lathe restrictions like the machining speed and power are entered. After assigning all the input data to the related fields, the queries found 16 inserts and 51 toolholders suitable for the purpose (Figure 3).

4. Conclusion

In this study, a modular submodule is developed to select cutting tools for cylindrical turning operations by using 'Microsoft Access Professional 1997' software. The cutting tool/toolholder systems and grades given in Sandvik Coromant Turning Tools Catalogue are used in the preparation of the database.

The Access Basic Modules are prepared to link the cutting tools database and the forms. The developed database properties, fields features in the tables, forms, queries, reports, macros and modules can be analysed from the Tools Menu of Access using 'Microsoft Access Professional 1997' software. The sample runs indicated the effectiveness of the developed software and the database in tool selection. More features can be added to the database by introducing new tables for turning and the other machining operations such as milling, threading, profiling for the cutting tool selection and for determining the machining conditions.

Report 1: Selected Inserts

Insert System	**T-MAX P**
Insert Type	*-QF*
Ordering Code	***Type***
CNMG 09 03 04-QF	Double Sided Insert
CNMG 09 03 08-QF	Double Sided Insert
CNMG 12 04 04-QF	Double Sided Insert
CNMG 12 04 08-QF	Double Sided Insert
CNMG 12 04 12-QF	Double Sided Insert
DNMG 11 04 04-QF	Double Sided Insert
DNMG 11 04 08-QF	Double Sided Insert
DNMG 11 04 12-QF	Double Sided Insert
SNMG 12 04 04-QF	Double Sided Insert
SNMG 12 04 08-QF	Double Sided Insert
TNMG 16 04 04-QF	Double Sided Insert
TNMG 16 04 08-QF	Double Sided Insert
TNMG 16 04 12-QF	Double Sided Insert
TNMG 22 04 04-QF	Double Sided Insert
TNMG 22 04 08-QF	Double Sided Insert
TNMG 22 04 12-QF	Double Sided Insert

Report 2: Toolholders

Coromant System	**T-MAX U**				
Clamping System	***Screw***				
Ordering Code	***Entering Angle***	***Degree***	***Chip Flow***	***Indexing Time***	***Accessibility***
BT25-SDJCR/L-2532	93	5	5	2	5
BT25-STDCR/L-2532	45	5	5	2	5
BT25-STFCR/L-2532	91	5	5	2	5
BT25-STGCR/L-253	91	5	5	2	5
SVJBR/L 2020K16	93	5	5	2	5
SVVBN 1212F11	72.5	5	5	2	5
SVVBN 2020K16	72.5	5	5	2	5
Summary for 'Clamping System Field' = Screw (51 detail records)					
Max	*5*	*5*	*25*		
Summary for 'Coromant System Field' = T-MAX U (51 detail records)					
Max	*5*	*5*	*25*		

Figure 3 Sample reports for insert and toolholder selection

References

[1] Sandvik Coromant, 1994, "General Turning", *Turning Tools Catalogue 93/94.*

[2] Ashaf, M., 1993, Automatic *Tool Selection For Turning*, M.Sc. Thesis, Middle East Technical University, (METU), Ankara

[3] Çubukçu, F., 1997, Microsoft Access 2.0/4.0, Alfa Publishers, İstanbul.

[4] Karagülle, İ. and Pala, Z., 1997, *Visual Basic Pro 5.0*, Türkmen Publishing, İstanbul.

[5] Microsoft Access 97, 1997, "Help: Content and Index".

[6] Uysal, M., 1998, Database Management with Access 97, Beta Publishing Comp, İstanbul.

[7] De Garmo, E.P., Black, J. T. and Kosher A. R., 1998, *Materials and Processes in Manufacturing*, Prentice Hall Int., London.

[8] Tokuroğlu, N. T. and Anlağan, Ö., 1990, 'Design of Machinibility Database Systems for Turning Operations', *4. Mach. Des. and Prod. Conf.*, METU, Ankara, 357-367.

[9] Tokuroğlu, N. T. and Anlağan, Ö., 1992, "Turnbase 1.0: Introduction and Representation of General Purpose Machinibility Database Software Prepared for Turning Operations", *5. Mach. Des. and Prod. Conf.*, METU, Ankara, 101-111.

MANUFACTURING SYSTEMS AND CONCURRENT ENGINEERING

A Concurrent Engineering Approach Using FEM and CAM for the Development of Scroll Compressors

Z. Jiang and D.K. Harrison*
* Correspondence: Department of Engineering, Glasgow Caledonian University
Cowcaddens Road, Glasgow G4 0BA, U.K Email: dha2@gcal.ac.uk

K. Cheng
School of Engineering, Leeds Metropolitan University
Calverley Street, Leeds LS1 3HE, UK Email: k.cheng@lmu.ac.uk

Abstract: Scroll compressors are a relatively new and advanced type of power equipment. To design and manufacture scroll compressors, however, is not easy for the reasons that their key parts have complex shapes with involute curves. The authors have developed a design approach that can cope with simultaneous design activities concurrently within one single environment. The approach is implemented by using Pro/ENGINEER and Visual C^{++} programming. The work presented is focused on the integration of Constructive Solid Geometry (CSG) solid modelling, FEM and CAM to effect the design of scroll compressors and with this particular approach a correct and precise solid model can be achieved, which leads to an optimal design in terms of specification adherence, quality, delivery time and cost.

1. Introduction

In recent years, scroll compressors have become increasingly popular in refrigeration and air conditioning products due to their high efficiency, low noise, etc. compared with other types of compressor. However, to design and manufacture scroll compressors is difficult because of the complex structure. The accuracy of assembly is high and the parts difficult to manufacture. The structure of a typical refrigeration scroll compressor is illustrated schematically in Figure 1. The conventional design of scroll compressor has tended to use a serial engineering approach, which hampers the design and manufacturing engineers' ability to deliver the products effectively and efficiently in the increasingly competitive global marketplace. In recent years, the concept of Concurrent Engineering (CE) has been proposed to overcome such serial design and manufacturing problems. Sturges and Kilani [1] presented work towards an integrated design for an assembly evaluation and reasoning system which was built on existing solid-modelling packages. The aim was the evaluation of the assemblability of the design, and the recommendation of design modification that decrease the assembly time and the level of difficulty. Kovacs et al [2] developed a CE approach to support a Computer Integrated Manufacturing (CIM) reference model based design system for Flexible Manufacturing Systems (FMS). The paper introduced a reference model that was developed from a functional point of view. Details are given on how this model can be used for the CE of products for the FMS. Lee et al [3] presented research on the development of a computer-based framework that supports concurrent mould manufacturing process-planning. Now although many researchers have presented significant applications of CE, few papers [4] are related to the integration or the CE approach of CAD/CAM used in the production of scroll compressors. This work attempts to address this shortcoming.

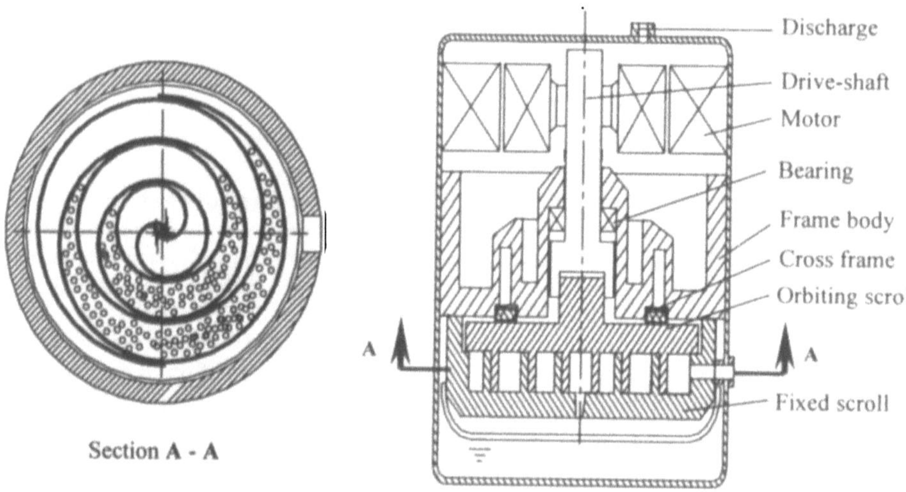

Figure 1 The structure of a refrigeration scroll compressor

2. The Integration of FEM and CAM for Scroll Compressors

The concurrent engineering approach requires that designers, manufacturing engineers and all the members related to the new product development team work simultaneously or co-operate closely to avoid redesign and other problems at a later stage. The approach shown in Figure 2 allows the designers of the scroll compressor to design the components and create features of the model in the CAD environment using the database for thermodynamic, dynamic and structural properties. At this stage the manufacturing engineers can evaluate the features of the part since the CAD system shows the main features of the solid model including the mass properties. The model can also be checked in any orientation and cross section. The FEM is the kernel of the approach. It enables the design output to be optimal, robust and efficient and so to achieve high quality products at low cost in the shortest delivery time. The scroll components work in a high pressure and temperature environment, which results in the deformation of the scroll wall. If the deformation is too great the compressor would fail and the compressor has to be redesigned and remanufactured. As the stress and deformation of the part are the factors that affect the working process of the compressor, they must be assessed. If the manufacturing engineers consider the model is suitable to implement the

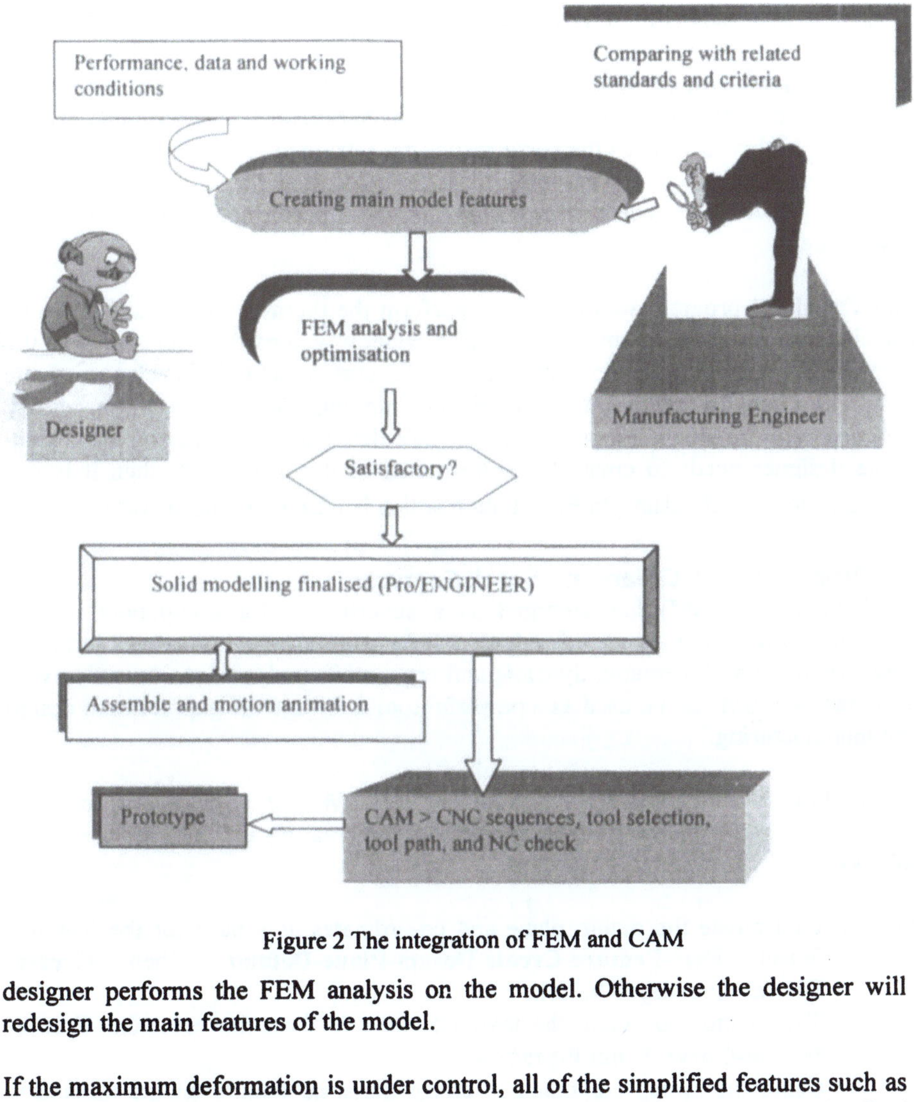

Figure 2 The integration of FEM and CAM

designer performs the FEM analysis on the model. Otherwise the designer will redesign the main features of the model.

If the maximum deformation is under control, all of the simplified features such as round corners, fillets etc will be added to the model and this then becomes the final model. The designer can seek optimisation under the limitations imposed by the displacement or stress of the scroll wall. The manufacturing engineers check the final model again. The designer can assemble the parts to be a set of components and animate the movement of the parts. The motion process may be checked and evaluated. If the manufacturing engineers and designers consider that the model is relatively good then the model is passed to the CAM environment. As the solid model of the scroll parts are evaluated for their structure, deformation, assembly, motion, material, manufacturing sequences, tool path codes and NC simulation, the models designed via this approach have much more reliability and producibility.

3. Solid Modelling of the Scroll Parts

3.1 Determination of Scroll Parameters in the Visual C^{++} Environment

The required performance of the refrigeration compressor is determined by the customer. In this case study the performance eg. refrigeration capacity, condensing temperature and so on are assumed. It is assumed that the compressor has no discharge valve. The calculation of the design is performed in the Visual C^{++} environment.

The Visual C^{++} program was created to perform the thermodynamic and dynamic calculations. In the program an instruction window guides the users through the software. A dialogue window appears in the system and the designer of the compressor can fill in the data interactively. Running this program the necessary parameters of the scroll component will be calculated and displayed on the screen. If the designer needs to create the solid model of the scroll parts then it is just necessary to click the Run_ProENG button which is located on the screen.

3.2 Procedures to Create the Scroll Features

Solid Modelling (SM) has emerged as a superb tool for component design, especially when there is added value in linking geometry to various forms of structure, thermal, kinematic, dynamic and ergonomic analyses, etc. It is also very important that SM can be used as a powerful tool to bridge the gap between design and manufacturing.

In the Pro/ENGINEER CAD environment the **SWEEP** method was used to generate the scroll features. The procedures for creating involute features is as follows:

1) First create the datum plane and co-ordinates as a base for the features. Select **Part-Feature-Create-Datum-Plane-Default.** Then **Create-Datum-Coordsys-Default.**
2) The method to create the involute curve is from the equation. **Create-Datum-Curve-From Equation.**
3) Using the sweep command to create the solid feature on the trajectory. Select **Solid-Protrusion-Sweep.**
4) The remaining features of the scroll part are simple ones. Using an **Extrude** method the scroll model can be built.

4. The Finite Element Method for Scroll Part Design

4.1 The FEM Analysis Process

In the process the solid model of the scroll component was created in the PART environment first and then in the FEM environment the solid model was modified to aid the analysis. Several kinds of elements such as bar elements, tetrahedron elements or mass elements can be selected to mesh the model. The element size is required and this should be much smaller than that of the solid model.

Procedures in the Pro/MECHANICA environment

i. Select the **Application** menu and click **Mechanica** and then choose **Independ Mec**. The system will change to the independent Pro/MECHANICA system.

ii. *Defining material.* The material property is specified for all elements in the model. Select **Model-Properties-Material-Part-All-Return**.

iii. *Adding constraints.* Before carrying out an analysis it is necessary to define the constraints acting on the scroll model. The constraints are placed on the bottom of the scroll wall. Select **Model-Constraints-Surface**.

iv. *Adding loads.* Loads act on the inner surface of the scroll wall. The different types of the loads are placed on the different regions of the model. Select **Model-Loads-Pressure-Surface**.

v. *Defining analyses and Running the program.* Select **Analysis** and enter the name of the analysis, change the convergence method to **Multi-Pass** adaptive. Then select stresses for calculation. Set the design study as standard. After finishing the setting of study the FEM program can now run.

4.2 Results and Analysis of the FEM

The calculation and analysis of the FEM was implemented in the Pro/MECHANICA environment. In the program the displacements and stresses were analysed. The displacement results are as shown in Figure 3. It shows that the displacement increases gradually along the scroll wall from the inside to the outside and from the bottom to the top. The reason is that the radius of the scroll wall is larger on the outside and hence the force increases as does the bending moment. This displacement result is very important for the scroll assembly clearances. If this displacement exceeds the allowable clearance the compressor will not run smoothly and in extremis, seize up.

5. CAM for Scroll Components

Manufacturing sequences and tool path codes can be created in Pro/MANUFACTURE, which is the same CAD/CAM environment that the solid model was created in. MANUFACTURE comprises mainly **MFG model, MFG set up** and **Machining.** In the **MFG model** environment, the solid model created in the PART environment was modified to become the manufacturing model and a workpiece model can be created based on the solid model. In **the MFG set up** environment, the operations, fixtures tooling and workcell can be selected. In the **Machining** environment the machining approaches and the sequence details can be created or selected. The CNC sequences created can be displayed on the system separately or serially as shown in Figure 4 which is the simulation of the CNC sequence to machine the scroll wall. The diagram shows that the tool is just at the half way point of the sequence.

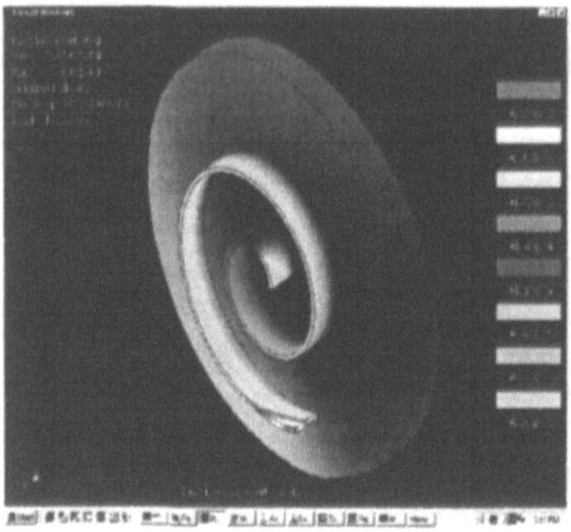

Figure 3 Displacement of the scroll wall

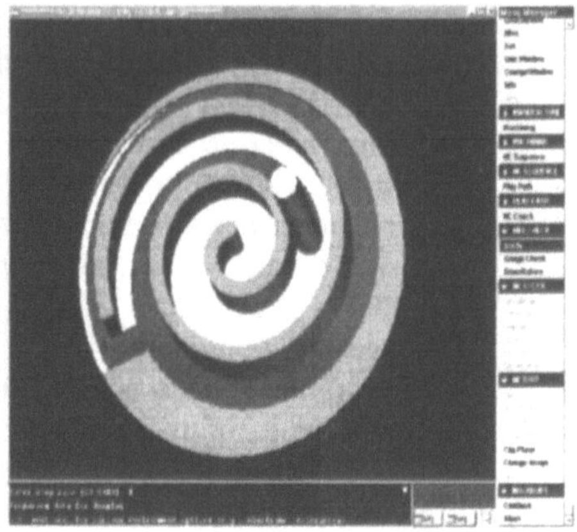

Figure 4 A simulation of machining the scroll wall

6. Concluding Remarks

An integrated approach using the FEM and CAM of scroll compressor components has been demonstrated based on concurrent engineering. The integration strategy has been developed to a full system. The system implemented is from customer requirements to the completion of manufacturing. The finite element method and optimisation are used in the system to assist the integrated approach, which are essential for the design of high quality products and to reduce the possibility of rework during the design and manufacture cycle.

References

[1] Sturges, R.H. and Kilani, M. I., 1992, Towards an integrated design for an assembly evaluation and reasoning system, *Computer-Aided Design*, Vol. 24, No. 2, pp. 67-79.

[2] Kovacs, G.L., Mezgar, I. and Nacsa, J., 1994, Concurrent engineering approach to support CIM reference model-based FMS design, *Computer Engineering Manufacturing Systems*, Vol. 7, No. 1, pp17-27.

[3] Lee, R., Chen, Y. and Cheng, H., 1998, A framework of a concurrent process planning system for mould manufacturing, *Computer Integrated Manufacturing Systems*, Vol. 11, No. 1-2, pp171-190.

[4] Jiang, Z.C., Cheng, K. and Harrison, D.K., 1999, Integration of solid modelling and FEM analysis for the design of scroll compressors, DETC99/DAC-8685, *Proceedings of the 1999 ASME Design Engineering Technical Conferences*, Las Vegas, Nevada, USA, CD-ROM.

Process Improvement in Mechanical Design Using a Derivation of The Systems Engineering Capability Maturity Model

I. Egan[1], I. Black[2], J. M. Ritchie[2], P.D. Gardiner[1]

[1] School of Management, Heriot-Watt University, Edinburgh, EH14 4AS, UK. i.egan@hw.ac.uk; p.d.gardiner@hw.ac.uk
[2] Department of Mechanical & Chemical Engineering, Heriot-Watt University, Edinburgh, EH14 4AS, UK. i.black@hw.ac.uk; j.m.ritchie@hw.ac.uk

Abstract: This paper outlines the results obtained from applying a derivation of the Systems Engineering Capability Maturity Model^{SM} to the mechanical engineering design process within a large, product-based, electromechanical engineering company. This new model, termed the Process Capability Model for Mechanical Design (PCM-MD), was applied via two separate studies to establish the capability level of the company's current mechanical design process and associated support areas. Results from an initial study clearly exposed deficiencies caused by the 'citadel' approach that the project-orientated company structure engendered whilst a second study indicated additional and more fundamental problems arising from the cost-accounting focus of the management.

1. Introduction

The Software Capability Maturity Model® [A] (SW-CMM^{SM}) [B] [1] was created and developed to tackle the scheduling and cost problems that permeated large, software-intensive contracts held by the United States Department of Defense in the mid-1980s. The principles on which the CMM® is based are consistent with the views expounded by Deming, Juran and others on TQM, i.e. views that foster the need for quality control of processes as well as products. Since the creation of the SW-CMM several other CMMs have been released which deal with related areas such as systems engineering and security engineering. It is the authors' view that the approach taken by CMM is sufficiently robust to stand exposure in other, more traditional engineering fields. In particular, the adaptation of the Systems Engineering CMM (SE-CMM^{SM})[B] [2] to a mechanical design and manufacturing environment was seen as a practical contribution to the management, control and strategic direction of the design processes involved with mechanical engineering. The model uses questions covering 18 different areas which relate to the engineering, project and support processes of an organisation. A key point is that not only are good working practices addressed but questions relating to the level to

---

[A] *Capability Maturity Model and CMM® are registered trademarks in the US Patent and Trademark Office.*

[B] SE-CMM^{SM} *and* SW-CMM^{SM} *are service marks of Carnegie Mellon University*

which the practices are carried out are also included, providing the evidence for different capability levels (from Level 0 to Level 5). This forms a crucial distinction between this type of assessment and those carried out for ISO 9000 type accreditation's. Work was carried out with a large electromechanical engineering company to develop the model and subsequently apply it to relevant areas in order to map out the actual capability levels of existing processes. This resulted in comprehensive recommendations for improvement to be made based on the resulting factual knowledge of the relevant strengths and weaknesses of the engineering, project and support areas of the business.

2. Objectives

The objectives of the study were to establish whether a 'CMM-type' model (Process Capability Model-Mechanical Design: PCM-MD©) could be used in the mechanical design environment to map the capability of existing processes and, from this, enable engineers to recommend improvements. This paper details the results obtained from an initial pilot study and highlights some relevant follow-up results from a second tranche of research which addressed the interaction between the company's mechanical design and manufacturing processes.

3. Methodology

The initial study was limited to the engineers and designers who provide the core input into the mechanical design process, i.e. those based either in the drawing office or situated geographically very near to the main function and easily accessible. The rationale for this approach was to test the feasibility of the technique in a new environment and to ensure that the results obtained would be of value to the company. The assessment methodology involved the Lead Mechanical Engineers and their team members answering questionnaires on eighteen different process areas associated with the company's mechanical design process and then participating in follow-up interviews. These eighteen areas were grouped under the general headings of 'Engineering', 'Project' and 'Support'. The follow-up interviews were carried out to further investigate areas of weakness, inconsistency and to corroborate practices that were stated in the questionnaires as taking place. A more detailed description of the methodology is given elsewhere in [3]. The answers provided by the questionnaire, amended following the interviews, allowed a capability level to be determined and allocated for each process area under investigation.

---

4. Analysis

The data collected from the questionnaires was entered into an Excel spreadsheet constructed to combine the answers given by individuals for each question (in the form of 'Yes', 'No', 'Not Applicable' or 'Do Not Know' responses) into an overall percentage. A minimum positive response of 90% [4] is required for each base practice before it can be accepted as being carried out. The objective of the questionnaire analysis was to establish the areas that needed further corroboration or required further probing to understand what features of the design process were giving rise to any contradictory or unexpected results from the questionnaires. This provided a general structure and focus for the interview sessions, the results of which provided an invaluable contribution towards understanding the true nature of both good points and problems alike.

5. Results

The results from the pilot study (Fig. 1, outer circle) indicated two main themes:

- Firstly, that although the traditional and mature engineering practices carried out by the mechanical engineers and designers were found to be mainly at Level 2, the level of communication across the company to give them a truly competitive advantage against other companies was low.

- Secondly, the drive to cut overheads had led to a decrease in the number of people available to implement changes which, ironically, the company had had the foresight to budget for in capital terms but not to provision sufficiently for in terms of implementation.

Importantly, no one interviewed stated that the lack of communication was felt to be a deliberate act by people prevaricating. Rather it was perceived to be a case of the parties involved being extremely busy sorting out their own individual areas and simply failing to appreciate the importance of the data they held to other members of the engineering 'team'. There were several instances of people citing the loss of large progress meetings as a cause of lack of communication; the meetings had been deemed to be a waste of time because each individual's input took up a relatively short part of the overall proceedings. Unfortunately, resorting to abandoning large meetings without recourse to an alternative form of mass communication, countered the ability for everyone to hear the same message.

Another point uncovered by the study was that although the reduction in overhead staff (defined as those not booking directly to a named project) had obvious attractions for the balance sheet, it induced knock-on problems when new tools and procedures were introduced. Typically, new processes do not immediately benefit a particular project. The company cost-accounting culture meant that no one wanted to pay for the introduction, least of all the programme managers on the projects who only saw inefficiencies and extra cost in the short term.

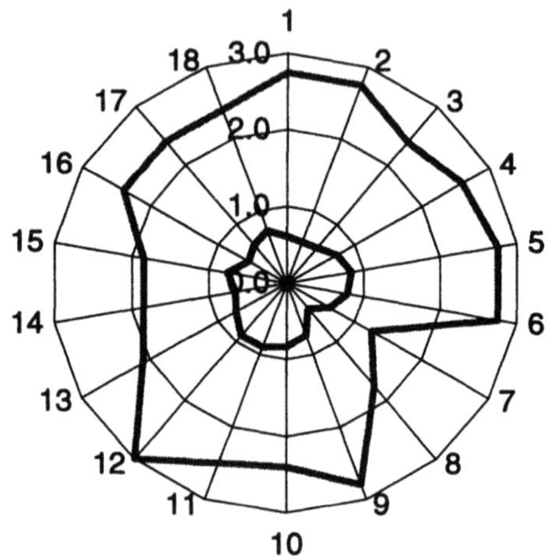

PROCESS AREAS		
ENGINEERING	**PROJECT**	**ORGANISATION**
PA 1 Understand Customer Needs and Expectations **PA 2** Derive and Allocate Requirements' Specifications **PA 3** Analysis of Possible Solutions **PA 4** Generate Design **PA 5** Integrate Product Modules **PA 6** Verify and Validate Product **PA 7** Integrate Disciplines	**PA 8** Ensure Quality **PA 9** Manage Risk **PA 10** Plan Product Development Effort **PA 11** Monitor and Control Product Development **PA 12** Manage Configurations	**PA 13** Define the Organisation's Mechanical Design Processes **PA 14** Improve the Organisation's Mechanical Design Processes **PA 15** Manage Mechanical Design Support Environment **PA 16** Provide Ongoing Skills and Knowledge **PA 17** Manage Evolution of Product **PA 18** Co-ordination With Suppliers

Figure 1: PCM-MD results showing process area versus level from the mechanical engineering study and mechanical design pilot assessment (outer circle) compared with the follow-up manufacturing interface (inner circle).

In addition to the more negative results, some very positive aspects were identified that highlighted the advantages of applying this tool to assess and discriminate between the state of the many processes taking place in such a dynamic environment. This ability to discriminate also highlighted common trends across many of the different process areas assessed, e.g. lack of resources and communication problems.

Prior to carrying out the second study on manufacturing's involvement in the mechanical design process, it was thought that the results would be of a similar nature to those found in the pilot study. However, the significance of the warning signs given by the low capability level for Ensure Quality (PA 8) and Integrate Product Modules (PA 5) found during the pilot had not been fully recognised. What was actually discovered during the second study was that small pockets of concurrent engineering were indeed taking place, where the funds allowed, but that the majority of practices were taking place in legacy programmes that had not budgeted for true concurrent engineering. An 'over-the-wall', serial product development, culture existed which had a large negative impact on the overall mechanical design process. Consequently, the overall capability level for the mechanical design process was substantially lower than the pilot had suggested (see Fig. 1, inner circle). On being made aware of this, one of the design engineers commented that, "you know we've been talking about concurrent engineering for the last fifteen years but this is the first time that something shows that that's all it's really been."

6. Discussion and Conclusions

The results derived from the pilot study provided useful pointers towards areas for improvement. The capability levels of the process areas were clustered around Capability Level 2; not surprising for a particularly mature process in a 'ring-fenced' area. The use of the scoring method gives a good focus on the eighteen different process areas, and clearly identifies positions of strength and weakness. The need for effective communication between people has been highlighted as a significant issue. Typical project structures lend themselves towards good communications within the confines of the main discipline. However, the need to work hard at communications across boundaries in a business environment is just as essential. The PCM-MD has been shown to be a very useful and practical tool in highlighting this.

Although the majority of the process areas in this first study were at Level 2, there was a batch of process areas below this level that related to the definition, improvement, quality and support of the mechanical design process. It was felt that the lower level did not reflect an absence of business intent in these areas but rather a lack of people to actually execute the many initiatives and improvements being pursued. A reduction in overhead-charging staff may be good for cash flow but the functions they carry out still have to be met if these types of process areas are to mature.

The need to cover a lot of ground is clear. An interesting comment from one of the interviews was that, "*there are enough engineers on site to cover the various tasks, it's just that they are not on the projects at the right time.*" If true, the ability to make people more efficient by enabling them to be available at the right time would allow their outputs to become the badly needed inputs to another's work. This would help the design process considerably. The positive results for the more 'traditional' engineering aspects covered by the model are encouraging. However, the next move towards improvement - concentrating on an organisation perspective rather than a project-centred solution - will not be an easy task. The benefits will be to have a business' central organisation continually learning from customer focussed programmes and knowing that any lessons learnt will seamlessly find their way to all other relevant people in the organisation.

The main point to come out of the second study was that the projects were being driven hard by a cost-accounting orientated project management approach, much to the detriment of engineering input. This was amplified during the manufacturing interaction with design assessment because of the point in the sequential project life-cycle where the bulk of manufacturing activities take place compared to those of design. Any 'extra' problems encountered and overcome at the design stages generally used up more time and capital than had been budgeted for with the consequence that there was less time and capital for downstream activities unless milestones and budgets were changed. This effect was further magnified by occasions of insufficient manufacturing input into the design before it became, "too expensive to change." In essence, concurrent engineering was talked about a lot but in practice the level of integration was much less than desired. This is another symptom of the lack of communication seen before but with the 'cost-orientated' cause being highlighted by the second phase of the study.

One of the key features of the PCM-MD is the ability to use it as a reference frame to balance the necessary process infrastructure that must be in place to deliver a company's business objectives. By developing a repeatable and easy-to-apply process assessment tool, engineering management have access to a valuable gauge with which to measure consistently the current and future capability levels of the company's design processes and associated support areas. As a result of the pilot assessment, the company's engineering management have constructed an action plan to take key processes, identified as deficient, and improve them to support their declared future business needs. Although the results from this type of assessment are typically more strategic in terms of implementation and timescale of business returns, the ability to take a tactical viewpoint to increase capability in specific targeted areas is viable.

Overall, this research has shown that CMM-based process improvement tools do have a useful and practical role to play in mechanical and manufacturing engineering environments. However, variants may have to be developed to suit the semantics of the function or specialist process being investigated.

References

[1] Paulk, M.C., Curtis, B., Chrissis, M.B. et al. 'Capability Maturity Model for Software V1.1' *CMU/SEI-93-TR-024*, Carnegie Mellon University. Pittsburgh, (February 1993).

[2] Bate, R., Kuhn, D., Wells, C., et al. 'A Systems Engineering Capability Maturity Model V 1.1' *CMU/SEI-95-MM-003*, (November 1995).

[3] Egan, I., Gardiner, P. D., Black, I., Ritchie J.M. Business Process Measurement - Adapting and Using a Capability Maturity Model in a New Domain. 15th World Congress on Project Management, London, May 2000.

[4] Kuhn, D. A. and Wells, C, et al. 'A Description of the Systems Engineering Capability Maturity Model Appraisal Method V 1.1' *CMU/SEI-96-HB-004*, 3-21 (March 1996).

CONSTRUCTION OF CONFIGURABLE VIRTUAL MANUFACTURING ENVIRONMENTS

Z. J. Xu[1], Z. X. Zhao[2] and R. W. Baines[3]

[1] School of Engineering, University of Huddersfield, Queensgate Campus, Huddersfield HD1 3DH, z.xu@hud.ac.uk
[2] School of Engineering, University of Derby, Kedleston Road, Derby DE22 1GB, z.zhao@derby.ac.uk
[3] School of Engineering, University of Derby, Kedleston Road, Derby DE22 1GB, r.w.baines@derby.ac.uk

Abstract: This paper reports a method of constructing configurable virtual manufacturing environments for knowledge representation, acquisition, and shop-floor simulation. The method uses a set of template virtual environments that are developed in a layered structure and stored in a database. The database is developed to store the environments and manage the virtual objects and control data within the each environment. A system is developed based on this method and a user interface is provided for the system to support rapid construction of large-scale and complex virtual manufacturing environments.

1. Introduction

Manufacturing planning and simulation are shifting from complex automated programming processes to a more intuitive user-machine interacting process. This has significantly reduced the effort of a manufacturing engineer in mastering complicated computing tools to solve specific problems. One driving force behind this shift is the availability of virtual reality (VR) or virtual environment (VE) technology.

Traditionally user-computer interactions are restricted by computer devices that are only suitable for 1-D and 2-D systems [1]. Engineering software that claims to facilitate 3-D computing are mostly working in a "static" mode. In contrast, VR has two distinctive features - real-time interaction and on-site "involvement" - which differentiates itself from conventional computing systems. Using VR in solving engineering problems is becoming a common practice. Cases such as employing VR for design, rapid prototyping, process simulation and virtual assembly have been frequently reported in the literature. Although VR technology has a great potential in various manufacturing areas, VR systems has not been widely implemented as a mainstream computing platform like other simulation packages. Perhaps this is due to the fact that constructing and implementing a virtual manufacturing environment is a time consuming and difficult modelling process. Also most virtual environments have been built around a particular system, which makes them difficult to configure and change by users. There are also difficulties in coding a large amount of manufacturing knowledge in a graphic-oriented programme.

2. Configuration Environment Concept

Currently VE based simulation is achieved by running predefined VR programmes, which work well with specific applications. However, manufacturing VE built in this style is generally lack of flexibility in terms of modification and configuration. It also offers little reusability of generic data that can be extracted from one VE to another due to (1) the construction methods of VE are currently in a great diversity and there is no industrial or commercial standard to follow and (2) the internal simulation code assigned to an virtual object is not compatible from different VR systems (only VRML standard exist for external data exchange).

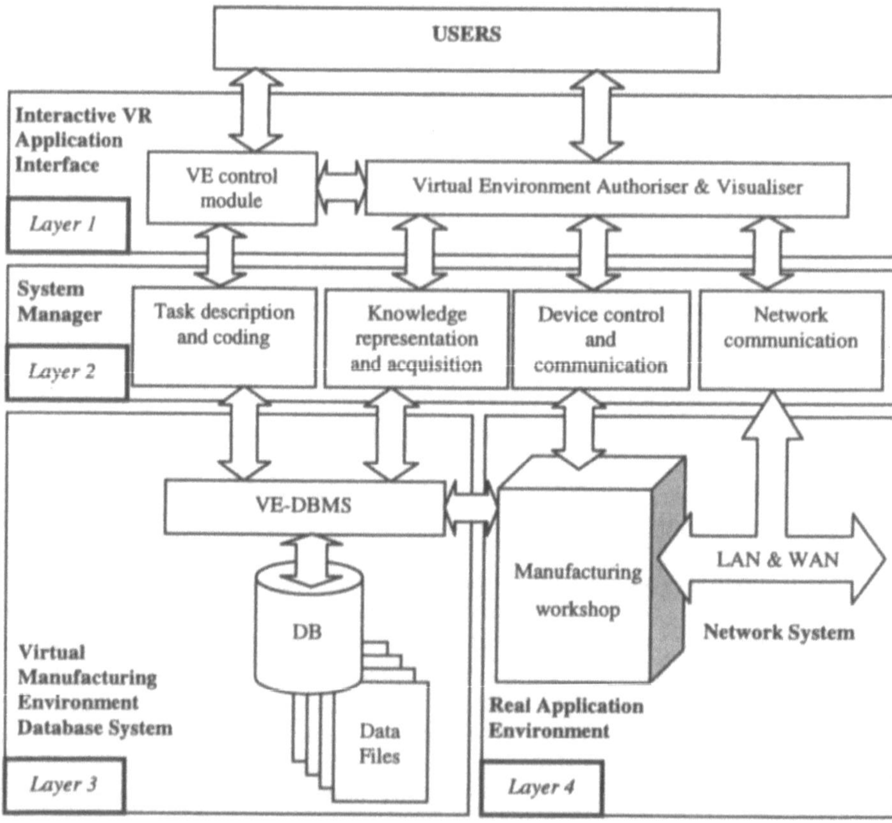

Figure 1 System structure

The configurable environments reported in this paper attempt to reduce the aforementioned difficulties by using a top-down domain-analysis approach to VE construction. By saying top-down, it means that, rather than building a VE from scratch, it allows user to built it by modifying an existing one that are similar to it in designated aspects. To do so, a set of template VEs – so called template environments - are constructed, classified, and stored in the database. Each template VE is constructed for a domain of manufacturing applications, thus it represents a range of similar VEs that are used in similar simulation tasks. In the database, each template VE is attached with a family of similar simulation tasks and the attached task family is indexed by a unique task code. This task code is a GT (group technology) like code generated according to a GT based coding and classification system.

When a specific VE needs to be constructed for a given simulation task (for example, the task of simulating a machining operation, a complex manufacturing process and a factory layout model), users can describe the task in a code and then use the code to search the database for a similar task family. The template VE attached to the retrieved task family will be a "base" VE on which users can use it straightaway if it is the same as the required VE or modify it if it is similar to the required one.

The system can also set a one-to-one relationship between a virtual object in the database and its physical counterpart in a real manufacturing environment. This enables a direct data acquisition and knowledge representation.

Figure 2 Constructed template environments

3. System Structure and Development

Figure 1 shows the structure of the system. The system is composed of four layers and each layer has its own functional modules. Due to the limit on the length of the text, this paper only describes two functions of this system: (1) the task description and coding and (2) knowledge representation and acquisition. These two functions are illustrated in the System Manager layer. The other functions of the system have been described in a separate publication [2]. The implementation of the specified system functions is described in the following sections.

3.1 Construction of template VEs and database management

The construction of template VEs is a one-off process that is carried out by the system developers not by the users. Figure 2 shows four template VEs built and stored in the database.

It must be pointed out that the data of a template VE can be stored and managed by the database, but it is not technically suitable to store every VE as an exact copy as it was constructed due to its similarity to other VEs. In this research, every template environment is stripped into its basic tree structure - called scene graph - before it being saved in the database. All the rendering information such as colour, texture, lighting, animation and simulation data are peeled off and stored as records in object property files, see Figure 3. These environment and object data are inter-linked by data files' key fields such as environment ID number and object ID number to form a relational database hierarchy, so that at environment run-time, data can be accessed for adding, deleting, and modifying object.

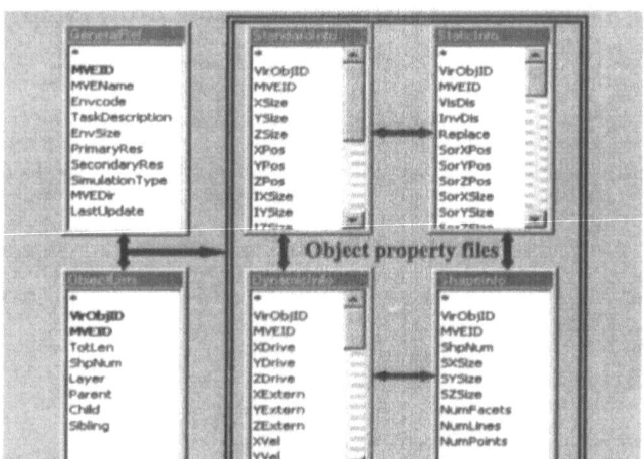

Figure 3 System database structure

3.2 Develop task-coding scheme for environment retrieval and modification

To retrieve a VE from the database, user needs to describe an application in a GT-like code that is so-called task code.

A task code is composite of three sections that describe part geometry, VE size and application type. The part geometry is defined by a 5-digit hybrid code. The VE size is defined by two digits, describing the scale of VE and the complexity of a manufacturing task the VE could be used for. The application type code is combined by three digits describing required object detail-level, simulation functions, and environment interactivity. The computing implementation of the task-coding and environment retrieval is through a virtual interface panel designed in an environment authoriser –Superscape VRT. Figure 4 has shown the snapshot of the interface.

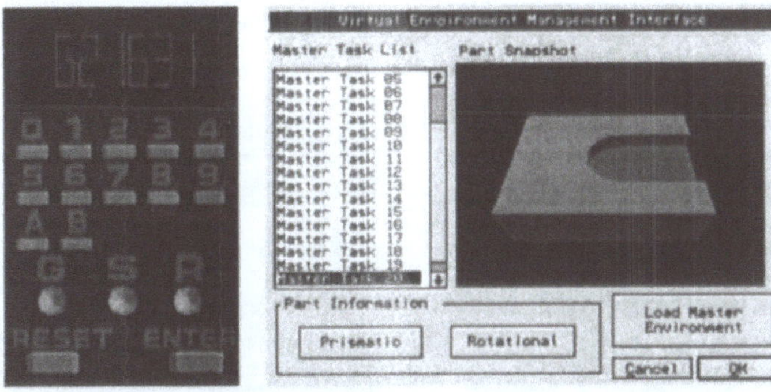

Figure 4 Virtual coding panel for environment retrieval

3.3 VE and database communication

A VEs at the run-time needs to communicate with the database both during its construction process and its life-time usage. The database is updated in accordance with the configuration or changes of the VE. This communication between the VE and the database is established by a set of dynamic-link-library (DLL) implemented between the VE's visualiser and the database query modules. It works as this: (1) search a template environment scene graph, (2) retrieve the base-shape geometry, (3) assign static information to objects, and (4) assign dynamic data to environment.

4. Using the System

The system modules have been integrated into a run-time platform. To use the system to create an environment, following example shows the simplicity of the construction process that the system offers.

For example, a virtual environment for machine operation training is to be constructed, where different parts with various features will be machined with specific quantities. The system works by following a few operational steps that are shown in Figure 5. These steps can be outlined as follows:

- The first step of using the run-time platform is to coding the application task using the virtual coding panel. A small help dialog explaining each digit's meaning is popped up as user required.
- The next step is to use the task code as task classification arguments for searching and starting a best suitable template environment.
- Then, the loaded environment is modified - using the information retrieved from environment database - and initialised.

- Now user can select any object to configure its specification. In this case, an virtual lathe has been configured using a dialog box where machine functional parameters such as maximum spindle speed is setup.
- Finally, user can explore the environment and start the machining simulation.

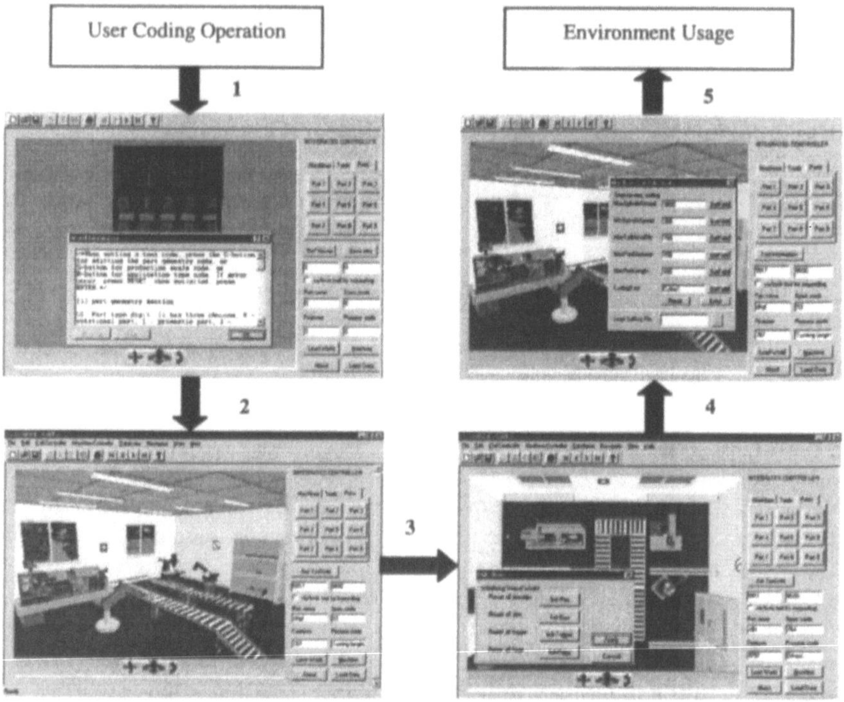

Figure 5 Using configurable environment run-time platform

5. Conclusions

The research reported in this paper has provided an integrated process of easy construction, configuration, and control of complex virtual environments. Future research will be conducted to develop an internal environment database to avoid the latency occurred from using the object-linking type environment and database connections.

References

[1] Zhao, Z. X, 1998, *A variant approach to constructing and managing virtual manufacturing environments*, International Journal of CIM, Vol.11, No.6.

[2] Xu, Z. J., Zhao, Z. X., and Baines, R. W. 1999, *A Virtual Environment For Manufacturing Simulation*, the 15th International Conference of International Foundation for Production Research, University of Limerick, Ireland, ISBN 1-874653-56-9.

Enhancement of Virtual Manufacturing through Web-centred technology

Qingjin Peng and F. R. Hall
School of Engineering and the Built Environment
University of Wolverhampton, Wolverhampton, WV1 1SB, UK

Abstract: Internet-based techniques provide a magnificent support to the sharing of resources among dispersed organisations. It can enhance the applications of Virtual Manufacturing (VM) in the development of products, which enables experts from different areas to join and evaluate a product simultaneously. This paper discusses the current applications and techniques of Internet-based VM, and proposes a user interface to support the Web-based VM applications.

1. Introduction

The global marketplace is forcing industrial organisations to develop new products using new technologies because the traditional approach to production design and manufacturing hampers the ability of many companies to compete effectively. Virtual Reality (VR) is a new technology that provides a visualised design and manufacturing environment at all stages of a product development. A VR-based product development process can speed up the total development of a product with the higher quality than traditional sequential methods. It provides exciting new ways of human involvement in the product development and allows a user to be a part of the computer generated model without a hindrance between them. Immersion, interaction and real-time computational issues from VR can be achieved and benefit to the whole manufacturing. Therefore, VR is an efficient tool to enhance the competitive capability of companies. However, the capital expenditure on VR facilities can run to many millions of pounds. A Web-centred interface can share the VR facility and distribute virtual design and manufacturing among experts at different geographical located place over a computer network. It is a valuable tool for the interactive communication within a product development process. A Web-based sharing environment will enable industries to utilise the VR technology cost-effectively, and will encourage industries to use the new technology, especially, for SMEs (Small and Medium sized Enterprises).

The University of Wolverhampton has recently been awarded a £3M European Regional Development Fund for Technology Transfer. A new Innovative Product Development Centre (IPDC) has been installed. This centre contains the latest

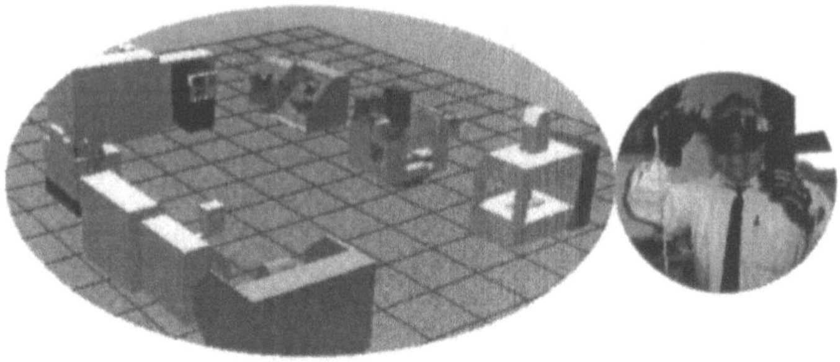

Figure 1The VR model of the Innovative Product Development Centre

equipment in VR design and manufacture. Figure 1 shows the VR model of the IPDC. As the manufacturing process is highly dependent on the user environment, such as specifications and conditions of the machine tool, a design can not be perfect without constant customisation. This paper will discuss how to share the VR facilities with industrial partners by developing Web-centred techniques. The objective of this research is to solve the current problems of VR applications, and to develop a user-friendly software cross-platform interface to enable virtual manufacturing to be utilised more efficiently in industries.

2. The background of VR and Web-based applications

VR and Internet techniques are widely recognised as very promising ways to meet today's challenges. They provide a rapid approach to deliver the industrial products into market. VR has been widely used in a design visualisation and a prototype evaluation [1]. The time required by all tasks in a design and manufacturing can be shortened through the use of VR. The developments of new products have shown encouraging results for the use of VR [2]. Internet techniques provide a cheap way to link and interconnect with industrial organisations and resources. Many barriers in industries have been broken down through Internet. By means of the network, a Web can link production, sale, management, supply and customer together. This allows a gain in competitiveness by shortening time to market and lowering costs.

Many current commercial CAD/CAM systems tend to make their products 'Net-enabled' with data sharing, Web-based 3D viewing and teleconferencing. Some of them claim that their products allow disparate groups to share 3D models and to be viewed and edited in real-time over the Internet [3].

The Internet is currently a 2D graphical environment. A Virtual Reality Modelling Language (VRML) has been developed to support 3D interactive simulation within Web applications [4]. A VRML file, which consists of descriptions of 3D scenes and promises to realise three-dimensionality image display, can be stored on the Web, and can be downloaded and run by users under Web browsers. The VRML viewer can receive instruction from the server, it is a self-contained environment. However, the files of the VRML in size can be many Megabits, to view such a file over the Web will have a very long wait because the

VRML requires to load the entire model before it can be displayed and this is limited by the bandwidth of today's network.

Many researchers are working for Internet-based developments, such as, Tam *et al* [5] developed a Web-based virtual environment for operator training, SEDA-VRML, which is a user interface incorporating VRML, Java and expert system. Burdea [6] reviewed WWW-based VR and robotics applications, and pointed out 'the recent introduction of high-bandwidth Internet2 communication will make possible real-time Web-mediated robot teleoperation applications'. The authors have used VR to enhance the design and manufacturing through sharing data via the Internet [7]. However, there are still some problems in current Web-based applications, such as issues in effective data support, navigation speed, collaborative tool and user-effective access. Other present limitations relevant to Internet include the absence of an effective global network management, questionable security, not actually a real time network and not deterministic. The following parts of the paper will discuss some techniques that can overcome above-mentioned problems, and provide the authors' solutions to support the Internet-based VR applications effectively.

3. The techniques to support Web-centred applications

The following techniques are proposed by the authors in developing a Web-centred virtual manufacturing system.

3.1 A layer structure of the application and database which can collect and update the user data easily

A virtual manufacturing environment can be created in three levels of details based on different demands in a product development process. The workshop layout is at the top level, which deals with the arrangement of machines, transportation control and scheduling; the intermediate level is a work/group based unit, in which there is a group of machines or products with the similar features; and the bottom level is based on the single machining process.

The user data comes from the existing/potential industrial partners. The user model and data can be linked and controlled to access the services according to the different applying levels. A unique identify of all participants is required and organised by a user management system. The user model contains the identification and authentication, user profiling, and the product and document model. The user has the ability to view and modify the model by changing the schedule or layout of agents within the virtual environment.

3.2 A universal interface based on Internet browsers for integrated applications

The VM applications using the Web server are available on different hardware platforms. The remote exploitation of data, simulation of a design and manufacturing process and interactive operation form a distributed environment for a Web-based virtual manufacturing system. The interface can be customised to suit the different users by letting them download the user window with their parameter configuration as a remote interactive application. The fundamental components on the interface window include drop-down menus, push button, check box, display windows, applets and scrolling lists. The actions of the dynamic interface behaviour in the 3D world are implemented using Java in the form of call-back functions and process methods. The user window provides an interactive approach to the design and construction of virtual prototypes, allows a product model to be erected and judged. It can show the geometric information, manufacturing information, and ergonomics. Figure 2 shows the interactive relationship between the IPDC and users.

3.3 An effective file transition approach to support real-time applications

Up-to-data information in real-time is the practical need in VR applications via the Web. To overcome the current limitations of transfer speed and bandwidth of Internet, data compression and decompression techniques are used. The transmission of jobs and data are divided into those of master and update, the master data contains a lager quantity of data and is transferred off-line. The update data is used to modify a design interactively and contain relevant less data that can be transferred on-line to satisfy the requirement in real-time. The CAD data

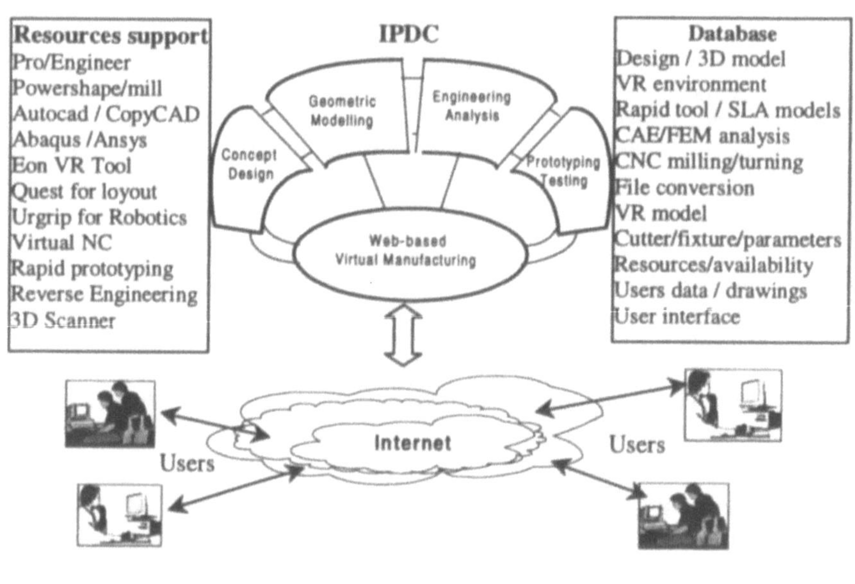

Figure 2 The interactive relationship between the IPDC and users

transmission is achieved via IGES and STEP standard data exchange formats, but some repair works are needed after data is received, a universal data exchange interface is developing by authors to overcome this problem.

4. An example and discussions

An example using the Web-based interface to support VR applications is shown in Figure 3. The interface, which contain all necessary information and interactive elements required for displaying the design or manufacturing process of products, is run under Web browsers. All data and models are stored on the Web server to be accessible by industrial partners. The evaluation and modification can be performed locally through the user interface plugged into the Web browser. The updated results are sent to IPDC through the Web. The IPDC Web site includes tools, resources, and techniques to support VR applications.

The significance of proposal tool is that it will allow the development process of products without having to make a real component until it is ready for satisfactory. Products will be produced in an economical and effective way by using Web-based virtual manufacturing. If there are more than one designer in the team, everyone can be presented in the same interface to evaluate a model, to make the desired changes and arrive at a consensus. Users can be geographically distributed and can interact with each other using real-time communication and the results can be synchronised with case. The hardware and software of VR facilities are shared

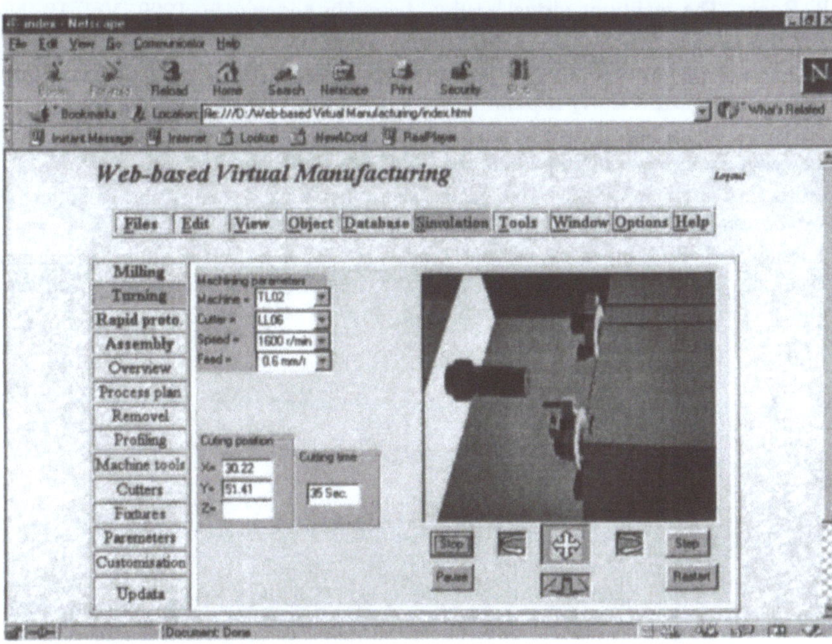

Figure 3 An example using the Web-based interface to support virtual manufacturing

within the university and industrial partners through the interface. Users can link their PC to the university for sharing facilities, or provide their demands, design to the IPDC for the virtual prototyping of their products.

5. Conclusions

The rapid development of Internet as the global network and Intranet as a corporate network creates a chance to meet the current challenge for industries in sharing resources and tools. The VM technique can be enhanced and distributed via Web-based applications.

Current progress in IT provides useful mechanisms and tools to support the Internet-based VR applications, but the limitations, such as in data transmission and use efficiency, have to breakthrough. To develop a user-effective interface can improve the VR applications based on Web.

The work presented in this paper is just the initial part of the whole project. The further work is to offer industrial partners better, more flexible ways of working, managing and viewing design and manufacturing through Web.

References

[1] John Vince, *Essential Virtual Reality fast*, 1998, Springer-Verlag London Lomited.

[2] B. Rooks, The reality of virtual reality, *Assembly Automation*, 1999, Vol. 19, No. 3, pp.203-208

[3] *CADD for Engineering Designers and AEC Professionals*, Vol. 20, No. 02, Feb. 2000.

[4] John Vacca, *VRML Clearly Explained*, Second Edition, 1998, Academic Press.

[5] Tam E. K., Badra F, Marceau R. J, et al. A Web-based virtual environment for operator training, *IEEE Transactions on Power System*, Vol. 14, No. 3, August 1999.

[6] Burdea G. C, Invited Review: The Synergy Between Virtual Reality and Robotics, *IEEE Transactions on Robotics and Automation*, Vol. 15, No. 3, June 1999.

[7] Q. Peng & F. R. Hall, Using Virtual Reality as a Tool to Enhance the performance of Manufacturing Industry, *Proceedings of the 15th National Conference on Manufacturing Research*, p.113-118, September, 1999, Bath, UK.

RELIABILITY, QUALITY AND MAINTENANCE ENGINEERING

Modeling Condition Based Maintenance Decision Support

W. Wang [1], F. Gu [2]

[1] Centre for OR and Applied Statistics, University of Salford, Salford, United Kingdom
w.wang@salford.ac.uk

[2] School of Engineering, University of Manchester, United Kingdom

Abstract: Condition monitoring is growing in popularity in industry with considerable sums now being spent on condition monitoring hardware and software. This paper focuses upon the prognosis aspect of condition monitoring and associated condition based maintenance decision making. The key element in such decision making is the concept of the conditional residual time, where the residual time of the component depends upon the age but is also influenced by the condition information obtained to date. A recent development in predicting the residual time distribution of a set of six bearings experimented on a test rig is introduced and the associated decision model is discussed

1. Introduction

Today, there exists a large and growing variety of forms of condition monitoring techniques for machine condition monitoring and fault diagnosis. However, irrespective of the particular condition monitoring techniques used, the working principle of condition monitoring is the same, namely, condition data becomes available which needs to be interpreted and appropriate actions taken accordingly. There are generally two stages in condition based maintenance. The first stage is related to condition monitoring data acquisition and its interpretation. There have been numerous papers contributing to this stage, as evidenced by the proceedings of COMADEM over recent years, Rao *et al* [1]. This first stage is characterised by engineering skill, knowledge and experience. The second stage is maintenance decision making, namely what to do now given available condition information. The decision at this stage can be complicated and entails consideration of cost, downtime, production demand, preventive maintenance shutdown windows, and most importantly, the likely survival time distribution of the item monitored. Compared with the extensive literature on condition monitoring techniques and their applications, relatively little attention has been paid to the important problem of the modeling decision making stage in condition based maintenance. Scarf [2] surveyed the papers on modeling condition based maintenance and highlighted the need for more theoretical and case studies in the area.

In this paper we present a case study of predicting the residual life distribution of items monitored based upon the condition information obtained. Once such a prediction is available, a model for condition based maintenance decision making can be readily established, Wang and Christer [3]. The model used in this paper is a special case of the general model developed in Wang and Christer [3] using stochastic filtering theory. The items concerned in this study are six bearings set up in a laboratory fatigue experiment where the overall vibration signal of the bearings is monitored on an irregular basis. The bearings were under accelerated life testing, and therefore, complete life data is available over a relatively short period of time. This is a rare case where the time to failure is available since the bearing with abnormal vibration signals will not be allowed to run to failure in practice. We used this data set and fitted a model to it. Since the life data is available this enabled us to compare the prediction with the actual data. The fitted results are satisfactory compared with observations.

2. Residual time Distribution Prediction

It is defined in Wang, Christer and Sharp [4] that the time lapse from any point that the operating system is checked by condition monitoring to the time that it may fail is called the conditional residual time (CRT). This means that the residual time is not only dependent upon the age or time of the item has survived, but also upon the condition information obtained to date. If at time t we are interested in how long the item monitored can possibly survive given the age and condition information obtained, the problem reduces to determining the residual time distribution conditional upon the time t and available condition information. We establish in this section the conditional residual time distribution based upon stochastic filtering theory, Wang and Christer [3].

2.1 Modeling assumptions and notation

According to the nature of the data and condition monitoring practice, we propose the following modeling assumptions.

1. Plant items are monitored irregularly at discrete time points.
2. There is no maintenance to the bearings.
3. There is a stochastic relationship between the item's residual time and condition monitoring information.
4. The condition information obtained at time t_i, say the current time, is a random variable which is a function of the current residual time and a random noise factor.

The notation used in this paper is as follows;

t_i denotes the ith and the current monitoring time since the bearing was new and still operating.

x_i denotes the residual time at time t_i.

y_i denotes the condition information obtained at time t_i.

$Y_i = \{y_i, y_{i-1}, \ldots, y_1\}$ denotes the history of cumulative condition monitoring information obtained up to t_i, the current monitoring decision point.

$p(x_i \mid Y_i)$ denotes the *pdf.* of x_i conditional upon Y_i.

$p(x_i \mid x_{i-1})$ denotes the *pdf.* of x_i conditional upon x_{i-1}.

$p(y_i \mid x_i)$ denotes the *pdf.* of y_i conditional upon x_i.

Since the residual time at t_i is the residual time at t_{i-1} minus the interval between t_i and t_{i-1} provided the item has survived to t_i and no maintenance action has been taken, it follows that

$$x_i = \begin{cases} x_{i-1} - (t_i - t_{i-1}) & \text{if } x_{i-1} > t_i - t_{i-1} \\ \textit{not}\text{ defined} & \textit{else} \end{cases}. \tag{1}$$

The relationship between y_i and x_i is yet to be identified. From assumption 4 we know that it can be described by a distribution, say, $p(y_i \mid x_i)$. We will discuss this later when fitting the model to the data.

We wish to establish the expression of $p(x_i \mid Y_i)$, and therefore a consequential decision model can be constructed on the basis of such a conditional probability. It can be shown that, Wang and Christer [3], $p(x_i \mid Y_i)$ is given by

$$p(x_i \mid Y_i) = \frac{p(y_i \mid x_i)\int_0^\infty p(x_i \mid x_{i-1})p(x_{i-1} \mid Y_{i-1})dx_{i-1}}{\int_0^\infty p(y_i \mid x_i)p(y_i \mid x_i)\int_0^\infty p(x_i \mid x_{i-1})p(x_{i-1} \mid Y_{i-1})dx_{i-1}dx_i}. \tag{2}$$

It can be shown that since $p(x_i \mid x_{i-1})$ can be determined from equation 1, then if $p(x_0 \mid y_0)$ and $p(y_i \mid x_i)$ are known, equation 2 may be determined recursively.

2.2 Fitting the model to the data

The very first question is obviously what distribution forms for $p(x_0 \mid y_0)$ and $p(y_i \mid x_i)$. Since there is no condition monitoring reading available when the bearings were first installed, then $p(x_0 \mid y_0) = p(x_0)$ which is just the conventional *pdf.* of the bearing life. For the bearing life distribution, an appropriate choice is the Weibull distribution which has been widely used in bearing fatigue experiments, Chen [5]. The distribution of y_i conditional upon x_i is chosen as the normal distribution with mean μ_i and a constant variance σ^2. The conditional relationship between y_i and x_i is established by letting $\mu_i = A + Be^{-Cx_i}$ where *A, B* and *C* are parameters to be estimated from the data.

The reason for choosing a normal distribution is arbitrary. However, by setting $\mu_i = A + Be^{-Cx_i}$, we established a negative correlation between y_i and x_i as expected. Other distributions could be selected, we have tested using exponential and gamma distributions, but the normal gives the best fit. The variance can be time dependent too, but obviously it adds more parameters to be estimated. Since the sample size is small, (6 bearings), we choose a constant σ for the sake of convenience.

The model parameters are estimated in two steps. The first is to estimate the parameters in $p(x_0)$, that is the scale and shape parameters of the Weibull distribution. We used the maximum likelihood method based upon the six bearings life data , but failed to find the reasonable values of them due to the small sample size (sample size=6). We then fixed the shape parameter at value 2 which is suggested in, Chen [5], from the bearing life test, we found the estimated value for the scale parameter is 0.0023. Accepting these two values, the next step is to estimate A,B, C and σ, which is nontrivial.

We wish to estimate A, B, C and σ based upon the available condition monitoring readings collected at discrete points and the age (operating hours) at each checking point using the maximum likelihood method. However, the classical theory of maximum likelihood is based upon the situation in which observations are independently and identically distributed. It should be noted in our case that the observations are neither independent or identical. Here the definition of a condition probability density function is used to write the likelihood function as

$$L = \prod_{j=1}^{m}\left(\prod_{i=1}^{n_j} p(y_{ij} \mid Y_{j,i-1})\, p(x_{j,i-1} > t_{ji} - t_{j,i-1} \mid Y_{j,i-1})\right) p(x_{n_j} = t_{jf} - t_{n_j} \mid Y_{n_j}), \qquad (3)$$

where n_j is the number of condition monitoring checks for the *j*th bearing, y_{ji} is the vibration reading of the *j*th bearing in the *i*th monitoring check, t_{ji}, x_{ji-1} is the residual time of the *j*th bearing at its *(i-1)*th monitoring check point, t_{ji-1}, and Y_{ji-1} is the monitoring history of the *j*th bearing up to the *(i-1)*th check. The last term in equation 3 is the pdf. of the time to failure of the jth bearing at time $x_{n_j} = t_{jf} - t_{n_j}$, where t_{jf} is the failure time and t_{n_j} is the last monitoring point of the *j*th bearing. Minimizing equation 3, we have the estimated values of *A, B, C,* and σ shown in table 1.

Table 1 Estimated parameter values

A	B	C	σ
3.364	28.795	0.046	2.447

The functional form of $p(x_i \mid Y_i)$ can be obtained recursively from equation 2. If x_0 is Weibull distributed and $y_i \mid x_i$ follows a normal distribution, it follows that

$$p(x_i \mid Y_i) = \frac{(x_i + t_i)^{\beta-1} e^{-\frac{1}{2\sigma^2}\sum_{k=1}^{i}(y_k - A - Be^{-C(x_i+t_i-t_k)})^2 - (\alpha(x_i+t_i))^\beta}}{\int_0^\infty (z+t_i)^{\beta-1} e^{-\frac{1}{2\sigma^2}\sum (y_k - A - Be^{-C(z+t_i-t_k)})^2 - (\alpha(z+t_i))^\beta} dz}. \tag{4}$$

Note the rule played by y_k where all the historical information of y has been taken into account.

It has been observed that the predicted residual times are well within our expectation, which give us confidence of the model established.

3. The decision model

The fundamental decision to make at various monitoring points is whether we should replace the bearing or not given all information available. If the answer is yes then what is the best time for such a replacement and whether we should wait until a suitable production window such as a scheduled shutdown arrives. Suppose that the mean cost of a failure is C_f and C_p denotes a mean preventive replacement cost, t_i is the current monitoring check point and $t - t_i$ is some planning period from t_i onward, if monitoring cost is negligible and no preventive replacement in (t_i, t), we have the total expected cost over (t_i, t) is given by

$$C_f P(x_i < t - t_i \mid Y_i) \tag{5}$$

If a preventive replacement is scheduled in (t_i, t), say at t_i or as earlier as possible, the expected cost is C_p, then a preventive replacement is recommended if

$$C_p < C_f P(x_i < t - t_i \mid Y_i) \tag{6}$$

that is $\frac{C_p}{C_f} < P(x_i < t - t_i \mid Y_i)$.

This model is dynamic and updated at each monitoring point once new information becomes available. This implies that t is bounded within (t_i, t_{i+1}).

4. Conclusion

This paper reports on a study of the application of a model in prognosis of bearing residual time distribution given available condition monitoring information. The model is established under a special case where life information is available. The fundamental concept behind the model is the conditional residual time which differs with the conventional concept of the residual time in that it not only depends upon the current age of the item but also the condition information available to date. The condition information we used is the overall vibration readings of six bearings used in a lab fatigue experiment.

Since the vibration analysis is the most popular condition monitoring means used in industry, the model developed has its potential to be applied in a wider situations where similar condition monitoring processes exist. It should be pointed out that this model is dynamic in that whenever a new vibration signal is available, this model is updated itself.

Acknowledgment

The research reported here is partly supported by EPSRC under grant number GR\M96582.

References

[1] Rao B.K.N. *et al* eds, 1988-1999, proceedings of COMADEM international congresses and exhibitions.

[2] Scarf. P., 1997. On the modelling of condition based maintenance, proceedings of ESREL97, June, Lisbon, Portugal, (C. Greder Soars Ed.), Pergamon.

[3] Wang W and Christer A.H., 2000. A general of condition based maintenance model for a stochastic dynamic system, in press, *J. Opl. Res. Soc.*

[4] Wang W., Christer A.H., and Sharp J.M., 1996. Stochastic decision modelling of condition based maintenance, proceedings of COMADEM96, Sheffield, (Rao B.K.N, Smith R.A., and Wearing J.L., Eds.), Sheffield Academic Press, Sheffield, 1175-1184.

[5] Chen J., 1993, Survival analysis and reliability (in Chinese), Anhui Education Press.

Towards a Decision Support System for Maintenance and Reliability Models using a Multimedia Based Approach

Paul A. Shorrocks and Ashraf W. Labib
Mechanical Engineering Department, Manufacturing Division,
University of Manchester Institute of Science and Technology (UMIST), UK,
Email - mcjiipas@ fs4.umist.ac.uk

Abstract. The requirements of maintenance have demanded the creation and development of a modular designed Decision Support System (DSS). This tool is capable of selecting the most appropriate Decision Making Model (DMM) for a specified company. It uses established mathematical procedures such as Analytical Hierarchic Process, Fuzzy Logic and Group Decision Making. This electronic package will make its knowledge available to the largest number of businesses through a multimedia web based approach in a manner that will offer some guarantee of best practices by the use of feed-back and the application of continual improvement. All these criteria will be met at the most economical cost whilst the final decision on choice of DMM will remain under the control of the participating organisation.

1. Introduction

The pressures imposed on modern manufacturing systems can only be met through effective and efficient use of both manpower and plant. The productivity of the manufacturing process can therefore be drastically reduced by a loss of core plant through unavailability, unreliability and difficulties associated with poor maintainability. This is in contrast to older and more traditional manufacturing systems where a significant amount of redundancy of both equipment and personnel was always maintained by system design in order to buffer resources. Availability is the operational capability of the production facility at the start of the production run. Reliability is the guarantee that equipment will survive for the lifetime of the production run. Maintainability is the ease of repair, replacement or refurbishment when failure occurs. Maintainability is usually the domain of the original equipment manufacturers and will therefore not be considered further in this document. Whereas availability and reliability can be significantly improved by using the correct maintenance policies and the most appropriate scheduling regime.

2. History

Traditional maintenance uses corrective or repair maintenance practices that restore the effectiveness of parts and equipment after both catastrophic and chronic failure. This policy is based on reactive management where little or no forward planning is made and very little data is collected or evaluated. Although there are some advantages to this type of maintenance policy in terms of reduced clerical activity, its disadvantages make it unsuitable for modern production systems that require maximum overall effectiveness. These systems require the use of proactive policies that conserve the effectiveness of equipment by replacing, repairing or refurbishing them before failure. The advantages of proactive maintenance are;

- Equipment can be repaired without stopping the production process (off-line and idle time refurbishment).
- There is no lost production cost (lost production costs three times the resources to replace than original production, this is because of);
    - The loss of the original production.
    - The loss of future production time whilst the original order is being replaced.
    - The loss of production whilst future orders are being replaced.
- There in no secondary damage (when parts fail catastrophically, associated parts may also be damaged).
- Forward planning allows more efficient use of physical resources (proactive planning reduces the burden of unexpected costs).
- Lower staff levels may be possible (because of effective use of man/hours).
- Specific training can be given for planned jobs (if records detect a skill deficiency).
- Inventory is maintained at lower levels (because only the immediately required parts are held in stock).
- The stores department can be reduced in size (in line with its lower spares stock).
- Continual improvement is possible (because the system is better managed).

There are however some problems with proactive systems. These problems are often connected with the frequency or scheduling of maintenance activities.

3. Scheduling systems

Scheduling of maintenance tasks is a multi criteria problem that aims to find the most appropriate number of maintenance activities required for optimum efficiency of the production system. The least number of maintenance activities often results in the maximum number of catastrophic failures. On the other hand the maximum number of maintenance activities results in loss of production performance and high maintenance overhead costs because the machine is always unavailable due to maintenance activities. Therefore the appropriate scheduling of maintenance tasks is of benefit to both the maintenance and production systems. The settings of maintenance schedules can typically be determined in one of three ways;

- Using the manufacturers' estimated life for a part e.g. a rolling element bearing may be rated at 10,000 hours of continuous running.

- The historic data collected by Computerised Maintenance Management Systems (CMMS's) shows how long a previously installed part survived in the form of mean time between repair.
- The final method of scheduling uses heuristical processes, (these are not the optimum methods, but produce useful results for difficult situations) these may be based on time, manufacturing cycles or algorithms, e.g. major parts will be changed or refurbished as part of an annual shutdown.

All three methods of determining part longevity are flawed. The manufacturers' estimate of life is based on component quality tests that use constant loading and unidirectional motion. Even with this restricted set of test parameters the sample parts fail over a range of times and therefore the published longevity can only be a statistical mean. Part life determined from actual operational data may have two fundamental faults. It can only be accurately estimated if many years of data are provided and secondly there is an assumption that the operational conditions remain the same. This may not be true if the general state of the equipment is deteriorating or there is a changing trend in the type of work undertaken. Heuristical planning is a multi-criteria process based on the convenience of the production system. This leads to maintenance activities being carried out annually or during bank-holidays as opposed to maintenance related intervals. Because of the limitations of the three previous methods of scheduling, a number of DMM's and maintenance systems have been developed that will dictate a maintenance regime which is more reliable. These include Failure Mode and Effect Analysis, Weibull Analysis, Root Cause Failure Analysis, Reliability Centred Maintenance and Total Productive Maintenance. These techniques are considered superior because of their mixture of real data, mathematical logic and constant analysis of the effectiveness of any previous decisions. Because there are many differing types of DMM's, it is difficult to select one in reference to any particular manufacturing system. The proposed model in this paper addresses this specific problem.

4. The Proposed model

A DMM is a tactical tool that dictates scheduling and sequencing of maintenance activities after consultation with historic records and manufacturing data, the difficulties outlined in the previous section show that help is required if the most appropriate model is to be determined. This selection can be made by a higher decision selection device known as a Decision Support System [1]. Regrettably these systems are small in number, not easily obtained and not widely known beyond the academic establishment. A good DSS will have the following attributes. It will be easy to obtain and use (available and user friendly) and will select a DMM with respect to the needs of a specific company. It will be suitable for the company that uses it, without excessive modifications or staff training and the decisions made by the DSS will be based on the most up to date analysis. The DSS will only advise on the most suitable models, it will not dictate which model to use, therefore allowing the manufacturing organisation to retain control. Finally, a full description of how the DSS reached its conclusion and a description of the DMM's will be available. Ongoing research has identified a possible method of satisfying these criteria through a modular multimedia computerised tool, the remainder of this document outlines the general construction of this system.

5. Methodology

As this is a prototype tool that may require a number of future upgrades, the tool has been designed and constructed from a variety of modules that have been built and modified as autonomous units within high level language of C++. Where possible, standard computer programmes have been utilised in order to reduce any requirements for user training programmes. A diagram of the modules and their linkage is shown in Figure 1.

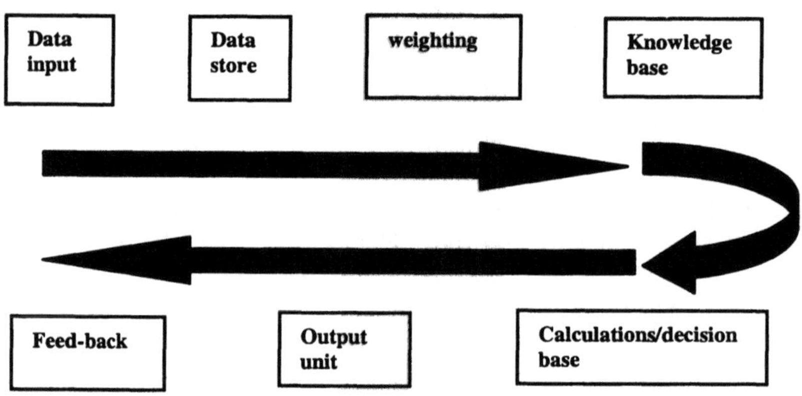

Figure 1. How the DSS modules are configured.

The importance of multimedia tools is now realised globally, these tools use the power of the internet or world wide web as a rapid information transfer media. The front end of this web tool (data input) is based on a reactive on-line questionnaire built using Microsoft® FrontPage. This questionnaire is constructed from a number of interlinked pages that ask relevant but simple closed questions. These questions have been selected because they categorize and define establishments. In the eventuality of a question requiring a more comprehensive explanation, auxiliary pages of supplementary information have been provided. The purpose of the questionnaire is to determine the ability of any company to install and effectively operate a new process. The numerical values that the computer assigned to each response have been determined using the power of fuzzy Delphi [2] and triangular numbers. Delphi is the process of obtaining advice from a number of human experts who operate in isolation. This isolation removes any outside influence a group member may experience. The system of triangular numbers is a mathematical process based on a compromise between three possible outcomes, the minimum of the most likely outcomes, the maximum of the most likely outcomes, and the stated most likely outcomes. The questionnaire will rank the responses to each answer in terms of a common set of parameters. There are a number of criteria that are commonly used to measure and classify a companies activities, costs, and benefits are the two favoured by accountants and financiers, but because they are both retrospective measures it was considered good policy to include some futuristic parameters as a balance. These extra values are opportunity and risk. The captured data is then stored and manipulated so that it can be used as both a measure of company description and DMM ranking. When all data has been collected and summed in terms of the four parameters, it is weighted in order to give it relative importance. In order for these parameters to be correctly compared they must be in commensurate units [3] before the weighting procedure. This tool uses the universally recognised units of money. The chosen method of weighting is the Analytic Hierarchic Process (AHP) [4]. This will give numeric output values to standard financial management accounting. A principle of financial accounting is that cash now is more important than cash in the near future and cash in the near future is worth more than cash in the distant future. This gives the ranking of Figure 2;

Cost > Risk > Benefit > Opportunity.

Figure 2 The ranking hierarchy.

AHP uses a squared matrix to produce eigenvectors that in turn produce simple numerical values as weightings for the four chosen parameters. The DSS has an extra data base module that is used as a knowledge base for all the characteristics of the proposed DMM's and reliability models. This knowledge-based approach [5] holds a mathematical description of each model and allows a match to be made between any specified company and the most suitable of the DMM's. This calculation is at the heart of the selection process and is determined by a rule-based process shown in the following equation.

$$\text{Ranking number} = \frac{\text{Benefit (B)} \times \text{Opportunity (O)}}{\text{Cost (C)} \times \text{Risk (R)}} \quad (1)$$

Where $C + R + B + O = \text{Unity}$.

In the final selection, the ranking number refers to a specific DMM in respect of a specific company's attributes. This numerical value will be used by both the web tool and the human operator in the participating company when determining the most suitable match between a decision-making model and the specified company. When the questionnaire has been completed and all the calculations resolved, the output information will be made immediately available on the same computer screen. This information takes the form of a hierarchical list of DMM's and the numerical ranking of these models in respect of the specified company. The user will not be pressured into accepting any given ranking. In order to assist in their final decision making, there is a supplementary page that gives a description of the models and the most appropriate methods of implementing them. Therefore the end user will always be in control of any decision-making process or new system selection. In order for the system to continually offer the highest standard of decision-making, an interactive system of feed-back is available, this will take the form of an email reply system built into the tool.

6. Conclusion

As a response to both internal and external evaluation, the tool will undergo a range of assessments during its development and use, the mode of these investigations will depend on the stage of completion of the model. Initially the tool will be subjected to simulation using stochastic data and latter will be field tested through participating organisations. Throughout its operational life the tool will be monitored with a view to detecting and rectifying any abnormalities. The success and longevity of this tool can only be maintained through continual improvement. The tool is expected to be available commercially in the early part of 2002 as part of an internet or intranet package. The tool's success or otherwise will be measured by company acceptance and individual company performance, this is possible because the selection of a suitable DMM will enhance maintenance proactive planning and be reflected in production figures.

References

[1] Kobbacy K.A.H., Proudlove N.C., Harper M.A., 1995, *Towards an intelligent maintenance optimisation system*, Journal of Operational Research Society, V46, No7, pp 831-853.

[2] T-H. HSU., 1999, *Public transportation system evaluation using the analytic hierarchy process: A fuzzy delphi approach,* Transportation Planning and Technology, V22, No1, pp 229-246.

[3] Wedley W.C., 1999, *Magnitude adjustment for AHP Benefit/cost ratios.* Simon Fraser University, Burnaby, Canada, unpublished paper.

[4] Saaty T.L., 1980, *Analytical Hierarchical Processes.* Mc Graw Hill, New York.

[5] Kobbacy K.A.H., 1992, *The use of knowledge based systems in evaluation and enhancement of maintenance routines,* International Journal of Production Economics, V24, No1, pp 243-248.

Detection of Gear Failures in Helical Gears by Using Wavelet Transforms

Naim Baydar , Maintenance Engineering Research Group, Manchester University, Manchester - UK, Email: mbgmgnb1@fs1.eng.man.ac.uk

Prof. Andrew Ball, Maintenance Engineering Research Group, Manchester University, Manchester – UK, Email: adball@fs1.eng.man.ac.uk

Abstract: In this paper, the use of a Continuous Wavelet Transform (CWT) and its modified version (Scalogram) is proposed to detect a growing local fault in a two stage helical gearbox. A broken tooth was simulated as a local fault and small fault conditions were tested. The vibration signal was used to monitor the fault conditions. It is found that the Scalogram offers interesting and more easily interpretable results for the fault detection process. It displays the fault features more distinctively when compered to the amplitude and phase map of the CWT.

1. Introduction

In rotating machinery such as gearboxes, the vibration signal is very commonly used for fault diagnostics. This is because a gearbox with faults has different vibration characteristics when compared to a healthy state of the same gearbox. The conventional techniques for fault detection, such as spectrum, cepstrum, time domain averaging and demodulation analysis are based on the assumption of stationarity of the vibration signals. They are [1-4] well established and have been proved to be effective in machinery diagnostics. However, in many cases these methods are not sufficient to reliably detect different types of faults in the early stage and give detail information about their conditions.

During the 1990's some time-frequency methods gained considerable recognition in the field of condition monitoring. The early applications of the methods to gear faults in particular began with Forrester's work [5]. He applied the Wigner-Ville distribution (WVD) to averaged gear vibration signals and showed that different faults such as a tooth crack and pitting could be detected in the WVD plot. Later, McFadden and Wang applied a weighted version of the WVD to gear failure in order to improve the detection capability of the method [6]. Staszewski and Tomlinson applied the Wavelet Transformation (WT) and Wigner-Ville distribution to a spur gear to detect a broken tooth. They also used statistics and neural networks for the classification of fault conditions [7]. More recent work conducted by Yesilyurt and Ball applied wavelet analysis and Instantaneous Power Spectrum (IPS) to gear faults and identified incipient faults in the gears [8].

In machinery diagnostic, it is crucial to detect the fault conditions in the earliest stage possible. This paper presents an investigation carried out concerning a detection of small progressing local faults in gearboxes. Initially, an ordinary wavelet transform based on modulus and phase will be given and then a modified version of the wavelet, Scalogram (a square modulus of the wavelet) will be presented.

2. Fault Simulation and Test Rig

The fault was simulated by the removal of a percentage of the tooth face width on the pinion gear. Gradual fault advancement was studied by removing face width in 10% increments. Fault conditions are represented from 10 % to 50 % removals of the single tooth. The torque on the output shaft was 260Nm and other specifications of the gearbox are given in Table-1. The vibration signal generated by the gearbox was picked up by an accelerometer. A reference signal obtained from an optical pick-up was then used to synchronise time domain averaging of the vibration data. The signals were sampled at 6.4 kHz.

Table 1. Specification of the gearbox

	First Stage	Second Stage
Number of teeth	34 / 70	29 / 52
Speed of shafts	24.33 rev/sec.	6.59 rev/sec.
Meshing frequency	827.73 Hz.	342.73 Hz.
Contact ratio	1.359	1.479
Overlap ratio	2.89	1.478

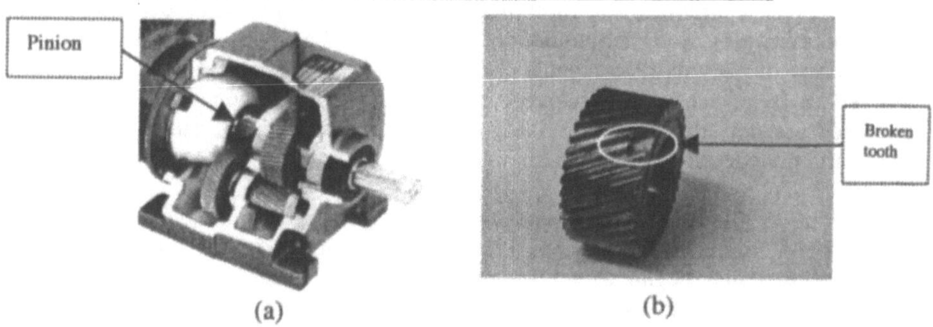

Figure 1. Gearbox and fault simulation a) sectioned view of the gearbox b) broken tooth

3. The Wavelet Transform (WT)

One of the important distinctions between wavelet transform and other time-frequency methods is that the wavelet transform is more capable of treating transient events in the signal. The wavelet transform moves data from a space to a scale domain with the wavelets as the basis function giving the localised features of the original signal. Wavelets occur in sets or a family of functions and each is defined by a dilation, which controls the scaling parameter, and translation, which controls the position of the kernel in time. In Continuous Wavelet Transform

(CWT), time t and time-scale parameters vary continuously. The CWT of a signal x(t) is defined as :

$$CWT_x(a,b) = W_\psi(a,b) = \frac{1}{\sqrt{a}} \int_{-\infty}^{\infty} x(t)\,\psi^*\left(\frac{t-b}{a}\right) dt \qquad (1)$$

where a(a>0) is dilation factor, b is the translation factor and ψ(t) is the mother wavelet or analysing wavelet. The WT is expanded by a family of kernel functions, which are translated and dilated from the mother wavelet and expressed as:

$$\psi_{ab}(t) = \frac{1}{\sqrt{2}} \psi\left(\frac{t-b}{a}\right) \qquad (2)$$

where ψab and ψ are called the daughter and mother wavelet respectively. Since the wavelet has a property of preserving energy it can be expressed as [9,10]:

$$\iint \left|CWT_x(a,b)\right|^2 \frac{da\,db}{a^2} = E_x \qquad (3)$$

where $E_x = \int |x(t)|^2 dt$ is the energy of the signal x(t). Similar to the Spectrogram, Equation.3 leads to definition of the modified wavelet or Scalogram, as the square modulus of the CWT.

4. Results and Interpretation

Figure.2 displays the phase and magnitude map of the CWT for various conditions. In the phase plot of the WT for healthy condition, the fundamental and second toothmeshing frequencies are visible, though the energy of the vibration signal mainly concentrated around the fundamental meshing frequency. Although some changes occur in the phase plot of 10 and 20 % fault conditions, most of the signal energy is still concentrated over the fundamental meshing frequency. Therefore, the fault conditions are not clearly identifiable at this stage. From 30 % fault conditions onwards, most of the signal energy appears to be (500 Hz) outside the fundamental meshing frequency and this indicates the presence of a fault in the gearbox. Similarly, the amplitude plot of the WT illustrates that when there is no fault (or a very small fault) the energy of the signal concentrates over the fundamental meshing frequency. With the growing fault the energy of the signal shifts significantly outside the meshing frequency region. The amplitude map seems to be easier to interpret though both the phase and amplitude of the WT reveals the fault conditions. When the results from the scalogram (Figure.3) is compared to the results from Figure.2b, it is quite clear that the scalogram produce finer and easily interpretable plots to identify the fault conditions. Signal energy outside the fundamental meshing frequency (about 500 Hz.) indicates the presence of fault in the gearbox.

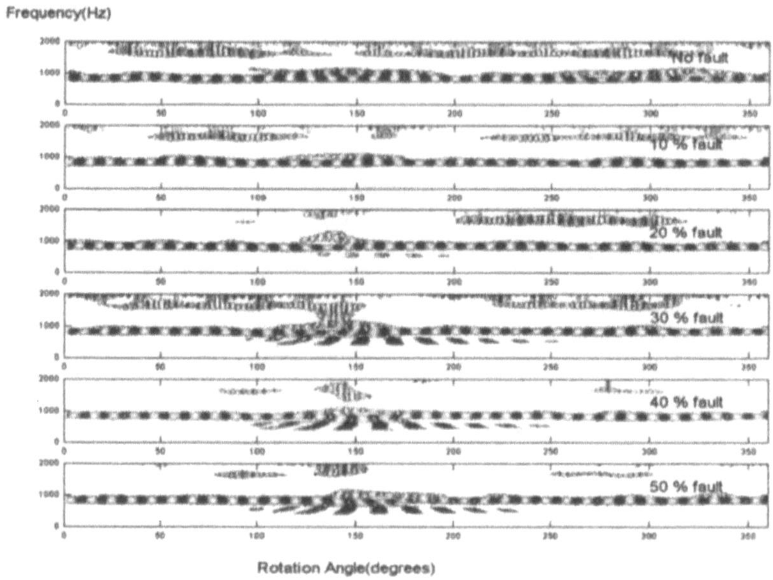

(a)

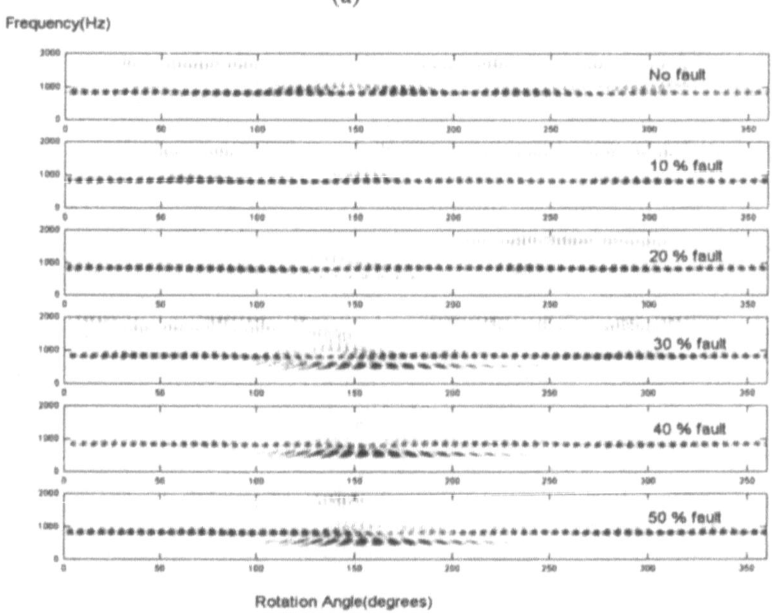

(b)

Figure 2. Detection of small fault conditions by CWT a) phase b) amplitude

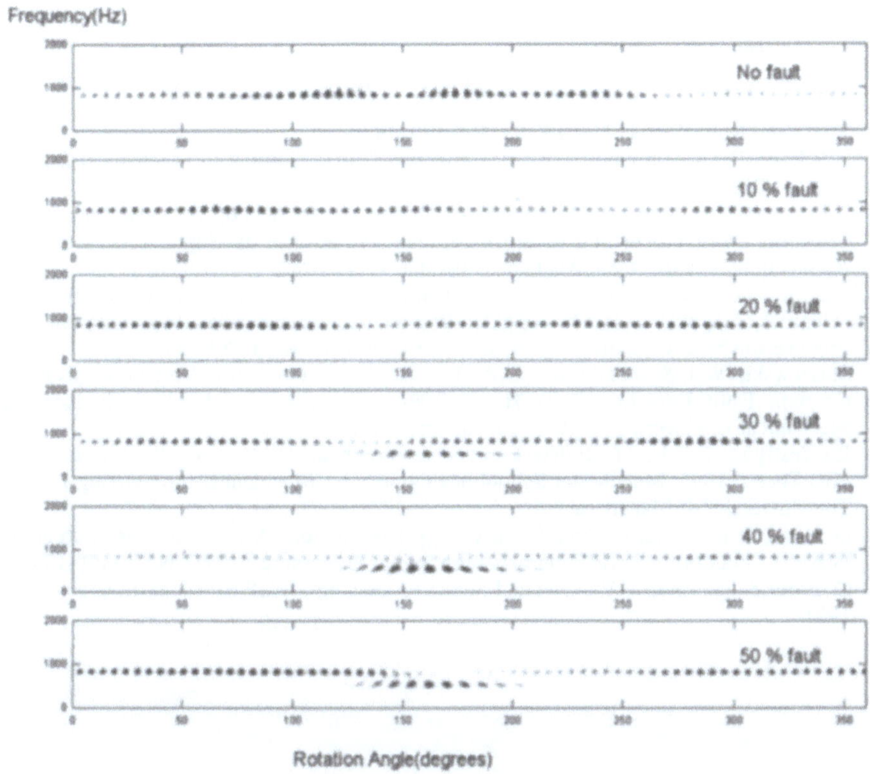

Figure 3. Scalogram results for small fault conditions

5. Conclusion

The result of the present work showed that the wavelet transform on its own is a very useful tool in gear diagnostics and can indicate progressing fault conditions in gearboxes. However, the scalogram provides a finer and a more easily interpretable contour plots to detect and identify early fault conditions.

Acknowledgement

This work has been supported by EPSRC in U.K.

6. References

[1] R.M.Stewart, Some useful data analysis techniques for gearbox diagnostics, Institute of Sound and Vibration Research, Southampton University, 19-22 Sept 1977

[2] R.B.Randall, Bruel & Kjaer, Application of cepstrum analysis to gearbox diagnosis – ImechE-1980

[3] P.D.McFadden, Detecting fatigue crack in gears by amplitude and phase demodulation of the meshing vibration. Transaction of the ASME, Journal of Vibration, Acoustic, Stress and Reliability in Design 108, 165-170, 1986

[4] R.B.Randall , A new method of modelling gear faults – Journal of Mechanical Design, Vol.104/259, April 1982

[5] B.D.Forrester, Use of the Wigner-Ville distribution in helicopter transmission fault detection, 1989 Proceeding of the Australian Symposium on Signal Processing and Applications – ASSPA-89, Adelaide,Australia, 17-19 April

[6] P.D.McFadden & W.J.Wang, Time-frequency domain analysis of vibration signals for machinery diagnostics(2), The weighted Wigner-Ville distribution, 1991 Department of Engineering Science, University of Oxford, Report No. OUEL1891/90

[7] Staszewski, W.J & Tomlinson, G.R, The Application of Time-Variant Analysis to Gearbox Fault Detection, PhD Thesis, 1994 - Manchester School of Engineering, Mechanical Engineering Dept. - The University of Manchester

[8] Isa Yesilyurt, Gearbox fault detection and severity assessment using vibration analysis, PhD Thesis, 1997, Manchester School of Engineering, Mechanical Engineering Dept. – The University of Manchester

[9] O. Rioul & P. Flandrin, Time-Scale Energy Distribution: A General Class Extending Wavelet Transforms. IEEE Transaction on Signal Processing Vol.40, No.7, July-1992

[10] O. Rioul & M. Veterli, Wavelet and Signal Processing , IEEE SP Magazine, October – 1991

Cost Estimating and Risk Analysis in Manufacturing Processes

Rajkumar Roy and Aitor Palacio

Department of Enterprise Integration, School of Industrial and Manufacturing Science, Cranfield University, Cranfield, Bedford, MK43 0AL, UK.
Tel: +44 (0)1234 754072, Fax: +44 (0)1234 750852,
Email: r.roy@cranfield.ac.uk

Abstract: The paper develops a methodology to identify and manage the risks involved in cost estimating a manufacturing process. A Case study is presented based on an automotive manufacturing company. The research starts by developing an 'AS-IS' process model of cost estimating for the manufacturing process. The causes of cost estimation inaccuracy provide the basis for a costing risk analysis within the company. The research has identified where the risk is, what are the different categories of risks and obviously how to control and manage them.

1. Introduction

The world manufacturing industry is becoming increasingly cost conscious. Any cost estimating process involves a risk and it is necessary to be aware of where this risk lays. Currently the manufacturing cost estimating is mostly dependent on experts' knowledge and judgement and this is another big challenge for the risk management. The research has developed a methodology to reduce risk in manufacturing cost estimating, through risk identification and then by developing Cost Estimating Relationships (CER).

It is observed that currently the estimates of the manufacturing costs of a sheet metal product are performed mostly from experience, and estimator's judgement. The research particularly looks into the areas of process time (sheet metal forming) related costs associated to sheet metal forming. The study has identified the most important features an estimator would look for while doing an estimate of the manufacturing costs. A list of possible sources of inaccuracy and their control mechanism is also identified.

This risk management technique is enhanced by complementing the 'managerial issues' with a more objective cost estimating process: the development of CERs. The CERs are a relationship among the characteristics of the parts and their final cost of production, and they are developed based on past data. The relationships are developed using a statistical analysis approach. The paper will outline a step-by-step procedure to develop CERs.

2. Manufacturing Cost Estimating

Cost estimating is the process of predicting or forecasting the cost of a work activity or work output [1]. By following a process and interpreting historical data it is possible to develop a probable cost for an activity or output. The accuracy of this forecast is dependent on the quality of the data and the relevance of the costing method. In this section the most common costing procedures are discussed: traditional, Activity Based Costing, Feature Based Costing and Parametric Costing.

In traditional costing there are two main estimates "first sight" estimate, an estimate done early in the cost stage, and a detailed estimate, done to calculate costs precisely [2]. These cost estimating methods are largely based around the experience of the estimator. Activity based cost (ABC) [3] is a tool for cost management. Activity based management seeks to portray a company as a series of activities which are related to customer desires and cost. This is a popular method of estimating detailed manufacturing costs.

Features based models offer considerable advantages and have the potential to close the gap between design on the one side and process planning and manufacturing cost information on the other side [4]. Parametric cost estimating is generally used in the earlier stages of a project such as concept and early design phases [5]. The technique, through statistical analysis, develop relationships called CERs (cost estimating relationships) with cost as a dependant variable and performance or physical dimensions as independent variables [6].

3. AS-IS Cost Estimating Process within the Company: The Context

The process of an estimate starts when the cost of a part sent from the Sales Department needs to be estimate (part or assembly). The Sales Department is in touch with the customer and that is who requires the cost estimate to be able to propose an offer to the customer. The estimators need to know the manufacturing process plan for the part to be able to estimate the cost of it. The Engineering Department the planning information to the estimators. Following are the stages of the estimation process [2]: *variable input lists, operations descriptions, piece price, maintain other investment, maintain other material.* The final piece price is calculated as the sum of three different prices: the price associated with the cost code centre and the operation, the price related to the material, and the price related to other material option. It is observed that the manufacturing cost estimation is dependent on experience of the estimators, semi-computerised, and is prone to errors. It is identified that the process time is a major manufacturing cost driver (Figure 1). Therefore, the rest of the study is focused on Risk Analysis of the process time for pressed and assembled parts. The study is based on the manufacture of body systems for automotive industry. The parts are made of either aluminium or steel, and are produced in low volume.

2. Risk Analysis

The risk analysis [7] is performed on the process time associated with the press sheet metal items. Semi-formal interviews, case studies are performed with the estimators to extract relevant cost estimating information about this area and to identify which are the main features they look at when they are doing an estimate. The risk analysis study is performed in four stages, as follows:

- Identification of the most important characteristics to estimate the cost of a part.
- Identification of the main causes of the error in the estimates: a list of some potential sources of error.
- Capturing estimator's knowledge on how to control the risks involved in the estimates.
- Define ways to improve the actual manufacturing cost estimating system.

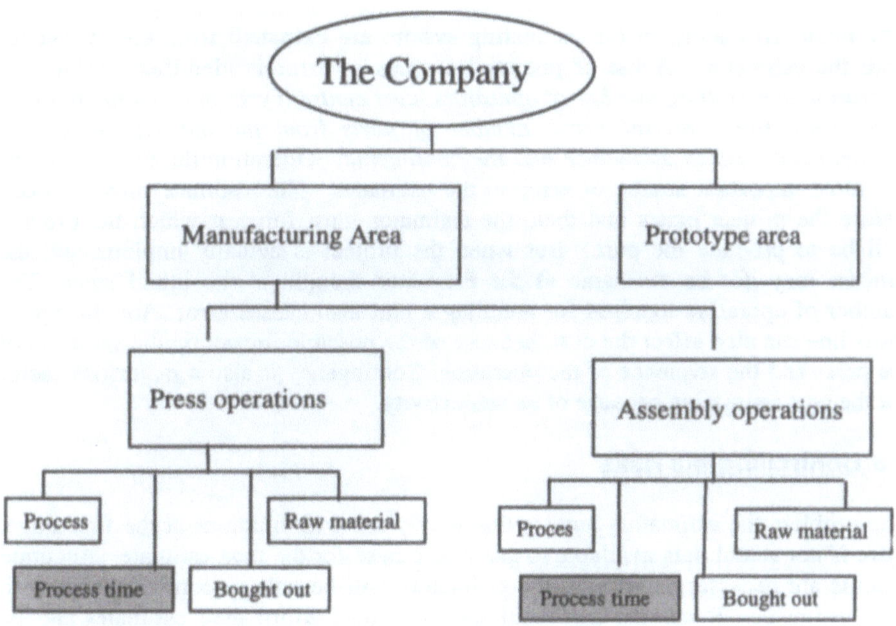

Figure 1: The most risky area for the cost estimating.

2.1 Identification of the most important characteristics

As a result from the first part of the study a list of those features the estimators look at in a pressed part to do an estimate is established. The list includes: *type of the part, material of the part, size of the part, operations type during the production, shape of the part, depth of the draw, material thickness, dimensions of the blank size, and batch quantity.*

There are two types of part, one is called skin part and the other one is called structure part. This has a very high influence in the cost because if the part is a skin part that means that it will be in the outside of the final product. Within the company there are three main types of materials used: aluminium, steel and coated steel. The material has not only influence in cost because its own price as raw material, it has also influence in the production process. The steel can be pressed quicker than the aluminium, then, this another factor why an aluminium part is more expensive than a steel one, because it needs more time to be processed. Similarly, the part size, shape, types of processes, material blank size and the batch quantity has direct influence on cost. The depth of the draw needed to produce the part is also a very important factor on the cost of the part. The higher the draw depth is the more expensive the part will be, because the process will be complex and slower and this means that the cost will be higher.

2.2 Main causes of error

The main risky areas in the estimating system are extracted from the interviews with the estimators. A list of potential sources of error is identified as follows: *operation description, number of operators, cost centre, cycle time, maintain other investment, raw material price, number of parts from the material sheet, the synthetics, the scrap allowance and the contingency*. Operation description can be the most important source of error in the estimates. The estimate must be done before the project exists and then, the estimator must forecast which the process will be to produce the part. But when the project is actually implemented the process may not be the same as the estimator thought at the initial stage. The number of operators required for handling a part also causes error. Any change in press line can also affect the cost, because of the possible impact on the selection of the press and the sequence of the operation. Contingency is also a major risk factor for the cost estimating because of its subjectivity.

2.3 Controlling the risks

The problem the estimators have is that they work with opinions at the first stage, there is not actual data available to use it as a base for the next estimate. Accurate records are necessary and then, the estimators will be able to compare them with their estimates. Receiving that feedback they could adjust their estimates and be more accurate.

2.4 Improving the System

The main suggestion to improve the current method is the need of receiving feedback from the shop-floor with the actual time. It would be possible to put that data in the synthetic, and this would help to improve the system significantly. It is proposed that historical data is utilised to develop cost estimating relationships (CER), that can be used for more objective estimation.

3. Cost Estimating Relationship (CER) Development

Cost Estimating Relationship development will reduce the subjectivity within the cost estimating. This section describes the procedure followed to get that relationship and the limitations found during the process. After initial investigation, the following cost drivers are selected for data collection: material thickness, blank size, length, width, height, number of operators, and number of operations. The actual time taken to produce the part is also noted. Most of the necessary information to fill in the table is provided by the MRP system implemented within the company. The only information that is not obtained from the MRP system is the one related with the overall dimensions: length, width and height. These data is collected manually by measuring the parts. The parts are categorised as large, medium and small. A large part has at least two dimensions above or around one metre. Medium parts are with only one of their three dimensions around or above one metre. And the small parts are the rest, with all three dimensions smaller than one metre. For this case study only large parts are considered, because they constitute 80% of the total manufacturing cost. The next stage of CER development is the data collection. The data collection is time consuming and labour intensive. Within the constraints of the project only eight large parts' data is collected for the relationship development.

4.1 Statistical Analysis and Results

Once all the data is captured, the next step is to go through a statistical analysis. The software used for this study is SPSS. Before the final analysis is done, estimators are consulted to reduce the number of independent variables (cost drivers) for the regression analysis. This is necessary to reduce the complexity of the regression analysis, assist in the data collection for future development and to work with the most important variables. Finally five cost drivers are selected for the analysis: material thickness, length, width, height and number of operations. Linear multivariate regression analysis is performed on the data to develop a CER for the large parts. The CER developed from the data is shown below:

COST OF A PART=12.34143*(Material thickness) + 0.011335*(Length) - 0.000767539*(Width) + 0.015808*(Height) + 0.805702*(No of operations) - 20.687356

It is worth noting that this equation is reached with a very small sample size, only eight large parts are used. This makes it quite difficult to assign a level of accuracy to this equation. One trial is carried out by excluding a part; and then carrying out the analysis with the other seven parts. The analysis produced similar equation, and when applied to the excluded part, the estimated cost is 14% lower than the actual manufacturing cost of the product. The CER is also validated in consultation with the estimators, and it matched with their general understanding about the cost estimation.

5. Discussion and Conclusions

The research has developed a systematic approach to analyse risks in cost estimating of pressed parts. The approach is based on capturing estimator's knowledge in this area. Within an organisation where plenty of relevant past data is available, a data centric, more objective approach can be used. It is observed that communication of feedback among different departments within the company is not very strong, as a result the estimates are mostly based on estimator's experience, and thus causing a lot of errors in the estimates. Development of CERs can introduce some objectivity in the estimating process, and thus making the process less risky. Due to non-availability of data within the time frame of the project, only eight part data is utilised for the CER development, this is very small for any statistical analysis. Therefore the CER developed should be used as a guide rather than a tool to estimate cost.

The approach developed by the research identifies the major characteristics of a pressed part that an estimator considers before developing the estimate, and also discusses the possible sources of errors in the cost drivers. The paper also presents some comments to control the risk and improve the estimation process. Use of CERs is proposed to reduce risk at the early stage of the cost estimating. It is observed that good quality data collection is a major challenge, and that can affect the quality of CERs produced. The cost estimators have validated the approach presented in this paper. In the future the research should concentrate in improving the communication between the departments in relation to the cost data. It is necessary to develop a best practice procedure to collect data for CER development in the future.

References

[1] Stewart J., 1995, *Cost estimator's reference manual.* Wiley Interscience.

[2] Palacio A., 1999, *Cost Estimate and Risk Analysis in Manufacturing Processes*, MSc Thesis, Cranfield University, UK.

[3] Yahya-Zadeh M., 1998, *Product –mix decisions under activity based costing with resource constraints and non-proportional activity costs*, Journal of applied business research.

[4] Wierdam L. S., 1991, *Linking design, process planning and cost information by feature based modelling*, Journal of Engineering Design.

[5] Department of Defence, 1999, *Parametric Estimating Handbook*, 2nd Ed., DoD, http://www.ispa-cost.org/PEIWeb/cover.htm .

[6] Roy R., Bendall D., Taylor J.P., Jones P., Madariaga A. P., Crossland J., Hamel J., Taylor I. M., 1999, *Development of Airframe Engineering CERs for Military Aerostructures.* Second World Manufacturing Congress (WMC'99), Durham (UK), 27-30th Sep., pp. 838-844.

[7] Dean E. B., 1986, *Cost risk analysis based on perception of the engineering process*, Proceedings of the Annual Conference of the International Society of Parametric Analysts.

Improve Availability by Enhanced Maintenance Strategies

G. Günther[1], A. Seufzer[2], H. Ahlers[3]
Institute of Production Engineering and Machine Tools,
University of Hannover, Schosswender Str. 5, 30159 Hannover
[1]guenther@mail.ifw.uni-hannover.de
[2]seufzer@mail.ifw.uni-hannover.de
[3]ahlers@mail.ifw.uni-hannover.de

Abstract: Knowledge about the condition of processing equipment is getting more and more important due to growing requirements for reliability and availability. This article presents a general method for the building of condition models. This method is explained in more detail for the modelling of the wear condition and a computer aided maintenance assistant is presented in which these methods have been integrated.

1. Introduction

Shorter innovation periods, a high number of product variations, small stock amounts and decreasing lot sizes lead to increased application of highly automated machines and plants in the manufacturing techniques. This trend results in a growing subjection to the dependability and availability of the system components.

Within the collaborative research centre 384 which is sponsored by the DFG (Deutsche Forschungsgemeinschaft), different methods and procedures are being developed at the University of Hannover, Germany to insure the availability of production systems (http://www.sfb384.uni-hannover.de).

The availability is defined as the ratio of the average service life between two failures and the sum of the average service life between two failures and the average downtime (DIN 40041 [1]). Consequently, the following requirements are needed for high availability: A long mean time between failures (MTBF) and a short mean down time (MDT). A long mean time between failures can be achieved by:

- Perfection: long-life highly reliable components (to avoid early failures)
- Tolerance: failure tolerance due to redundant structures
- Efficiency: complete utilisation of the usage reserves

A short mean down time can be achieved by:

- Fast and reliable failure detection and diagnosis (inspection during operation)
- Quick and perfect corrective maintenance (if possible during times of no production)
- Short maintenance times (if possible during operation, otherwise in times of no production)

Condition monitoring can make important contributions to ensure availability in several aspects. On the one hand condition monitoring leads to a faster and more reliable failure detection and diagnosis. Knowledge of the wear condition obtained by the monitoring process on the other hand enables complete utilisation of the usage reserves. The monitoring supplies in addition important data for the organisation of the maintenance by providing knowledge of the current wear condition and predicting its further trend. In this way, maintenance measures can be planned before the defect occurs. Spare parts and maintenance capacity for example can be provided in good time. Therefore, the monitoring process also supports a quick and perfect corrective maintenance. Figure 1 illustrates the correlation.

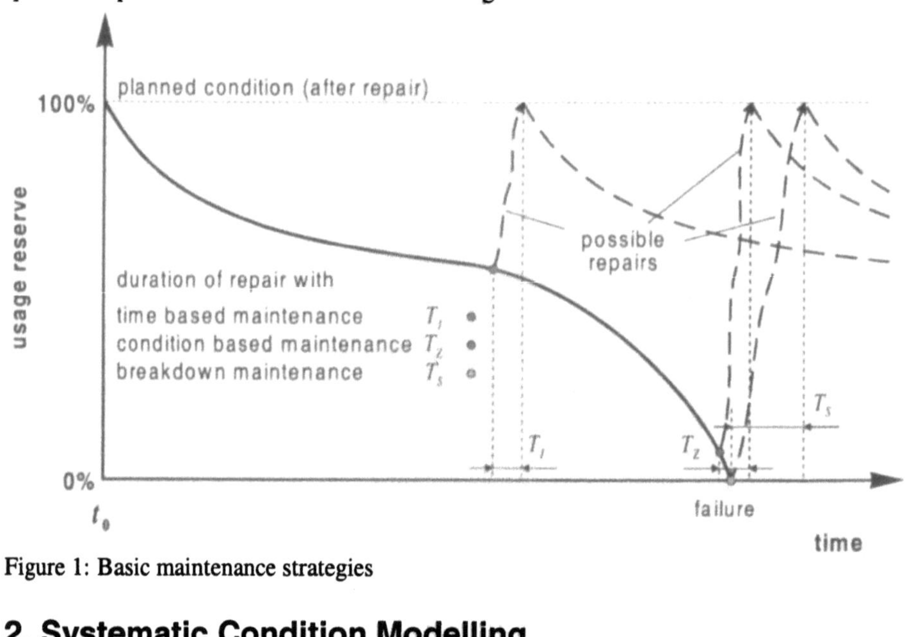

Figure 1: Basic maintenance strategies

2. Systematic Condition Modelling

It is necessary to have the production plants in an optimal conditional range to keep the output, production quality and the processing equipment on a high level. It is therefore important not only to consider the wear of the machines, as it is still mainly done within maintenance processes, but also the output of the production process as well as the quality of the fabricated product.

If the term of "condition" is extended according to this view, it can be seen that all measures which are needed to reach the aimed goals are deduced from current conditions. Wear conditions of components for example are used for optimal planning of maintenance measures. Necessary adjustment parameters which are needed to reach an aimed optimal condition as fast as possible can be deduced from the current operating condition. This indicates the importance of knowing the machine conditions to efficiently ensure availability.

Detailed researches of condition models show that their development follows a certain schema. The condition of a unit is according to DIN 40041 and DIN 55350-11 [2] the entirety of features of a unit at inspection time. Features are properties which can be used to detect or to distinguish units. A condition is always related to the regarded object. In general not all properties of an object are relevant. Conditions are therefore mostly determined under a certain aspect of inspection. Only the properties which influence further action are relevant. For decisions about maintenance schemes e.g. the wear condition, of the considered units is critical.

A condition can adopt different values. A wear condition for example can adopt the values "in working order", "damaged but in working order" or "not in working order". Conditions can be determined by suitable features which have to be fixed dependant on the aim of the condition determination.

The time needed for the determination of a condition can become problematic, since the condition always refers to a certain point in time per definition. The calculation of the condition however needs a certain time period. If the condition is not constant within this period, it is defined as "mean condition".

The first step for the development of condition models is to fix the targets.

Apart from the targets, the boundary conditions to be fulfilled by the condition model are another important precondition for systematic model development. Afterwards, the output quantities are determined. The possible input quantities of the model result from a detailed examination of the influence factors on the output quantities. Up to this point of the model development a linear procedure without loops is possible with thorough consideration.

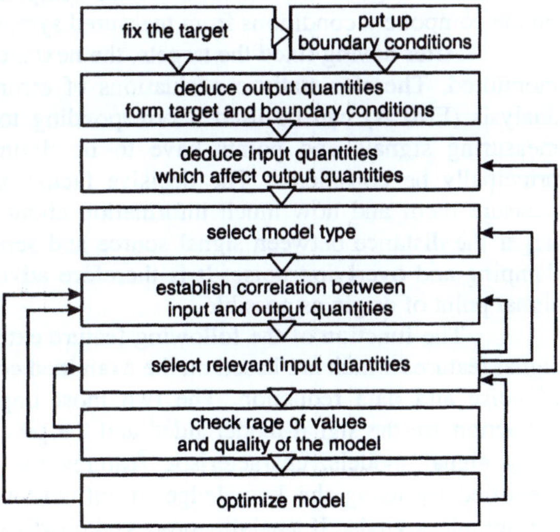

Figure 2: Universal approach to establish a condition model

The next step is to choose a suitable type of model. This depends on the type of information which are available about the identified input and output quantities. Most used model types are mathematical, rule based and neural network models.

The heart of the model building is to construct the correlations between the input quantities regarded as relevant and the output quantities deduced from the target, based on the chosen type of model. Detailed theoretical considerations or extensive series of tests are required identify the relations.

Correlations between input and output quantities being constructed, in a next step the input quantities determined at the beginning of the model development can be reduced to the relevant ones. Afterwards, the state model is checked for range of values and exactness.

The last step of the development is to optimise the constructed condition model in regard to the criteria robustness, closeness to reality and expenditure. In [3], the modelling of the operating condition of a complex assembly line is described in more detail. This article mainly concentrates on the modelling of wear condition.

3. Methods to Gain Knowledge about Wear Condition

For condition monitoring, methods are in demand which enable measuring the wear condition of components in an automatic, online and non-destructive way. It is seldom possible to measure the wear directly under this restrictions. There are basically two strategies. If a model is available which links stress with lifetime it is possible to measure the stress and calculate the remaining lifetime. Lifetime models have to be established by extensive wear experiments. Examples are the crack propagation concept or the L_{10} model for ball bearings and many other components. The other strategy is to establish so-called symptom models with allow conclusions on the component conditions from measured symptoms.

After having fixed the targets, the next step is to select the conditions to be monitored. Therefor statistic evaluations of errors or a failure mode and effects analysis (FMEA) can be used. Corresponding to the methods mentioned above, measuring signals and points have to be defined. All physical quantities can principally be considered. The decisive factor is how much effort is needed to measure them and how much information about the condition they contain. The larger the distance between signal source and sensor, the more is the influence of damping and overlying noise. It is therefore advisable to measure as close to the signal point of origin as possible.

The function of the following feature extraction is the amplification of the signal features which are based on the examined condition variation, the attenuation of noise and data reduction. The two most important used methods of feature extraction are the signal-model aided and the process-model aided methods. When using signal models, characteristic features for the wear condition have to be identified by using the knowledge of influences of the wear conditions on the measurable signals. If analytic process knowledge in the form of mathematical models does exist, it is often advantageous to use process-model aided methods for the feature extraction [4]. Condition variations are often reflected in variations of the normally not measurable process parameters, as for example stiffness, damping and friction coefficients, resistance or capacities. But these parameters can be determined with a parameter estimation method.

The causal (wear) conditions are inferred from the extracted features. They are compared with the symptoms of hypothetical errors to decide whether a certain condition variation exists. The condition-symptom relation is frequently determined in a heuristic experiment.

Finally a prognosis for further prediction of the condition may be added. In this way there is enough time to plan and initiate respecting measures timely before an outage.

4. Integration of the Condition Monitoring into a Computer aided Maintenance Assistant

Within the scope of the collaborative research centre, a prototype of a computer aided maintenance assistant has been developed and implemented, which assists the maintenance in all its phases [5]. Starting point was a thorough analysis of maintenance in several companies. The phase of maintenance was examined in regard to lacks and needs. All phases of the maintenance, starting with the failure detection about the diagnosis, the maintenance work planning, the control incl. planning all necessary resources (spare parts, tools, staff), the support of the maintenance workers at the actual repair up to the final documentation, are supported by the computer aided maintenance assistant. All necessary information are available to the maintenance worker in each phase.

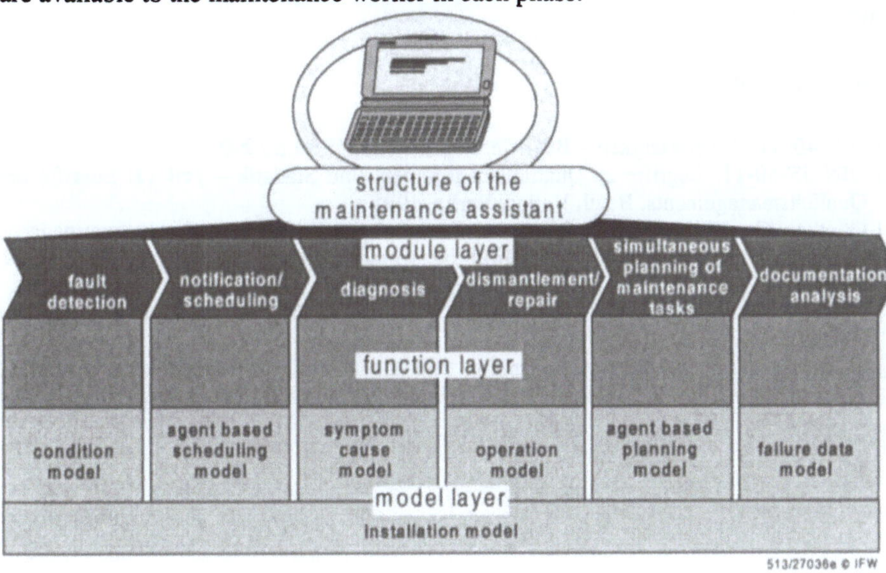

Figure3: Structure of the computer aided maintenance assistant

The installation structure is first modelled an programmed in the maintenance assistant. Lifetime models as well as the symptom-cause models described in the previous chapter can be build into the maintenance assistant for the detection of the remaining useful lifetime and error diagnosis. Critical components are preferably monitored with the help of the symptom-cause models and additional sensors. Less critical components are considered in the system with their lifetime models. The stress in the form of working cycles is determined via signals present in the machine control. Therefor the information about the remaining lifetime of all machine parts and components is always available. These data can be used for the planning of maintenance measures. When replacing a defect motor, it may for example be useful to exchange also further components whose estimated remaining lifetime is only short and which would otherwise cause an additional outage. The lifetime models and the documentation of actual system behaviour in the maintenance assistant enable the optimal planning of the time for inspections and maintenance and the optimising the stock keeping of spare parts.

5. Summary

The interdisciplinary collaborative research centre 384 at the University of Hannover develops procedures and methods to improve the availability of production systems. As part of the work the systematic modelling of conditions has been presented which can be used in different tasks to ensure availability. Methods for the determination of conditions of machine parts and components have been presented. These methods have been implemented into a computer aided maintenance assistant which has also been developed. Practical tests of the maintenance assistant in an industrial environment resulted in significantly reduced down times.

References

[1] DIN 40041: Zuverlässigkeit – Begriffe. Beuth Verlag, Berlin 1990

[2] DIN 55350-11: Begriffe zu Qualitätsmanagement und Statistik – Teil 11: Begriffe des Qualitätsmanagements. Beuth Verlag, Berlin, 1995

[3] Günther, G.; Köhrmann, C.; Niemeier, F.: Verfügbarkeit sichern durch systematische Zustandsmodellierung, ZwF (Zeitschrift für wirtschaftlichen Fabrikbetrieb) 94, Nr. 9, 1999, S. 482-485

[4] Natke, H. G.; Cempel, C.: Model-Aided Diagnosis of Mechanical Systems. Springer Verlag, 1997

[5] Tönshoff, H.K.; Seufzer, A.: Planning and Controlling of Maintenance Activities. Proceedings of the 2nd International Conference Design and Automation, USA, 1998, S. 234-238

AUTOMATION AND ROBOTICS

A Micro-controller Based Control System to Rejuvenate Old Robots

P.Brunn[1], C.Trigonis[2]

[1]Total Technology and [2]Mechanical Engineering Departments, UMIST

Abstract The paper describes the design, development and testing of a micro-controller and PC based control system which was used to repair and enhance an ASEA IRB6 welding robot in the authors' laboratory. The principles described can be applied to any similar age robot and provide a low cost route to revitalise any working robot hardware that is limited by an outdated control system.

1. Introduction

As part of a MPhil project [1] to investigate the integration of machine vision with robotic devices for use in manufacturing systems, it was decided to repair a non-functioning ASEA robot at UMIST. It had been previously investigated by another student, who found that it could not easily be brought back into use, however there was no obvious damage and the major elements of the devices still seemed to be in place and thus it was considered sensible to invest the time and effort to repair it.

2. The ASEA IRB6 Welding Robot

The ASEA IRB6, the original design of which dates from 1973, is a five axis robot, with three major axes and two wrist axes, all of which are electrically driven by d.c. servo motors. Initial attempts at fault finding established that the robot could not be started into any form of controlled motion because the start up and initial calibration procedures would not complete successfully. It was possible to gain access to the rear of the control cabinet without entry into the restricted workspace of the robot itself, so special connectors were developed to allow access to components and test points on the circuit boards of the original controller to investigate further. This established that each axis could be powered up and moved at a defined speed, but only under open loop control. The safety circuits were all working correctly. Within each axis servo system both the tachometer and the resolver indicating position were also working correctly. All the faults could thus be traced to the circuit boards containing the main control computer system. It became apparent that it would not be practical or possible to repair the main circuit boards since many of the components used in the original design were now obsolete and direct replacements were no longer available. It was doubtful if pin-compatible replacements would be available in more modern forms or if these replacements would be compatible with the timing requirements of the remaining original parts. The processor system (Intel 8008) used in the original controller was very slow and of very limited capability. Also it seemed likely that the control code for the processor, stored in 7Kb of PROM was corrupt and following a general enquiry the manufactures agents were unable to provide replacement PROMS. They were however most helpful in providing documentation on the system, [2] without which the project would have been much more difficult or even impossible within the time scale.

3. The Chosen Repair Philosophy

The majority of the parts of the robot were both in good condition and of fairly traditional design. The main control computer however was both obsolete in design and performance and very definitely non-functional. The program and data storage facilities were similarly obsolete. The chosen approach was therefore to salvage as much hardware as possible from the cabinet but to replace and enhance the entire control system, with the exception of the main power supplies, motor control circuits and safety interlocks. This decision required a further choice as to the structure of the new controller, in particular with regard to the relationship between the individual axis controllers and the overall control system. In the original design the 8008 micro-controller had been responsible for the management of all functions, from inverse kinematics calculations to axis speed setting and position comparison. It was thus both slow and limited in its capacity to interface with the outside world. partly because of its specialist operating system.

One alternative, was a ruggedised PC to carry out all the control functions and provide data storage and interfaces. Other single computer options were investigated but none had the price and power advantages of an Intel PC. The strengths of a PC based system are in its ability to provide overall control and system optimisation but the weaknesses are in the real time control of individual axes. It was therefore decided to use single chip micro-controllers to provide low level control and simple I/O for each axis and a conventional PC, with a standard parallel interface, to control and monitor the axis controllers. Additional circuitry would be required to provide the necessary "housekeeping" functions, such as the sinusoidal excitation signals for the resolvers. This solution meant that a modular axis controller board could be designed, built and tested and then duplicated for each axis and for spares. The wide variety and low cost of the many simple chip micro controllers currently available would fit within the limited budget of the project. The circuit boards for these controllers would be relatively simple and thus could be designed, made, populated and tested in house.

4. The Axis velocity and position control modules

The control philosophy chosen was to add intelligence to each axis controller to allow the PC time for overall system control and other functions within the cell. The PIC family of micro-controllers from Arizona Microchip was chosen for the axis controllers partly because of its impressive technical specification but also because both the Mechanical Engineering and Electrical and Electronic Engineering Departments at UMIST had considerable experience of their use. Other devices may have had better cost or performance specifications, but the experience already gained using the PIC family's Harvard architecture and Reduced Instruction Set (RISC) approach meant that development time would be kept to a minimum.

Six identical axis control modules were produced, one to act as a spare, based on the PIC 16C74A which used a 14 bit instruction bus and a separate 8 bit data bus.

The device has 33 I/O pins which can be configured as digital I/O ports or multiplexed with other internal modes of the micro-controller such as inputs to the onboard ADC. A separate DAC was used to provide an output analogue voltage for the velocity setting for each axis motor drive unit. The resolver feedback voltage was converted to a series of digital pulses for input to the micro-controller.

Each axis controller was designed to operate in three modes. The first, particularly suited to joint interpolated motion, required that the supervisory components sent to each axis controller a velocity and position value and a command to initiate its movement. The axis controller would then start to move the axis, using the drives inbuilt circuits to ramp up to the set speed, continue the motion at constant speed until close to the final position and then decelerate the axis by reducing the speed setting until the final position had been reached. The second mode of operation is very similar to the first except that the reduction in speed towards the end of the motion does not occur. This mode is ideally suited to controlled motion in a straight line and the use of 'via' points, typical of several second generation languages. The third mode allows the supervisory computer to set the speed voltage directly and control the axis open loop

All five axis controllers are normally synchronised to move together and return digital values of their current velocity and position data to the supervisory computer, while continuously checking the overall safety circuits. These are linked with the existing systems that monitor hardware performance, such as the motor drive thermal cut outs, axis travel limit switches etc. The axis control circuit boards were mounted in the existing control cabinet, on a common back-plane which also carried a new card that provided the excitor signals for the resolvers and other standard signals that were previously included in the processor boards of the original control computer. A vero-board back-plane, within a card cage, with sockets for each axis control board kept electrical noise to a minimum and allowed the use of nearly all the cables and connectors that made up the original system. As much as possible of the original hardware was reused, except the magnetic tape drive and the teach pendant since these functions could be more easily duplicated and enhanced in the GUI on the supervisory computer. It is interesting to note that while the design, development and testing of the new circuit boards were not particularly straightforward, it was carried out by a research student whose first degree was in Computation, with a supervisor whose main subject area was Mechanical Engineering. Thus while some training in electronics was essential it was not necessary to employ a specialist to design and test the circuits or to produce and populate the circuit boards that were used. It should therefore be equally possible for others who are primarily robot users rather than robot designers to undertake such an exercise to replace elements of a similar controller.

5. The Supervisory Computer

The system design includes considerable intelligence in each axis controller and therefore requires less continuous effort from the supervisory control computer. For playback control of simple taught programs the supervisory system simply needs to send sets of position and speed information for each axis and commands to initiate motion. If sensor input and tool and/or gripper control are required the supervisory system has simply to interface with the I/O channels of the robot and respond accordingly. Overlaid on these requirements is the need to monitor for alarm signals or incorrect position information. However, this does not need to be done in real time, since each axis micro-controller is also monitoring its own performance and the overall safety circuits. For the system described herein an IBM PC with a Pentium processor (running at 166 MHz) and 32 Mbytes of RAM was used for overall control. The system had ISA bus slots and thus could contain a relatively low cost 48 bit programmable digital interface card based on two 8255 PIOs. This computer had more than sufficient speed and processing power for this application but was available and served as an ideal platform not only to control the robot but also upon which to develop the software for both robot control and the vision related aspects of the overall research project. It was also the most cost effective way to provide programme and data storage and standard communications with other devices. It also meant that a standard operating system could be used (Windows 95) which had advantages both in terms of cost and operator training. In this particular project the development language and programming environment chosen was Borland's Delphi since both research student and supervisor had experience in its use. The object oriented nature of Delphi allowed quite complex and imaginative Graphical User Interfaces (GUI's) to be developed. Delphi also has the advantage that assembler code can be integrated with the high level language where necessary for direct access to the hardware.[3]

A GUI that duplicated the original teach pendant and some of the front panel switches was developed to allow the robot to be started, moved to its calibration position and then controlled to move through a series of taught points while accepting and sending signals from a range of I/O devices. The geometry necessary to perform forward and inverse kinematic calculations was encoded in a linked software unit. It was thus possible to write simple programs that moved the robot between predefined points in cartesian space and subsequently to specify the path along which the robot should move. Additional, application specific features could easily have been added to the system controller at this stage.

The main objective of the particular project for which the ASEA robot had been repaired was to integrate a vision system with robot control. Consequently some considerable time was spent selecting and configuring a vision system and writing software that carried out image analysis and then selected control strategies prior to converting these to actual robot motion commands. The PC used was quite capable of using the majority of its processing capabilities to control the image grabbing and analysis system while also generating all the necessary control codes to move the robot in response to visual events in the workspace. Given the compute intensive nature of vision analysis it is fairly safe to assume that the supervisory PC would therefore be capable of handling almost any task in a real manufacturing cell or similar situation.

6. System Testing and Use

System testing was split into two main areas namely hardware testing and software validation after the detection and removal of all code 'bugs'. Obviously, there was only a very limited amount of hardware testing that could be accomplished without the supervisory computer and none that related to the system without the individual axis controllers. Standard test instruments and a variable voltage source had been used during the early stages of the project to establish that the motor drives, resolvers and tachometers were all working correctly. A simple GUI, that provided the learning and motion commands for taught points, was used to test repeatability. Using a pair of LVDTs and a simple reference frame [4] the repeatability of the new system at a range of points in the workspace was shown to be as good as that expected from the controller it replaced, based on the manufacturers specification. However it was not possible to establish the accuracy of the new controller because the exact link parameters were unknown since they were not provided in the extensive literature supplied for the robot. It was possible to estimate these, based on measurements taken from the outside of the assembled robot, but not with sufficient accuracy to make it worthwhile to then use this data as part of a programme of position accuracy testing. The estimates could however be used to demonstrate that cartesian motion could be specified and controlled by a series of inverse kinematics calculations.

In this project it was considered more important to demonstrate that the control program in the supervisory computer could keep up with or outpace the robot and that communication with the axis controllers was not an excessive overhead. A simple demonstration was written to use the video system, integrated with the robot control software in the PC, to recognise when a new object, in this case a small disk, appeared in the work area, which was a white table top, and then to move the robot to point directly at that object wherever it landed. Recordings of processor usage, from the Windows system monitor utility, showed that even a relatively slow Pentium processor had spare capacity when tracking moving objects, establishing when they had stopped moving and then controlling the robot to point to each new disk as it arrived in the workspace.

This MPhil project had a planned time scale of only one year. In practice this extended to nearly eighteen months in total on all aspects of the work. Seven man months were spent on the design, programming, manufacture and testing of the individual axis controllers. A further four man months were spent in programming the supervisory computer and subsequent testing of robot motion control and the various systems. The remainder of the time was required for the initial project feasibility study, selection of vision hardware and software and final thesis preparation. The axis control hardware development was delayed at one stage because of a lack of facilities to manufacture printer circuit boards to the required quality within the department and earlier by problems with interpreting the resolver output to the correct resolution. With the experience gained so far, and equally importantly, knowing that it has already been done successfully, it should be possible to repeat the exercise of upgrading the controller of a robot of a similar age within a much shorter timescale.

7. Conclusions

The approach used in this project, has been shown to be successful and relatively inexpensive. It employs a distributed processing system, whereby a supervisory computer manages overall control by co-ordinating the collective operation of a set of intelligent axis controllers. The supervisory PC is responsible for performing all the computationally intensive, but not time–critical tasks, such as inverse kinematic calculations and trajectory generation. The axis controllers have the form of micro-controller based single board computers whose main function is to carry out the time-critical tasks of regulating the position and motion of their respective axes. The position control loop is based on a client–server architecture where the systems supervisor, the client, requests this service of the axis control units, the servers, via a series of low frequency messages in order to execute the assignment appointed by the user. This offers a range of benefits over the more traditional architecture. Technically the new system makes more effective use of the supervisory computers resources. Communication is easier, the system is easily expandable and upgradeable and can be safer in operation since additional features can be added such as collision avoidance either through software or hardware. Commercially the new controller has cost advantages, insofar as it makes use of low cost micro-controller and PC hardware. It uses standard interfaces and communication protocols outside the robot cell and it uses standard data storage and retrieval devices and operates with an industry standard computer operating system. Code to control the system can be written in any suitable high level language, e.g. C++ or in this case Delphi, by a programmer who does not have to be an expert in machine level code. Most significantly this approach allows the owner of an existing robot to update and enhance the control system of that robot, at relatively low cost while retaining and re-using the expensive hardware which even after many years of productive use is probably still in reasonably good working order. A further consequence of this work was that all involved obtained a much better understanding of the operation and control of that robot. It was possible to build relatively simple diagnostics into the software of each axis micro-controller and it would have been even easier to build extensive diagnostics into the supervisory programs had that been considered a significant element of the project. Thus there are considerable long term benefits that could be gained from a project of this type by any organisation which uses even a limited number of robots in terms of staff training, system maintenance and ongoing product and process improvement.

References

1. The selection of Hardware and Software Control Element for Robot Dynamics
   C. Trigonis - MPhil Thesis 1998 UMIST
2. Programming and Operating ASEA's Industrial Robots
   ASEA (Now ABB) Documentation 1975
3. Thurrot P. Delphi 3 Superbible
   Waite Group Press 1997
4. Low Cost Robot Performance Assessment
   P. Brunn, F. Hildago
   International Workshop on Advanced Robotics and Intelligent Machines
   University of Salford, March 1997

Structural identification and calibration of kinematic robot models by genetic search

Jens- Uwe Dolinsky, I.D.Jenkinson, G.J. Colquhoun

Liverpool John Moores University, School of Engineering,
Byrom Street, Liverpool L3 3AF
Email: {engjdoli, I.D.Jenkinson, G.J.Colquhoun}@livjm.ac.uk

Abstract: Accurate robot modelling is of great importance to the application of enhanced robot programming tools such as Offline Programming systems. This paper describes a prototype of an automated kinematic modelling environment, which is primarily based on evolutionary computation. A genetic algorithm herein attempts to find an optimal model structure of the forward kinematic of an industrial robot based on measurements reflecting individual characteristics. Finally it will be reported on results obtained from simulation experiments.

1. Introduction

Modern robot programming methods such as Offline Programming require accurate robot models sufficiently capturing the robot kinematics. Since every robot, even within the same series, has individual deviations of kinematic properties which is due to e.g. manufacturing tolerances and tear, the controller model has to be updated e.g. by calibrating its kinematic parameters with measurement data taken from that specific robot. In Offline Programming robot programs are developed and validated within a simulated environment assuming an accurate match of robot and workcell models with their physical counterparts.

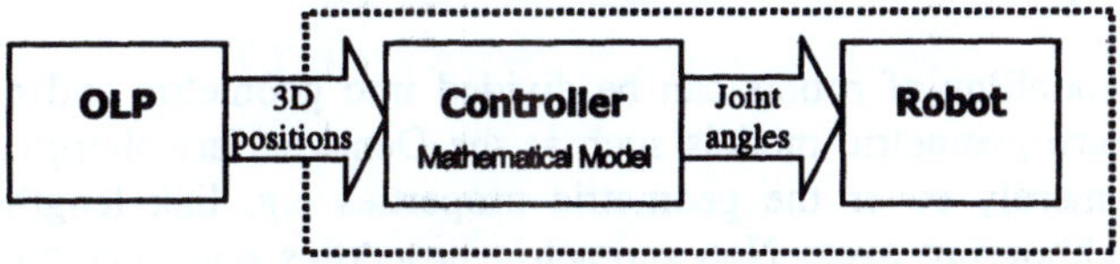

Figure 1: Relations OLP ↔ physical robot

Since information about the robot controller model is usually not available (the black box in Figure 1) and in order to create portable programs the common approach of an Offline Programming System (OLP) such as Workspace [5] to achieve higher absolute accuracy of a particular robot is to employ path error compensation. A calibrated kinematic model acts as a filter, which alters target positions in the robot program to direct the robot to the desired targets.

Kinematic Calibration in robotics is a well-established field of research that has delivered several models and calibration methods addressing special types of robots and numerical stability issues of the parameter identification procedure [2]. Some approaches utilise black box models based on artificial neural networks (ANN) with little or no relation of the connection weights used to physical robot parameter. In general the approach has been to determine a fixed model structure and to subsequently fit its parameters using measurement data.

This paper presents a new automated general hybrid modelling method to generate a forward kinematic robot model based primarily on evolutionary computation and gradient search. The genetic algorithm proposed has the task to create an appropriate model structure whose parameter will be numerically identified by subsequent gradient search. Unlike other hybrid search heuristics, e.g. genetic algorithms on artificial neural networks, the objective of this approach is not to derive a black box model. Instead knowledge of the nominal kinematic model, e.g. the serial links and their order, will be integrated and preserved.

2. Genetic modelling of robot kinematics

Genetic Algorithms (GA) are search procedures inspired by concepts of natural evolution. In principle a population of potential solutions (termed individuals) is searched or explored for the best performing or fittest individuals. To keep and exploit their characteristics those individuals will be selected for reproduction or recombination in further generations eventually undergoing mutational changes. In that way over a number of generations the algorithm attempts to gradually breed highly fit solutions to a problem. An introduction to GA's can be found in [1]. Genetic Algorithms have been found to be useful for solving many different types of problems across different disciplines. This independence from the actual problem domain and the underlying concepts, which are relatively easy to implement in computer programs, has contributed to their success.

2.1 Model representation

Kinematic Modelling of robots can be divided into geometric and non-geometric modelling. Pure geometric models such as the Denavit-Hartenberg (DH) (see e.g. [3]) model, merely cover the geometric properties e.g. link lengths and angles between neighbouring axes. Non-geometric properties e.g. gear compliance and gear backlash however contribute considerably to the accuracy of a robot and need to be taken into account (see e.g. [4]).

In this approach a genetic algorithm attempts to model the influence of non-geometric properties by inserting additional co-ordinate transformations into a DH model. The fact that forward kinematic models can be built from sequences of elementary homogenous transformations makes them an appealing representation for genetic algorithms permitting effective application of genetic operations such as crossover or mutation. The principle idea is to view the entire kinematic robot model as an ordered sequential concatenation of fundamental homogenous transformations (DH- transformations with nominal parameter values provided by the robot manufacturer) together with additional co-ordinate transformations (such as translations along and rotations about the x, y and z-axis) inserted by the genetic algorithm. Those sequences make up the kinematic genome[1] on which the genetic algorithm operates. Figure 2 shows the genome of the 6-link PUMA 762 robot used in our experiments.

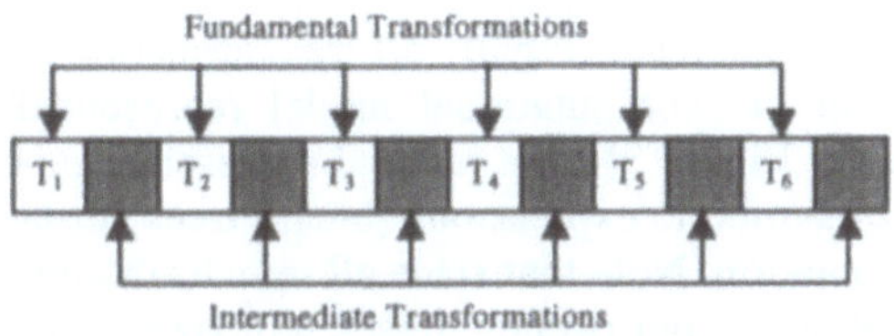

Figure 2: Kinematic Genome

The initial population of genomes is seeded with each genome containing the fundamental transformations (T_1 to T_6) in the order according to the links of the robot. The genome of the kinematic model in Figure 2 consists of 12 gene sections, 6 of which (the fundamental transformations) must not be altered. This creates a primary schema, which expresses the initial knowledge (that of the provided serial link design and specifications) which is to be preserved. Random numbers of different additional transformations will be placed between these fundamental transformations by the genetic algorithm. The inserted transformations can be single translations, rotations or composed transformations (e.g. building blocks).

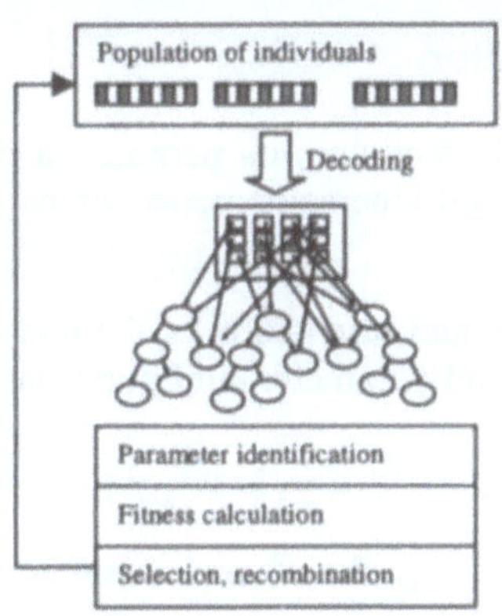

Figure 3: Genetic Algorithm

Thus the genetic algorithm explores the space of sequential homogenous transformations for the optimal kinematic design of a particular robot. It evaluates, selects and recombines the transformation strings (genomes) within a population of a given size.

[1] Genome or chromosome (string of genes) in this context means the entire information required to construct a kinematic model

2.2 Decoding and evaluation

In order to evaluate a population of models, all genomes have to be decoded and prepared for evaluation. For each genome the corresponding homogenous transformation matrix (phenotype) is created. Each gene in the section of intermediate transformations represents one single transformation for which a parameter needs to be generated. Finally all transformation matrices are multiplied symbolically to form the computational model represented by its homogenous transformation matrix (Figure 3). The internal representation of this transformation matrix is an array of symbolic expressions (binary trees) instantiating the equations for position and orientation. Note that since all transformations are homogenous the genetic algorithm, although performing drastic structure changes, cannot violate the orthonormality constraint of the rotation matrices. Also, the application of an analogous crossover technique prevents the genetic algorithm from changing the order of the fundamental transformations, which preserves the predefined serial link design. This means only genes within the same gene section can be exchanged.

It is possible during the course of evolution that the algorithm creates gene sections with several consecutive equal transformations e.g. 2 translations along the x- axis. This would obviously cause a parameter redundancy, which is avoided immediately during decoding by ignoring those multiple equal transformations. However, those redundancies are kept in the genome in order to preserve a diversity of genetic material for further generations.

2.3 Parameter identification

Once the model has been instantiated, its parameters need to be identified using measurement data. The forward kinematic model can be written as:

$$y = f(\theta;\beta), \tag{2.1}$$

where f returns the position and orientation of the end-effector tool depending on the given joint angle set θ and parameter configuration β. Identification is carried out by minimising:

$$\sum_{i=1}^{n} \| y_i - f(\theta_i;\beta) \|^2 \tag{2.2}$$

with subject to β, where y_i is the ith measurement and n is the number of measurements. Since the model equations are nonlinear in rotational parameters the identification has to be carried out iteratively e.g. by nonlinear least squares. For this the model equations have to linearised by first order Taylor expansion around the current parameter estimate. The functional (2.2) to be minimised can then be rewritten as:

$$\| \Delta y - C \Delta\beta \|^2 \quad \text{with } C = \begin{pmatrix} C_1 \\ \vdots \\ C_n \end{pmatrix} \text{ and } \Delta y = \begin{pmatrix} \Delta y_1 \\ \vdots \\ \Delta y_n \end{pmatrix} \tag{2.3}$$

subject to $\Delta\beta$ and **C** being the Jacobian of f approximated by finite differences at the current parameter estimate. To solve this linear minimisation problem for the parameter update (Gauss-Newton update) vector yields:

$$\Delta p = (C^T C)^{-1} C^T \Delta y \tag{2.4}$$

Hence the parameter values are iteratively obtained by:

$$p_{k+1} = p_k + \Delta p_k \tag{2.5}$$

The Moore-Penrose generalised inverse $(C^T C)^{-1} C^T$ is not formed explicitly in order to avoid numerical instabilities e.g. due to round-off errors. Instead orthogonal decomposition is applied to C in (2.3) by Householder Transformations. However $C^T C$ may become singular and hence not invertible. In that case the algorithm uses the Levenberg-Marquardt update $\Delta p = (\delta I + C^T C)^{-1} C^T \Delta y$ with δ being a positive scalar constant to be determined and I being the identity matrix.

2.4 Fitness calculation

The fitness value of an identified model is the residual error (2.2) computed on a second set of measurements. Thus the fitness is a function of the evolved kinematic structure and the identified parameters. The genetic algorithm exploits the Baldwin effect e.g. the learned information (identified parameter values) only affects the fitness function and is not backcoded or retained in the genetic description of an individual model.

3. Experiment

For our experiments we implemented a steady state GA[2]. Individuals (kinematic models) are selected from the population by Tournament Selection (see [1]). This type of selection was chosen because it is easy to implement, and by setting an appropriate tournament size it allows convenient adjustment of the selective pressure and thus the convergence property of the genetic algorithm. To create a new model, two tournaments were run and the best performing model of each was chosen to be parents for the new model.
The initial population was seeded with genomes with each intermediate gene section containing up to 5 random elementary transformations. Fundamental transformations were initialised with the nominal DH parameters of the PUMA 762 robot. The parameters of elementary transformations in the intermediate gene sections were initialised with the value of 0.01 (translational and rotational parameter). The number of measurements used for identification and evaluation was 50. The probability of applying crossover was 0.8 and mutation 0.3. The evaluation of 50 generations (population size was 15) took about 20 min on a PC (with a 500 Mhz Intel Celeron processor) running Linux. The results of a typical run of the genetic algorithm are shown in Figure 4.

---

[2] Offspring created by the GA replaces bad or worst performing individual in a population.

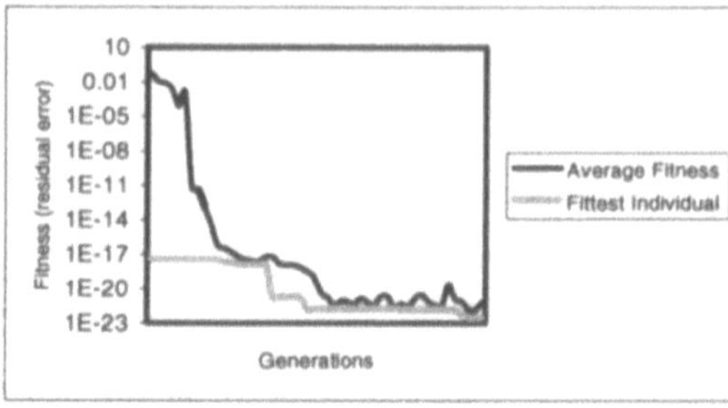

Figure 4: Results obtained from a GA run over 50 generations (population size: 15)

As the evolution progresses, fitter individuals appear in the population, the fitness of individuals improved almost gradually. However, because of the limited population size of 15 individuals the GA settles relatively quickly. To keep the evolution going i.e. to explore other genome configurations additional mutation was applied to some individuals in generations, the spikes in the average fitness curve (Figure 4) illustrate this effect. This method proved useful because this perturbed genetic material contributed to slightly fitter individuals.

4. Conclusion

The genetic algorithm presented appeared to be well capable of fitting a kinematic robot model (structure and parameter) to measured data. Despite compromises made with the choice of the genetic algorithm parameter (e.g. number of generations and population size) due to the high computational complexity, it performed robustly and delivered parametric models, which performed even better on our measured data than an ordinary DH model calibrated with these data. The algorithm works with simple mechanisms (e.g. estimation of parameter values), which need to be refined for further support of the results. Further work planned in this context is to reduce the number of parameters to be estimated by applying variable projection, which requires an initial estimate for nonlinear parameters only.

References

[1] Beasley, D., D.R. Bull and R.R. Martin, (1993), *An Overview Of Genetic Algorithms*. University Computing, Part 1 "Fundamentals", 15(2) 58-69, Part 2 "Research Topics", 15(4) 170-181.

[2] Bernhardt, R. and S.L. Albright, editors. (1993). *Robot Calibration*. Chapman & Hall, London.

[3] Hollerbach, J.M., (1999), *Introduction to Robotics*, Lecture Notes CS 5310/6310 & ME 5220/6220, Chapter 4, University of Utah.

[4] Whitney, D.E., Lozinski, C.A. and Rourke, J.M., (1986), *Industrial Robot Forward Calibration Method and Results*, ASME Journal of Dynamic Systems, Measurement and Control, Vol. 108, No.1.

[5] Workspace 4, (1998), User Manual Robot Simulations Ltd, Newcastle Upon Tyne, UK.

Structured Programming Languages for Industrial Robots with Backwards Execution and Editing Running Programs

Günther Martinez-Dreyer, Gerhard Schreck, Cornelius Willnow
Fraunhofer IPK, Pascalstr. 8-9, D-10 587 Berlin, Germany
e-mail: cornelius.willnow@ipk.fhg.de

Keywords: Industrial Robots, Programming Languages, Man Machine Interfaces, On-line Programming, Off-line Programming, Simulation, Robot Applications

Abstract: The article presents a method for implementing high level, structured programming languages for industrial robots that realizes the robot specific requirements of editing programs under execution and executing programs backwards. Additionally, the method allows the user to choose among text based, icon based and path based program display.

1. Introduction

Since precisely predicting robot behavior mentally is hard, teach-in programming includes a considerable share of try and error, especially in tight work space. Backwards program execution is therefor an important means for undoing robot motions. After some steps of backwards execution, the program has to be modified in its current context. This concerns the current program state, as well as the state of the robot's environment. Therefor the program may be stopped but has to be still under execution when it is modified.

For programming the first generations of robot controllers, interpreter systems for assembler style, line oriented languages have been provided. Although being simple, these interpreters fulfilled the robot specific requirements of backwards program execution and editing programs under execution. With increasing complexity of robot tasks and robot programs, structured programming became important. First controllers with such high level languages appeared on the market in the end of the 1980ies. However, the compiler technology used for realization considerably reduced the abilities of program backwards execution and, especially, of editing program under execution.

Within the EC sponsored ESPRIT 22 646 project AMIRA (Advanced Man-Machine Interfaces for Robot System Applications), a program representation has been developed that combines high-level language features like structured programming with the teach-in requirements for backwards execution and editing running programs. In addition to these fundamental requirements, the developed system supports different presentations of programs to users: They may choose among conventional textual programming, as it is preferred by programming experts, and icon based flow diagram style programming [1] as it is often preferred by expert workers. Furthermore, a geometrical, path based presentation as it is known from off-line programming systems can be provided what is a step towards closing the gap between on-line and off-line programming.

2. System Overview

The core of the language system is formed by an Internal Program Representation that is used for storing programs in the robot controller's working storage. Its structure is the key to fulfilling the above requirements and will be described in more detail in the following sections [2]. Around this Internal Program Representation, the language related processes and devices are grouped, as illustrated in figure 1.

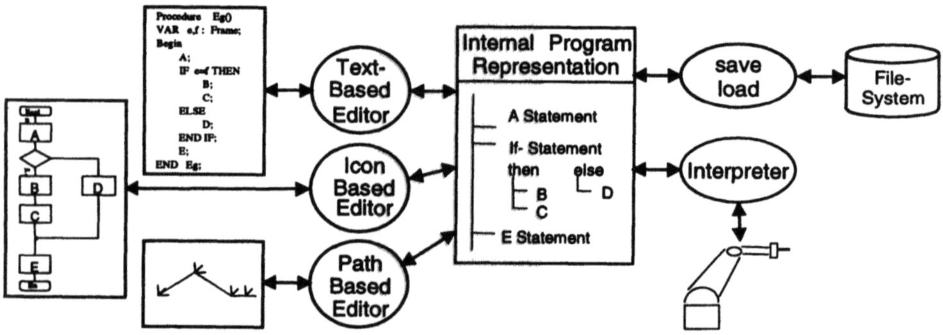

Figure 1: Program representations

The three different methods of program display are illustrated on the left by short example programs. For each of them an editor with two major tasks is provided: For displaying the Internal Program Representation the method for accessing the Internal Program Representation is uniform for all three display modes, while the mapping to the user representation and the process of displaying are specific to each user presentation. When modifying the Internal Program Representation, the user commands modifications by pressing buttons on the teach pendant. For convenience, the assignment of buttons to functionality should be uniform for all three user presentations, while the displayed labels and menus should conform to the display mode, i.e. text based menus for text based programming and icon based menus for icon based programming. The respective modification of the Internal Program Representation is uniform for all three display modes.

After program modification, the display is updated by mapping the Internal Program Representation to the user presentation again. For textual programming, this mechanism gives the user the impression that he is working with a text source code that is later on somehow translated and executed. However, actually the Internal Program Representation is the source and the displayed text is the translation.

For executing programs, the Internal Program Representation is processed by an interpreter. The interpreter affects the robot, as illustrated at the right of figure 1. Furthermore it interacts with I/Os and other devices of the robot controller. Since variable values are stored directly in the Internal Program Representation, the interpreter also modifies it up to a certain degree. The program statements, however, are in the current implementation not modified by the interpreter.

For archiving programs, the Internal Program Representation can be transferred to a file format and retrieved from a file format. For debugging purposes, the current implementation uses a text based file representation. Therefor loading programs is actually a compilation.

3. Internal Program Representation on Parse Tree Level

The backbone of the Internal Program Representation is given by the grammatical structure of structured programming languages, i.e. the hierarchical decomposition of a program by the grammar's production rules as it is used for parse trees of compilers. Such parse tree level representations enable a reasonable reconstruction of user representations as illustrated in figure 2.

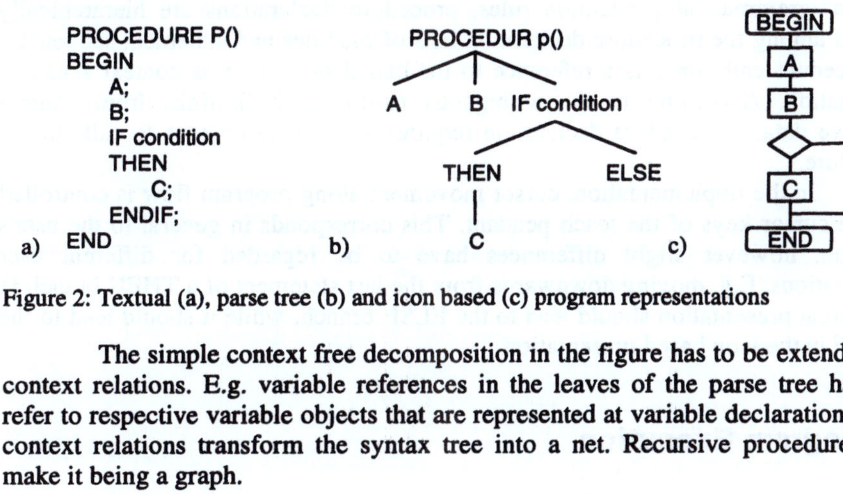

Figure 2: Textual (a), parse tree (b) and icon based (c) program representations

The simple context free decomposition in the figure has to be extended for context relations. E.g. variable references in the leaves of the parse tree have to refer to respective variable objects that are represented at variable declarations. The context relations transform the syntax tree into a net. Recursive procedure calls make it being a graph.

4. Program Editing

A parse tree level representation, equipped with a suitable systems of pointers as illustrated in figure 3, can be efficiently edited.

Sequences of statements are represented by double linked lists what enables moving an editing pointer, or cursor, forwards and backwards. Flow control statements like IF-THEN-ELSE and WHILE are organized as single statements that refer to the statement sequences that they control. Vice versa, each statement sequence is linked to the controlling flow statement.

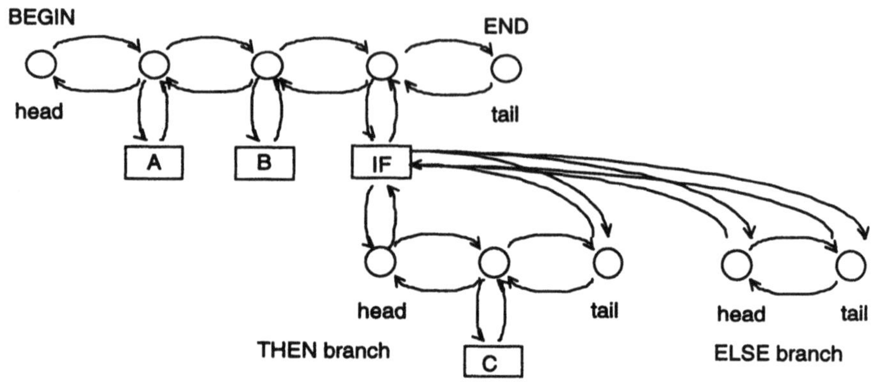

Figure 3: Pointer structure of the parse tree representation of the program in figure 2

An example for more complex relations are procedures. From the point of view of grammatical production rules, procedure declarations are hierarchically located among the procedure declaration lists of modules and statement sequences. A procedure call contains a reference to the called procedure as context sensitive information. According to the ambiguous nature of back links from context sensitive data, a procedure declaration requires a list of references to calls to the procedure.

In the implementation, cursor movement along program flow is controlled via the cursor keys of the teach pendant. This corresponds in general to the user's intuition, however slight differences have to be regarded for different user presentations. E.g. moving downwards from the last statement of a THEN branch in the textual presentation should lead to the ELSE branch, while it should lead to the ENDIF in the icon based presentation.

5. Program Execution

For forwards program execution, the already existing pointer structure for moving the editing cursor can be used for moving a program execution pointer or program counter.

For backwards execution, the same pointer structure could be used, however, this is often ambiguous. E.g. when backwards execution arrives at an ENDIF, it is not clear if the THEN or the ELSE branch has to be executed. This information can not be reliably obtained from the branching condition, since its value might have been modified when executing the IF-THEN-ELSE statement. For loops this is always true, since the condition for leaving the loop is always true after its execution.

Furthermore, backwards execution requires that variable assignments are undone. The previous variable value can, however, not be reconstructed from the data of the Internal Program representation. Therefor, a different mechanism is used, that consists of an undo-list. The undo-list stores modifications in a queue like manner. When the program is executed backwards, items in the undo-list are taken from it in a first-in last-out manner. This technique is used not only for variable values, but also for program counter values. The latter enables to uniquely retrace the behavior of flow control statements.

Since the Internal Program Representation can be executed as well as edited, both can take place in parallel. However, it has to be observed that backwards execution should refer to the unmodified program, since this was what has been actually executed, while the modifications have not been executed yet. This forms no principle obstacle for added or modified statements since the undo-list does not refer to the additions or modifications. However, deletions have to be handled carefully. For maintaining backwards execution consistent with the program, deleted statements could be removed from the undo-list. This, however, may result in differences among forwards and backwards execution. For enabling proper backwards execution, deleted statements have to remain in the undo-list. Therefore, they are removed from the pointer structure of the Internal Program Representation and will be actually deleted with the last reference to them in the undo-list. This method may, however, be irritating for the user. When such zombies create backwards execution program behavior may be not reflected by the displayed program.

6. Implementation

A test implementation of the above described principles has been performed in the ESPRIT 22 646 project AMIRA (Advanced Man-Machine Interfaces for Robot System Applications). Figure 4 shows the icon based user presentation in the

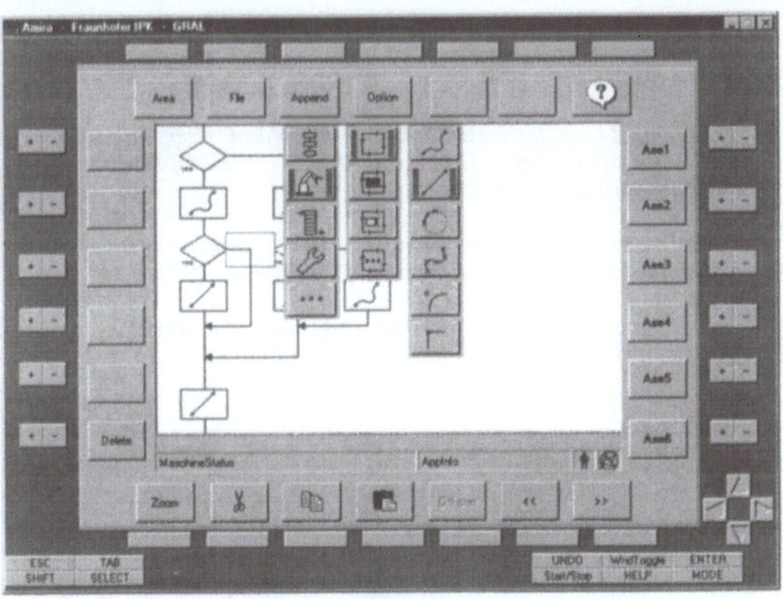

Figure 4: Icon based editor in the AMIRA reference teach pendent

AMIRA reference teach pendent [3] with a pull-down menu for inserting robot statements.

Currently the implementation of the Internal Program Representation covers program flow structures, including procedure calls, as well as robot commands and enables editing programs under execution. For verifying the different user presentations, the icon based editor and saving/loading programs to a textual presentations are realized. For executing motion commands, a simple interpolator and a VRML-based visualization are used.

7. Summary and Outlook

A method for combining structured high-level robot languages with the teach-in requirements of backwards program execution and editing running programs has been designed and feasibility has been verified by implementation. The method also enables different modes of program display, and the implementation showed the feasibility of textual and icon based user presentations.

A next step consists in realizing also a path based user presentation as it is known from off-line programming. Analyses performed during the AMIRA project point out that a considerable gap exists among the user's mental presentation of a robot task, that normally corresponds to some spatial imagination of robot motions, and the task presentation given by on-line programming devices today. Icon based programming adds a further presentation towards visualization, a path based representation, however, would fit the user's mental model best.

References

[1]ISO/DIS 15 187, *Manipulating Industrial Robots - Graphical User Interfaces for Programming and Operation of Robot (GUI-R)*, Frankfurt/Main 1998, (http://www.irf.de/auto_grp/guir/gui-r.htm)

[2]Günther Martinez-Dreyer, *Konzeption einer zur Laufzeit editierbaren Programmrepräsentation für Industrie-Roboter*, Diplome Thesis, Berlin, 1999

[3]*Style Guide for Robot System Applications*, AMIRA (Advanced Man Machine Interface for Robot system Applications) ESPRIT Project 22 646, AMIRA Consortium, Brussels 1999

Neural Network Adaptive Control For Servo Systems with Nonlinear Disturbances

X. Ouyang[1], C. Morgan[2], C. Nwagboso[3]
Transportation and Automotive Systems Research Centre
School of Engineering and the Built Environment, University of Wolverhampton,
Priorslee, Telford TF9 2NT
Email: [1]e9718841@wlv.ac.uk, [2]me1957@wlv.ac.uk, [3]in6079@wlv.ac.uk

Abstract: In this paper a new model-reference adaptive control method for a specific nonlinear servo system is proposed by using neural network. The controller consists of a conventional PID control law and a neural network inverse model compensator. It is shown that the nonlinear dynamics of the real plant can be well compensated and excellent dynamic performance can be obtained. Computer simulation results show that the proposed model-reference adaptive controller is very effective for servo systems with nonlinear disturbances.

1. Introduction

Conventional controllers, such as PI or PID controllers, are widely used in systems control due to their simplicity, robustness and low-cost. However, the effectiveness of such controllers will deteriorate when applied to nonlinear systems since linear control methods rely on the key assumption of small range operation and the nonlinearities in the system cannot be properly compensated by using a conventional methods [1]. Artificial neural network (ANN) technologies have the potential to provide effective methods for solving the problems of dynamical system identification and control due to their capabilities of learning, nonlinear mapping and parallel distributed processing [2]. ANNs have been widely used to detect faults in control systems, solve nonlinear optimisation and nonlinear system identification and control [3-9].

In this paper a new neural network model-reference adaptive control system (NNMRACS) for servo system with nonlinear disturbances is proposed. The control system consists of a conventional PID controller and a neural network inverse dynamics compensator. With the neural network inverse dynamics compensator, the dynamics of the servo system with nonlinear disturbances can be well compensated and excellent dynamic adaptive performance can be obtained. Experimental simulation results show that the proposed NNMRACS is very effective for servo systems with nonlinear disturbances.

2. Dynamics of Servo Systems with Nonlinear Disturbance

The design specifications for control systems normally include several time-response indices for a specified input command (usually a step signal) as well as a desired steady-state accuracy. The PID controller is widely use in servo control. The controller provides a proportional term, an integral term, and a derivative term. Integral action is able to eliminate long persistent errors. Proportional action responds immediately to changes in deviation but is insensitive to how rapidly the deviation is changing. Derivative action adds an extra term which is proportional to the rate of change of deviation and counteracts the de-stabilising effects of the integrator. The closed-loop digitised control structure of dc servo system (DDCS) with PID controller is shown in Figure 2.1 where d_1, d_2, c_1, and c_2 are corresponding difference equation coefficients.

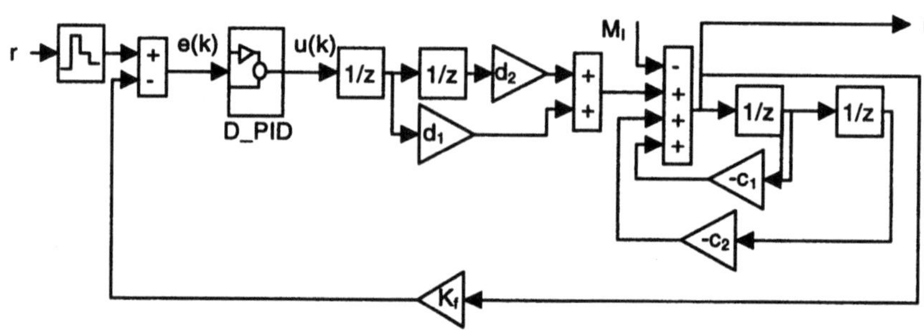

Figure 2.1 Closed-Loop Control Structure of DDCS

The parameters of an industrial DCS are as follows

$$L_a = 0.19H, \;\; R_a = 3.8\Omega, \;\; K_a = 0.12Nm/A,$$
$$K_e = 0.12Vs/rad, \;\; K_f = 0.1, \;\; J = 6.6667 \times 10^{-4} Kgm^2$$

and the sampling period is set to $T = 0.04s$. Under the condition that the load disturbance is equal to zero, the dynamic performance of the closed-loop DDCS is excellent. However, performance of such control systems will deteriorate if there is nonlinear load disturbance since the nonlinearities in the system cannot be properly compensated by using conventional controller [1].

In this paper, we will consider a parabolic nonlinear load disturbance. In practice, load disturbances caused by centrifugal pumps, fans, and propellers give rise to this type of disturbance. In these cases, the load torque varies directly with the square of the velocity

$$M_l = \lambda n^2 \tag{2.1}$$

For the servo system under consideration, the corresponding discrete-time dynamic equation is

$$n_{k+1} = (-c_1 n_k + m_1 n_k^2) + (-c_2 n_{k-1} + m_2 n_{k-1}^2) + d_1 u_k + d_2 u_{k-1} \tag{2.2}$$

Figure 2.2 shows the simulation result for the unit step response of the closed-loop DDCS with $\lambda = 3 \times 10^{-6}$. Obviously, the nonlinear load disturbance causes the system performance to deteriorate significantly. In order to properly compensate for the effect of nonlinear load disturbance, methods other rather than conventional linear controllers should be found since linear control methods rely on the key assumption of small-range linear operation.

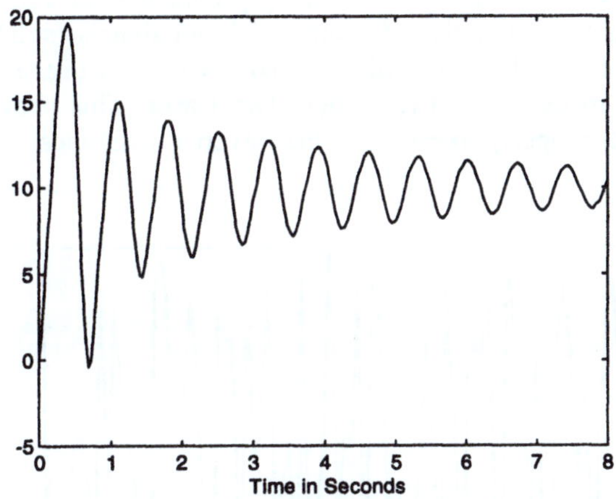

Figure 2.2 Step Response of DDCSwith Parabolic Load Disturbance

3. Neural Network Identification

A three-layer neural network will be trained to represent the inverse dynamics of the nonlinear servo system under consideration. Inverse models of dynamical system play a crucial role in a range of control structures [10]. Conceptually the simplest approach is *direct inverse modelling* as shown schematically in Figure 3.1. Here, a synthetic training signal, u, is introduced to both the plant and the neural network. The plant output is also used as input to the neural network (the series-parallel structure). The network output is compared with the training signal (the system input) and this error is used to train the neural network. This structure will tend to force the network to represent the inverse of the plant.

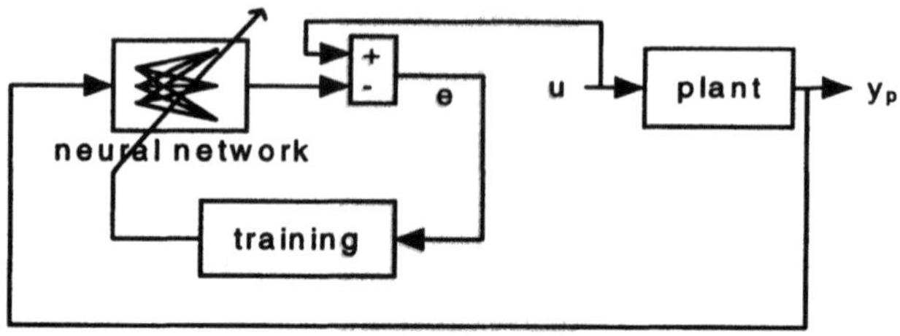

Figure 3.1 Direct Inverse Modelling by Neural Networks

One hundred pairs of stochastic input-output data are generated to train the inverse neural network. It is important to note that the function of the inverse neural network is to give correct input signals after being provided with a set of output signals. Figure 3.2 gives the comparison for the real input signals (solid line) and the outputs of the inverse neural network (dashdotted line). From Fig. 3.2 it is seen that there is very small difference between these two traces. This shows that the inverse neural network properly represents the inverse dynamics of the real nonlinear servo system.

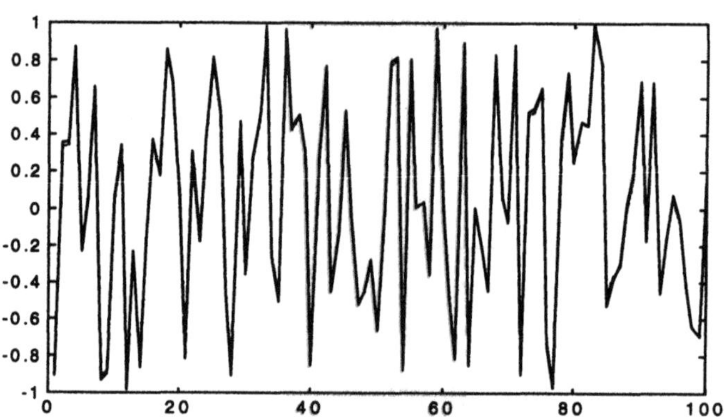

Figure 3.2 Real Input vs. Output of Neural Network Inverse Model

4. Neural Network Model-Reference Adaptive Control

The basic structure of the proposed neural network model-reference adaptive control system (NNMRACS) is shown in Figure 4.1. The controller consists of a conventional PID controller and a neural network inverse dynamics compensator. In Figure 4.1, e is the error between the outputs of the reference model and that of the real plant; u_1 is the control action produced by the PID controller; u_2 is the compensating control produced by the neural network inverse model. The inverse model will automatically adjust the control action according to the output difference and produce a compensating control action that will bring the error to zero.

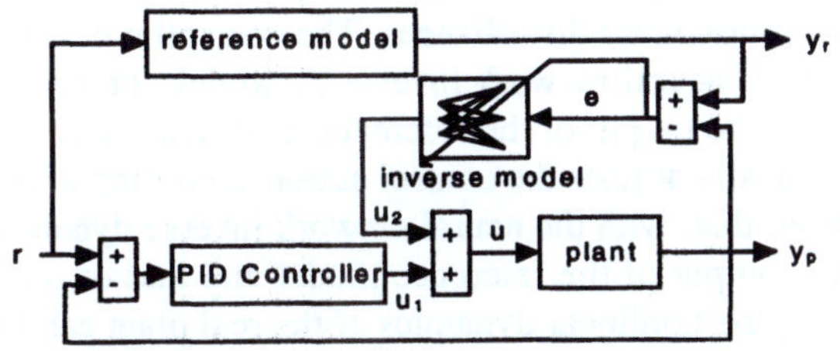

Figure 4.1 Structure of NNMRACS

The unity step responses of both the reference model and the controlled nonlinear servo system are given in Figure 4.2 where the solid line represents the response of the reference model, and the dotted line represents the response of the controlled nonlinear servo system. As comparison, the step response for the nonlinear servo system without neural network inverse model compensation is also given by the dashdotted line. The simulation results show that with the neural network inverse model adaptive compensator, the nonlinear dynamics of the real plant has been well compensated and excellent dynamic adaptive performance obtained.

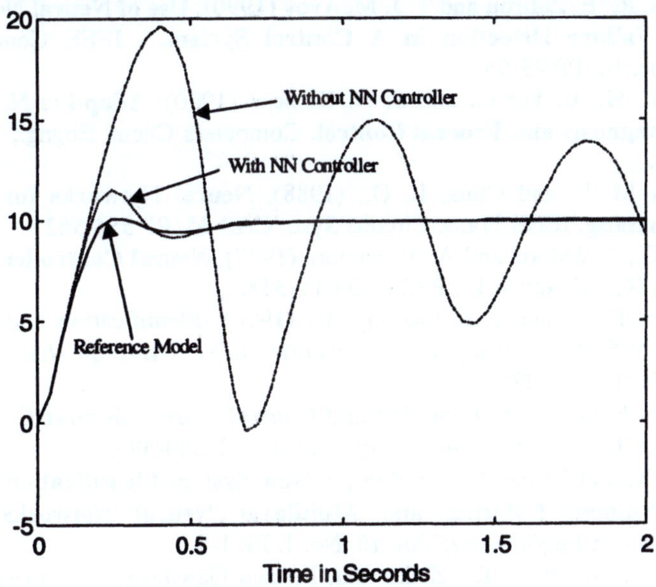

Figure 4.2 Step Responses

5. Conclusion

In this paper a new neural network model-reference adaptive control is proposed for servo systems with nonlinear disturbances. The controller consists of a conventional PID controller and a neural network inverse dynamics compensator, which uses the difference between the output of the reference model and that of the real plant as its input and automatically adjusts the control action according to the output difference. It has been verified that, with the neural network inverse dynamics compensator, the error between the output of the reference model and that of the real plant will tend to zero. Therefore, the nonlinear dynamics of the real plant can be well compensated and excellent dynamic adaptive performance can be obtained. Computer simulation results show that the proposed NNMRACS is very effective for nonlinear servo system identification and control.

References

1. Slotine, J.-J. E. and Li, W. (1991), **Applied Nonlinear Control**, Prentice-Hall, Inc.
2. Ishii, K., Ura, T., and Fujii, T. (1994), **A Feed Forward Neural Network for Identification and Adaptive Control of Autonomous Underwater Vehicles**, Proceedings of 1994 IEEE International Conference on Neural Networks, Vol. 5, PP 3216-3221.
3. Naidu, S. R., E. Zafirou and T. J. McAvoy (1990), **Use of Neural Networks for Sensor Failure Detection in A Control System**, , IEEE Control System Magazine, 10, PP 49-55.
4. Ungar, L. H., B. Powell and S. N. Kamens (1990), **Adaptive Networks for Fault Diagnosis and Process Control**, Computers Chem. Engng., Vol. 14, PP 561-572.
5. Kennedy M. P. and Chua, L. O., (1988), **Neural Networks for Nonlinear Programming**, IEEE Trans. Circuits Syst., CAS-35, PP 554-562.
6. Psaltis, D., A. Sideris and A. Yamamura (1987), **Neural Controllers**, IEEE Int. Conf. on Neural Networks, PP IV-551-IV-558.
7. Narendra, K. S. and Parthasarathy, K. (1990), Identification and Control of Dynamical Systems Using Neural Networks, *IEEE Trans. on Neural Networks*, vol. 1, No. 1, PP 4-27.
8. Ydstie, B. E. (1990), **Forecasting and Control Using Adaptive Connectionist Networks**, Computers chem. Engng., Vol. 14, PP 583-599.
9. Lou, K. N. and Perez, R. A. (1996), **A New System Identification Technique Using Kalman Filtering and Multilayer Neural Networks**, *Artificial Intelligence in Engineering*, Vol. 10, No. 1, PP 1-8.
10. Hunt, K. J., Sbarbaro, D., Zbikowski, R., and Gawthrop, P. J. (1992), **Neural Networks for Control Systems – A Survey**, Automatica, Vol. 28, No. 6, pp. 1083-1112.

CO-ORDINATE
MEASURING MACHINES

A Multiple Sensor, Geometrically Corrected Co-ordinate Measuring Machine System

Allan D. Spence[1], Robert V. Fleisig[2] and Nael A. Barakat[3]

McMaster University, Mechanical Engineering, Hamilton, Ontario, Canada
[1]adspence@mcmaster.ca, [2]robert@mcmaster.ca , [3]barakat@mcmaster.ca
www.mech.mcmaster.ca

Abstract: This paper describes a modular and open Co-ordinate Measuring Machine architecture that merges quasi-static thermally compensated geometric error compensation with multiple data collection sensors. The complete system consists of thermocouples attached to the CMM and part, a motion control computer, trigger and analogue touch probes, and a non-contact laser digitiser. To simplify error correction algorithm implementation, motion control is performed directly on a Pentium based microcomputer. Linear scale quadrature signals are used to synchronise the sensor data with the CMM position, and ethernet provides data interconnection amongst the components. Use of the laser digitiser reflected light intensity with calibrated artefacts is proposed as a promising and practical self calibration method.

1. Introduction

Today's accelerated manufacturing pace demands rapid part inspection, analysis, and process feedback. Rather than bring parts to a 20 degree Celsius environmentally controlled room, industry seeks a practical inspection solution that can be located nearer to the associated production site. Because of thermal expansion at elevated temperatures, geometric error compensation of the Co-ordinate Measuring Machine (CMM) volume is essential. To maximise data collection rates, high point density analogue touch probes and non-contact digitisers are preferred over touch trigger probes. Output from these sensors must be corrected using the geometric error map to achieve the true part dimensional co-ordinates. Since, for a particular industrial need, equipment from several vendors will likely be required, a modular and open architecture system is most suitable.

This paper describes such a CMM system. Subsequent sections describe the CMM systems components and geometric error compensation strategy. Use of the laser digitiser reflected light intensity and calibrated artefacts for self calibration is proposed. A final summary concludes the paper.

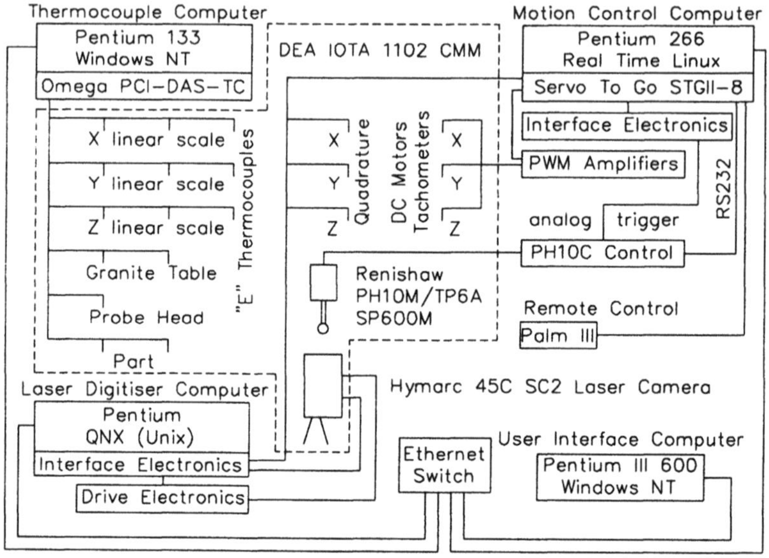

Figure 1. CMM system architecture.

2. CMM System Components

Components of the system are shown in Figure 1. The Co-ordinate Measuring Machine in the laboratory is a DEA IOTA model 1102, with a standard bridge Z on Y on X on granite table configuration. The working volume is 900 mm (X) by 550 mm (Y) by 600 mm (Z). The X axis has dual rack and pinion drives and scales, but the slave pinion is free to slide axially. That is, the overall machine has no kinematic redundancy that would bend the structure due to thermal expansion. The linear scales are interpolated to a 1 μm resolution. Three type "E" Chromega-Constantan thermocouples are attached to each axis scale. Additional thermocouples are attached to the top and bottom of the granite table, to the probe head, and to the part being inspected. The Pentium 133 MHz based thermocouple computer includes an Omega [1] PCI-DAS-TC interface card, including ice bath Cold Junction Compensation (CJC). Open thermocouple circuits are detected using 20 MΩ pull up resistors. An uninterruptible power supply is used to prevent data loss during short duration electrical outages. The thermocouple data is time stamped and recorded. The CMM motion control and user interface computers obtain error compensation model coefficients on demand via private switched ethernet circuits.

Motion control is provided using a Pentium 266 MHz based computer. A Servo To Go STGII-8 [2] motion interface card provides scale quadrature counting (including probe switch latching), digital control system interval timing, analogue to digital (A/D) tachogenerator input and digital to analogue (D/A) output to the Pulse Width Modulated (PWM) DC motor servo amplifiers. This approach was chosen over the usual Digital Signal Processor (DSP) based motion control method

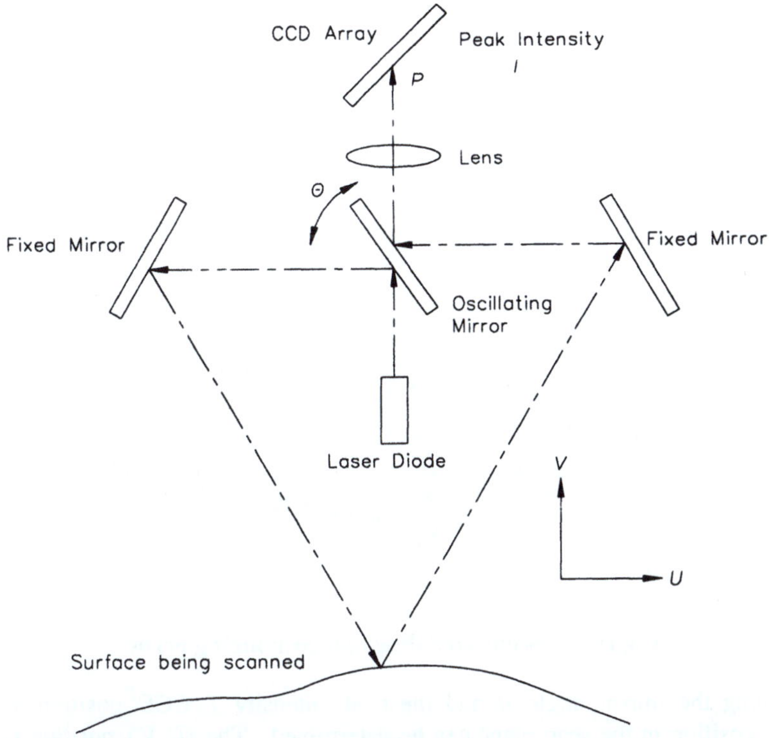

Figure 2. Hymarc laser camera operating principle [10]

to simplify the Continuous Compensation to the Feedback (CCF) [3] error compensation algorithm.

The hand held remote control [4] includes a 3 axis potentiometer, a Peripheral Interface Controller (PIC) based A/D to RS232 serial interface (Ontrak [5] model AD101), and an Emergency Stop (servo amplifier disable) switch. A Palm III [6] provides a Digital ReadOut (DRO) display, and user input facility for choosing axis speed at full joystick deflection, etc. As opposed to CMM controllers that provide joystick commanded motion using a direct analogue input to the servo amplifiers, this implementation feeds the joystick deflection through the geometric error compensation algorithm and the position control loop. Hence, when an axis is "locked", an improved geometric axis lock, as opposed to a servo axis lock, is achieved. The motion controller software is written using the RealTime extensions [7] to the Linux kernel [8].

For touch probing, the CMM is equipped with a Renishaw [9] PHC10 controller, PH10M articulating probe head, a TP6A touch trigger probe, and an SP600M analogue probe with AC2 A/D interface card. Non-contact inspection is performed using a Hymarc Hyscan 45C SC2 [10] laser digitizer. This device is controlled using a separate QNX [11] operating system based Pentium computer. The laser camera consists of a laser diode light source, a mirror that oscillates at 3.3 Hz, and a Charge Coupled Device (CCD) light detector array (Figure 2). By

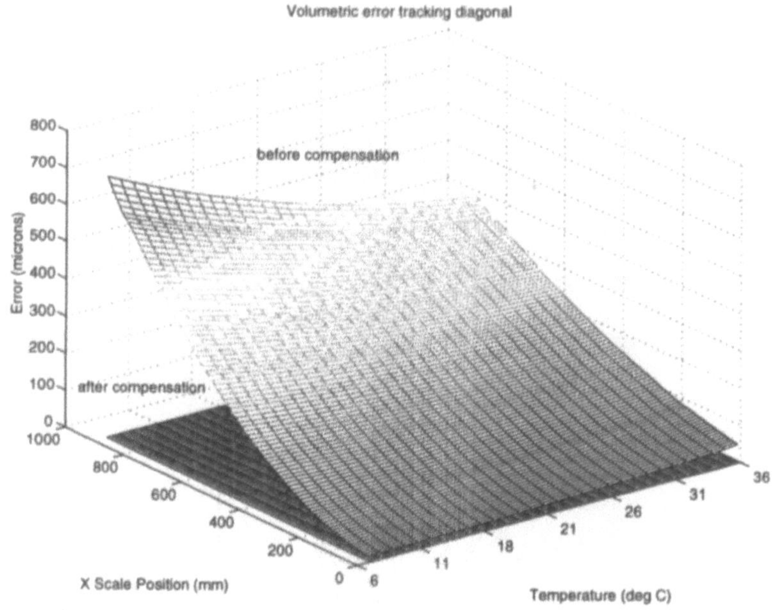

Figure 3. Simulated diagonal contouring error

combining the mirror angle θ and the peak intensity I CCD position P, the (U,V) position in the scan plane can be determined. The (U,V) position is then combined with the CMM axis scale positions to obtain the (X,Y,Z) part surface position. This approach provides the necessary synchronisation without the need to specifically design to the motion controller architecture, as would be required using the Next Generation Inspection Systems (NGIS) II interface [12].

The final component of the CMM system is the user interface computer. This provides an interface to the remaining systems, and performs the bulk of the data processing. Hymarc supplied software, with custom McMaster geometric extensions, is used for real time data acquisition.

3. Geometric Compensation

3.1 Calibration

The McMaster CMM was enclosed in an isolated room to facilitate the research. Room temperature was varied, in 1 deg C steps, from 20 to 27 deg C using electric resistance heaters and fans. At each step, a laser interferometer was used to measure the length, straightness, pitch, and yaw errors. A double ball bar, placed along the 3-D machine diagonals and measured using the touch trigger probe, was used to calculate squareness errors. Using regression analysis, the linear scales coefficient of thermal expansion was determined to be $\alpha_{\text{scales}} = 9.5\ \mu\text{m/m/deg C}$.

The roll errors remained near zero regardless of temperature. There was a constant Y with respect to X squareness error of $6\ \mu m / m$. This is due to the difficulty in equally spacing the air bearings and bridge magnets with respect to the granite table mounted guides. At the highest room temperature and position $X = 900$ mm, an X axis pitch error of $-3\ \mu m(Z) / m(X)$ was observed. This was caused because of the 1 deg C / m thermal gradient from the bottom to the top of the CMM table, and is in approximate agreement with the second order model $Z(X) = -0.5\alpha_{\text{granite}} X^2 \nabla T_Z$, where $\alpha_{\text{granite}} = 7\ \mu m$ and $\nabla T_Z = 1$ deg C / m. With the short probe tips used, the elevated probe head temperature did not result in a measurable error. Using this calibration data, a graph of the simulated contouring error from the CMM home position diagonally to the opposite corner, both before and after compensation, was prepared (Figure 3). Additional experimental results are available in [13,14].

3.2 Real time compensation

As stated earlier, the Continuous Compensation to the Feedback (CCF) [3] error compensation algorithm is used. The scale thermocouples are used to correct for linear thermal expansion, and the granite table thermocouples are used to correct for the X axis pitch error. Data from the thermocouple computer is transmitted on demand, via ethernet, to both the motion control computer and to the user interface computer. In the motion control computer, the updated thermocouple data is stored in a lookup table. A RealTime Linux kernel extension is used to compute the geometrically corrected target at the position loop closing frequency of 1250 Hz. This approach is used for all motions, as well as for correction of the actual measured touch trigger and analogue probe data.

Thermal geometric compensation for the Hymarc system is applied via the user interface computer. First, the camera is calibrated using a special inclined plane target to establish the relationship between the mirror angle and CCD peak intensity I position (θ, P), and the scan plane co-ordinates (U, V). Research [15] indicates that best accuracy occurs when the part surface normal is tilted 25 degrees about the U axis. During active point collection, the QNX based control computer transmits (U, V) and CMM axis scale positions via ethernet to the user interface computer. Data from the thermocouple interface computer is used to update the compensation model coefficients, and the laser digitiser points are corrected to yield the final (X, Y, Z) co-ordinates. To test the implementation, a calibrated aluminium metre rule, with dark 1 mm graduations, was located at different orientations within the CMM volume. The minimum peak intensity I was used to determine the centre of the graduation mark. Observed results [16] verify that the thermal geometric correction procedure yielded improved results to within the $25\ \mu m$ laser camera uncertainty. Future research will concentrate of reducing the camera uncertainty, and using low coefficient of thermal expansion (e.g. Zerodur® [17]) artefacts for self calibration.

4. Summary

This paper has presented a modular, open CMM architecture that includes thermal geometric compensation and multiple data collection sensors. Ethernet is used for data communication. Synchronisation is obtained using the scale quadrature signals. Use of the laser digitiser reflected light intensity for self calibration is a proposed future work.

Acknowledgements

The authors gratefully acknowledge financial and in-kind support from Omni-Tech CMM Services (Canada) Inc., Hymarc 3D Vision Systems, Ltd., and Materials and Manufacturing Ontario.

References

[1] Omega Engineering, Inc., Laval, PQ, Canada, www.omega.ca.
[2] Servo To Go, Inc., Redmond, WA, USA, www.servotogo.com.
[3] Sartori, S., 1995, "Geometric Error Measurement and Compensation of Machines", *Annals of the CIRP*, Vol. 44/2, pp. 599-609.
[4] Fisher, A.D., Fleisig, R.V., and Spence, A.D., 1998, "A Distributed Object Architecture CMM Interface Pendant", *IFIP PROLAMAT '98*, Trento, Italy.
[5] Ontrak Control Systems, Inc., Sudbury, ON, Canada, www.ontrak.net.
[6] Palm, Inc., Santa Clara, CA, USA, www.palm.com.
[7] Yodaiken, V., *RealTime Linux*, Department of Computer Science, New Mexico Institute of Technology, USA, www.rtlinux.org.
[8] Linus Torvalds, Linux Kernel Archives, www.kernel.org.
[9] Renishaw plc, Gloucestershire, UK, www.renishaw.com.
[10] Hymarc 3D Vision Systems, Ltd., *Hyscan 3-D Laser Digitizing System - Operator's Manual*, Nepean, ON, Canada, www.hymarc.com.
[11] QNX Software Systems, Inc., Kanata, ON, Canada, www.qnx.com.
[12] Rippey, W.G., Michaloski, J.L., Herman, M., and Szabo, S., 1998, *NGIS SIM Specification*, NISTIR 6116, National Institute of Standards and Technology, Gaithersburg, MD, USA.
[13] Barakat, N.A., Elbestawi, M.A., and Spence, A.D., 2000, "Kinematic and Geometric Error Compensation of a Coordinate Measuring Machine", *Int. J. Mach. Tools Manufact.*, v. 40, pp. 833-850.
[14] Barakat, N.A., Elbestawi, M.A., and Spence, A.D., "Adaptive Compensation for Errors in an Intrinsic Machine", *Int. J. Mach. Tools Manufact.*, submitted for review.
[15] Liu, Y., Feng, H.Y., and Xi, F., 1999, "Error Analysis of a High-speed Laser Scanner Digitizing System", 1999, *Proc. ACMC Annual Meeting*, London, ON, Canada.
[16] Spence, A.D. and Chang, D.W., 2000, "Laser Digitizer Based Curve Detection on Sculptured Surfaces", *Proc. ACMC Annual Meeting*, Niagara-on-the-Lake, ON, Canada.
[17] Schott Glass Technologies, Duryea, PA, USA, www.schottglasstech.com.

Thermal Compensation for a CMM Based on Interferometer Measurements and Testing its Applicability with a Thermal Stable Artefact.

Jean-Pierre Kruth, Paul Vanherck, and Christophe Van den Bergh

Katholieke Universiteit Leuven,
Division PMA, Celestijnenlaan 300B, 3001 Leuven, Belgium
Jean-Pierre.Kruth@mech.kuleuven.ac.be

Abstract. The use of CMMs outside a strictly controlled environment necessitates the management of the temperature dependent geometry of the machines. The paper presents a thermal error compensation based on physical relations between temperature and deformation, rather then on an empirical model linking inputs and outputs.

Measurements on a thermal stable artefact verify the applicability of the model. The model reduces the uncertainty of length measurements at various temperature levels to the level of the stochastic errors at 20 °C.

keywords: CMM accuracy, thermal behaviour, software correction

1 Introduction

Traditionally coordinate measuring machines (CMM) are located in temperature controlled rooms. The growing demand for quality control on the workshop floor forces the manufacturers to come with solutions to deal with the environmental issues.

This paper deals with measurement errors caused by non-standard temperature conditions. The focus lies on near steady state temperature distributions within the range of 17 to 25 °C. The goal was to give an alternative for the empirical black-box models, like neural networks or regression techniques, commonly used to deal with thermal problems [1,2,4]. The presented model is based on physical machine behaviour. Such models are less obscure than the black box approaches, thus more easily adapted for a commercial implementation by industry. Extrapolation with this type of models is not a step in the dark. Special attention was given to the possibility of extending the developed techniques to other machines of the same type, with limited efforts.

2 The test set up

The subject of this research is a small sized, numerically controlled CMM. It has a L-shaped bridge configuration (see fig. 1). The machine structure has a very homogeneous construction. The guides, the slides as well as the base structure are made out of aluminium. A granite worktable is placed in the base, but can easily be removed. The main machine properties can be summarised by:

- work volume: $500 * 400 * 300$ mm^3
- $u_3 = 3 + 3,5 * \frac{L}{1000}$ µm at 20 ± 1 °C with L in mm

The CMM controller is equipped with a geometrical compensation that corrects the values read out by the scales for the geometrical imperfections in the CMM. A test set up is build up to efficiently monitor the thermal

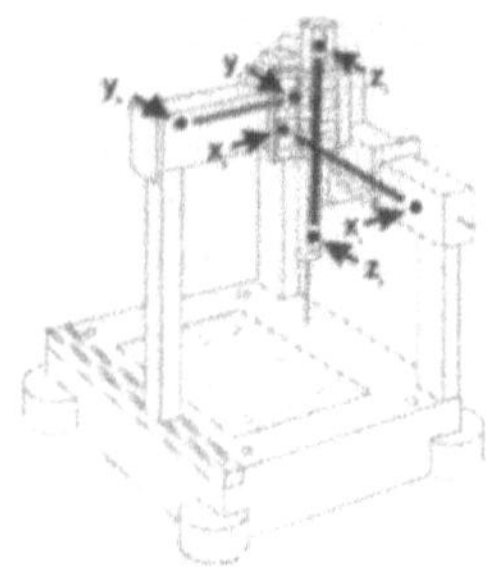

Fig. 1. Temperature sensor placement on the CMM.

deformations [3]. Temperatures are recorded with a set of negative temperature coefficient (NTC) thermistors. Shielding the sensors for environment influences ensures that only the temperature of the CMM structure is measured.

The main issue in thermal deformation research is the way the machine structure responds to a thermal load. Practically the displacements of the measurement probe — the end of the Z-axis — are examined with a laser interferometer that is equipped with temperature influence correction. During the verification phase the machine performs actual measurements on an artefact.

3 Temperature dependent error compensation

The thermal compensation is based on different positioning accuracy measurements. During this data collection phase the CMM's internal geometrical correction was disabled. Different sets, i.e. at different locations in the CMM's work volume, were recorded for each axis at three different temperature levels (18, 21, 25 °C).

These data are used to calculate from an uncompensated set of coordinates (X_U, Y_U, Z_U) and the temperature distribution of the machine, the correct values (X_C, Y_C, Z_C) axis per axis. The system uses a total of six temperature sensors, all placed at the scale ends (fig. 1). The different steps will be discussed for the Z-axis by means of fig. 2.

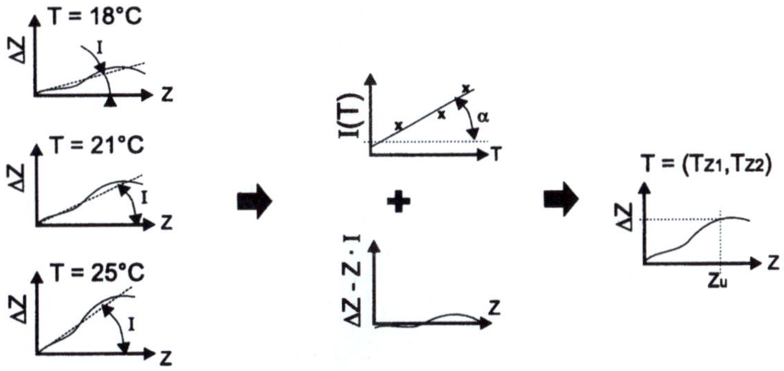

Fig. 2. Compensation approach for the Z-axis.

A set of positioning accuracy measurements is recorded along the Z-axis near the centre of the CMM work table. The set includes curves at different temperatures. In a first step, the least-square-best-fit-line through each curve is calculated. The difference between each measured curve and the least-square line can be determined. The obtained difference values are identical on the micrometer level within the temperature range under consideration and thus considered to be temperature independent.

The expansion coefficient α links the inclination (I) of each best-fit line and the corresponding temperature value T. This latter relation allows to determine I at any temperature. The incremental expansion ΔZ along the scale can be calculated when a linear temperature gradient over the scale is assumed. Based on the measured α and the temperatures ($T_{z1, z2}$) at the outer edges of the Z-scales, the correction curve corresponding with the current situations can be determined. The correct value Z_C can then be calculated.

The described procedure can be repeated for the Y-axis, as well as for the X-axis. For these coordinate directions a spatial interpolation is added at the end of the procedure. For the Y-axis two data sets measured at two different Z-values are required (different ABBE offsets to the scale). For these two ΔY is determined for the T = (T_{y1}, T_{y2}). Interpolation is done between these two, based on Z_u. In the case of the X-axis, three data sets are required and spatial interpolation is done based on (Y_u, Z_u).

4 Verification of the thermal compensation approach

To verify the applicability of the thermal compensation, measurements were done on a thermal stable artefact (see fig. 3). For these tests a hole plate made out of Zerodur was used. This material has a very low coefficient of expansion, $|\alpha| < 0{,}05$ µm/m·°C.

The standard geometrical correction is compared with the full correction (geometrical correction complemented with the developed thermal compensation). The artefact was measured under different near steady-state environmental conditions, the temperature levels varied from 17,5 °C to 25 °C. Based on the recorded coordinates different relative distances — lengths — and angles can be defined. The evaluation of the model is done based on these items.

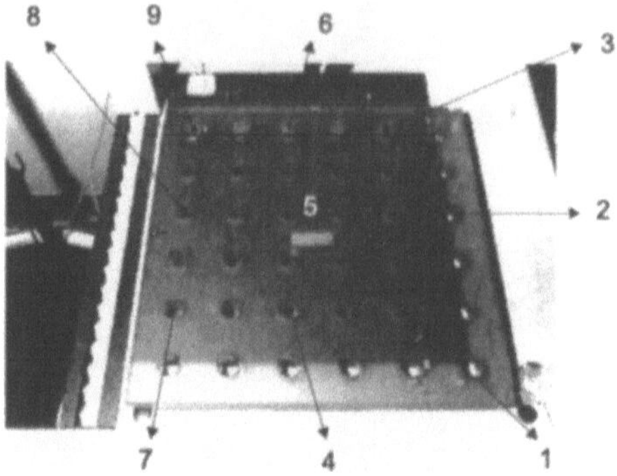

Fig. 3. Identification of measured points on the holeplate.

4.1 Correction while measuring lengths

Table 1 summarises the length measurement results. Different lengths can be defined between the measured 9 points. The first column of table 1 identifies the different lengths by their end points (see fig. 3), for example (4-6) indicates the length defined by point 4 and 6. For the evaluation, the standard deviations (Std) on the measured lengths are compared. The lengths measured at different temperature levels are treated as one set. Because the standard correction does not use the temperature information the values at the various conditions will be different. This leads to a high Std value, and as a consequence to a high uncertainty.

Table 1 indicates that without the knowledge of the temperature — thus without temperature control or thermal compensation — the uncertainty of a measured length is relatively large. The Std is smaller for lengths measured parallel (//) with the machine axes than for diagonal measurements (><). The first is a one-dimensional error (u_1), the second is a two-dimensional value (u_2).

The thermal correction leads to a significant reduction in the Std. For situations involving uniform temperature distributions, the uncertainty interval — which depends on the Std value — reduces by more than 80%. The temperature dependent measurement error is eliminated, and the uncertainty for the temperature range in focus is reduced to the stochastic errors at 20 °C.

The values recorded at different near steady-state spatial gradients were also combined and treated as one set. Again all measurements were processed with both the geometrical correction and the thermal correction. The temperature of the room was set at 20 °C. Halogen and conventional filament lamps were used as heat sources to create various spatial gradients.

The fact that the Std values with the standard correction are much lower then in the test with uniform temperatures is due to the overall room temper-

Table 1. Length measurements under non-standard temperature conditions.

	Uniform temperatures				Spatial gradients			
	Standard		Full correction		Standard		Full correction	
	Mean (mm)	Std (μm)	Mean (mm)	Std (μm)	Mean (mm)	Std (μm)	Mean (mm)	Std (μm)
(1-3)	319.498	14	319.509	1.0	319.494	4.6	319.508	1.0
(4-6)	319.325	14	319.337	1.3	319.322	5.0	319.336	1.3
(7-9)	319.594	14	319.606	1.5	319.590	5.1	319.606	1.5
(1-7)	319.468	15	319.489	1.5	319.464	6.6	319.479	0.8
(2-8)	319.347	15	319.358	1.1	319.342	6.7	319.358	1.0
(3-9)	319.607	15	319.618	1.1	319.602	7.0	319.618	1.1
(1-9)	451.874	21	451.892	1.1	451.867	9.2	451.889	3.6
(3-7)	451.926	20	51.940	1.5	451.922	7.8	451.942	3.3
Global		16		1.3		6.5		1.7
// axes		14		1.3		5.8		1.1
>< axes		20		1.3		8.5		3.5

ature of 20 °C during the spatial gradient tests. The correction is better for parallel then for diagonal orientations, because the spatial gradients have an influence on the interaction between the axes. Although the model can deal with temperature differences between two sensors on an axis, changes in axis orientation remain uncompensated for. Thus measurements of 2D-distances (><) are only partially corrected.

The consistency of the correction can be verified by a comparison of the results of the full correction for both uniform temperatures and spatial gradients.

4.2 Correction while measuring angles

Apart from the distances, the angles between lines defined on the holeplate can be examined. The results listed in table 2 prove that in uniform steady-state situations, the interaction between the different axes does not change. The measured angle values do not change and the Std values are low (<0.6 arcsec).

Spatial gradients imposed by external heat sources do influence the measurement of angles. Here the Std values are higher, in the order of 4 arcsec (see table 2). The spatial gradients influence the orientation of the axis. The correction has no influence on the Std value, because the system contains no knowledge about the interaction of the temperatures and axis orientation.

This last section indicates the limitations of the developed correction technique. There is room for improvement based on the experience gained.

Table 2. Angle measurements under non-standard temperature conditions.

	Uniform temperatures				Spatial gradients			
	Standard		Full correction		Standard		Full correction	
	Mean (°)	Std (")	Mean (°)	Std (")	Mean (°)	Std (")	Mean (°)	Std (")
(2-8)×(4-6)	89.983	0.1	89.983	0.4	89.984	4.0	89.983	4.0
(1-9)×(3-7)	90.001	0.6	90.002	0.2	89.997	3.4	89.997	3.4

5 Conclusions

A basic compensation approach was presented. The compensation system eliminates the temperature dependent errors on length measurements. The uncertainty for the temperature range and conditions in focus is reduced to the level of the stochastic errors under standard conditions by means of only six thermometers. Based on the temperature readings the otherwise unknown environment influence is identified and the resulting errors become systematic and therefore manageable.

Further research in the field of thermal deformations at the K.U.Leuven has already shown promising developments towards non steady-state situations with spatial gradients.

Acknowledgements

This research work is supported by the European Brite-Euram program under project number BE-3545: MEDECOTHER. COORD3 (It.) kindly provided the CMM under consideration for the duration of the research. CRIF/WTCM kindly provided the laser interferometer and PTB the thermal stable artefact. Special thanks goes to ir. B. Timmers and ir. P. Van de Perre for their significant contribution to this study [3].

References

1. A. Balsamo, D. Margues, and S. Sartori. A method for thermal-deformation corrections of CMMs. In *Annals of the CIRP*, volume 39/1, pages 557–561, 1990.
2. P. A. McKeown, M. Weck, R. Bonse, et al. Reduction and compensation of thermal errors in machine tools. In *Annals of the CIRP*, volume 44/2, pages 589–597, 1995.
3. B. Timmers and P. Van den Perre. Correctie van thermische fouten bij gebruik van een 3D CMM in een niet-geklimatiseerde omgeving. Master's thesis, K.U.Leuven, 1998. Dutch.
4. Q. D. Yang, C. Van den Bergh, P. Vanherck, et al. Study of the thermal deformations on a coordinate measuring machine and compensation via a neural network. *International Journal of Flexible Automation and Integrated Manufacturing*, 7/1&2:129–147, 1999.

CMM ACQUISITION CHAIN

I. Amara [1], J.M. Linares [2], J.M. Sprauel [2]
[1] IGM, Université de Constantine, route de Aïn El Bey, 25000 Constantine, Algeria, amara.idriss@caramail.com
[2] Laboratoire Mécasurf, ENSAM, 2 cours des Arts et Métiers, F13617 Aix en Provence, linares@iut.univ-aix.fr

Abstract :. To calculate a CMM uncertainties we have studied the CMM acquisition chain. We propose to model the CMM. The initialisation procedure is very important for the control of the measurement quality. We show the main operations required for a use of CMM. This work is the roots of our study on CMM uncertainties.

1. Introduction

The coordinate measuring machine (CMM) extensive automation has largely reduced the measurement acquisition and processing time in the take up measurement operation. This is to explain the widely use of the CMM in the mechanical industry for ten years. Nowadays automation advances facilitate the measurement of the classical or complex surfaces. By using a very ergonomic human machine interface the operator is free of data processing programs. Then he can not observe all the computation phases. The good control of this measurement process relies on a great knowledge of its functional mode. Several works propose the study of the kinematics model for CMM [1]. But these studies focus on the identification of geometric errors [2][3]. In this paper we would like to model the CMM acquisition chain. This work is the roots of our study on CMM uncertainties [4].

2. Model of CMM Acquisition chain

The portico of the CMM is studied to illustrate this model. This analysis should be the same for other kinds of structure. This study is focused on the following elements :

Measuring machine : It is constituted by sliding joints X, Y, Z, equipped by optical rulers. The marble is the element that supports the reference sphere and the piece support device. The surfaces that assure the mounting of probe system move in 3 directions by three sliding joints in relationship with the fixed marble. The displacement is measured by optical rulers.

Processing software : It uses the coordinates of a current point arbitrary chosen on the last mobile part (vertical sliding translation) as the axis origin. The software acquires the Mj coordinates point measured in terms of the axes of the piece.

Piece support device : It is represented by the support surfaces on the piece and the contact surfaces with the marble.

Piece : It is defined by the measured point and the contact surfaces with the piece support device.

Probe head [5]**:** It is constituted by the head and the touch trigger probe system (5 rubies). The crucial elements are the contact points between the ruby and the piece, the ruby's centre and the linking surfaces with the CMM.
Reference sphere : The centre of the reference sphere and its linking surfaces with the fixed marble represent the most important characteristics.
Pre-processing system : It is an acquisition and counting electronic board. During the displacements the incremental optical rulers supply a square signal. When the ruby is touching a surface, the pre-processing system saves the probe head position on the three axes and sends it to the software (current coordinates point in terms of the axes of the machine).
Man : He acts on the software. He measures and controls the piece via the metrology software.
The model obtained is presented in the figure 1. It allows to represent the acquisition process finely enough for our study. An analysis with a higher level of decomposition is however possible but not necessary.

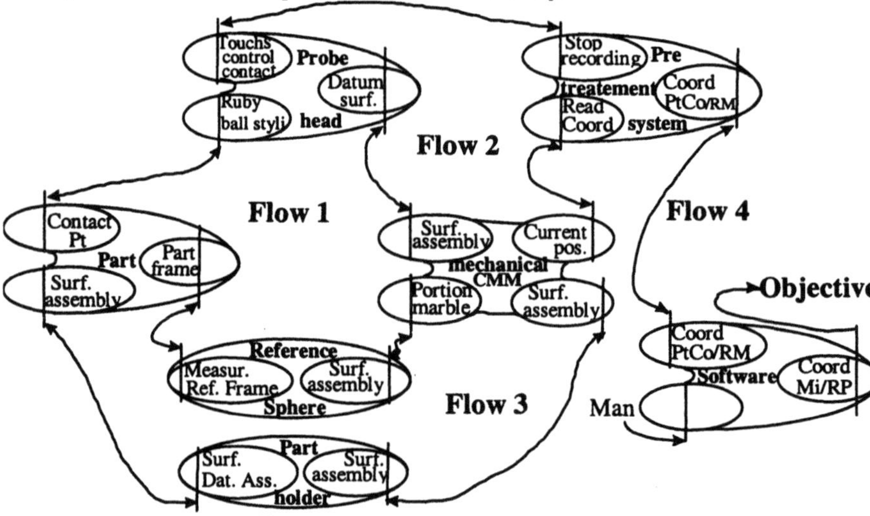

Figure 1 : Acquisition chain modelled

2.1 Main flows

There are four main flows :
Flow 1 : Positioning or locating loop (CMM, Probe system, Piece, Reference sphere). It is the central flow of the process. Other flows depend on it (acquisition flow) or influence it (the flow 3 disturbs it).
Flow 2 : Acquisition loop (CMM, Probe system, Pre-processing). When the stylus ball is touching a surface of the piece a front down signal is provided. That orders the processing system to acquire the position value by counting the number of pulses generated by the three axes optical rulers (X, Y, Z). The result is the current point position in terms of the axes of the machine (initialised to zero by contact on three stops).

Flow 3 : Mechanical Loop (CMM, Support piece, Piece, Sphere). It influences the flow 1 only when the system of reference of the piece is not straightened for every piece. Variation of temperature should be considered as negligible.
Flow 4 : Processing loop (Pre-processing, software). This flow gets pre-processed data constituted of the Mj points coordinates in terms of the axes of the machine. Then the loop provides the coordinates of the Mj points in terms of the axes of the piece.
From this model we can extract the central flow 1 that permits to locate the various parameters involved in the initialisation.

2.2 Initialisation vectors

First all parameters included in the flow 1 are described. In the considered configuration the CMM is placed on its stops (Origin of the axes of the Machine, O_M). The behaviour of the flow 1 permits to bring to the fore vectors that contribute on the value of the desired vector.

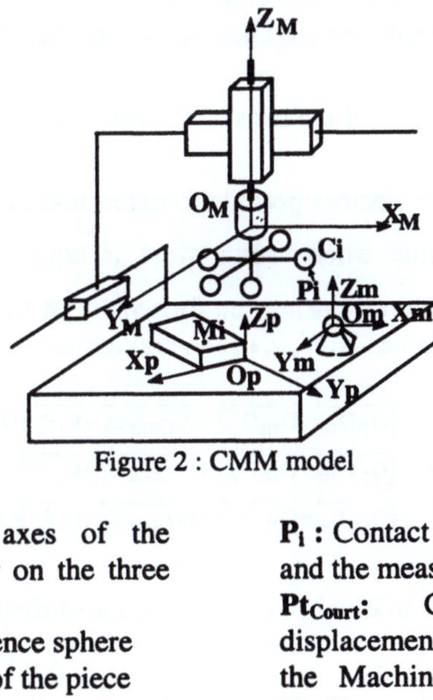

Figure 2 : CMM model

Note :
O_M: Origin of the axes of the Machine (Zero counter on the three axes)
O_m: Centre of the reference sphere
O_P: Origin of the axes of the piece
M_J: Point belonging to the measured surface j
C_i : Centre of the ruby i
P_i : Contact point between the ruby i and the measured surface
Pt_{Court}: Current point after displacement in terms of the axes of the Machine O_M (not pictured at Fig.2).

The objective is to find the vector $\overrightarrow{O_PM_j}$ in the axes of the piece (O_P, X_P, Y_P, Z_P). To perform this mission we should identify vectors that influence vector $\overrightarrow{O_PM_j}$. The measuring machine moves according to a displacement estimated in the axes of the machine (O_M, X_M, Y_M, Z_M). By using Châles theorem we obtain the following vectorial equation written in terms of the axes of the machine (O_M, X_M, Y_M, Z_M).

$$\overrightarrow{O_PM_j}_{/RM} = (\overrightarrow{O_PO_m} + \overrightarrow{O_mO_M} + \overrightarrow{O_MPt_{Court}} + \overrightarrow{Pt_{Court}C_i} + \overrightarrow{C_iP_i} + \overrightarrow{P_i\,M_j})_{/RM}$$

Several assumptions are proposed to simplify the issue. Dimensions are considered as constant between the process initialisation and the Mj point acquisition (static equilibrium, no temperature variation, ...). The measuring machine kinematics is considered as perfect.

Under these assumptions, we can identify the initialisation vectors :

$\overrightarrow{P_iM_j}$: Displacement vector required to put in contact P_i and M_j.

$\overrightarrow{C_iP_i}$: Projection of ruby radius along the contact normal n_j to the point M_j in the measuring reference system.

$\overrightarrow{Pt_{Court}C_i}$: Offset of the C_i ruby centre in relationship with an unspecified point of the mobile system.

$\overrightarrow{O_MPt_{Court}}$: Displacement realised by the current point from the origin of the axes of the machine O_M(value X,Y,Z of the counter).

$\overrightarrow{O_mO_M}$: Position of the sphere reference centre to the origin of the axes of the machine.

$\overrightarrow{O_PO_m}$: Position of the origin of the axes of the piece in relationship with the centre of the reference sphere.

The knowledge of all these vectors permits to determine the vector $\overrightarrow{O_PM_j}$ in terms of the axes of the machine when vector $\overrightarrow{P_iM_j}$ is equal to zero (contact). This transformation can be written by using the formalism of matrix with homogeneous coordinates by using the values X, Y, Z of the counter as an input vector.

$$\overrightarrow{O_PM_j}_{/RM} = \begin{pmatrix} 1 & 0 & 0 & \left(\overrightarrow{O_PO_m} + \overrightarrow{O_mO_M} + \overrightarrow{Pt_{Cour}C_i} + \overrightarrow{C_iP_i}\right)\bullet\vec{X}_M \\ 0 & 1 & 0 & \left(\overrightarrow{O_PO_m} + \overrightarrow{O_mO_M} + \overrightarrow{Pt_{Cour}C_i} + \overrightarrow{C_iP_i}\right)\bullet\vec{Y}_M \\ 0 & 0 & 1 & \left(\overrightarrow{O_PO_m} + \overrightarrow{O_mO_M} + \overrightarrow{Pt_{Cour}C_i} + \overrightarrow{C_iP_i}\right)\bullet\vec{Z}_M \\ 0 & 0 & 0 & 1 \end{pmatrix} \times \overrightarrow{O_MPt_{Cour}}_{/RM}$$

A change of axes is needed to obtain $\overrightarrow{O_PM_j}$ in terms of the axes of the piece.

$$\overrightarrow{O_PM_j}_{/RP} = \begin{pmatrix} \vec{X}_p\bullet\vec{X}_M & \vec{X}_p\bullet\vec{Y}_M & \vec{X}_p\bullet\vec{Z}_M & 0 \\ \vec{Y}_p\bullet\vec{X}_M & \vec{Y}_p\bullet\vec{Y}_M & \vec{Y}_p\bullet\vec{Z}_M & 0 \\ \vec{Z}_p\bullet\vec{X}_M & \vec{Z}_p\bullet\vec{Y}_M & \vec{Z}_p\bullet\vec{Z}_M & 0 \\ 0 & 0 & 0 & 1 \end{pmatrix}^{-1} \times \overrightarrow{O_PM_j}_{/RM}$$

Both transformations can be defined by only one matrix of homogeneous coordinates. That allows to easily identify the desired vector. After the initialisation vectors are known, the process can calculate the M_j point coordinates in terms of the axes of the piece.

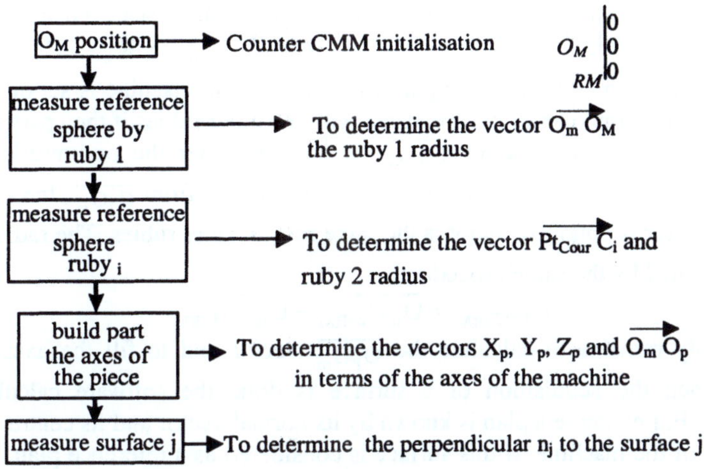

Figure 3 : CMM initialisation procedure

Now our study is focussing on the preparation phase of the CMM. A flow chart of initialisation phase for CMM is proposed by using the modelling and the vectorial equation. This flow chart could be used for any kind of MMC. The counter initialisation is done by touching the stops. A point of the mobile part composed of the vertical sliding joint Z and the probe system is chosen as the machine origin. For the MetroSoft's Metromec software the origin of the axes of the machine is defined during the choice of the first calibration stylus ball.

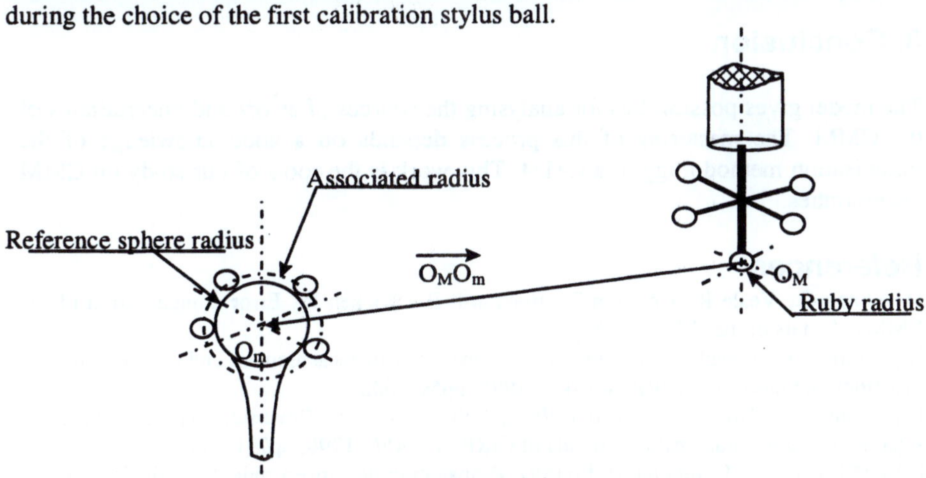

Figure 4 : Acquisition of $\overrightarrow{OmOM}$ and ruby's radius

Secondly it is necessary to define the reference sphere position in relation with the machine origin. The ideal sphere is associated to the measured points on the reference sphere by a least square method. The software provides the $\overrightarrow{O_M Pt_{Cour}}$ vector (Pt_{Cour} is confused with O_M if no displacement) and the associated sphere radius R_{ass}. The real value of the diameter of the reference sphere R_{real} is available in the setup file. The radius of the ruby that has measured the reference sphere is immediately deduced by subtraction (Fig.4): $R_{ruby} = R_{ass} - R_{real}$

Once this operation is done the origin of the axes of the machine is set on the first stylus ball. The same operation is repeated with the other rubies if they participate in the checking phase. The vector $\overrightarrow{C_2O_m}$ is obtained after the optimisation phase realised by the software. By subtracting the vector $\overrightarrow{O_mO_M}$ from $\overrightarrow{C_2O_m}$, the $\overrightarrow{O_MC_2}$ or $\overrightarrow{Pt_{Cour}C_2}$ vector is obtained, and it is the same for the other rubies. The radius of the ruby is obtained by the same procedure.

$$\overrightarrow{O_MC_2}_{/RM} = \overrightarrow{O_mC_2}_{/RM} - \overrightarrow{O_mO_M}_{/RM}$$

Now, it only remains to calculate the $\overrightarrow{O_PO_m}$ vector and to fill the axes change matrix. When the acquisition of a surface is done the software calculates its parameters. For example a plan is known by its normal vector and its centre in terms in the axes of the machine. If this surface is considered as a reference plan, the user commands through a software function to put the normal in the change matrix previously defined. A second vector is used by the same mode and fills a second column of the change matrix. The third vector is obtained from the vectorial product. Then the change matrix is completed (Fig. 10). The $\overrightarrow{O_pO_m}$ vector is obtained according to two phases :

- Geometric constructions realised by the user.
- Its acquisition by a specific function of the software.

When this procedure is completed the CMM is initialised to start a checking phase.

3. Conclusion

The model gives possibilities for analysing the sources of errors and uncertainties of the CMM. The mastering of this process depends on a good knowledge of the initialisation methodology of a CMM. This work is the roots of our study on CMM uncertainties.

References

[1] Zhang G., Veale R., Charlton T., Borchardt B., Hocken R., Error compensation of the CMMs, Annals of the CIRP, 1985, pp445-448.

[2] Satori S., Colonnetti G., Zhang G.X., Geometric error measurement and compensation of machines, Annals of the CIRP, Vol 44/2,1995, pp599-602.

[3] Evans C.J., Hoken R.J., Estler W.T., Self calibration: Reversal, Redundancy, Error separation and absolute testing, Annals of CIRP, Vol 45/2,1996, pp791-794.

[4] J.M Linares, J.M. Sprauel, P. Bourdet, Propagation de l'incertitude d'acquisition dans la chaîne de traitement d'une machine à mesurer tridimensionnelle, Métrologie 99, 9è Congrès International de Métrologie, Bordeaux, pp.185-188, Octobre 1999.

[5] Renishaw, CMM Touch-Trigger Probe System, User guide, 1990.

The Dynamic Performance of a CMM during the Probing Procedure

Zhu Xijing,(Scientific Advisor)

Brown & Sharpe Qianshao Technology Co. Ltd., China
11 Luoyang Road, Qingdao, China. Email: zhuxijing @cn.bnsmc.com

Abstract: This paper describes an analysis into the dynamic performance of a CMM (Coordinate Measuring Machine), specifically during its Probing Cycle. By the use of a laser interferometer, the whole probing procedure is observed and the effects of positioning, velocity and acceleration curves graphically analysed.

1. Introduction

The repeatability, accuracy and the 3D positioning accuracy are very important for a CMM. Many standards, including ISO10360, have specified the error of indication of a CMM and probing error. All of these specifications are based on the measurement of a Gage block or a standard sphere measurement. They indicate the result but do not answer *why* things happen or *how* to improve the performance.
Many articles have analyzed the performance of a CMM from control theory and gave the transfer equations, including PID parameters but it could not comment on any typical CMM.

The main purpose of many papers, which involved error compensation, has been to form the database used to compensate the 21 items static error.

This paper attempts to find a method by using a laser interferometer to observe the whole probing procedure and record the positioning, velocity and acceleration curve. On the basis of analyzing the curve, then give the reasons why the repeatability, accuracy and 3D positioning are good, and how to improve it. Also using the curve for comparison, and to ascertain whether it is possible to improve the design, or adjustment of controller or driver system.
This method is simple, and straightforward. It is very useful for research or develop purpose.

1.1 On a CMM, measurement takes place during movement, i.e. it is dynamic. So the positioning, velocity and acceleration curves are very important and explain the whole dynamic story.

1.2 The probe performance, especially for touch trigger probe is influenced by the uniformity of the velocity and acceleration.

1.3 If there is the backlash, then upon reversing the direction of movement, there is a pause (for position curve there is a horizontal section and velocity equals zero).

1.4 If moving along the diagonal direction, the positioning curve keeps a straight line and no vibration for velocity curve. That means good synchronization for X, Y and Z movement. The 3D positioning are stable and good.

1.5 The whole procedure of the probing

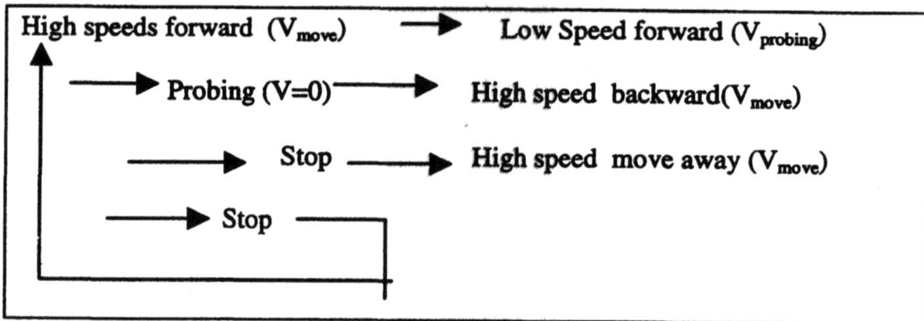

Fig A: Diagram of the whole procedure of the probing

2. Typical positioning, velocity and acceleration curve

Now we show three different curve files, which happened along the diagonal direction here. You could compare each other.
Fixed Bar driver without intermediate point (Fig.1, 4,7)
Metal Belt without intermediate point (Fig.2, 5,8)
Metal Belt with intermediate point (Fig.3, 6,9)
If we check it according to positioning curve only, we could not see the big difference (Fig. 1,2,3). But if we check it according to the velocity curve. You will see the difference (Fig. 4,5,6).

Conclusion:
2.1 The Repeatability would be good if the $V_{probing}$ is uniform (Fig. 4).
During the speed change period. The vibration happened when using a metal belt driver for a very heavy CMM and controller is not suitable adjusted. (Fig.5)

2.2 When the positioning curve of Low Speed forward (measuring speed) could keep straight line ($V_{probing}$ keeps constant). The 3D positioning keeps well (Fig.1, 4).

2.3 The curves of positioning, velocity and acceleration measured by laser interferometer could be used to comment on the performance of a CMM.

2.4. Adding an intermediate point (Fig. 5), controller suitable adjusted and good mechanical structure could be used to improve the CMM performance.

3. Typical probing curves

For the accuracy and repeatability for the form measurement the most important thing is the system stiffness (including mechanical and controller's stiffness) and adjustment which influence the behavior after first time probing,

If the vibration is at low frequency, after the first time probing there is vibration (Fig. 11). The measuring velocity changes the sign (Fig.13), the CMM goes forward and backward. The circuit could not block this first signal (signal period is 0.1s) so that CMM could get the wrong position when it is at the second or the third peak signal.

The procedure for probing:

Low Speed forward ($V_{probing}$) → Probing → Deceleration (forward)

→ Stop(V=0) → Acceleration (backward)

→ High speed backward(V_{move})

Fig.B: The procedure for probing

Observe the curve at Fig. 10. It is one ideal CMM, which has constant speed (small acceleration Fig.14), before probing (Fig.12) and one probing point (Fig.10). There is very sharp deceleration stage (Fig.12). After back off at stop stage there is a small speed change. No second or third peak signal at positioning curves(Fig.10).

Fig. 16 shows another good adjusted CMM, Not only the positioning curve but also the velocity curve is smooth. The acceleration curve near the straight line. There is less impact than the other CMM. It is important for high speed CMM.

Fig. 17, There is very clear second and third peaks at the positioning curve therefore there are obvious change of velocity (Fig 19) and variety on acceleration. This CMM performance is poor.

Conclusion:

From above we can see that a CMM like in Fig. 1,4,7 Fig. 10,12,14.or like Fig. 16,18,20 is a good one whose position curve keeps straight line and velocity curve has no vibration.

From microphenomenon, during probing time, only one touch point, no second and third peak of position curves, no sign change of velocity or like Fig.16, 18,20 all curves are smooth.

A CMM that could meet the requirements mentioned above would be a good CMM both on accuracy and repeatability. If not then you should check the reason until getting above mentioned ideal curve.

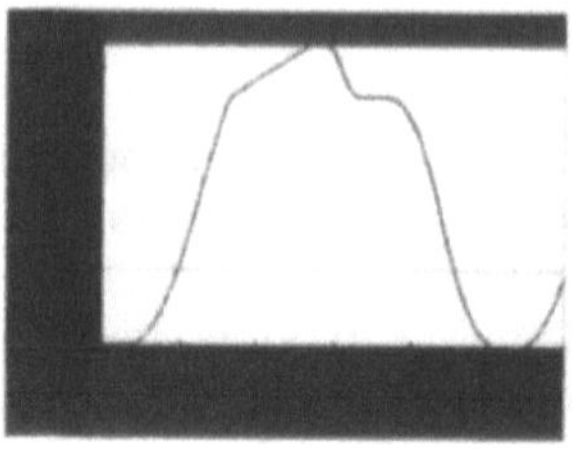
Fig.1

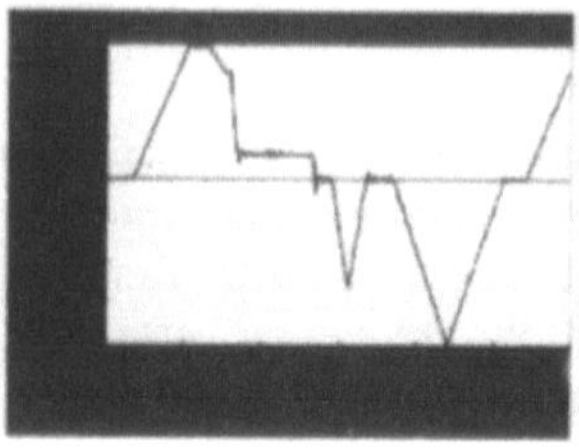
Fig.4

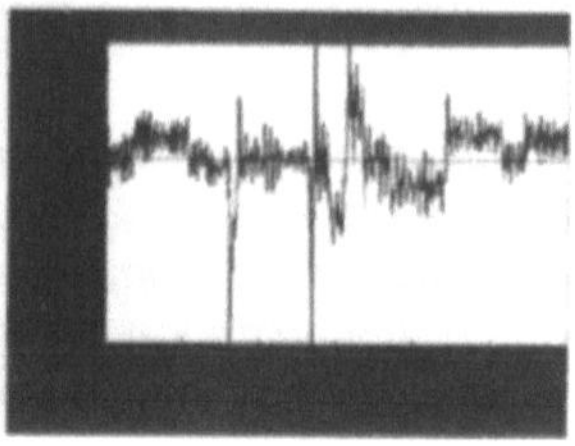
Fig.7

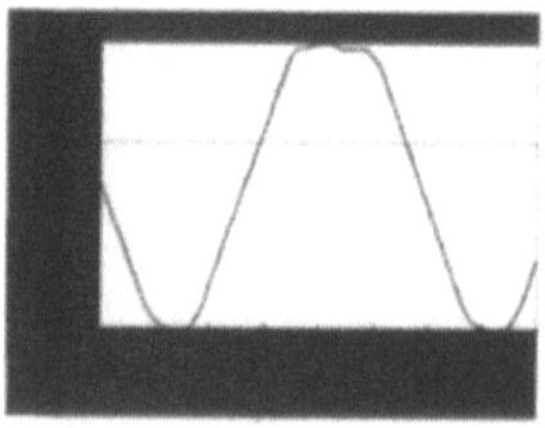
Fig.2

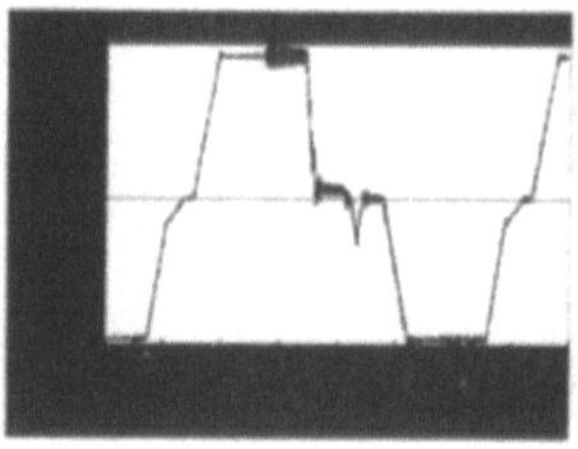
Fig.5

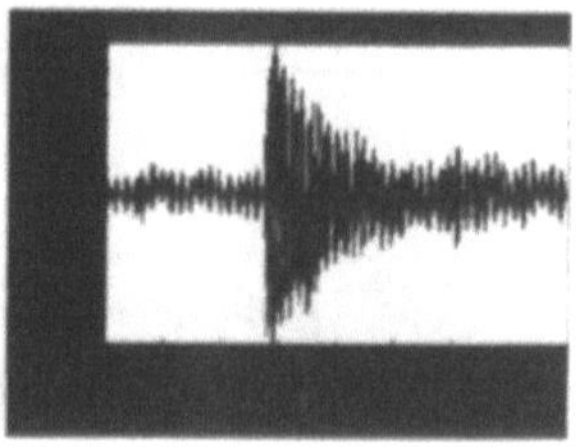
Fig.8

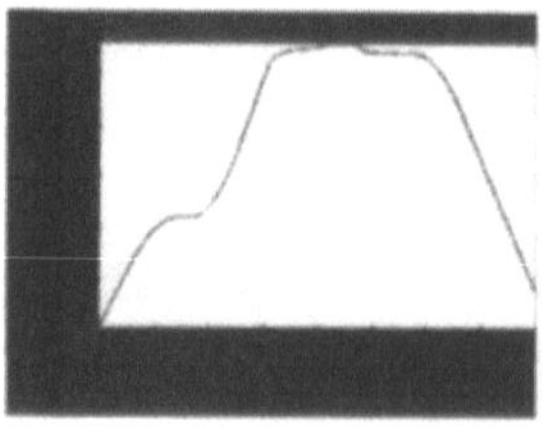
Fig.3

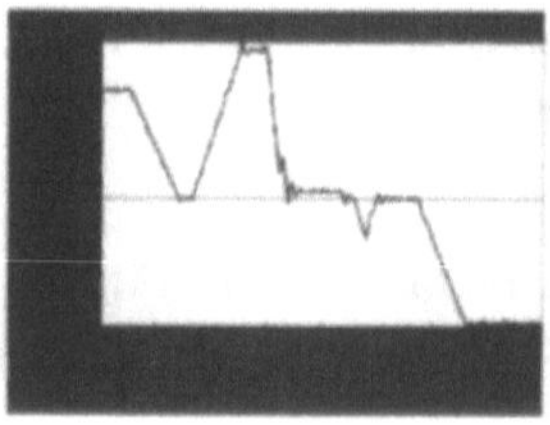
Fig.6

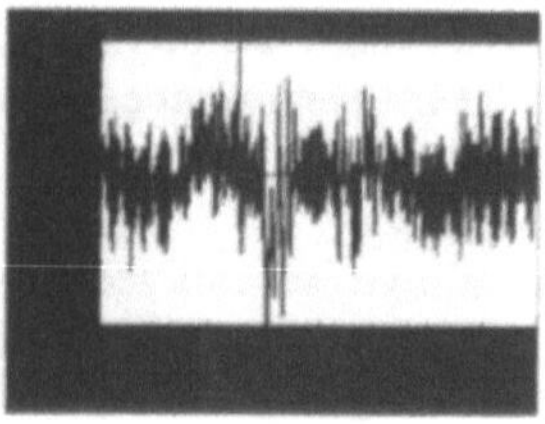
Fig.9

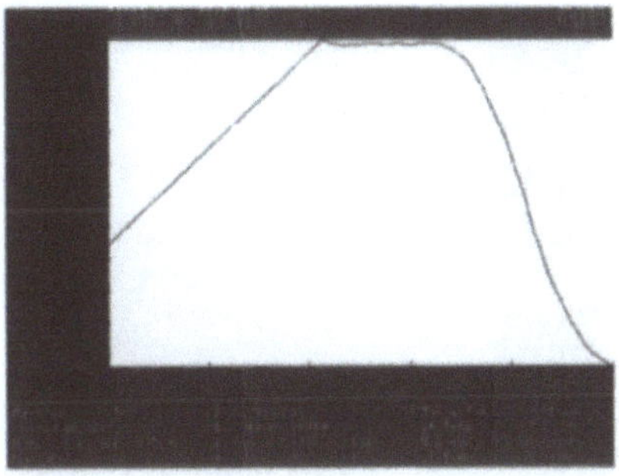

Fig.10

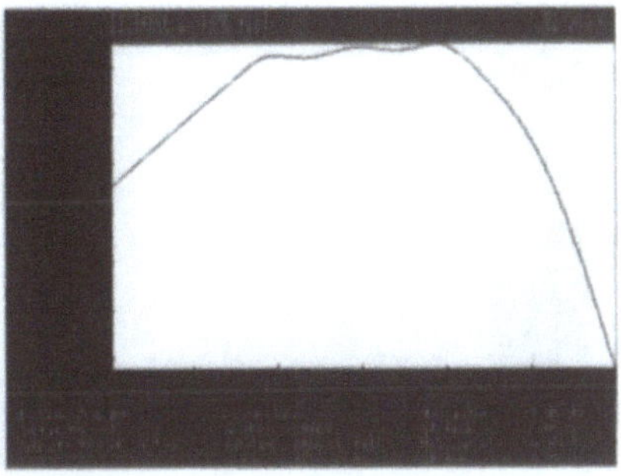

Fig.11

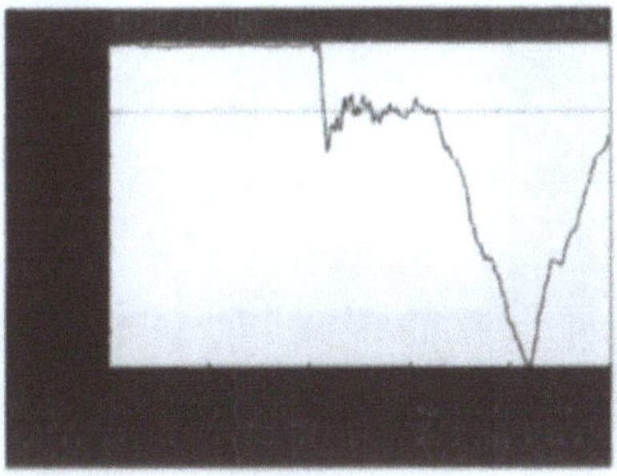

Fig.12

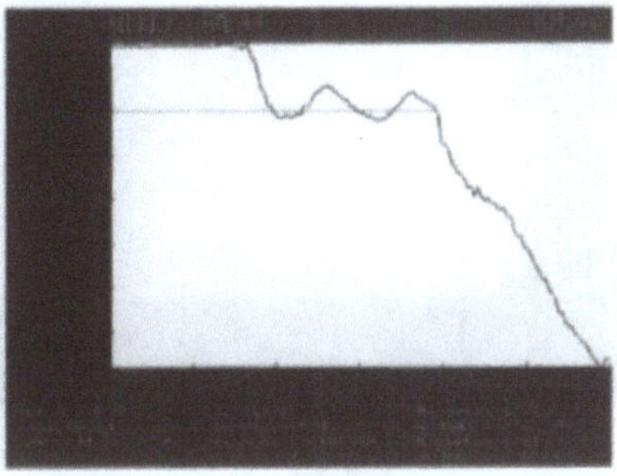

Fig.13

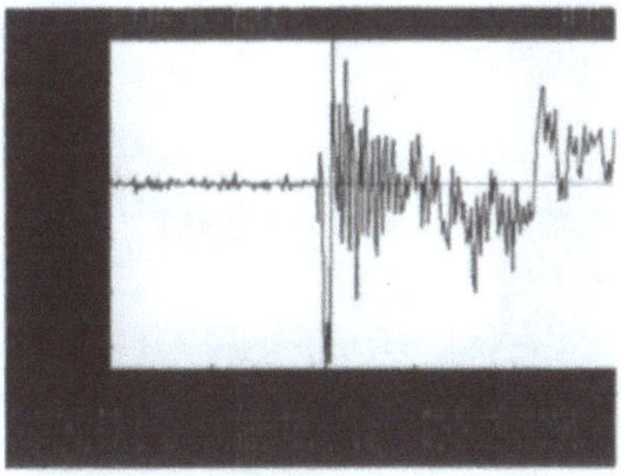

Fig.14

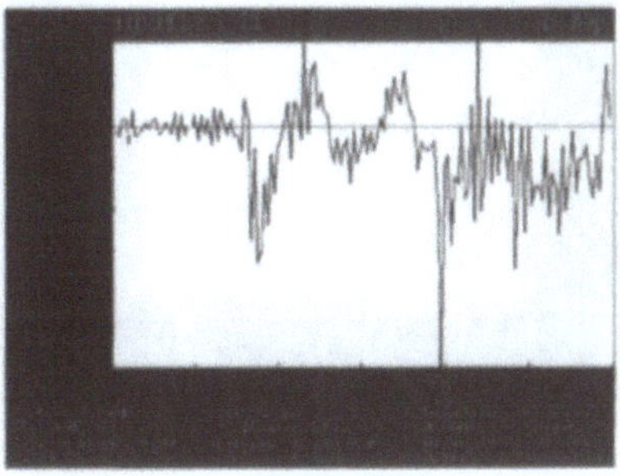

Fig.15

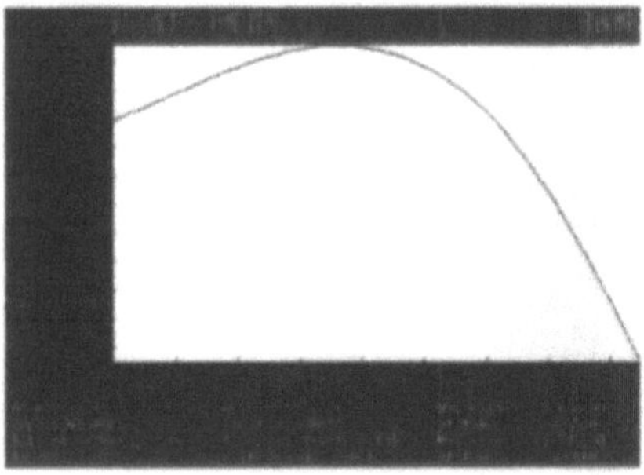

Fig.16

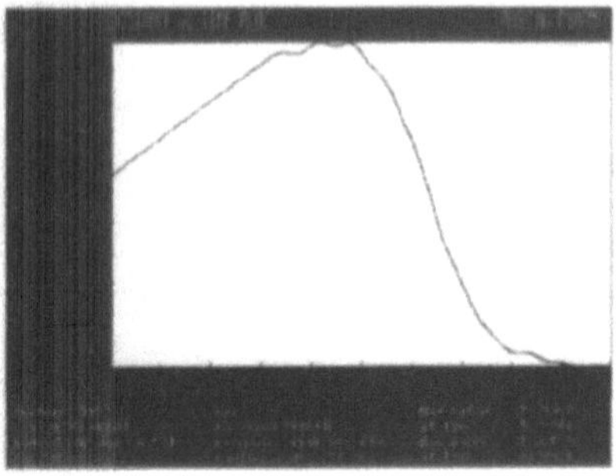

Fig.17

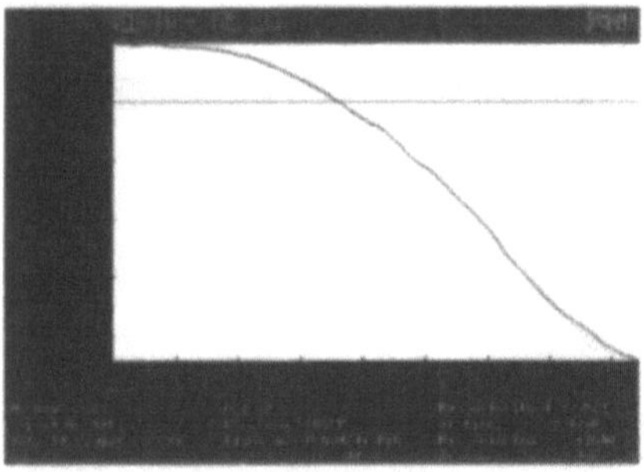

Fig.18

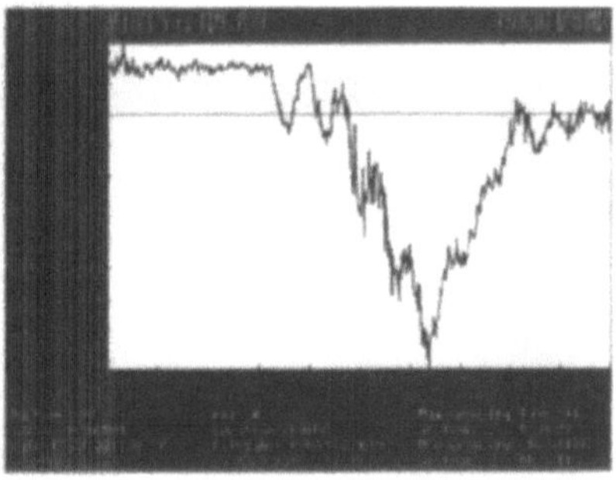

Fig.19

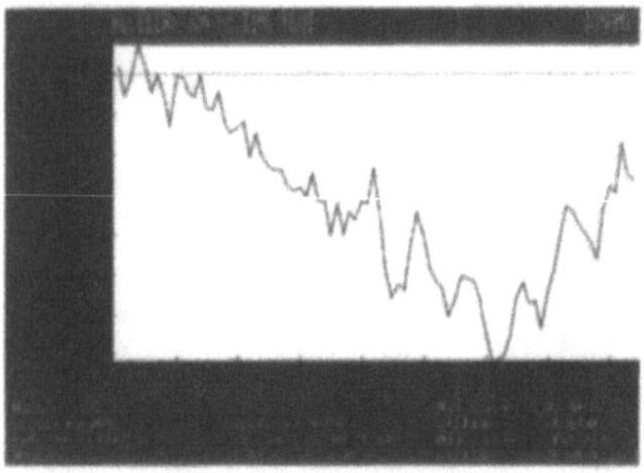

Fig.20

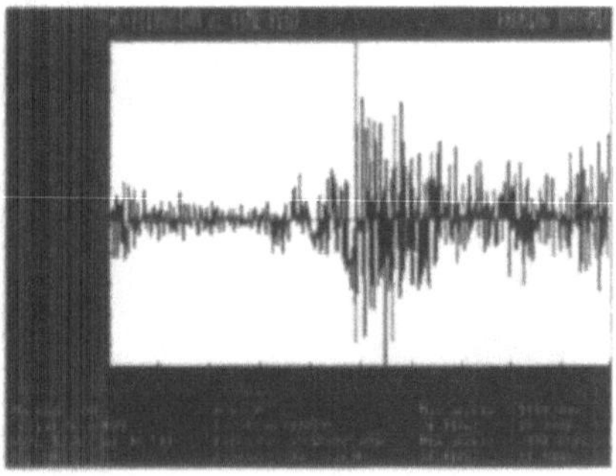

Fig.21

MACHINE TOOL ACCURACY I

A Fast and Low-cost Machine Tool Thermal Error Compensation System for Industrial Use

Kuang-Chao Fan and Ling-Zen Meng

Department of Mechanical Engineering, National Taiwan University
Taipei, Taiwan, ROC, E-mail: fan@ccms.ntu.edu.tw

Abstract: A measurement and compensation control system for the spindle thermal expansion of machine tools has been developed in this study. The low cost but accurate thermal sensor, AD 590 IC, and tool setting probe was developed for the temperature and spindle expansion measurements respectively. With the proper selection of sensor locations, a linear error model with respect to only one or two temperature terms could always be obtained, and then compensated to within 10 μm by an 8051-based single board computer. As the error model can be generated in one day under the real cutting condition and the whole system is low cost, this system is suitable for industrial use.

1. Introduction

Thermal error is the most important source, which significantly influences the machining accuracy of the machine tools. Although there are already many excellent research reports in this subject over the past years [1-3], most of them, however, still remain in the laboratory level utilizing various sophisticated models, which do not meet industrial requirements in terms of the cost and simplicity. Meanwhile, the current ISO 230-3international standard still specifies to experimentally generate the thermal error model of the machine tool under non-cutting (air-cutting) condition [4]. It, in fact, does not exactly reflect the real phenomenon in industrial practice, where any machine tool must cut the metal to produce the part. Therefore, thermal error of the machine tool should be paid with more attention under the real-cutting condition [5]. Major considerations are not only to the effectiveness and applicability but also its cost and simplicity in the model generation and error compensation within a very short time.

In this system, the thermal sensor was developed using an AD 590 IC, which performed high accuracy for a long period. The thermal displacement of the spindle was measured using a tool setting probe, which is a standard accessory installed in most machine tools. The thermal error model was generated under real-cutting conditions on steels, which are most used in machine shops. With the proper selection of sensor locations, the thermal error of any particular machine can be modeled by a linear function with least temperature terms. The diagnosed error model could be stored in a developed single board computer from which the compensation commands could drive the CNC controller to implement real-time error compensation without interfering the existing NC operation. Details are described below.

2. The Measurement System

2-1 Temperature measurement

In general, thermal couples and thermistors are widely used for the temperature detection on the machine structures. Its characteristics of epoxy bonding, however, brings disadvantages to the machine tool applications since the actual temperature rise is small and the best sensor locations are unknown in the early stage. This research selected a low cost IC-AD590 (US$2 per piece, range to 150•) and packaged it into a magnetic ring so as to become a portable temperature detector. An 8-channel A/D converter board was also designed to simultaneously detect eight temperatures located around the machine, as shown in Fig. 1. It is noted that the ambient temperature (T8) will also influence the thermal behavior of the machine tools. Calibration results of eight AD-590 sensors using the HP 2804A Quartz Thermometer have proved that all sensors performed very good linearity with accuracy all within 0.7 • , and resolution to 0.1• .

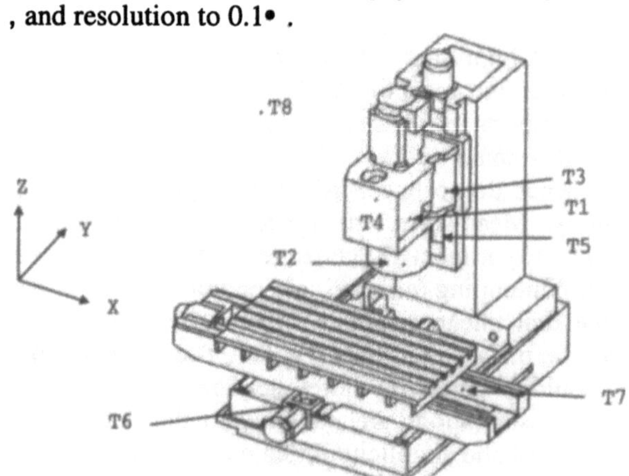

Figure 1: Initial locations for temperature sensors

2-2 Thermal displacement measurement system

There are several means of spindle thermal drift measurement vs. work table, such as 5-channel displacement transducer equipment [4] and the laser ball bar [6]. Such equipments are only suitable for the measurement under air cutting condition. This research, therefore, directly utilized the function of the tool-setting probe (MP4 of Renishaw Co.), which is a common accessory for the settings of the tool length and the cutter radius. Fig. 2 shows if we detect the touch-triggered positions of spindle in Z direction at cold start and after a machining cycle the positional change will be the amount of spindle expansion. This technique can actually apply to the thermal drift in three dimensions, as shown in Fig. 3. The repeatability of the touch-triggered probe was found only 1 μm. It is deemed as a very useful tool for this purpose.

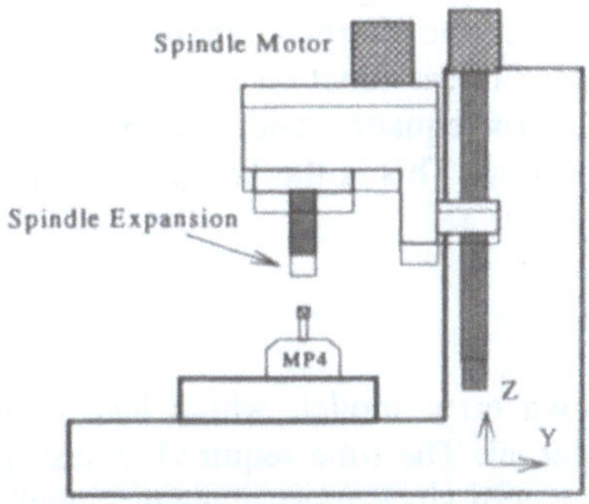

Figure 2: Spindle thermal expansion

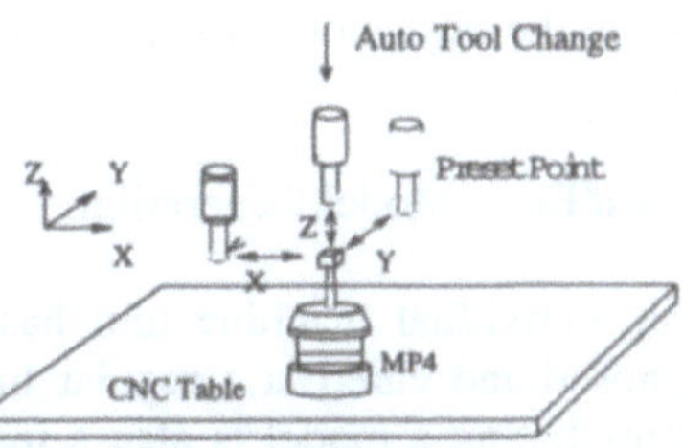

Figure 3: 3-D touch triggering

3. Thermal Model Generation

3-1 Experimental setup for real cutting tests

The procedure of temperature and spindle expansion measurements under real cutting condition is illustrated in Fig. 4. A DNC command is sent to the machine tool controller via RS232 interface by a personal computer. The real cutting process will be carried out on a steel block, such as S50C or the mold steel SKD61. These steels are commonly used in the machine shops. After a prescribed time the cutter will be replaced by a standard bar at the tool changer. An in-cycle gauging is implemented by touching the bar to the MP4 probe. The triggered position in Z-direction together with the current temperature detected by each AD590 sensor will be automatically recorded. It then continues to the next cutting cycle. The NC tool path can be as simple as a zig-zag planar motion.

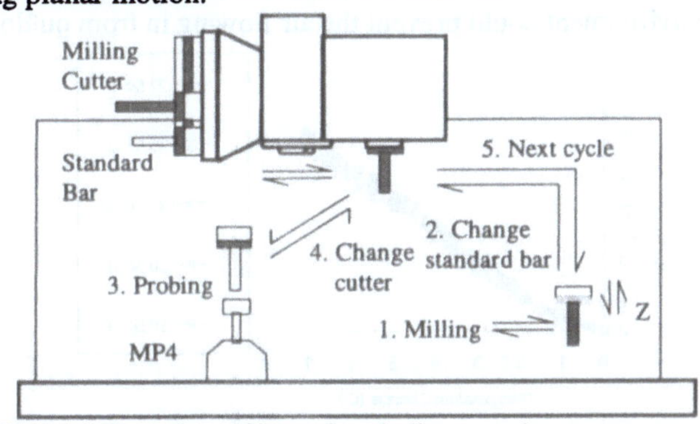

Figure 4: Procedure of in-cycle spindle expansion measurement

3-2 The best conditions for error model generation

(1) Optimum Temperature Sensor Locations

During machining all sensors respond the temperature rises in different rates. Proper selection of effective temperature terms is very essential to represent an accurate error model. Ni [7] pointed out the thermal displacement to temperature change could be as simple as a linear function if the optimum sensor locations could be found. A linear error equation with least variables must be the best model for error compensation. This is the key point of our researches.

(2) Fast Error Model Generation

Every individual machine tool has its own error model, which has to be diagnosed and cured at a regular time interval. The time required to testing and predicting a model is also a major concern since most managers would prefer as short as possible. Consequently, any acceptable solution of error model generation must be as fast as possible. This research has found that to increase the depth of cut is the most effective way to reduce the experimental time while maintaining the linearity of the error model. Moreover, due to the weather change all the year round the working environment of the investigated machine also induces serious disturbance to the generated model. Having had a long time experiments, this study has also found that if the room temperature could be controlled to a small variation the thermal error model of the investigated machine tool could still be stable.

Combining the efforts of the above two best conditions, Fig. 5 shows a very satisfied results. The work material is S50 carbon steel. The temperature rise at the spindle front housing (T2 in Fig. 1) has a very good linear relationship with the spindle expansion. Experiments were carried out in different days with room temperatures controlled at ±0.5• . It is found the repeatability is very good. Since the total time for each experiment lasted only 3.5 hours, it was not difficult to control the room temperature as long as the testing environment could prevent the air flowing in from outdoor.

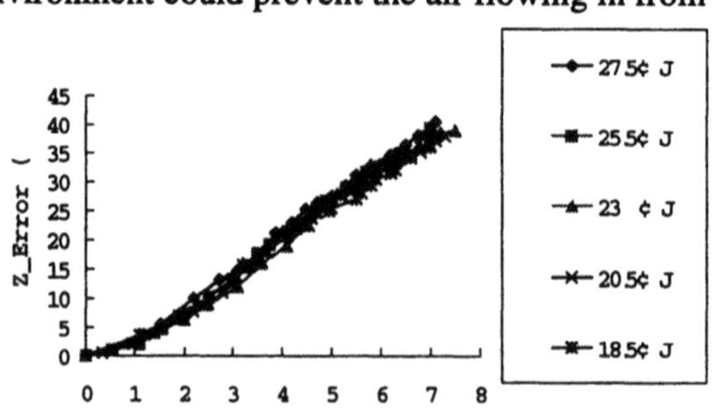

Figure 5: Linear error model generated at different room temperatures

4. Error Compensation Strategy

An Intel 8051-based single board computer was developed to implement the error compensation process. As shown in Fig. 6, the linear error function was initially stored in the 8051 processor. During the real cutting process AD590 sensor kept reading the T2 temperature for every 5 minutes and sending data to 8051 IC. This processor then output a compensation command via the I/O B2 interface card into the CNC controller so as to automatically shift the origin of the machine coordinate. The special feature of this strategy is that the compensation board is a stand alone one which can be placed in the CNC controller box, and the NC commands for the part machining will not be interrupted. The machine tool does not need to equip an external computer as required by most of other researches. Meanwhile the cost of the compensation board is rather cheap, which is the primary requirement.

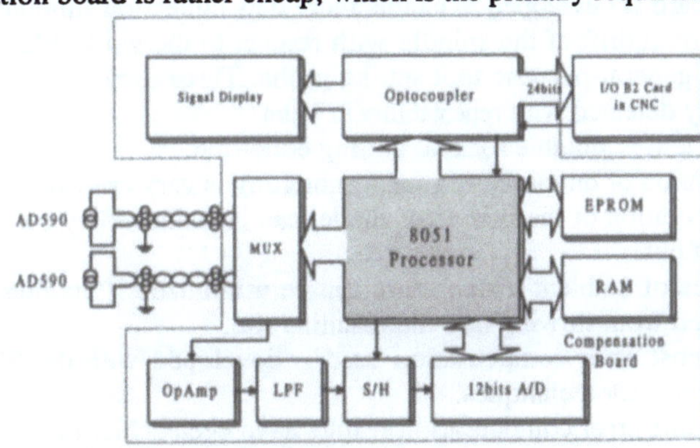

Figure 6: Block diagram of the error compensation

5. Error Compensation Experiments

In the experiments, both in the laboratory and factory site, thermal error model of the investigated machine tool was fast generated under various real cutting conditions on steel blocks, namely S50C and SKD61. The effort of error model generation took within 4 hours under proper room temperature control. Fig. 7 shows an example of the spindle position drift with and without error compensation. It is clearly seen that the errors caused by spindle thermal expansion can always be successfully controlled to within 10 μm during an eight -hour run.

6. Conclusions

A very effective method for the thermal error measurement and compensation of CNC machine tools is presented in this article. Many special features of this system are summarized as follows.

(1) The temperature sensor is made by low cost IC-AD590 with magnetic

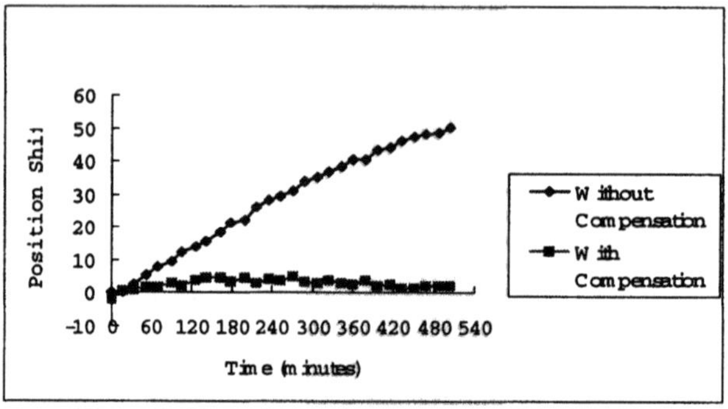

Figure 7: Experimental results, material: SKD61

mount so that the placement location can be changed. A multi-channel sensor card is developed to enable more data collection simultaneously.

(2) The thermal drift of the spindle with respect to the worktable is detected using a general-purpose tool setting probe. Three-dimensional shifts can be easily detected with repeatability to 1 μm.
(3) This system is suitable for real cutting condition.
(4) The location of temperature sensor mounting is very critical.
(5) Fast generation of thermal error model can be achieved by increasing the depth of cut.
(6) Influence of ambient temperature can be minimized if the outdoor air is prevented from flowing into the machine site.
(7) A low cost error compensation card is developed with the 8051-based microprocessor techniques.
(8) Remarkable error compensation results have been achieved.

References

[1] Jun Ni, 1997, "CNC Machine Accuracy Enhancement Through Real-Time Error Compensation," Int. J. of Manufacturing Science and Engineering, Vol. 119, pp.717~725.

[2] S. Yang, 1997, "Improvement of Thermal Error Modeling and Compensation on Machine Tools by CMAC Neural Network," International Journal of Machine Tools and Manufacture, Vol.36, No.4, pp.527~537.

[3] N. Srinivasa and J.C. Ziegert, 1996, "Automated Measurement and Compensation of Thermally Induced Error Maps in Machine Tools," Precision Engineering, Vol. 19,No. 2-3, pp.112-132.

[4] ISO 230-3, 1994, "Acceptance Code for Machine Tools--Part 3".

[5] K. C. Fan and K. Y. Hong, 1995, "Error Compensation of Spindle Expansion by Cutting Model on a Machining Center," 31st MATADOR Conf., England, pp.269~275.

[6] N. Srinivasa, J.C. Zigert and C. Mize, 1996, "Spindle Thermal Drift Measurement using the Laser Ball Bar," Precision Engineering, Vol. 18, No. 2, pp.118-128.

[7] C.H. Lo, J. Ni, and J. Yuan, 1996, "Thermal Sensor Placement Strategy for Machine Error Compensation," Proc. of the ASME Dynamics Systems and Control Division, DSC-Vol. 58, pp.341~348.

Thermal Error Measurement and Real Time Compensation System for the CNC Machine Tools

Heui Jae Pahk*, Suk Won Lee*
*School of Mechanical and Aerospace Eng.
Seoul National University, Seoul, Korea, 151-742

Abstract: Thermally induced errors have been significant factors affecting the machine tool accuracy. In this paper, the spindle thermal error and thermal feed axis error have been focused, the measurement/compensation of thermal error is introduced. In order to compensate the thermal error in real time bases, the coordinates of the CNC machine tool are modified in PC-based environment. After the real time compensation, the machine tool accuracy has been improved about 4-5 times typically.

1.Introduction

Thermally induced errors have been significant factors affecting the machine tool accuracy. The thermal errors generally come from the thermal deformations of the machine elements caused by heat sources that exist within the structure, ball screws, bearings, nuts, axis drive motors, friction on the way surfaces, cutting processes, the flow of coolant/lubricating oil, and the ambient temperature. Those thermal errors have been reported as much as about 70% of the total positioning error in the machine tool (e.g. [1]), and among them, the spindle thermal errors or spindle drifts have been considered as the dominant error components (e.g. [2]). In this paper, a measurement system for spindle thermal error and feed axis thermal error is developed and three methods of thermal error modeling are implemented for the spindle thermal error: multiple linear regression, neural network, and system identification. In order to obtain the compensation values, the algorithm of the volumetric error map is programmed using the models. The coordinates of the machine tool controller are modified to compensate in real time in PC interfaced environment program. After the real time compensation, the machine tool accuracy has been improved about 4-5 times typically

2. Spindle Thermal Error Measurement/Modeling

2.1 Measurement system for the 5 DOF spindle thermal errors

There are 6 degree of freedom (6 DOF) components in the spindle error motion in machine tools [6]: two radial error motions, one axial motion, two tilting motions, and one indexing error motion. In view of the machine tool accuracy, the indexing error motion of spindle can be ignored and thus 5 DOF spindle thermal errors are considered in this paper.

Fig.1 shows the measurement setup of the 5 DOF spindle errors, where the jig of two master balls and 5 gap sensors are arranged around the PC interfaced environment. Temperature variations are also measured with the thermocouples around the machine tool structure.

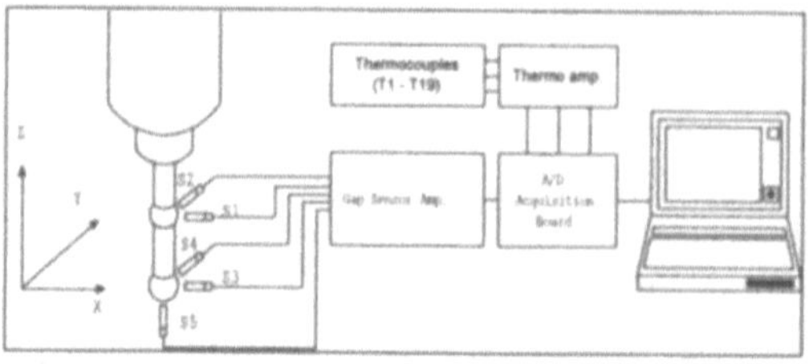

Fig.1 Sensors for measuring spindle drift errors.

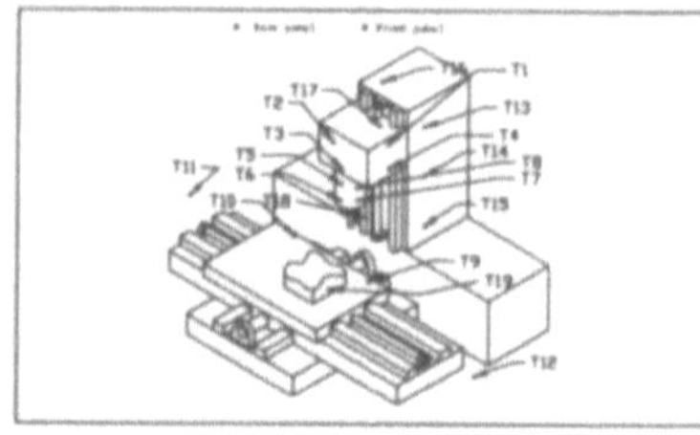

Fig.2 Thermal sensor location

Fig.2 shows the thermocouples located around the machine tool structure. The on line measurement feature has been implemented, in order to measure the spindle thermal error with efficiency. When the total measurement time and the measurement interval are input, the PC controls the CNC machine to operate according to the programmed speed via the RS232C serial interface connection. At the time of measurement, the PC sends the command "M19"(typically for the FANUC controller) to the machine tool such that the machine spindle stops at the angle of reference pulse signal (zero degree) of the spindle encoder. Then the machine spindle stops, and the spindle error measurements as well as the temperature measurements are performed. After the measurement operation, the PC sends the command of spindle rotation to the machine tool, and then the machine spindle rotates until the next command of measurement is given.

2.2 Spindle thermal error measurement and modeling

The developed thermal error measurement system has been applied to the three operation conditions of the CNC machine tool: (1) constant running condition (running at constant 3000 rpm) (2) progressive running condition (progressively increasing/decreasing rpm at 10 min interval) (3) random running condition (running at random rpm). The machine tool is running at 3000rpm constantly during 4 hours, then machine stops during another 4 hours in constant running condition. In progressive running condition, the spindle rotation is progressively increasing from such as 0 rpm, 1000 rpm, 2000 rpm, and 3000 rpm, then the spindle rotation is progressively decreasing from such as 3000rpm, 2000rpm, 1000rpm, and 0 rpm. In random running condition, the machine tool is running at random. In fig.3, temperature variations around machine tool and thermal spindle errors are shown over time in each spindle running condition. From the above three spindle thermal error measurement cases at the three typical running condition of machine tool, a strong relationship can thus be suggested between the temperature data and the spindle thermal error data. Therefore the relationship can be obtained by the thermal error modeling procedures which will be explained in the next section.

2.3 Spindle thermal error modeling

There have been several methods of the thermal error modeling for the machine tool, and three different methods of thermal error modeling are implemented and tested in this paper: multiple linear regression model, neural network model, and the system identification model.

- Multiple linear regression model

The relationship between the temperature data and the spindle error can be modeled as the multiple linear regression model. Let t_I(I=1,2...N) be the temperature data at several locations around the machine tool, then the thermal

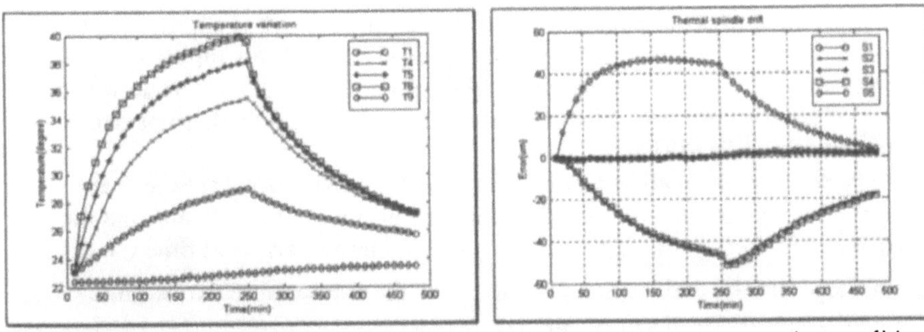

Fig.3a Temperature variations and thermal spindle variation in constant running condition

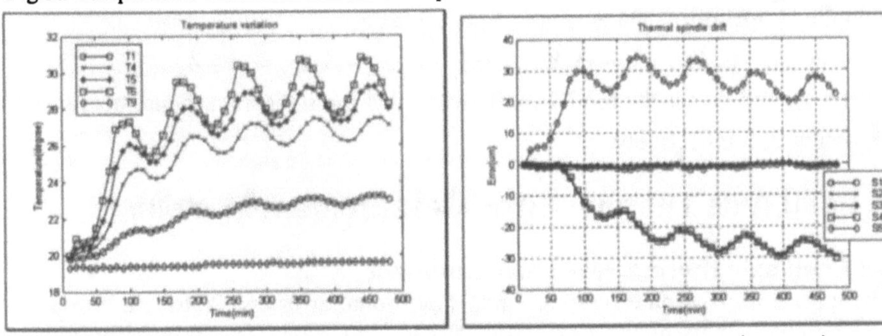

Fig.3b Temperature variations and thermal spindle variation in progressive running condition

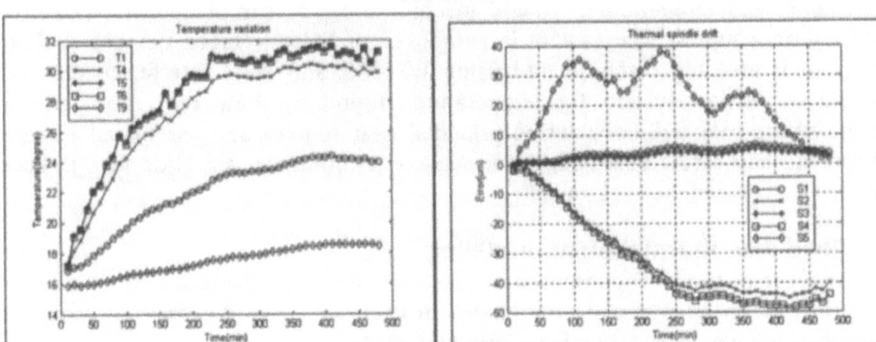

Fig.3c Temperature variations and thermal spindle variation in random running condition

error component, y_1 can be modeled as linear relationship as follows.

$$y_1 = a_{11}t_1 + a_{12}t_2 + a_{13}t_3 + a_{14}t_4 + \dots + a_{1n}t_n + b_1 \qquad (1)$$

where $a_{11}, a_{12}, a_{13}, a_{14} \dots a_{1n}$ are coefficient for temperature, b_1 is the constant for the thermal error model. Eq(1) can be extended to the 5 DOF spindle thermal errors, and thus can be represented in matrix form as follows

$$Y = AT \qquad (2)$$

Therefore the relationship between the thermal error and the temperature data can be found from equation(2), and the multiple linear regression model is completed.

- Neural network

The neural network is a kind of multiple nonlinear model in which the coefficients are called weights; and the coefficients are evaluated by training with an iterative technique called back propagation. Thus the neural network is appropriate to a system having the nonlinear relationship between the multiple inputs and the multiple outputs. For the thermal error application, the input layer is designed to consist of temperature data, output layer of the 5 DOF spindle thermal errors.

- System identification

In the system indentification model, the present thermal errors(at time t) are influenced by the present temperature data, the past temperature data, and the past thermal errors. Let $X_t, X_{t-1}, X_{t-2}, \cdots X_{t-n}$ be the thermal errors at time t, t-1, t-2, … t-n, respectively, and let $a_t, a_{t-1}, \cdots a_{t-n+1}$ be the temperature data at time t, t-1, … t-n+1, respectively. Then the thermal errors $X_t, X_{t-1}, X_{t-2}, \cdots X_{t-n}$ can be related with the temperature data $a_t, a_{t-1}, \cdots a_{t-n+1}$. That is,

$$X_t - \phi_1 X_{t-1} - \phi_2 X_{t-2} - \cdots - \phi_n X_{t-n} = \theta_1 a_t - \theta_2 a_{t-1} - \theta_3 a_{t-2} - \cdots - \theta_m a_{t-m+1} \quad (3)$$

where, ϕ_n and θ_m are the model coefficients, and the integers n, m are heuristically chosen as 3, 2 repectively.

3. Feed Axis Thermal Error Measurement/Modeling

3.1 Feed axis thermal error measurement

There are 3 translation errors(including 1 postional error and 2 straight errors) and 3 rotational errors(pitch, yaw, roll) along feed axis. The thermally induced translational errors, pitch, yaw errors are measured using the laser interferometer over time; and temperatures around machine tool are simultaneously measured with thermocouples. The system is composed of PC interfaced devices such as program interfacing machine tool using RS232C and laser interferometer; and having acqusition system for temperature around machine tool. Temperature locations are selected, considering principal heat sources and associated thermal distortion; ball screw bearings, linear scale, environmental air, slide face in each axis.

3.2 Feed axis thermal error modeling

Feed axis error model is composed of geometric term and thermal term.Geometric term has factor of nominal coordinates in each axis and the thermal term has factors of temperatures around machine tool, that is,

$$Error(P,T) = Error_g(P) + Error_t(T) \quad (4)$$

Where each term is defined as following:

$$Error_g(P) = a_0 + a_1 P + a_2 P^2 + \ldots \quad (5)$$

$$Error_t(T) = m_1(T) * P + m_2(T) \quad (6)$$

Temperature variation and positional error including thermal/geometric error in X axis is shown in fig.4. It has been found that the longer feed axis, the more thermal error machine tool has. The first term in equation(6) has physical meaning that thermal error is related to the length and temperature. The other thermally induced feed axis error such as straightness and angular errors are also modeled similarly. But angular error and straightness error have less variation than positional error over time as shown fig.5 and fig.6, while the angular errors are usually amplified by the Abbe's offset principle.

3.3 Volumetric error map and real time compensation

The volumetric errors are calculated using the homogeneous transformation matrix(HTM) including spindle thermal error and the feed axis thermal error. The HTM is defined as follows,

$$T_r^n = \begin{bmatrix} O_{11} & O_{12} & O_{13} & P_x \\ O_{21} & O_{22} & O_{23} & P_y \\ O_{31} & O_{32} & O_{33} & P_z \\ 0 & 0 & 0 & 1 \end{bmatrix} \qquad (7)$$

where **O** is related to angular terms and **P** is related translational terms. The actual errors at every nominal position are calculated in each axis as follows:

$$S_{AP_t} = T_A^D T_D^E P_t(p) \quad S_{AP_w} = T_A^C T_C^B P_w(p) \quad Error = S_{AP_w} - S_{AP_t} \qquad (8)$$

Real time compensation system is developed using the modified PLC program in CNC controller. The compensation data in each axis calculated using the HTM, and are sent to the controller. The developed system has been applied to a practical CNC machine tool, and the diagonal error is measured with laser interferometer for verification. The result is shown in fig.7, where the error was greatly reduced typically about 4-5 times.

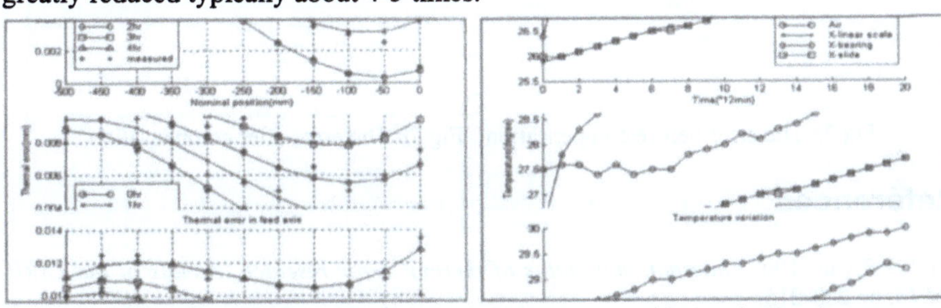

Fig.4 Temperature variation and positional error with time in X-axis

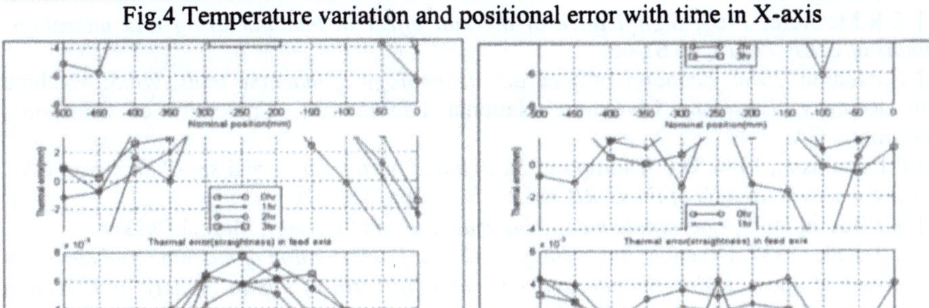

Fig.5 Thermal error in y-straightness and z-straightness in X-axis with time

4. Conclusion

Thermal error measurement and real tome compensation system has been successfully developed. The spindle thermal error and feed axis error including thermal and geometric error are modeled respectively. The real time compensation system is developed modifying the PLC program in CNC controller interfacing the PC which has algorithm calculating the volumetric error. After compensation, about 85% of error is removed. This result shows that the error modelings suggested have good performance.

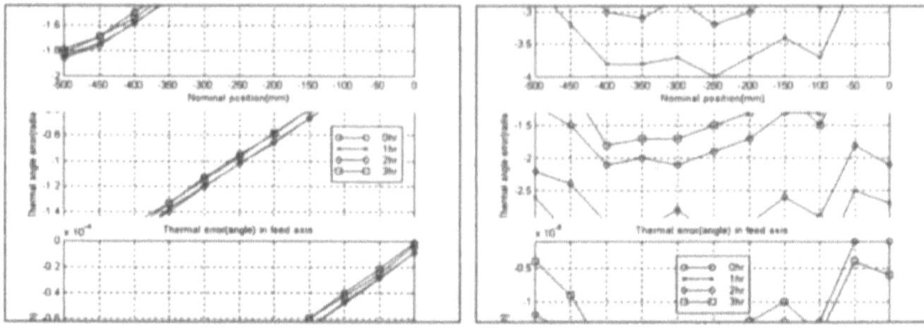

Fig.6 Themal pitch error and yaw error in X-axis with time

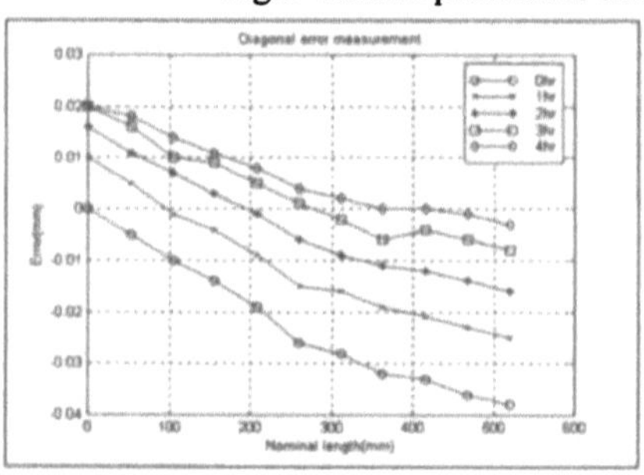

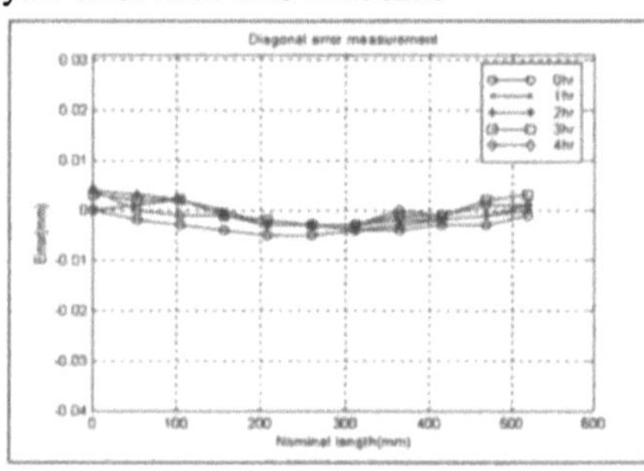

Fig.7a The error before compensation Fig.7b The error after compensation

References

[1] J.B.Bryan, 1968, *International status of thermal error research, Annals of the CIRP, Vol.16*, pp.203-215.

[2] E.R.McClure, 1967, *Significance of thermal effect in manufacturing and metrology, Annals of CIRP, Vol.15*, pp.61-66.

[3] G.P.Sutton, 1980, *Economy of accuracy, technology of machine tools, Vol.5*: Machine tool accuracy, Lawrence Livermore National Laboratories, University of California, Livermore, CA.

[4] R.Donaldson, 1986, D.C.Thompson, *Design and performance of a small precision CNC turning machine, Annals of the CIRP, Vol.35, No. 1*, pp. 373-376.

[5] A.J.Bosch, 1995, *Coordinate measuring machines and systems*, Marcel Dekker.

[6] K.Takada, 1987, I.Tanabe, *Basic study on thermal deformation of machine tool structure composed of epoxy resin concrete and cast iron, Bulletin of Japan Society of Precision Engineering, Vol.21*, No.3, pp. 173-178.

Closed Loop Dimensional Error Correction for 5 Axis Flexible Part Milling

Allan D. Spence[1] and Kevin P. Peffers

McMaster University, Mechanical Engineering, Hamilton, Ontario, Canada
[1] adspence@mcmaster.ca
www.mech.mcmaster.ca

Abstract: Because of deflection due to cutting forces, the surface of machined thin, flexible parts will be dimensionally incorrect. Although offline simulation can be used to predict a corrected tool path, significant data must be collected in advance to calibrate the process model. This paper presents a self-calibrating, closed loop correction strategy. The initial, climb milling pass is executed assuming no tool or part deflection. An inspection touch probe is then used to measure the dimensional errors. The process model is then used to calculate a second, 5 axis corrective pass. For subsequent parts, parameters are incrementally adjusted with the goal of achieving in tolerance dimensional errors after one pass. An RS232 serial connection between the machine tool controller and the process model microcomputer allows the system to run without operator intervention.

1. Introduction

In industry, machining of thin, flexible parts is frequently unavoidable. This is particularly true in the production of aircraft engine and turbopump blades, and aircraft frame components. Conventionally, cutting the part and subsequent inspection have been completed on separate machines. This lengthens the time delay before a process correction, and frequently produces parts that must be reworked or scrapped. Initial machining, inspection, and subsequent correction using a single setup reduces this time delay, as well as costly part handling. It also offers the opportunity to implement a self-calibrating process model and dimensional error correction strategy. Such a system is described in this paper. The remaining sections review existing work, describe the proposed closed loop strategy, and present some early experimental results. A final summary completes the paper.

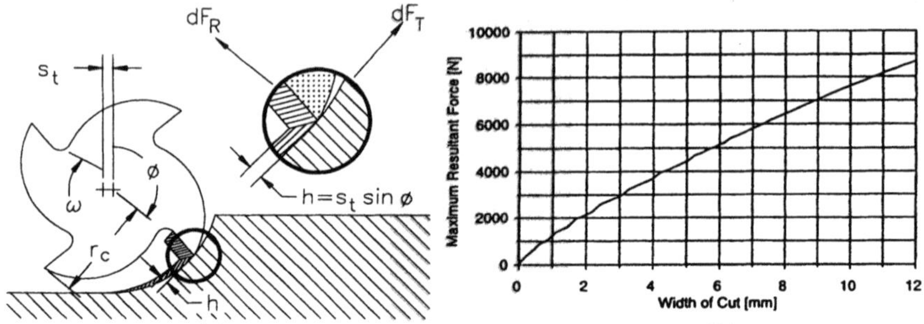

(a) **(b)**

Figure 1. Mechanistic milling process model

2. Previous Work

The metal cutting process has been extensively studied. A summary of process models was contributed by Smith and Tlusty [1]. In the mechanistic model used by Kline et al. [2], cutting forces are separated into tangential F_T and radial F_R differential components (Figure 1(a)) governed by the equations

$$dF_T = K_T h dz; \quad dF_R = K_R dF_T \tag{1}$$

where $h = s_t \sin\phi$ and dz is the differential element thickness along the z (spindle) axis. The value s_t is the feed per tooth. Note that the tooth is in cut only during the immersion segment $0 \le \phi_{st} \le \phi \le \phi_{ex} \le \pi$. The specific cutting pressure K_T is determined from cutting tests specific to the tool and part material combination, as is the radial/tangential force ratio K_R. Both quantities are expressed as exponential functions of the average chip thickness $\bar{h}$

$$K_T(\bar{h}) = M_T h^{P_T}; \quad K_R(\bar{h}) = M_R \bar{h}^{P_R} \tag{2}$$

where M_T, P_T, M_R and P_R are calibration constants, and

$$\bar{h} = s_t \frac{\cos\phi_{st} - \cos\phi_{ex}}{\phi_{ex} - \phi_{st}} \tag{3}$$

The vector sum of all differential force elements is calculated to determine the total force. A representative graph of the resultant cutting force vs. width of cut is shown in Figure 1(b). It is assumed that neither the tool nor the part deflects.

For cases where only the tool deflects, it can be represented as a cylindrical cantilever [3], with the force distributed over the dz slices. In this case the uncut chip thickness is expressed as $h = s_t \sin\phi - \delta_{normal}$, where δ_{normal} is the tool deflection component normal to the part surface. A representative paper is Sutherland and DeVor [4]. For modelling of chatter, time domain simulation is required [5]. When both the tool and part are flexible, the uncut chip thickness

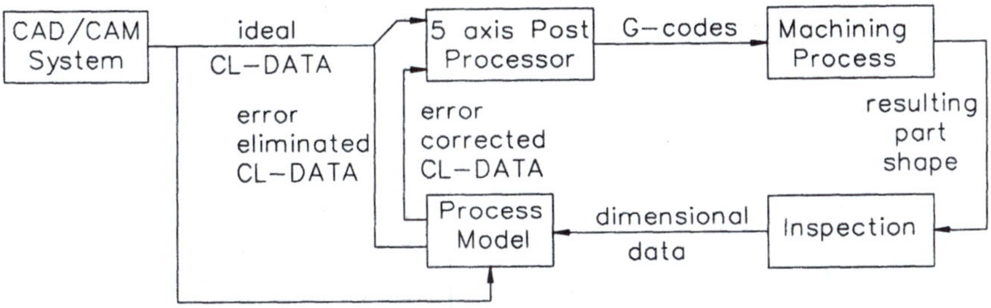

Figure 2. Proposed closed loop machining strategy

must be adjusted to account for both deflections [6]. Other influences, such as radial runout [7], must also be included where relevant.

These milling process models progressively increase in complexity and computation requirements. Calibration of the model becomes difficult, and is very sensitive to changes such as clamping force of the tool and part, material hardness variations, etc. For these reasons, a simple, self-calibrating closed loop machining strategy, that can be implemented using a standard 5 axis machine tool, touch probe, and microcomputer, was investigated.

3. Proposed Closed Loop Strategy

The proposed closed loop machining strategy is shown in Figure 2. The ideal, part co-ordinate system based Cutter Location Data (CL-DATA) tool path, assuming no flexibility, is planned using commercial, microcomputer based Computer Aided Design and Manufacturing (CAD/CAM) software. The post processor converts this to a machine tool specific co-ordinate system and controller G-code language. The G-codes are transmitted from the microcomputer to the Computer Numerical Control (CNC) controller via an RS232 serial connection to the controller, and the part is machined. A touch probe is then loaded into the machine tool spindle, and a grid of part surface dimensional data is collected. This data is transmitted back to the microcomputer, using the same RS232 serial connection. The microcomputer transforms the inspection data from machine tool to part co-ordinates (inverse post processing), and, using a simplified process model, calculates a new, 5 axis, error corrected CL-DATA tool path. This is then post processed, transmitted to the CNC controller, and a second machining pass is made. To minimise the overall machining cycle time, computations are performed in parallel with inspection data collection using multiple threads. A spoke style fixture is used to permit carousel style machining and inspection [8]. For subsequent parts, an error eliminated CL-DATA tool path is substituted. As the self-calibrating process model learns, the error eliminated CL-DATA tool path is expected to converge such that an in tolerance part is produced after only one pass. Cutter runout is accommodated by initially planning a material safe pass that is offset by the largest expected runout

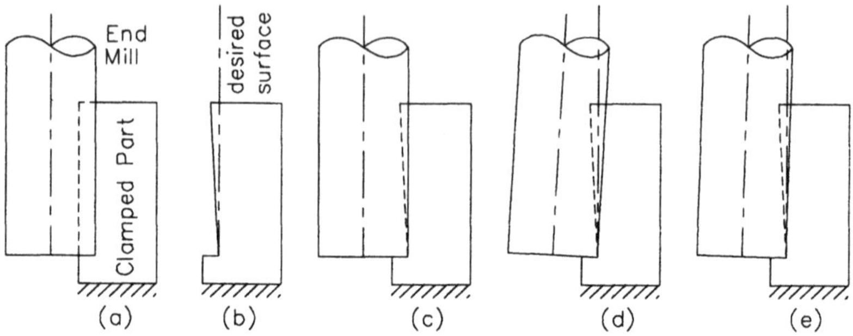

Figure 3. Error profiles and correction strategies.

radius. Compensation for tool wear is automatically accomplished by re-calibrating the model after each machining pass.

The primary goal of the system is to correct dimensional surface errors. During machining of flexible parts, deflection will cause the machined surface to be generated both translated and oriented away from the desired location (Figure 3(a,b)). The simplest second pass strategy is to cut again with the original tool path (Figure 3(c)). For very flexible parts, this method is not sufficiently aggressive, and significant dimensional error will remain. Conversely, offsetting the second pass by the total measured error will lead to overcutting (Figure 3(d)). A simple compromise strategy, based on fundamental knowledge of the machining process, and that estimates the part deflection (Figure 3(e)), was therefore investigated.

Assuming no deflection, the first machining pass is planned with a width of cut w_{command}, but produces a surface dimensional error e. The achieved width of cut is therefore $w_{\text{actual}} = w_{\text{command}} - e$. Defining the compliance factor $\mathrm{d}\delta = e / w_{\text{actual}}$, the corrected width of cut, to be used for the second pass, is

$$w_{\text{command,corrected}} = e\mathrm{d}\delta \tag{4}$$

This correction is applied for each inspected point on the part surface, and the tool path translation and orientation data is smoothly interpolated using B-spline curves. Linear lead-in extensions are added at the beginning and end of cut. A new, error corrected CL-DATA file results. The 5 axis post processor converts the CL-DATA file to a closely spaced series of linear motions that are executed on the machine tool using inverse time continuous contouring.

For subsequent parts, the same procedure is used for the error eliminated CL-DATA file, except that the new width of cut is calculated as

$$w_{\text{command.eliminated}} = (1 + \mathrm{d}\delta) w_{\text{command}} \tag{5}$$

A second error correction pass is used only if inspection detects that the surface is out of tolerance.

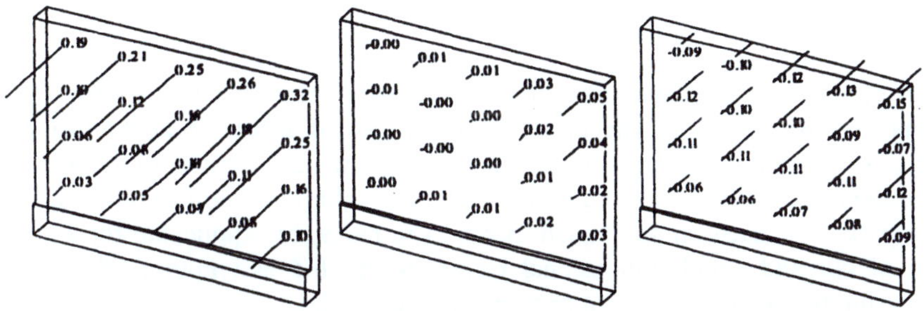

(a) (b) (c)

Figure 5. Representative experimental results. Units: mm

4. Experimental Results

The closed loop machining strategy was assessed by wet machining aluminium parts with three different geometries [9]. The machining centre was a Fadal [10] VMC 4020 with table mounted tilt/swivel 4th and 5th axes. A 25.4 mm diameter four flute helical end mill was used as the cutting tool. A low spindle speed of 600 rpm was used to avoid chatter. The feed was 0.152 mm per tooth. Radial runout (highest to lowest tooth) was 8 μm. The axial depth of cut was 40 mm. The touch trigger probe was a Renishaw [11] model MP12. Geometric modelling and data processing was performed with the aid of the Spatial Technology Inc. [12] ACIS solid modeller. Full results are available in [9].

Figure 5 shows representative results for a Clamped-Free-Free-Free (CFFF) part, with an initial width of cut linearly decreasing from 1.5 mm to 0.5 mm (down milling, over a 60 mm path). The error corrected CL-DATA pass reduced the maximum dimensional error from 0.32 mm (Figure 5(a)) down to 0.05 mm (Figure 5(b)). When the second part was cut, using the error eliminated CL-DATA pass, an overcut of 0.15 mm (Figure 5(c)) resulted at the corresponding location on the part surface. This occurred because the actual relationship between width of cut and cutting force is non-linear (Figure 1(b)), but a linear relationship at $w_{command}$ is assumed (Equation (4)). For the error corrected CL-DATA pass, the part is more flexible due the removal of material from the initial pass. Actual cutting forces, however, are lower than predicted by the linear force model, and these effects tend to cancel. With the error elimination CL-DATA pass, the part has the same flexibility as with the initial pass. The actual cutting forces are again lower than the linear force model predicts, the part bends less than predicted, and overcutting occurs. No overcut was observed with a constant 0.5 mm initial width of cut, suggesting that this problem can be reduced by adjusting the error eliminated CL-DATA pass more slowly. As more parts are cut, sufficient information can be gathered to improve the model beyond a simple linear relationship.

5. Summary

This paper presented a closed loop dimensional error correction strategy for 5 axis milling of flexible parts. The initial pass is executed assuming no tool or part deflection. After inspection, a second, corrected pass is made to reduce the dimensional error. The model automatically self-calibrates and calculates a new initial pass to be used on subsequent parts. The method is applicable for a small width of cut and small dimensional errors, where the assumption of a linear relationship between width of cut and error remains valid.

Acknowledgements

This work was supported by the Natural Sciences and Engineering Research Council of Canada (NSERC). Dr. M.A. Elbestawi co-supervised the work.

References

[1] Smith, S. and Tlusty, J., 1991, "An Overview of Modeling and Simulation of the Milling Process", *Trans. ASME, J. Eng. Ind.*, v. 113, pp. 169-175.

[2] Kline, W.A., DeVor, R.E., and Lindberg, R., 1982, "The Prediction of Cutting Forces in End Milling with Application to Cornering Cuts", *Int. J. Mach. Tool Des. Res.*, v. 22, n. 1, pp. 7-22

[3] Kops, L. and Vo, K.T., 1990, "Determination of the Equivalent Diameter of an End Mill based on its Compliance", *Annals of the CIRP*, v. 39/2, pp. 92-96.

[4] Sutherland, J.W. and DeVor, R.E., 1986, "An Improved Method for Cutting Force and Surface Error Prediction in Flexible End Milling Systems", *Trans. ASME, J. Eng. Ind.*, v. 108, pp. 269-279.

[5] Smith, S. and Tlusty, J., 1993, "Efficient Simulation Programs for Chatter in Milling", Annals of the CIRP, v. 42/1, pp. 463-466.

[6] Kline, W.A., DeVor, R.E., and Shareef, I.A., 1982, "The Prediction of Surface Accuracy in Milling", *Trans. ASME, J. Eng. Ind.*, v. 104, pp 272-278.

[7] Kline, W.A. and DeVor, R.E., 1983, "The Effect of Runout on Cutting Geometry and Forces in End Milling", *Int. J. Mach. Tool Des. Res.*, v. 23, pp. 123-140.

[8] Fleisig, R.V. and Spence, A.D., 1997, "Integrated Digitizing, Path Planning and Five-Axis Machining for the Refurbishment of Compressor and Turbine Blades", *Proc. ASME 1997 Int. Mech. Eng. Conf., Symposium on Sensor-Fused Intelligent Machining Systems*, MED-Vol. 6-1, pp. 31-38, Dallas, TX.

[9] Peffers, K.P., 1997, *Closed-Loop Five Axis Milling of Flexible Parts*, Master's Thesis, McMaster University, Hamilton, Ontario, Canada.

[10] Fadal, Chatsworth, CA, USA, www.fadal.com.

[11] Renishaw plc, Gloucestershire, UK, www.renishaw.com.

[12] Spatial Technology Inc., Boulder, CO, USA, www.spatial.com.

MACHINE TOOL ACCURACY II

A New Measuring Apparatus for Verifying The CNC Machine Tools Performance Using PSD

[1]Chieh-Li Chen and [2]Wen-yuh Jywe
[1]Institute of Aeronautics and Astronautics, National Cheng-Kung University, Tainan, Taiwan 701. Email: ckchen@mail.ncku.edu.tw
[2]Department of Automation Engineering, National Huwei Institute of Technology, Huwei, Yunlin, Taiwan, 632. Email: Jywe@sunws.nhit.edu.tw

Abstract: In this research, the implementation for measuring dynamic computer-numerically controlled (CNC) contour using the position silicon detectors (PSD) was investigated. The motivation behind this work is the desire to perform rapid, simplicity of setup, low cost and pre-process verification of CNC contouring path. This reduces scrap and increase process efficiency. The integrated Dual Laser PSDs System (DLPS) was designed, constructed and tested in this paper. Complete setups and procedures for dynamic contouring error and geometric error verification were also clearly presented. Experimental results showed that more information on analyzing the types of errors can be obtained and the performance of verification can be further improved.

1. Introduction

Machine uncertainty determinates the quality of the machined part mainly. It is of importance to check the performance of the machine tool systematically for direct quality control purpose or to compensate for this uncertainty. Schlesinger 1932 [1] first provided the systematic testing method for machine tools. This method was developed as the basis of ISO standard. Tlusty, Koenigsberger [2] and Burdekin [3] indicated new testing rules for machine tools. The tests for machine tool performance were then classified into direct cutting test and indirect cutting test. Ericson [4] described the first work zone of machine tools. Burdekin [5] provided the cutting tests with the circular path for accuracy assessment. J. B. Bryan [6] developed the first ball bar system for the contouring test. However, in this system, the uncertainty is high due to the friction between the master balls and the magnetic sockets and no accuracy contouring radius was given. Knapp's system [7,8] used the circle comparison standard disc mounted on the tested table of the machine tool and the 2D-probe. The problems on this system are the existence of the influence of friction between the 2D probe and the disc, its very small bandwidth which causes the system to be unusable for high speed contouring tests, and the high cost of the 2D probe. M. Burdekin and W. Y. Jywe [9] provided a method to diagnose the contouring error and to adjust the parameters of the CNC controller to optimize the performance of the tested CNC machine tool.

From the previous researches, some existence limitations must be further investigated and improved. The limitations are summarized as follows:

1. For the geometric error test:
   a. The sampling device is expensive.
   b. It takes much time to calibrate the 21 error components.
   c. The device for geometric error test is always not suitable for the circular test if the complete calibrating is considered.

2. For the circular test:
   a. It is difficult to analyze the individual error of single axis.
   b. The device for the circular test is always not suitable for the geometric error test.
3. The above developed systems or methods may not be suitable for a desktop CNC machine tools.

In this paper a simple measuring device is thus designed and developed to calibrate a desktop CNC machine tool and to check the contouring performance with its single axis output.

2. The Dual Laser PSDs System (DLPS)

The Dual Laser PSDs System is composed of two sub-systems, which are set vertically to each other. Fig.1 presents the principle of the DLPS, composed of two PSDs (Position Silicon Detector), two laser beams (provided by two laser sources). Each PSD receiving the laser beam decodes the position of each axis during a circular test. These two laser beams are located at the two PSDs. During the 2D circular test, the positioning data of each axis can be sampled by the PSD, interfaced with the computer by the A/D card and processed by the developed software to analyze either the geometric error or the contouring error.

The PSD, Model SL76-1 provided by UDT Sensors INC., provides a 76mmx 2.5mm working area. The voltage output of the PSD is proportional to the location, which the laser beam projects on.

2.1 Using the DLPS to calibrate the geometric error

The determination of the position error

Fig. 2 shows the measuring frameworks of measuring position error. The PSD_x was put to coincide with the X-axis of the setup working zone on the machine table. The laser beam perpendicularly lights on the PSD_x. When the table moves along the X axial direction, the changed displacement of light spot can make the PSD_x reflect the displacement. When the machine table arrived at a target point P_{ij}, the position error at the target point can be obtained by means of Eqn. (1).

$$\delta_x(x)_{ij} = P_{ij}(X,0,0|PSD_x) \bullet P_{ij} \qquad (1)$$

where $P_{ij}(X,0,0|PSD_x)$ = the displacement obtained from PSD_x when the table moves along the X direction.

$P_{ij}(X,Y,Z|PSD_n)$ = X,Y and Z are denoted as the direction of the machine table moving

P_{ij} = the displacement obtained from the machine tools at the jth target in the ith path.

The determination of the straightness error and the pitch and yaw angular errors

Fig. 3 shows the measurement instrument for straightness error, pitch and yaw angular error. As shown in Fig. 3(a), when the table moves along the X direction and the laser beam light on the PSD_y, the PSD_y samples the errors combined by straightness and yaw. The yaw angular error can be determined as the slope of the minimum zone line obtained from the data error fitting. The straightness can be determined by solving the minimum zone tolerance. The other frameworks for other straightness, pith and yaw errors are shown in Fig. 3(b)~(f). The measuring methods are similar to the above method.

The determination of the roll angular error

The measuring frameworks for roll angular error measurement are shown in Fig. 4. As shown in Fig. 4(a), two PSDs are vertically put on the machine table with an offset y_o in Y axial direction. Two parallel laser beams vertically light on these two PSDs. When the machine table moves along the X axial direction, the roll angular error, $\varepsilon_x(x)$, of X axis can be obtained by comparing the sample difference between these two PSD's. Then

$$\varepsilon_x(x) = [P(X,0,0|PSD_{z1}) - P(X,0,0|PSD_{z2})] / y_0 \tag{2}$$

As the similar frameworks shown in Figs. 4 (b) and (c), the roll angular errors of Y and Z axes can be obtained by means of Eqns. (3) and (4).

$$\varepsilon_y(y) = [P(0,Y,0|PSD_{z1}) - P(0,Y,0|PSD_{z2})] / x_0 \tag{3}$$

$$\varepsilon_z(z) = [P(0,0,Z|PSD_{x1}) - P(0,0,Z|PSD_{x2})] / y_0 \tag{4}$$

The determination of the squareness

Fig. 5 shows the basic diagonal measurement for squareness error measurement. Fig. 6 shows the framework for squareness error measurement using PSD. One PSD_{xy} is put and fixed on the machine table with an angle• between PSD_{xy} and X axis. As shown in Fig. 5, when the machine table move along the $\overline{OA}$ direction, the displacement $P_i(X,Y,0 \mid PSD_{xy})$ via PSD_{xy} can be obtained in each target points• 1st, 2nd,....Nth•. Then

$$P_i(X,Y,0 \mid PSD_{xy}) = X_p \cos\theta + Y_p \sin\theta \tag{5}$$

From Eqns. (1) and (2), then X_p and Y_p can be written as

$$X_p = X + E_x$$

or

$$X_p = X + \delta_x(x) + \delta_x(y) - Y\varepsilon_{xy} - Y\varepsilon_z(x) \tag{6}$$

$$Y_p = Y + E_y$$

or

$$Y_p = Y + \delta_x(x) + \delta_y(y) \tag{7}$$

Take the obtained $\delta_x(x)$, $\delta_x(y)$, $\delta_y(y)$ and $\varepsilon_z(x)$ into Eqns. (6) and (7). Then

$$\overline{X} = X + \delta_x(x) + \delta_x(y) - Y\varepsilon_z(x) \tag{8}$$

$$\overline{Y} = Y + \delta_x(x) + \delta_y(y) \tag{9}$$

Take the Eqns. (6),(7),(8) and (9) into Eqn. • 5 •, then

$$P_i(X,Y,0 \mid PSD_{xy}) = \overline{X}\cos\theta + \overline{Y}\sin\theta - Y\varepsilon_{xy}\cos\theta \tag{10}$$

The squareness error in each target can be solved by means of Eqn. (11).

$$\varepsilon_{xy} = -\frac{P_i(X,Y,0 \mid PSD_{xy}) - \overline{X}\cos\theta - \overline{Y}\sin\theta}{Y\cos\theta} \tag{11}$$

3. Verification of the contouring performance with the DLPS

As mentioned above, with the general circular test device only radius error can be obtained. To analyze the contouring error in each individual axis, the DLPS is thus developed.

3.1 The contouring test with DLPS

Fig. 1 shows the framework of the DLPS. Two PSDs, PSD_x and PSD_y, are orthogonally set on the machine table. The laser sources are set outside the table. The individual error in each contouring axis is sampled by each PSD respectively. During the contouring test, two orthogonal PSDs sample the error signals of the X- and Y- axis which are denoted as $P_i(X,Y,0 \mid PSD_x)$ and $P_i(X,Y,0 \mid PSD_y)$. These signals are combined together to obtain the complete contouring result. The combined signal, R, can be written as

$$R = \left(P_i(X,Y,0 \mid PSD_x)^2 + P_i(X,Y,0 \mid PSD_y)^2\right)^{1/2} \tag{12}$$

4. Experiment results of the contouring test using DLPS

The DLPS was herein used to check the dynamic contouring error of the CNC machine tools. The DLPS is simplicity of setup, and quick measurements without the problem of using the cable in the traditional contouring system. Thus, the multi-circles or non-circle test, such as an ellipse test, a rectangle test or a curve test can also be implemented. The experiment results for testing the CNC machine tools are shown in Fig. 7, which shows the X and Y path errors. The radius of the contouring path is 2.5 mm. The test results are the average of the sampling signals for 3 run. The feed rate is 100 mm/min.

5. Conclusion

A complete integrated system DLPS for verifying the performance of the CNC machine tools is presented in this paper. The DLPS system improves the traditional measurement method for checking the dynamic and static errors. In this system the error signals of each axis can be respectively obtained for the geometric error test and then combined together for the contouring test. By using the integrated DLPS, the measurement methods, characterized by low cost, simplicity of setup, and quick measurements are developed to directly verify the dynamic contouring error of the CNC machine tool.

Reference

1. G. Schlrsinger., "Inspection Test on Mechine Tools," *Machinery publishing Co. Ltd*, Lodon (1932)
2. J. Tlusty and F. Koenigsberger., "Specifications and Tests of Metal Cutting Machine Tools," *UMIST Manchester*, (1970)
3. G. Schlrsinger, F. Koenigsberger and M. Burdekin., "Testing Mechine Tools," *8th edition Machinery Publishing Co. Ltd.*, London (1978)
4. C. Ericson., "Machine Alignment-The First Step Towards Product Accuracy," *Tech. Paper MM66-171, ASTME* (1966)
5. M. Burdekin., "Cutting Tests for Accuracy Assessment", *Technology of Machine Tools*, Vol. 5, pp. 9-11 (1980)
6. J. B. Bryan., "A Simple Method for Testing Measuring Machines and Machine Tools," *Prec. Eng*, Vol. 4, No. 2, pp. 61-69.
7. W. Knapp., "Test of the three-dimensional Uncertainty of Machine Tools and Measuring Machines and its Relation to Machine Errors," *Annals of CIRP*, Vol. 32, NO. 1, pp. 459-464 (1983)
8. W. Knapp., "Circular Test for Three Coordinate Measuring Machine Tools," *Prec. Eng*, Vol. 5, No. 3, pp. 115-124 (1983)
9. M. Burdekin and Jywe W.Y., "Optimizing the Contouring Accuracy of CNC Machines Using the Contisure System," *29th Int'l MATADOR Conference*, April, (1992)

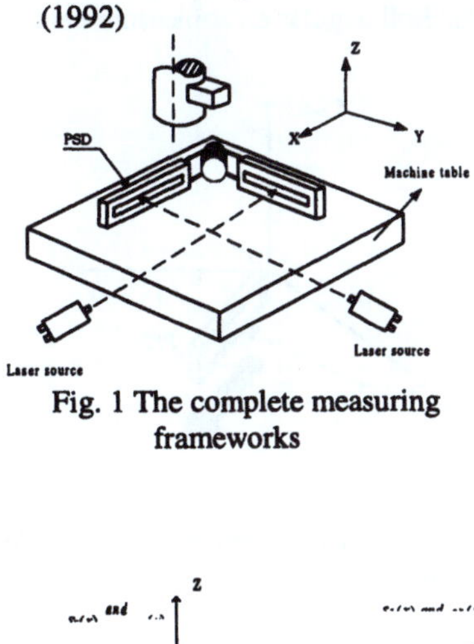

Fig. 1 The complete measuring frameworks

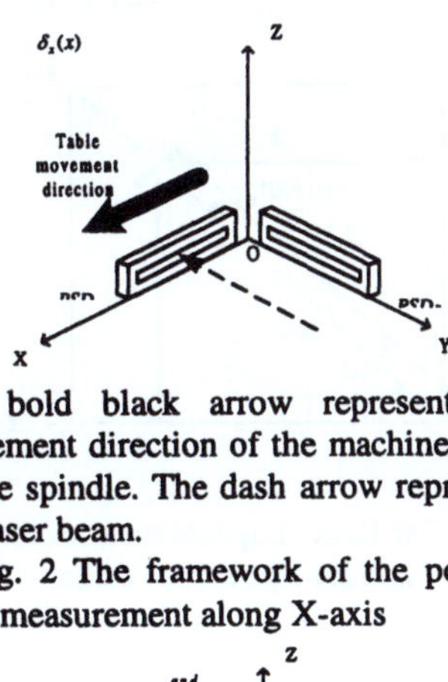

The bold black arrow represents the movement direction of the machine table/ or the spindle. The dash arrow represents the laser beam.

Fig. 2 The framework of the position error measurement along X-axis

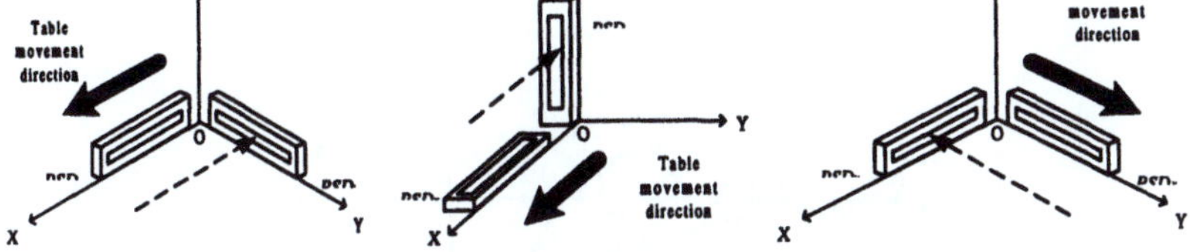

(a) (b) (c)

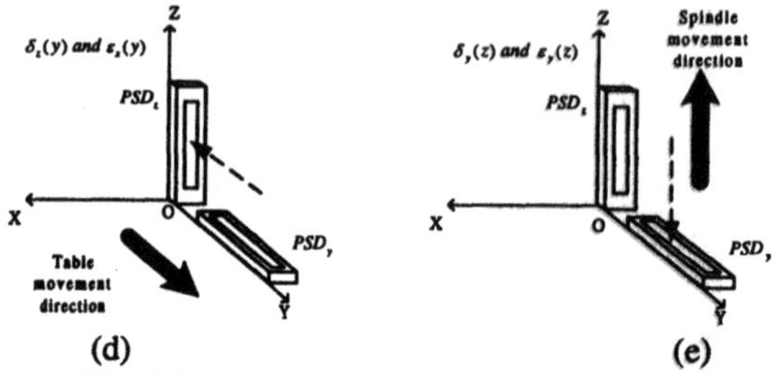

Fig. 3 The complete measurement frameworks of straightness error, pitch and yaw angular errors measurement

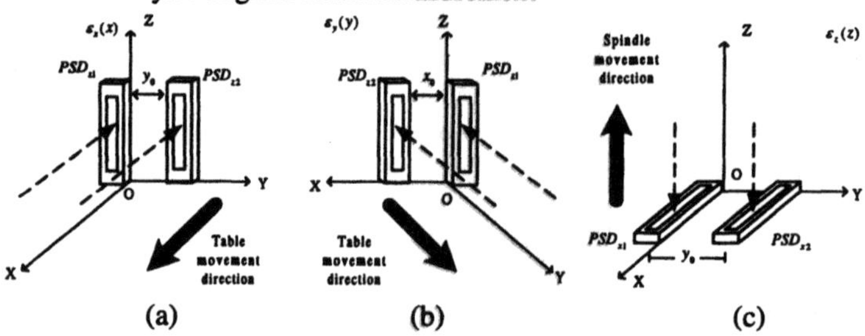

Fig. 4 The complete measurement framework for Roll angular error measurement

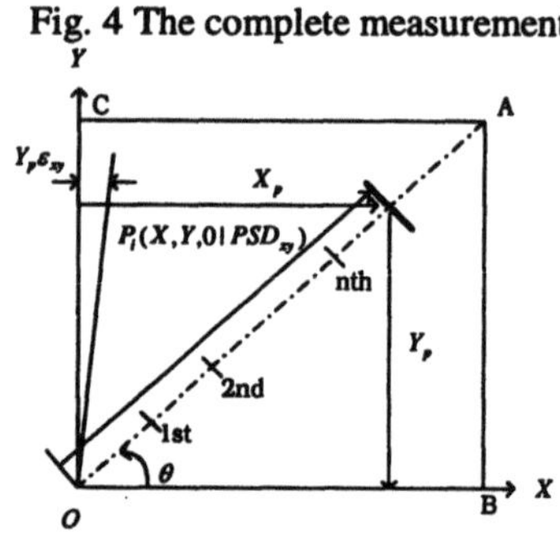

Fig. 5 The basic diagonal measurement principle for squareness error

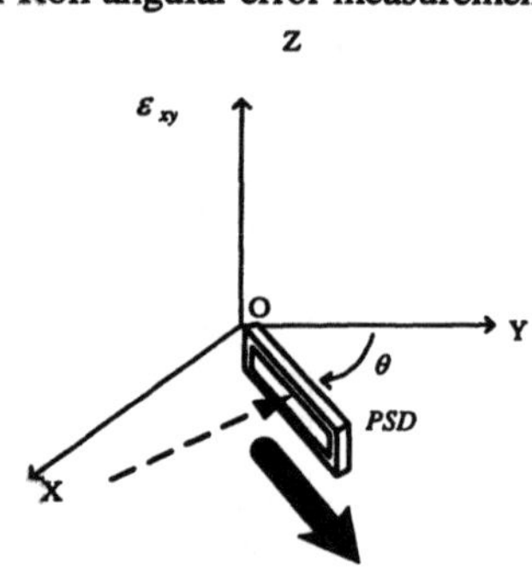

Fig. 6 The measurement framework for squareness error measurement

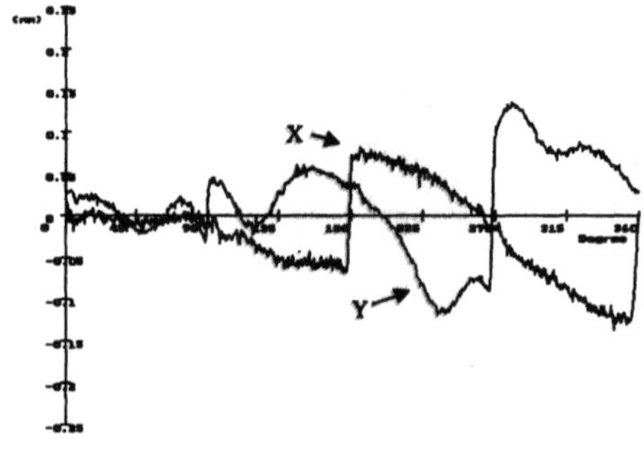

Fig. 7 The contouring results in the X and Y axis by using DLPSs

Developing contouring testing device for CNC machine tools

Jywe Wen-Yuh
Department of Automation Engineering, National Huwei Institute of Technology, Huwei, Yunlin, Taiwan, 632
Address: 64 WenHua Road, National Huwei Institute of Technology Huwei, Yunlin, Taiwan. ROC
Email: Jywe@sunws.nhit.edu.tw

Abstract: This paper presents a non-connect contouring system for CNC machine dynamic performance test. This system not only deals with the testing of the dynamic performance of CNC machine tools, but also can be extended to test the dynamic contouring performance under cutting situation. The experimental results show that this system can be employed for CNC machine tools test.

1. Introduction

Machine uncertainty determinates the quality of the machined part mainly. It is of importance to check the performance of the machine tool systematically for direct quality control purpose or to compensate for this uncertainty. Schlesinger 1932 [1] first provided the systematic testing method for machine tools. This method was developed as the basis of ISO standard. Tlusty 1980 [2] proposed a non-cutting testing method. The tests for machine tool performance were then classified into direct cutting test and indirect cutting test. Meanwhile circular tests were developed to check both geometric error and contouring error. Burdekin [3] provided the cutting tests with the circular path for accuracy assessment. J. B. Bryan [4] developed the first ball bar system for the contouring test. However, in this system, the uncertainty is high due to the friction between the master balls and the magnetic sockets and no accuracy contouring radius was given. Knapp's system [5,6] used the circle comparison standard disc mounted on the tested table of the machine tool and the 2D-probe. The problems on this system are the existence of the influence of friction between the 2D probe and the disc, its very small bandwidth which causes the system to be unusable for high speed contouring tests, and the high cost of the 2D probe. Y. Kakinov [7,8,9,10] provided a serious of method by using the ball bar system to calibrate a coordinate measuring machine and CNC machine tools. W. Knapp [11,12] described the rule to reduce the errors due to the stick-slip and others. M. Burdekin and J. Park 1988 [13] modified the original ball bar system by employing a four-rod linkage. M. Burdekin and W. Y. Jywe [14] provided a method to diagnose the contouring error and to adjust the parameters of the CNC controller to optimize the performance of the tested CNC machine tool. The proposed system in this paper can be used for CNC machine tools test and can be extended to test the contouring performance under cutting situation.

2. The developed Non-Contact contouring system (NCCS)

2-1. The testing device system of NCCS

The proposed NCCS is shown schematically in Fig.1. The system is composed of one high precision reference spheres rigidly mounted at spindle, four rods, a L-type fixed part, a fixed part, the magnetic base, a rotating axis and a non-contact sensor. A carbon fiber transducer link containing one non-contact sensor (eddy current sensor) is located between the reference spheres. The sensor output represents the change in centre distance of the sphere as the machine performs a circular contouring operation.

2-2 Sampling procedure

Fig. 2 shows the sampling procedure, which ensures that the sampled data are the reliable actual circular contouring data. At the first stage, sample are taken to check the whole condition, which includes the NCCS, the machine table and the environment are acceptable. It is necessary to calibrate the sampling speed in the various CPUs, A/D card and other sampling devices to control the proper sampling number during the auxiliary memory. During the sampling, the sampled data cannot be sent to the auxiliary memory such as the floppy or the hard disk because the sampling will be stopped during the data recording. This method let per sampling period to be the same.

3. Extend NCCS for cutting test

The NCCS are herein used for testing the dynamic performance of CNC machine tools during the cutting process. The NCCS can not only be used for testing CNC machine tools under non-cutting situation, but also can be extended to test the dynamic performance during cutting process.

3-1 Experimental result

NCCS was employed to test the dynamic performance of CNC machine tools without spindle rotating. After that, it was used again to test the dynamic performance of CNC machine tools under spindle rotating. Figure 3 shows the result at Feed rate 2000 mm/min ,clockwise-contouring and without spindle rotating. Figure 4 shows the result under Feed rate 2000 mm/min, anti- clockwise contouring and without spindle rotating. Figure 5 shows the result at Feed rate 2000 mm/min , clockwise contouring and with spindle rotating speed 1000 rpm. Figure 6 shows the result at Feed rate 2000 mm/min ,anti-clockwise contouring and with spindle rotating speed 1000 rpm.

The contouring error bands in Fig.5 and Fig.6 are much larger than those in Fig.3 and Fig.4. This presents significant information that in the actual cutting condition the eccentricity between the center of the cutter and the center of the spindle causes dominate contouring errors, while compared to that without cutting. It is more important to study the way to reduce the eccentricity than to optimize the contouring parameters without cutting.

4. Conclusion

The proposed system NCCS in this paper was successfully used for CNC machine tools test and extended to test the contouring performance under cutting situation. The test results can be further used to evaluate the cutting situation in practice for in-processing test.

Acknowledgments:

The work was supported by National Science Council, Taiwan, Republic of China Number NSC-86-2212-E-150-013 (Group: Precision Measurement).

References:

[1] G. Schlrsinger., "Inspection Test on Mechine Tools," *Machinery publishing Co. Ltd*, Lodon (1932)

[2] J. Tlusty., "Testing of accuracy of NC Machine Tools," T*echnology of Machine Tools, Supplement 1*.(1980)

[3] M. Burdekin., "Cutting Tests for Accuracy Assessment", *Technology of Machine Tools*, Vol. 5, pp. 9-11 (1980)

[4] J. B. Bryan., "A Simple Method for Testing Measuring Machines and Machine Tools," *Prec. Eng*, Vol. 4, No. 2, pp. 61-69.

[5] W. Knapp., "Test of the three-dimensional Uncertainty of Machine Tools and Measuring Machines and its Relation to Machine Errors," *Annals of CIRP*, Vol. 32, NO. 1, pp. 459-464 (1983)

[6] W. Knapp., "Circular Test for Three Coordinate Measuring Machine Tools," *Prec. Eng*, Vol. 5, No. 3, pp. 115-124 (1983)

[7] Y. Kakino., "Compensation of 3-Dimensional Coordinate Measuring Machine and Evaluation of Double Ball Bar Test by Error Vector Analysis," *Journal of JSPE*, Vol. 51, No. 6, pp. 1224-1250 (1985)

[8] Y. Kakino., Y. Ihara and Y. Nakatsu., "The Measurement of Motion Errors of NC Machine Tools and Diagnosis of Their Origins by Using Telescoping Magnetic Ball Bar Method," *Annals of CIRP*, Vol. 36, No. 1, pp. 337-380 (1987)

[9] Y. Kakino., Y. Ihara, A. Kamei and T. Ise., "Study on the Motion Accuracy of NC Machine Tools (1st Report) - The Measurement and Evaluation of Motion Errors by Double Ball Bar Test," *Journal of JSPE*, Vol. 7, pp. 1193-1198 (1986)

[10] Y. Kakino, Y. Ihara and Y. Nakatsu., "A Study on the Motion Accuracy of NC Machine Tools (2nd Report) - Diagnosis of Motion Error Origins by Using Double Ball Bar Test," *Journal of JSPE*, Vol. 52, No. 10, pp. 1739-1745 (1986)

[11] W. Knapp., "Circular Test on NC Machine Tools in Industrial Application," *Journal of Industrial an Production Engineering*, Vol. 10, No. 3, pp. 99-100 (1986)

[12] W. Knapp., "Accuracy of Length Measurement and Positioning: Statical Measurement and Contouring," *Annals of CIRP*, Vol. 37, NO. 1, pp. 511-514 (1988)

[13] M. Burdekin. and J. Park., "CONTISURE - A Computer Aided System for Assessing the Contouring Accuracy of NC Machine Tools," 27th Int'l MATADOR Conference, pp. 197-203 (1988)

[14] Burdekin and Jywe W.Y., "Optimizing the Contouring Accuracy of CNC Machines Using the Contisure System," *29th Int'l MATADOR Conference*, April, (1992)

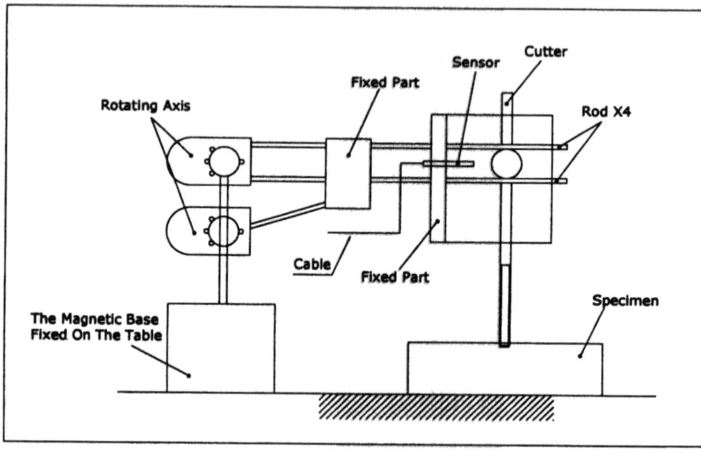

Fig. 1 The non-contact system set-up

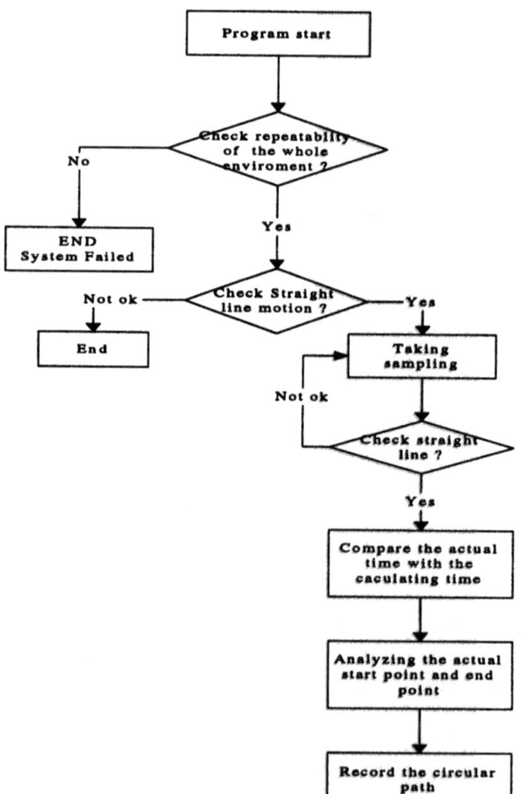

Fig. 2 Sampling procedure

Feedrate: mm/min	2000
Spindle speed: (rpm)	0
Contouring radius: (mm)	153
Direction:	Anticlockwise
Out of Roundness:(•m)	30

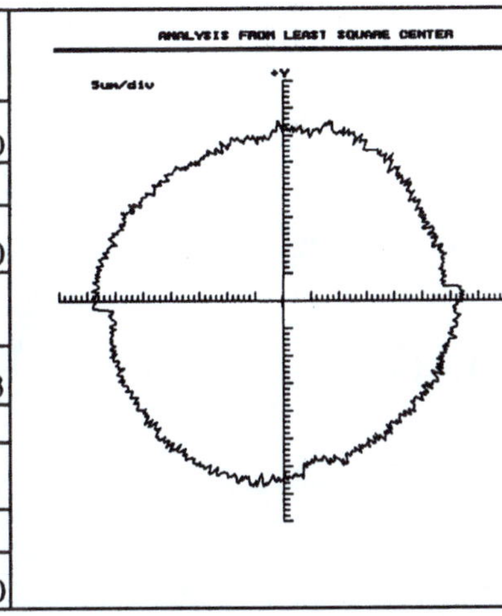

Fig. 3 The result under Feed rate 2000 mm/min and clockwise without spindle rotating.

Feedrate: (mm/min)	2000
Spindle speed: (rpm)	0
Contouring Radius: (mm)	153
Direction:	Clockwise
Out of Roundness:(•m)	28

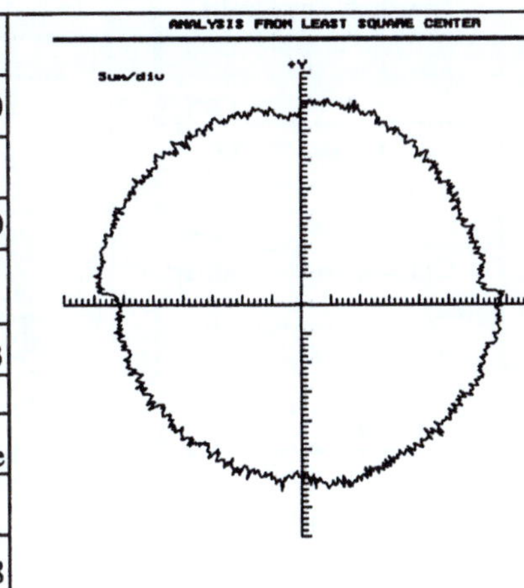

Fig. 4 The result under Feed rate 2000 mm/min and anti clockwise without spindle rotating.

Feedrate: mm/min	2000
Spindle speed: (rpm)	1000
Contouring Radius(mm)	136
Direction:	Clockwise
Out of Roundness:(•m)	228

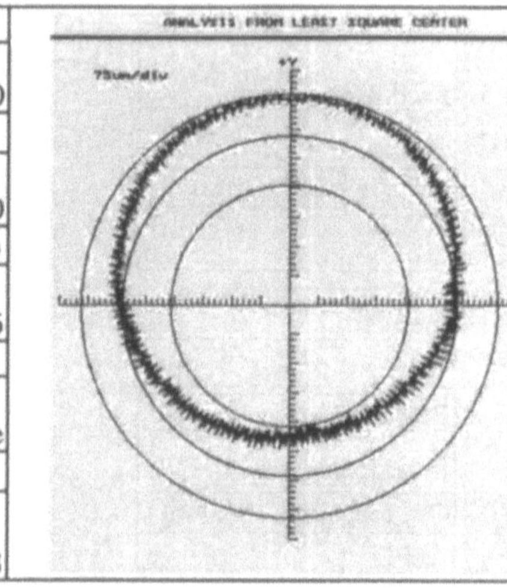

Fig. 5 The result under Feed rate 2000 mm/min and clockwise with spindle rotating speed 1000 rpm.

Feedrate: mm/min	2000
Spindle speed: (rpm)	1000
Contouring radius: (mm)	153
Direction:	Anti-Clockwise
Out of Roundness:(•m)	306

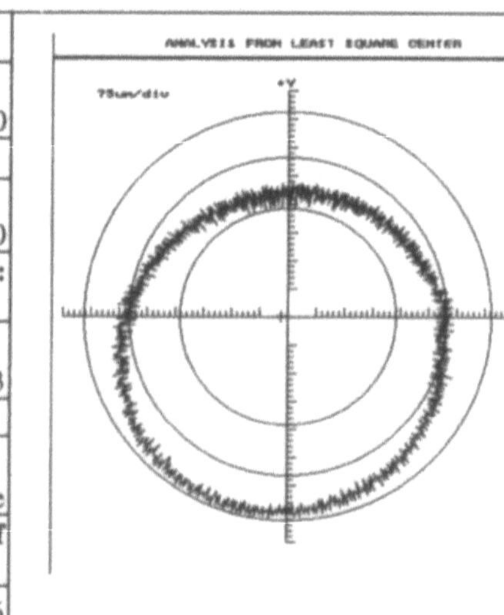

Fig. 6 The result under Feed rate 2000 mm/min and anti-clockwise with spindle rotating speed 1000 rpm.

Calibration of a Five-Axis Machine Tool for Position Independent Geometric Error Parameters Using a Telescoping Magnetic Ball Bar

J.R.R. Mayer[1], Y. A. Mir[1], C. Fortin[1]
[1]Ecole Polytechnique, Montréal, Canada, rene.mayer@meca.polymtl.ca

Abstract: Many parameters influence the quality of machined parts on five-axis machine tools. One important family of parameters are the geometric inaccuracies in the relative location of the machine axes. A method is proposed which uses a telescoping magnetic ball bar to identify these error sources. The methodology is applied to position independent geometric error parameters such as joint misalignments, angular offset and axes separation. Simulation results support the effectiveness of the approach.

1 Introduction

The accuracy of machine parts is affected by the geometric errors of machine tools. Two groups of geometric error sources causing tool positioning errors may be defined for a machine. The first group includes the position independent geometric error parameters (PIGEPs) which describe the relative location of the machine's successive rotary and prismatic joints. Examples of these are squareness errors, parallelism errors, joint offsets, and rotary axes separation errors. The second group includes the position dependent geometric error parameters (PDGEPs) also called motion errors describing the non-ideal motion of a joint or axis. Examples of these are positioning errors, straightness errors and angular errors such as yaw, pitch, and roll for prismatic axes, and, the angular, axial, radial and tilt errors for rotary axes.

Methods, instruments and setups exist and are well documented, for example in national standards (ASME B5.54)[1], to measure these parameters. However such procedures are relatively involved in terms of equipment and operator expertise and simpler means are thus sought. One of the most popular devices is the telescoping magnetic ball bar (TMBB) [2, 3]. This method has been extensively used for three prismatic axis machine tools [4] and allows the measurement of numerous quasi-static and dynamic parameters including all three squarenesses via circular test paths. More recently, attention has been given to the position dependent or motion errors. Pahk [5] represents the motion errors as polynomials and incorporates the straightness errors in those. The experimental data is gathered using a TMBB and circular test path in the XY, YX and ZX planes respectively. Ziegert [6] proposed the direct measurement of the x, y, and z volumetric positioning errors by a trilateration principle and a novel laser ball bar (LBB) system which provides high accuracy over a large range. Driels [7] used a fixed-length ball bar to calibrate the position independent geometric error parameters of an industrial robot made of a single serial kinematic chain of six rotary joints. He uses a modified Denavit-Hartenberg (D-H) formulation to avoid the problem occurring with D-H modelling in the presence of nominally parallel rotary axes.

In this paper a method is proposed which uses a TMBB to identify the PIGEPs of a five-axis machine tool. The modelling approach delivers an observability matrix used to judge the suitability of the proposed test procedure. Both the rank and condition number are used to assess the observability of the unknown parameters. Simulations are conducted using the Matlab language for a Matsuura five-axis NC machine tool.

2 Five-axis machine tool model

From the base frame F we define a tool branch and a workpiece feature branch using their homogenous transformation matrices (HTMs) ${}^{F}\boldsymbol{T}_{t}$ and ${}^{F}\boldsymbol{T}_{w}$ respectively. ${}^{j}\boldsymbol{T}_{i}$ denotes the 4×4 HTM representing the pose (position and orientation) of frame i with respect to frame j. Thus, the pose of t relative to w is given as

$$ {}^{w}\boldsymbol{T}_{t} = [{}^{F}\boldsymbol{T}_{w}]^{-1} \cdot {}^{F}\boldsymbol{T}_{t} \tag{1}$$

The target Matsuura five-axis machine tool has a CAXYFZ topology, see Figure 1. The joint coordinates are defined as $\boldsymbol{\theta} = [z\ y\ x\ a\ c]^{T}$. At $\boldsymbol{\theta} = [0\ 0\ 0\ 0\ 0]^{T}$ the reference frames of all joints coincide, see Figure 2.

The HTM for a non-ideal link and joint combination including the nominal state and the imperfections in the links (PIGEPs) can be modelled using 3 HTMs. Two link HTMs describe the nominal location and the locational error between two successive links followed by a joint motion HTM

$$\boldsymbol{D}_{s} = \boldsymbol{D}_{1s}\, \boldsymbol{D}_{2s}\, \boldsymbol{D}_{3s} \ . \tag{2}$$

$\boldsymbol{D}_{1s}$ is the nominal link HTM and $\boldsymbol{D}_{2s}$, the joint location error (PIGEPs) HTM.

$$\boldsymbol{D}_{1s} = \boldsymbol{trans}(\boldsymbol{i}, a_{x,S})\, \boldsymbol{trans}(\boldsymbol{j}, a_{y,S})\, \boldsymbol{trans}(\boldsymbol{k}, a_{y,S})\, rot(\boldsymbol{i}, \alpha_{x,S})\, rot(\boldsymbol{j}, \alpha_{y,S})\, rot(\boldsymbol{k}, \alpha_{z,S}) \tag{3}$$

where $a_{x,S}$, $a_{y,S}$, $a_{z,S}$, $\alpha_{x,S}$, $\alpha_{y,S}$ and $\alpha_{z,S}$ are the nominal location values and S represents Z, Y, X, A, and C and all nominal parameters are null for the Matsuura.

$$\boldsymbol{D}_{2s} = \boldsymbol{trans}(\boldsymbol{i}, e_{x,S})\, \boldsymbol{trans}(\boldsymbol{j}, e_{y,S})\, \boldsymbol{trans}(\boldsymbol{k}, e_{z,S})\, rot(\boldsymbol{i}, \gamma_{x,S})\, rot(\boldsymbol{j}, \gamma_{y,S})\, rot(\boldsymbol{k}, \gamma_{z,S}) \tag{4}$$

where $e_{x,S}$, $e_{y,S}$, $e_{z,S}$, $\gamma_{x,S}$, $\gamma_{y,S}$, $\gamma_{z,S}$ are the PIGEPs of the S link.

The joint HTM $\boldsymbol{D}_{3s}$ describes the nominal motion of the joint frame,

$$\boldsymbol{D}_{3s} = \boldsymbol{trans}(\boldsymbol{dir}, s) \text{ for prismatic joint; } \boldsymbol{D}_{3s} = rot(\boldsymbol{dir}, s) \text{ for a rotary joint} \tag{5}$$

where for prismatic joints ***dir*** is ***i***, ***j*** or ***k*** and s (the joint coordinate) is x, y or z, for axes X, Y and Z respectively. For rotary joints ***dir*** is ***i*** for A, and ***k*** for C, and *s* is a or c, the displacement angle (joint coordinate).

The tool and workpiece features are constant HTMs $\boldsymbol{T}_t$ and $\boldsymbol{T}_w$ respectively.

The tool versus workpiece feature location error ${}^w\boldsymbol{E}_t$ is calculated as follows:

$$ {}^F\boldsymbol{T}_t = \boldsymbol{D}_z . \boldsymbol{T}_t \tag{6} $$

$$ {}^F\boldsymbol{T}_w = \boldsymbol{D}_y . \boldsymbol{D}_x . \boldsymbol{D}_a . \boldsymbol{D}_c . \boldsymbol{T}_w . \tag{7} $$

$$ {}^w\boldsymbol{E}_t = [{}^F\boldsymbol{T}_w]^{-1} . {}^F\boldsymbol{T}_t . \tag{8} $$

The Jacobian matrix determining the analytical relationship between the tool versus workpiece feature pose and the error parameters may be obtained using the Newton-Euler equation for a specific machine tool topology, geometry and joint positions

$$ {}^w\boldsymbol{E}_t = \boldsymbol{J}\, \boldsymbol{\delta p} . \tag{9} $$

This Jacobian is referred to later as a sensitivity matrix when analysing the effect of parameter errors on the pose, or as an observability matrix when proceeding to the calibration of the unknown error parameters as with the ball bar procedure.

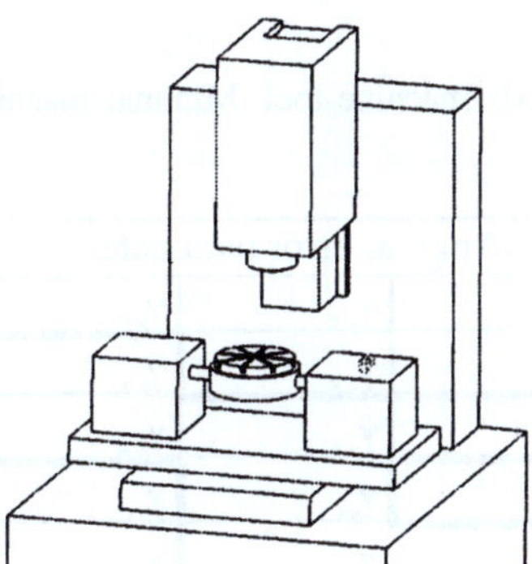

Figure 1 Matsuura machine

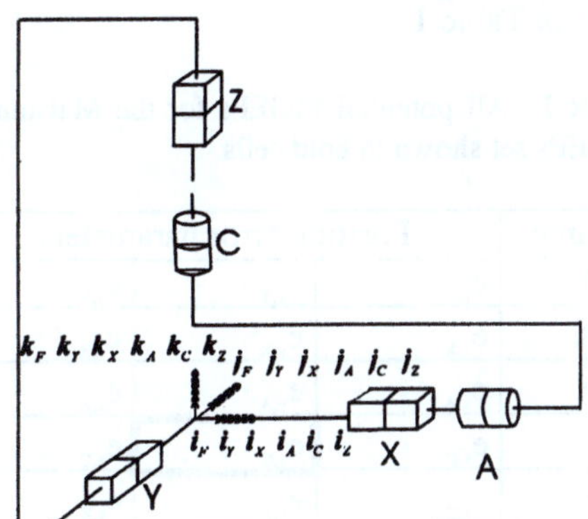

Figure 2 Joint frames at zero coordinates.

3 Complete and Minimal Set of PIGEPs

There are potentially 42 PIGEPs for modelling a five-axis machine tool in order to predict the pose error of the tool with respect to the workpiece feature. There are 6 PIGEPs per axis, 6 error parameters for positioning and orienting a part on the table and finally 6 error parameters for positioning and orienting the tool on the spindle. These error parameters are listed in Table 1. Pt_C and Pt_t define the workpiece feature frame with respect to the C table and the tool frame with respect to the spindle respectively.

The sensibility matrix, ***J***, calculated for a nominal machine has a size of $6m \times 42$ (where m is the number of joint coordinate sets considered) and a rank of only 20, indicating 22 redundant error parameters. Closer inspection reveals numerous identical columns (except for sign) thus confirming the parameter redundancies. One parameters (the one in parenthesis) of each of the following coupling groups $[e_{x,Y}, e_{x,X}, e_{x,A}, e_{x,C}, e_{x,Z}, (e_x(Pt\_t))]$, $[e_{y,Y}, e_{y,X}, e_{y,A}, e_{y,Z}, (e_y(Pt\_t))]$, $[e_{z,Y}, e_{z,X}, e_{z,A}, e_{z,Z}, (e_z(Pt\_t))]$, $[e_{z,C}, (e_z(Pt\_C))]$, $[\gamma_{x,Y}, (\gamma_{x,Z})]$ $[\gamma_{x,X}, (\gamma_{x,A}), \gamma_{x,C}]$, $[\gamma_{y,Y}, \gamma_{y,X}, (\gamma_{y,Z})]$ and $[\gamma_{z,Y}, (\gamma_{z,Z})]$ need be retained. This parameter reduction is easily confirmed by an elementary analysis of the sensibility matrix looking at its size, rank and condition number. There remain two redundant error parameters. A singular value decomposition provides a simple means of identifying the coupled parameters. The removal of one of those coupled parameters for both coupling groups eliminates redundancy and leads to a complete and minimal parameter set. The first groups is $[\gamma_{z,C}, e_x(Pt\_C), e_y(Pt\_C), \gamma_z(Pt\_C)]$ and we remove $\gamma_{z,C}$ which is often defined by a part setup procedure prior to machining. Also $\gamma_{z,Z}$ is not required as it would couple with the tool. Furthermore $\gamma_z(Pt\_t)$ is not relevant for a rotating tool. The remaining tool and part parameters, Pt_C and Pt_t, are application dependent and do not define machine PIGEPs. As a result we now have 8 machine tool PIGEPs shown in bold cells in Table 1.

Table 1: All potential PIGEPs for the Matsuura five-axis machine tool. Minimal machine PIGEPs set shown in bold cells.

Frame	Position error parameters			Angular error parameters		
Y	$e_{x,Y}$	$e_{y,Y}$	$e_{z,Y}$	$\gamma_{x,Y}$	$\gamma_{y,Y}$	$\gamma_{z,Y}$
X	$e_{x,X}$	$e_{y,X}$	$e_{z,X}$	$\gamma_{x,X}$	$\gamma_{y,X}$	**$\gamma_{z,X}$**
A	$e_{x,A}$	$e_{y,A}$	$e_{z,A}$	**$\gamma_{x,A}$**	$\gamma_{y,A}$	**$\gamma_{z,A}$**
C	$e_{x,C}$	**$e_{y,C}$**	$e_{z,C}$	**$\gamma_{x,C}$**	$\gamma_{y,C}$	**$\gamma_{z,C}$**
Z	$e_{x,Z}$	$e_{y,Z}$	$e_{z,Z}$	**$\gamma_{x,Z}$**	**$\gamma_{y,Z}$**	$\gamma_{z,Z}$
Pt_C	$e_x(Pt\_C)$	$e_y(Pt\_C)$	$e_z(Pt\_C)$	$\gamma_x(Pt\_C)$	$\gamma_y(Pt\_C)$	$\gamma_z(Pt\_C)$
Pt_t	$e_x(Pt\_t)$	$e_y(Pt\_t)$	$e_z(Pt\_t)$	$\gamma_x(Pt\_t)$	$\gamma_y(Pt\_t)$	$\gamma_z(Pt\_t)$

4 Single socket method for PIGEPs identification

The TMBB provides a subset of the complete pose data. Thus it is essential to verify that all 8 machine PIGEPs can be observed (calibrated) by the test procedure. The observability can be determined from the Jacobian matrix relating the effects of PIGEPs on the ball bar readings. 6 additional PIGEPs define the positions of the ball bar ends.

These errors have a cumulative effect on the actual relative position between the tool and the workpiece and so directly influence the readings from a MTBB connected between the spindle and the table. The data acquired by the ball bar procedure and the observability matrix are used to estimate the unknown error parameters by iteratively solving a linear system of equations.

5 Simulation results

Table 2 Results of simulations A and B for the values of the error parameters. The linear errors (e) are in inch and the angular errors (γ) are in radian.

PIGEPs name	Simulation A		Simulation B		
	True values	Identified values	True values	Identified values	Identification errors
$e_{x,Y}$	0.0040		0		
$e_{y,Y}$	0.0160		0		
$e_{z,Y}$	0.0010		0		
$\gamma_{x,Y}$	0.0040		0		
$\gamma_{y,Y}$	0.0080		0		
$\gamma_{z,Y}$	0.0020		0		
$e_{x,X}$	0.0070		0		
$e_{y,X}$	0.0030		0		
$e_{z,X}$	0.0020		0		
$\gamma_{x,X}$	0.0020		0		
$\gamma_{y,X}$	-0.0080		0		
$\gamma_{z,X}$	-0.0060	-0.0060	-0.0060	-0.0060	-3.3E-16
$e_{x,A}$	0.0060		0		
$e_{y,A}$	0.0020		0		
$e_{z,A}$	0.0015		0		
$\gamma_{x,A}$	0.0030	0.0050	0.0500	0.0500	-7.8E-16
$\gamma_{y,A}$	0.0060	0.0060	0.0060	0.0060	4.0E-16
$\gamma_{z,A}$	-0.0040	-0.0040	-0.0040	-0.0040	-5.0E-16
$e_{x,C}$	0.0020		0		
$e_{y,C}$	0.0030	0.0030	0.0030	0.0030	-2.8E-15
$e_{z,C}$	0.0070		0		
$\gamma_{x,C}$	0.0000		0		
$\gamma_{y,C}$	0.0040	0.0040	0.0110	0.0110	-7.6E-17
$\gamma_{z,C}$	0.0000		0		
$e_{x,Z}$	0.0020		0		
$e_{y,Z}$	-0.0010		0		
$e_{z,Z}$	0.0110		0		
$\gamma_{x,Z}$	0.0030	-0.0010	0.0300	0.0300	-2.3E-16
$\gamma_{y,Z}$	0.0030	0.0030	0.0350	0.0350	9.0E-17
$\gamma_{z,Z}$	0.0035		0		
$e_{x,(Pt\_C)}$	0.0070	0.0070	0.0070	0.0070	-1.2E-15
$e_{y,(Pt\_C)}$	0.0060	0.0060	0.0060	0.0060	1.4E-15
$e_{z,(Pt\_C)}$	0.0030	0.0100	0.0030	0.0030	-1.1E-15
$\gamma_{x,(Pt\_C)}$	0.0000		0		
$\gamma_{y,(Pt\_C)}$	0.0000		0		
$\gamma_{z,(Pt\_C)}$	0.0000		0		
$e_{x,(Pt\_t)}$	0.0040	-0.0131	0.0040	0.0040	2.0E-15
$e_{y,(Pt\_t)}$	0.0070	-0.0149	0.0070	0.0070	-1.2E-15
$e_{z,(Pt\_t)}$	0.0090	0.0156	0.0090	0.0090	-1.9E-15
$\gamma_{x,(Pt\_t)}$	0.0000		0		
$\gamma_{y,(Pt\_t)}$	0.0000		0		
$\gamma_{z,(Pt\_t)}$	0.0000		0		

30 machine configurations are used for the identification. No conventional circular path tests are needed. Another 60 configurations are used for verification of the tool versus workpiece position error prediction. Table 2 shows the identified parameters for test A and test B. In test A, error values are imposed to all retained machine PIGEPs and the others except $\gamma_{x,C}$, $\gamma_{z,C}$, and $\gamma_{z,Z}$ the effect of which could not be corrected. Errors are also added to the ball bar end positions. The identified PIGEPs are different to the true ones as expected due to redundancy in the complete parameter set of the true model. The ball bar residuals are practically null (10^{-14}) and the predicted tool error for the identification and verification groups is smaller than 10^{-13}. In test B, only the minimal error parameter set has non-zero true errors. In this case, and as anticipated, the identified error values are identical (within computer resolution) to the true ones. For both tests the Jacobian as a rank of 14 and a condition number around 100 is obtained which is very satisfactory.

6 Conclusion

A systematic approach has been proposed for the calibration of the position independent geometric error parameters of a five-axis machine tool. A minimal but complete set of error parameters is defined. Then the TMBB is used to gather length readings at a small number of machine configurations. Simulations show excellent results both for the parameter values and for the tool versus workpiece error prediction.

Acknowledgements

This work was performed with the support of Rene Mayer's NSERC Grant N° 155677-98 and Clément Fortin's NSERC Grant 36396-98. Dr. Guy Cloutier's help with Matlab libraries and with numerous discussions was much appreciated.

References

[1] ASME B5.54-1992, *Methods for Performance Evaluation of Computer Numerically Controlled Machining Center*, The American Society of Mechanical Engineers.

[2] Bryan J. B., 1982, *A simple method for testing measuring machine and machine tools Part 1: Principles and applications*, Precision Engineering, **4**(2), pp. 61-69.

[3] Bryan J. B., 1982, *A simple method for testing measuring machine and machine tools Part 2: Construction details*, Precision Engineering, **4**, pp. 125-138.

[4] Kakino Y., Ihara Y. and Shinohara A., 1993, *Accuracy Inspection of NC Machine Tools by Double Ball Bar Method*, Hanser Publishers.

[5] Pahk H. J., Kim Y. S. and Moon J. H., 1997, *A New Technique for Volumetric Error Assessment of CNC Machine Tools Incorporating Ball Bar Measurement and 3D Volumetric Error Model*, Int. J. Mach. Tools Manufac., **37**(11), pp. 1583-1596.

[6] Ziegert J. C. and Mize C. D., 1994, *The laser ball bar: a new instrument for machine tool metrology*, Precision Engineering, **16**(4), pp. 259-267

[7] Driels M. R., 1993, *Using Passive End-Point Motion Constraints to Calibrate Robot Manipulators*, Trans. of the ASME, Vol. **115**, 560-566.

Contribution to Software Compensation of Structural Errors in Machine Tools

Nunzio D'Addea[1], Luigi Giannotti[1], Enrico Gargano[1], Alessandro Bartolini[1]

[1]ITIA-GAD Gruppo di Affidabilita`e Diagnostica, Consiglio Nazionale delle Ricerche, Cinisello Balsamo (ITALIA), email nudaddea@tin.it

Abstract: The demand for improved machining accuracy has increased with the factory automation but thermal errors of machine tools are yet critical. These errors can be controlled from design either by hardware, e.g. geometry and structural anisotropy, or by software compensating them. In this area many applications were carried out, also from the Authors. In this paper they enhance some recent results obtained with the associated use of a traditional homogeneus coordinate transformation method and of new anticipatory algorithms.

1. Some Contributions to Structural Deformation Understanding and Compensation.

The growth in factory automation has led in the recent years to the demand for higher and better machine productivity and at the same time to the desire for increased levels of geometric accuracy, both in the work volume and over time.
Bryan [1] divides errors into six source types which directly affect the machine, the piece and the translation system. Cotta Ramusino (The 3th IMTEC) highlights the links between temperatures, heat sources and deformations.The internal sources of heat are more active and since they are linked to the working process itself, they are the main cause of structural deformation and product errors. A correct design is, according to Spur, of fundamental importance in reducing deformation, but insufficient and "compensation techniques are necessary to increase accuracy, in N.C. machines". The most important module among the active sources of heat is the spindle. There have been many attempts, Spur, Weck, Bryan and Sato [2] to reduce exchange with the structure and compensate errors. Cotta Ramusino sought to compensate lengthening of the spindle by comparison with a reference in Zerodur. The tests reported by D'Addea and Quaranta [3] indicate that lengthening is often prevalent but it is not the only spindle error. About the effect of the working parameters on the spindle deformation it was noted that maximum accuracy is sought in the finishing phase when the cutting load is lower, however dissipated power depends more on speeds than on cutting load.Over the years, techniques have been proposed to compensate structural errors, starting with deformations caused by one's own or additional mobile weights. D'Addea et Al. [4] divide mobile weights hardware corrections into the two techniques of "contrasting pre-tensioning" and "balancing" with opposite loads. Thermo-geometric design is not new: sensitivity to

symmetry has increased and a solution to the problem has been sought by in process metrology. The software compensation techniques involve auxiliary means of calculation to forecast errors due to deformations. The increase in work velocities has made the thermal problem even more difficult to solve but more instruments, e.g.computer based numeric control and so on, have also become available.
From the beginning of the 1960's, the work of Bryan, Ferreira, Tlusty, Stute, Sata, Koenigsberger, Wiele and Zhang [5] [6] [7] [8] has made important contributions for measuring machines and machine tools. The correction techniques move in two directions: the first, involves use of a pre-calculated and stored error map as a look-up table; the second uses on-line error calculation, since the computer holds a mathematical model rather than a map.
Dufour et Al.[9] have faced the problem of software based geometric error compensation on large machines using the error matrix method. The work of Donati et Al [10] offers the possibility of autocalibrating the measuring machine or machine tool and generating the mathematical model. Attia et Al.[11] underlined the importance of the joints in the study of deformations, and Donnez [12] investigated the method based on the transformation matrix of homogenous coordinates. Srinivasa [13] and Chen made some attempts to forecast thermal machine errors using a neural network approach, but results have not been wholly satisfactory, probably because of difficulties in categorising knowledge. More recently Revilla et Al.[14] have performed an industrial application by the same system. D'Addea et Al.[4] report an experimental approach: the algorithm relating temperature to deformation is based on the position in the work volume and on key temperatures. The proposal is based on the identification of key temperatures and even the authors admit that it is difficult to use. A recent study by Huang is based on the same methodology. Srivastava et Al. use the transformation matrix to calculate the errors dependent on the work position and on time (as thermal). Fraser et Al [15] recently have proposed a new concept of generalized stuctural modelling based on the real time extimation and control of intensity of heat sources coming from temperatures.

2. Using the Homogeneus Transformation Matrix and the Anticipatory Algrithms in the Errors Compensation

Machine tools have rigid modules which are linked by fixed or mobile linear and rotating restraints to form a closed kinematic chain from the piece to the tool. The transformation matrix describes the transfer via the translation and rotation joints of deformations in the modules to the tool tip. The approach has been extensively used in the past for the transfer of geometric errors. The innovation, supported by extensive field testing, which distinguishes the present approach from others, consists in considering 'a priori' the coefficients of the homogeneus transfer matrix as time and position variables. The correlation between temperature and deformations is sought identifying those temperatures which allow the deformation to be described over time with a sufficient level of approximation. To date, these key temperatures have been identified by means of extensive experimental analyses [4]. This process can now be integrated by using calculation methods and field testing. The machine tool used for the software compensation of thermal deformation has five axes, three for translation (X,Y,Z) and two for rotation, parallel to the Y and Z, Fig.1.

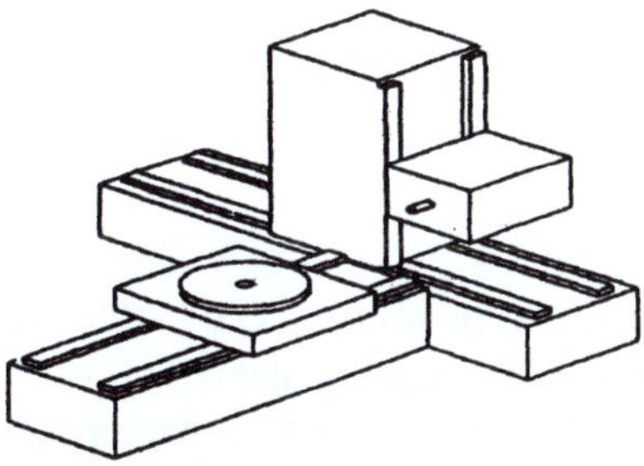

Fig.1 Five axes machining centre
(A is not here faced)

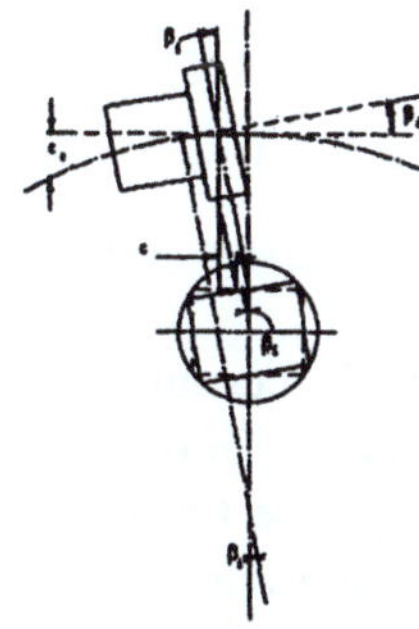

Fig.2 Orientation between tool and piece

Compensation was applied to the three principal axes; the rotating axes, as B axis parallel to Y, were used to re-establish the correct orientation between the piece and the tool, Fig.2. The compensation methodology is based on the concept that it is possible to identify on each axis of the machine one or more temperatures (key temperatures) which are representing the state of deformation: this procedure is an approximation but it can supply good results if the themperature checkpoints are selected regarding all thermical loads. The technical way consists in a number of steps. A thermo-deformational mathematical model of machine is proved by field experiences: a real thermal map coming from a working cicle and an equal map on the model have to give us approximately the same deformations to the tool point so it allows to simulate on the model many different thermal loads without difficult and long time experimental tests. The selection of temperature key points and the experience to seek the rules that links deformation state to temperature variation comes from this first data analysis. The homogeneus transformation matrix application with the use of syntetic description of machine joints, as in Fig.3 , is a further step:

$$[T]_{0,4} = [T]_{0,1}[\Phi]_{0,1}[T]_{1,2}[\Phi]_{1,2}[T]_{2,3}[\Phi]_{2,3}[T]_{3,4}[\Phi]_{3,4} \quad (1)$$

where $[T]_{i,j}$ and $[\Phi]_{i,j}$ are matrices for the translations of origin and rotation.
The experimental observation on the particular machine suggested as simplified hypothesis that the distortions of the base guide of the work table, axis Z, are very small. Following the scheme in Fig.3, the first coordinate transformation is that which moves the origin 0 (0,0,0) at the base of the upright. This point has the coordinates 1 (x+a, -b, -c), while the rotations are α_1, β_1, and γ_1, respectively. The origin of the second reference is given as point 2 (x+a, y, -c), and the rotations are α_2, β_2, and γ_2. The third reference is on the axis of the spindle sleeve 3(x, y, -c) with negligible rotations, and the fourth on the spindle tip, 4 (x, y, z), with rotations α_4, β_4,γ_4. The spindle excites a rotation around the axes but also a deformation Δx, Δy,

Δz [4], [7]. If the radial dimensions of the tool are overlooked and the rotations are small enough to disregard the second order terms, the corrections are:

$$\varepsilon_x = \varepsilon_{xx}(x,T_{xi}) + \varepsilon_x(y,T_{yi}) + (c+z)(\beta_1+\beta_2)+\Delta x + \text{superior order terms};$$

$$\varepsilon_y = \varepsilon_{yy}(y,T_{yi}) + \varepsilon y(y, Tyi) + (c+z)(\alpha_1+\alpha_2) + \Delta_y + \ldots\ldots \qquad (2)$$

$$\varepsilon_z = \varepsilon_{zz}(z, T_{zi}) + \varepsilon_z(x,T_{xi}) + \varepsilon_z(y,T_{yi}) + a(\beta_1+\beta_2) + \Delta z + \ldots\ldots$$

where x, y and z are the current coordinates in the work volume; $T_{x,i}$, $T_{y,i}$ $T_{z,i}$ are the key points temperatures; $T_{m,k}$ the temperatures on the spindle sleeve; $\varepsilon_{xx}(x, T_{xi})$, $\varepsilon_{yy}(y, T_{yj})$, $\varepsilon_{zz}(z,T_{zi})$ the linear errors along the axes X, Y and Z, respectively, in function of the position and the key temperatures; $\varepsilon_x(y,T_{yj})$, $\varepsilon y(y,Tyi)$, $\varepsilon_z(x,T_{xi})$, $\varepsilon_z(y, T_{yj})$ the cross errors in X, Y, Z axis as a result of the deflexion of the upright in the XY, YZ plane; $(c+z).(\beta_1+\beta_2)$ error along X caused by the effect of the rotation at the base and the torsion of the upright; $(c+z).(\alpha_1+\alpha_2)$ the error on Y caused by the effect of the rotation on the base and upright; $a.(\beta_1+\beta_2)$ the error along Z axis as a result of the rotation at the base and the upright of the machine;

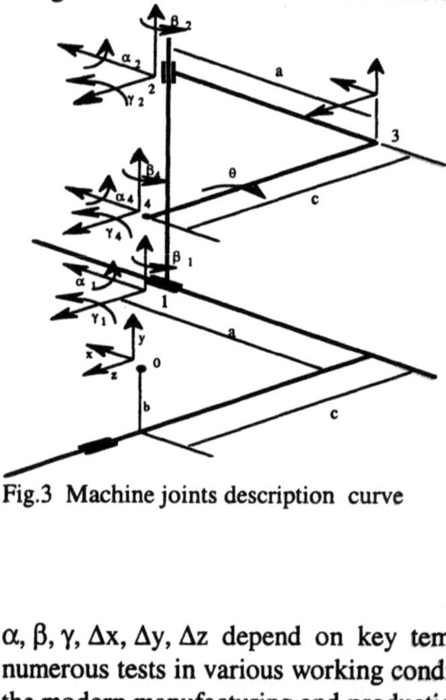

Fig.3 Machine joints description curve

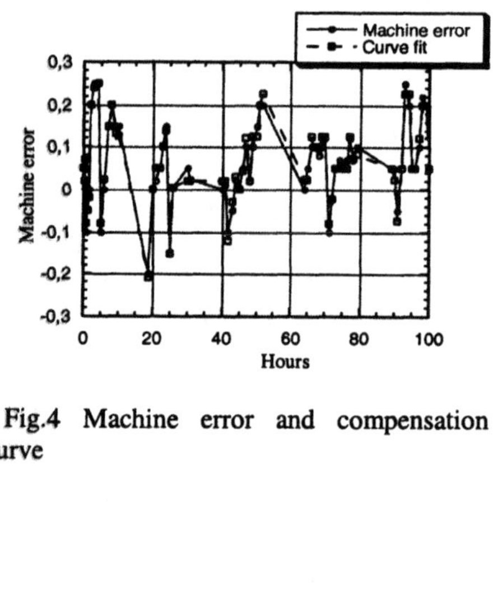

Fig.4 Machine error and compensation curve

α, β, γ, Δx, Δy, Δz depend on key temperatures. The problem solution demands numerous tests in various working conditions: a lead time wich is incompatible with the modern manufacturing and production. Research recently undertaken by the AA. has shown that it is possible to move from an initial FEM thermo-deformation model to one which approximates real machine behaviour. This result was obtained comparing the temperature and deformation maps in the model with real ones and modifying the model to obtain with some approximation the same performance. Such a model eases and accelerates recognition of other situations, e.g. transience and the overlapping of external and internal thermal variation, and allows the choice of the key temperature points. The methodology, already outlined over the past ten years [4], has been recently modified using new rules to process the temperatures and applied successfully in a number of cases, including the controlled five-axis machine described above and gear cutting machines. The compensation algorithm was sought with regard to the uncertainty wich defines the temperature key points, so the thermal data are processed by fuzzy systems[16], Fig.5, to try the coefficients of homogeneus transformation matrix. The terms of the matrix which associate the error at the point of contact between the piece and the tool to the direct or induced deformations in the chain are together a function of the position of the mobile bodies

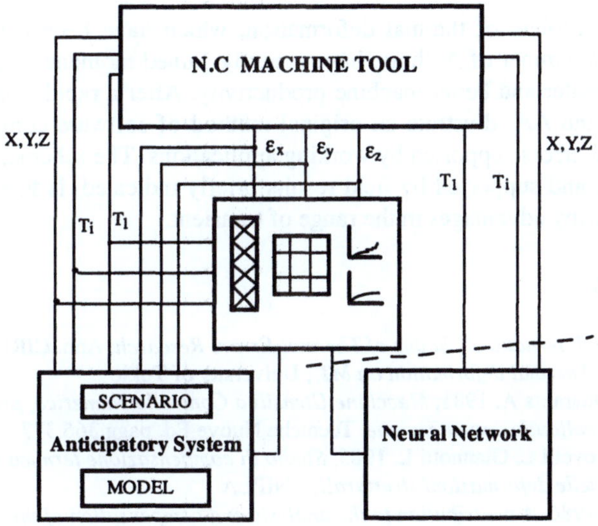

Fig.5 Fuzzy system and rules variation by neuralnetwork or anticipatory algorithm

in the machine and the distribution of temperature. The variation of the position of the source often has a double effect on the error: the endogenous heat is linked to the position of the sources, while changes in position often also modify the length of the transfer chain. The rules of the fuzzy matrices can be revised and varied to control the errors arising from the position of the thermal sources and from the structure thermal memory, that is a cause of higher deviations in the desired approximation. This rules revision can be obtained by a recoursive knowledge based algorithm as in a neural network, Fig.5. This method is far from a direct application of neural network to the total errors of axis [14] and more difficult but the results are already satisfactory. Fig.4 highlights the correspondence over long periods of time between the deformations calculated with reference to temperature and those actually found in the axis of a gear cutting machine.
Recently to model the fuzzy rules [16] systems the AA. were trying a new algorithm based on the anticipation of the probable states of the structure thermodeformations:

$$f(t+1)= F [f(t), F(f(t), f(t+1),p),p] \qquad (3)$$

where "f" represents the vector of state variables; "F" is a reactualization function based on the system model; "p" the probability associated with the vector of state variables and with the state transition. The decision strategy is defined as to find, in the scenario characterized by the reactualization function, the transitions probability of state variables. The system complexity is used to point out the fact that the problem treated here can t be expressed only in hard relations.

Conclusion

Studies of the problem of thermal deformation, which have been carried out for many years by a number of Authors, have recently gained momentum as a result of the needs for greater and better machine productivity. After a rapid overview of the present culture the AA. illustrate an original method of software compensation of thermostructural errors supported by working applications. The scheme, proposed in previous papers and supported by field results briefly indicated, is formulated as it seems to have many advantages in the range of transient.

References

[1] Bryan J, 1990, *International Status of Thermal Errors Research*, Ann. CIRP

[2] Sato H, 1988, *Thermal deformation on MT* , University of Tokio

[3] D'Addea N,.Quaranta A, 1981, *Macchine Utensili a Controllo Numerico, progetto, collaudo, manutenzione*, Tecniche Nuove Ed. pagg.366,377

[4] D'Addea N, Covelli L, Giannotti L, 1985, *Studio di compensazione termica via Sw delle deformazioni strutturali*, ANIPLA

[5].Ferreira F.M, 1986, *A contribution to the analysis and compensation of the geometric error of a machining center*, Ann. CIRP

[6] Tlusty J, Mutch G, 1973, *Testing and evaluating thermal deformations of machine tools*, MTDR

[7] Koengsberger F, Tlusty J, 1970, *Specification and tests of metal cutting machine tools*, UMIST

[8] Sata T, Takeuchi Y, Okubo N, 1977, *Improvement of working accuracy of machining centre by computer control compensation*, MTDR

[9] Dufour P, Groppetti R, 1981, *Computer aided accuracy improvement in large N.C. machine tools*", MTDR

[10] Donati D, Gorini I, Sartori S, 1987, *Coordinate measuring machines and machine tools selfcalibration and error correction*, Ann. CIRP

[11] Attia M, Kops L, 1979, *Computer simulation of nonlinear thermoelastic behaviour of a joint in M.T. structure and its effect on thermal deformation*, Trans of ASME

[12] Donnez A, 1990, Ph.D.Dissertation, Purdue University

[13] Srinivasa, Ziegert, Smith,1992, Japan/USA Symp on Flexible Automation,Trans ASME

[14] Revilla J, Arana R, 1997, *Thermal deformations compensation by neural networks:an industrial experience*, BRITE/EURAM, OPTIMA Proj.5251

[15] Fraser S, Attia M, Osman M, 1998, *Modelling,identification and control of thermal deformation of M T structures*, J of Manufacturing Science and Engineering

[16] D'Addea N, 1999, *Multisensor systems today and tomorrow: machine control, diagnosis and thermal compensation*, CASYS '99

Computer Vision for Automated Quality Inspection of Colour Printing Products

J Luo[1], Z Zhang[2] and H Ismail[3]

[1]University of Liverpool, L69 3GH, UK, ispgjjl@liv.ac.uk.
[2]University of Liverpool, L69 3GH, UK, zhengwen.zhang@liv.ac.uk.
[3]University of Liverpool, L69 3GH, UK, hsismail@liv.ac.uk.

Abstract: Quality inspection of colour printing products constitutes a very important QC task in printing industry. The current practice in industry is to inspect complex colour products manually. This process is labour-intensive and the results are not reliable. The types and causes of defects on colour prints are discussed in this paper. We present a novel algorithm to automate the printing inspection process, which incorporates colour histogram-based techniques for colour image processing and a neural network for image classification. The experimental results show that this algorithm is able to detect both registration and purity defects under varying illumination conditions.

1. Introduction

In printing industry, inspection of colour products such as labels constitutes a very important QC task. Such colour products are often supplied to food, drink, cosmetic, pharmaceutical packaging and chemical industries. In some of these industries, packaging or labelling is a critical part of the supply chain, and label identification is of paramount importance. Currently, commercially available vision systems mainly inspect miss/over-printed characters/texts, alphanumeric codes, or verify single colours on simple colour printings. However, many printing products are complex coloured and contain sophisticated texture information. Defects on such products are caused by mismatches of colour and texture information between the printed products and customer requirements, which can not be easily inspected by conventional vision systems. Two types of defects are commonly found on such products, the colour registration defects and the colour purity defects. The colour registration defects are the result of displacements of basic printing colour planes, caused by mechanical problems and set up errors of the production line. The colour purity defects are resulted from colour variations due to the impurity of primary colours, which are caused by errors on the quantity/quality of ink of the basic colours supplied to the printing heads.

The current practice in printing industry is to inspect the quality of complex colour products manually by human inspectors. This process is labour-intensive and the results of inspection are often subject to variations due to the fatigue and tedium of inspectors, as these vary with mood, time, experience and personal skills of inspectors. The printing production lines often run at very high speeds, inspection therefore has to be carried out off-line, resulting in significant financial losses for printing businesses in terms of wasted raw materials and resources. Furthermore, there is no quantitative report, as to how closely a product matches the requirements of customers. Also, illumination changes affect the colour appearances of products. These problems often result in disputes between suppliers and customers.

During past decade, colour image processing has received considerable research interest. However, most work in this area is carried out for applications in the media industry, such as multimedia data retrieval [1] [2] and video image segmentation [3]. Relatively, less work has been carried out for applications in printing inspection. The algorithms for colour printing inspection proposed by Yu et al [4] and Sodergard et al [5], which were both based on gradient operators, can only inspect some defects on simple colour prints (e.g. defects causing significant colour intensity edge changes). Another algorithm proposed by Mirmehdi et al [6] employed Hough transform to detect straight lines in an image caused by colour registration problems. This algorithm has been shown to be able to detect faults on the outlines of label products. All these algorithms attempted to deal with registration problems, however, they are only capable of inspecting some simple colour defects. In this paper, we propose an algorithm for complex colour inspection, which integrates a histogram-based approach to colour image processing with a neural network for subsequent pattern classification.

2. Colour Recognition and Neural Network

2.1 The histogram-based approaches

Approaches for colour recognition reported in the literature are mostly histogram-based [7] [8] [9]. The full colour 3D histogram of an image is the statistical measurement of colour distributions within the image. It is independent from variations such as a change in the orientation, a shift in viewing position, a change in scene background, or even a partial occlusion. Generally, if two objects have different colour information, their histograms would be different. Therefore, features extracted from colour histograms can be used to discriminate between different colour images, and for inspection purposes, between good and faulty products. Swain and Ballad [7] have shown that colour distributions can be used directly for recognition from a large database of colour objects, even without using corresponding geometric information. However, histogram-based approaches suffer from a severe limitation due to the fact that a histogram only describes the global colour distribution in an image and is very sensitive to varying illumination conditions. This limitation has led to a collection of algorithms for colour constancy [10]. Despite the progress in colour constancy, pre-processing for colour constancy presents even more difficult problems to be solved. Simply placing a few patches of known spectral reflectance in the scene will not work well, because illumination usually varies spatially both in intensity and spectral composition.

The ignored spatial information about the colour vectors in an image can also provide important information for colour recognition. Images with significantly different spatial structures can not be discriminated if they have similar colour distributions. For example, one can not discriminate an image of a random distribution of red and green spots from an image that is half red and half green only based on their colour histograms. For printing inspections, the histogram-based algorithms are only effective in detecting some of the colour purity defects, and will be unable to detect small registration defects since the colour distributions of good and faulty products in such cases are very similar.

2.2 Colour Image Filtering for Image Recognition

Spatial structure within a colour image that is not captured by a colour histogram provides useful information for colour recognition, especially in a textured image. Such spatial information can be captured by spatial filters that may be designed to selectively enhance or eliminate some structure features of an image [11] [12]. Therefore, the distortions of spatial information on a printed label due to registration defects or some of the purity defects may be detected from the colour histograms of the filtered label images. Such a method alone will however not be able to detect some of the purity defects that do not cause changes on the selected spatial information.

2.3 Neural networks as classifier

The measured colour distribution from colour histogram can be used for indexing colour objects into a data base of models or can be used as features to be fed into a classifier for pattern classification. We use neural network as the classifier to perform the discriminatory tasks.

There are many types of neural networks but, generally, they are composed of either analogue or logic nodes. The logical neural network that utilise single layer of deterministic logic nodes has been shown [9] to be able to provide good recognition performance for some colour objects. As logical nets utilise weightless nodes, they are faster and easier to train than conventional neural nets. However, the inputs of the logical nets have to be a set of binary statistics. As such, the histogram has to be sliced by multiple thresholds into a binary set. This results in a lot of relevant information to be lost and therefore, affects the recognition accuracy. The conventional neural networks utilise computer simulations of analogue summing nodes. These networks either the supervised neural nets[13] or the unsupervised neural nets (i.e. self-organising feature maps)[14] require training time for iterative adjustments of the weights at each node. As the training stage can be performed off-line, this does not add up execution time for real time applications. In the application for printing inspection, it is very difficult to obtain full set of defect samples. An unsupervised neural net is more desirable.

3. The Proposed Algorithm

In this paper, we attempt to integrate information from colour histograms with that from image filtering so as to be able to detect a wide range of different defects. As shown in Fig.1, firstly, the colour histogram of the image of a printed label is calculated. The image is then filtered by a specially designed spatial filter. The colour histograms of the resulting images are then calculated. From each of these histograms (both original and filtered), a number of features are calculated and the results are combined to form a feature vector, which is then fed to a neural network for classification.

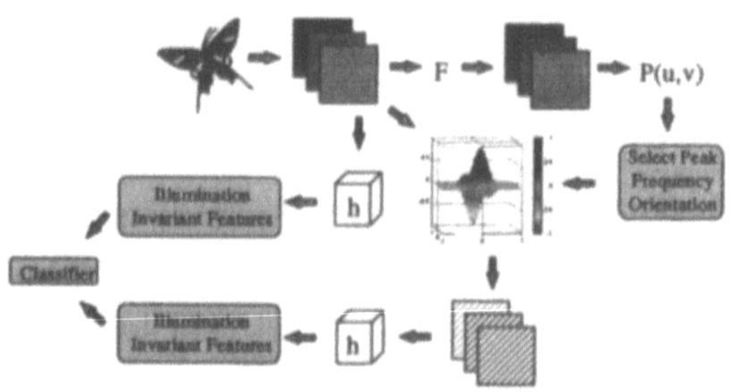

Fig. 1 Combined filtering and histogramming

As colour descriptors based on histograms are sensitive to varying illumination conditions, they can not be used directly for illumination-invariant inspection. Moment invariants have been used as features for pattern recognition [15], for image data annotation [16] and content-based retrieval of multispectral satellite images [17]. It has been proved [11] that moment invariants computed from the second-order centred moment matrix of a colour distribution are independent from illumination changes. Therefore, such distribution invariants from colour distributions (the original and the filtered distributions) can be used as feature vector for illumination-invariant recognition.

An important question here is how to design/select an optimised filter set that is able to extract the distinct spatial characteristics from an input colour image, which are likely going to be affected by registration and impurity defects. Gabor filters [13] that have been used widely in texture analysis as band-pass filters to detect spatial information at specific frequencies and orientations are used as spatial filters in this paper, because the image of a faulty print must be different from the reference image in certain frequencies and or orientations. A Gabor wavelets is a function that presents a cosine wave enveloped within a Gaussian function. The simplified 2-D GEFs is as follows:

$$h(x,y) = e^{[-(\frac{x'^2}{\sigma_x^2}+\frac{y'^2}{\sigma_y^2})]}\cos(2\pi\omega x'+\phi)$$

Where $(x',y') = (x\cos\theta + y\sin\theta, -x\sin\theta + y\cos\theta)$

There are five parameters σ_x, σ_y, ω, ϕ, θ.The first two determine the standard deviation of the envelop 2-D Gaussian; ω is the spatial frequency of the cosine wave; ϕ is the phase shift of the cosine wave; θ represents the rotation angle of the spatial - domain rectilinear coordinates

From above equation, we can see that GEFs can be designed to be highly selective in frequency with different spatial orientations. The selection of Gabor filters for a particular image are based on spectrum analysis. The Foriour transform and power spectrum of a colour image are first computed. From the spectrum, the principal frequency peaks and the spatial orientations of the structural information in the input image are determined. These parameters are then used to construct Gabor filters that emphasise these peak frequency characteristics in the image corresponding to their orientations.

The Feature vector of moment-invariants computed from an image colour distribution and from the colour distributions of the filtered images can be inputted into a neural network for final image classification. Because the inspection of colour prints is only concerned with the discrimination between the reference image and the test image, the "Golden template" image may not be unique, sometimes it has to be a set of reference images to allow for a range of tolerance, the Nearest Neighbour (Nearest Euclidean distance) classifier may not be suitable in this case. We use a neural network to perform the colour classification. This conventional neural net contains three layers with one hidden layer. The number of nodes in each layer has been optimised through experiments for each specific products order.

4. Experimental Results and Discussion

We first tested this algorithm on a group of synthetic images, which contain pairs of similar colour patterns. The correct classification rates were nearly 10/10. We then carried out the experiments on a range of different colour printing samples. Nearly 9 out of 10 sample images can be correctly identified. When we increase the number of Gabor filter, the miss-classified images can also be correctly identified. We compared the performances between this algorithm and other algorithms. The overall performance of this algorithm in terms of correct recognition, is superior to the algorithms based on either histogram or image filtering techniques alone. The performance of this algorithm in terms of speeds is slightly worse than another histogram-based algorithm proposed by us. These preliminary test results show that the algorithm proposed in this paper is able to detect both registration and purity defects in colour print inspection under varying illumination conditions. We expect that this algorithm will lead to a prototype automated visual inspection system with potential to be used both off-line and on-line. This algorithm may also be used for image retrieval and video segmentation tasks.

References

[1] Fountain S. R. and Tan T. N., 1998, *Efficient Rotation Invariant Texture Features for Content-based Image Retrieval.* Pattern Recognition, Vol. 31, No. 11, pp. 1725-1732.

[2] Colombo c. and Bimbo A. D., 1999, *Color-induced Image Representation and Retrieval.* Pattern Recognition, Vol.32, pp.1685-1695.

[3] Drew M. S.,Wei J. and Li Z. N., 1999, *Illumination-invariant Image Retrieval and Video Segmentation.* Pattern Recognition, Vol. 32, pp.1369-1388.

[4] Yu X. H., Sodergard C. and Juha Y. J., 1993, *On-line Control of Color Printing Quality by Image Processing.* Proc. IEEE Region 10 Conf. Comput. Comm. Contr. Power Eng. Pp.1039-1041.

[5] Sodergard C., Launonen R. and Aikas J., 1996, *Inspection of Color Printing Quality.* Int. J. of Pattern Recognition and Artificial Intelligence, Vol. 10, No.2, pp.115-128.

[6] Mirmehdi M., West G. A. and Dowling G. R., 1991, *Label Inspection Using the Hough Transform on Transputer Networks.* J. Microprocessors and Microsystems, Vol. 15, No. 3, pp.167-173.

[7] Swain M. and Ballard D., 1991, *Color Indexing.* Int. J. Comput. Vision, Vol.7, pp.11-32.

[8] Funt B. V. and Finlayson G. D., 1995, *Color Constant Color Indexing.* IEEE Trans. Pattern Anal. Machine Intell., Vol. 17, No. 5, pp.522-529.

[9] Wang Y. S., Griffiths B. J. and Wilkie B. A., 1996, *A Novel System for Colored Object Recognition.* Computer in Industry, No. 32, pp.69-77.

[10]Forsyth D. A., 1990, *A Novel Algorithm for Color Constancy.* International Journal of Computer Vision, Vol. 5, No. 1, pp.5-3.

[11]Healey G. and Slater D., 1997, *Computing Illumination-invariant Descriptors of Spatially Filtered Colour Image Regions.* IEEE Trans. Image Processing, Vol. 6, No. 7, pp.1002-1013.

[12]Plataniotis K. N., Androutsos D. Vinayagamoorthy S. and et al, 1997, *Colour Image Processing Using Adaptive Multi-channel Filters.* IEEE Trans. Image Processing, Vol. 6, No. 7, pp.933-949.

[13]Barnard E. and Holm J. E. W., 1994, *A Comparative Study of Optimization Techniques for Backpropagation.* Neurocomputing, Vol. 6, 19-34.

[14] Pal N. R., Bezdek J. C. and Tsao E. C. K., 1993, *Generalized Clustering Networks and Kohonen's Self-organizing Scheme.* IEEE Trans. Neural Networks, Vol. 4, No. 4, pp.549-557. 15]Hu M. K., 1962,*Visual Pattern Recognition by Moment Invariants.* IRE Trans. on Information Theory. Vol. 178, pp.179-187.

[16]Picard R. W. and Minka T. P., 1995, *Vision Texture for Annotation.* Multi-media System, Vol. 3, pp.3-14.

[17]Healey G. and Jain A., 1996, *Retrieving Multispectral Satellite Images Using Physics-based Invariant Representations* IEEE Trans. Pattern Anal. Machine Intell., Vol.18, pp.842-848.

CONSOLIDATION PROCESSES

Casting of Clad Strip Using a Twin Roll Caster

[1]Toshio HAGA and [2]Shinsuke SUZUKI

[1] Osaka Institute of Technology, 5-16-1 Omiya Asahi-ku Osaka city, Osaka 535-8585, JAPAN,
haga@med.oit.ac.jp
[2] Technical University Berlin, suzuki@hmi.de

Abstract: A twin roll caster to cast clad strip was designed and assembled. This twin roll caster could cast clad strips of aluminum alloys at the speed of 20 m/min. Thickness of the clad strip was about 2.0 mm. The clad strip had a clear interface between the two strips. The clad strip was connected strongly at the interface. Therefore, the clad strip did not peel off at interface after continuous bending until broken. Clad ratio of the thickness between overlay strip and base strip was 2:1•1:10. Alloy that melting point was lower than the other was suitable for overlay strip. Roll speed, cooling distance of base strip, temperature of the melt for overlay strip were important factors to cast sound clad strip.

1. Introduction

Roll casting of strips has some merits. For example, they are rapid solidification, low equipment cost, energy saving and process saving. Casting of clad strip adds new merits to the roll casting. In the present study, a roll caster for the clad strip was designed and assembled. And fundamental conditions for clad strip casting were investigated.

2. Twin Roll Caster and Experimental Conditions

Figure 1 shows a twin roll caster and an enlarged view of the roll bite. The twin roll caster was obtained by combining a melt drag twin roll caster and a downward melt drag twin roll caster [1,2]. Strip cast by an upper roll was called overlay strip, and strip cast by a lower roll was called base strip in this paper. Separating force was very small (10•50 N/mm width). Therefore, strips could not be connected by separating force. Diameter of copper rolls were 250 mm, and width of the nozzle for the overlay strip and the base strip were 20 mm and 50 mm, respectively.

As a first step of the study about the casting of the clad strip using the twin roll caster, fabrication of over top lay type clad strip was tried. Effect of the casting conditions that was thought to be concern to the connection of the overlay strip with the base strip was investigated. The conditions were cooling distance, roll speed, melt temperature and melting point of the overlay strip.

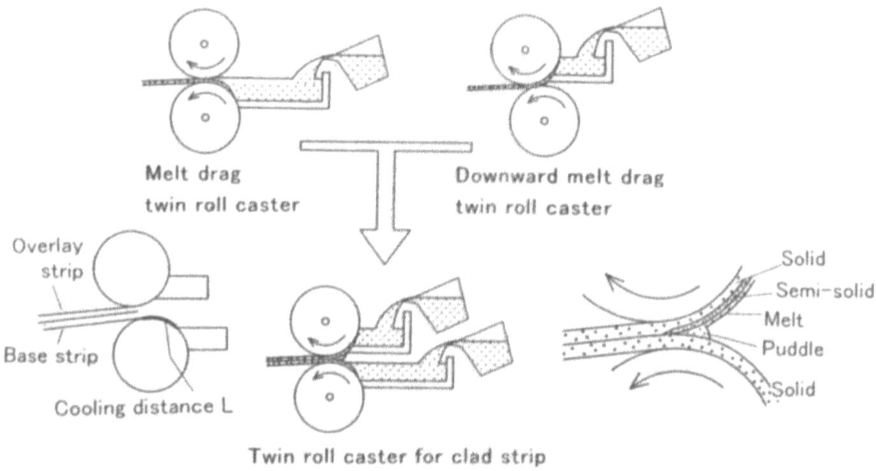

Fig. 1 A twin roll caster for clad strip and enlarged view near the roll bite

3. Result

Figure 2 shows relationship among the cooling distance, melt temperature of the overlay strip, ability of connection of the two strips, and condition of interface between the two strips. When the cooling distance was 150 mm and melt temperature was 600 •, strips could not connect. Temperature of the upper surface of the base strip was lower than proper temperature for connecting. The temperature of the base strip was raised by a puddle of the melt. When the melt temperature of the overlay strip was higher than 700 •, the temperature of the base strip was raised up to the proper temperature for connection of the strips. Therefore, the overlay strip could be connected with the base strip. The shorter the cooling distance, the higher the temperature of the base strip at the connecting point. When the cooling distance was shorter than 100 mm, the overlay strip could be connected to the base strip.

The interface between the overlay strip and the base strip was classified into A, B and C as shown in Fig. 2. A shows that the interface was not clear. The upper side of the base strip was melted. C shows that there were gaps at the interface. The temperature of the upper side of the base strip was too low. B shows that the interface was clear and gaps did not exist at interface. When the cooling distance of the base strip and the melt temperature of the overlay strip were proper, the clad strip that belonged to B was obtained.

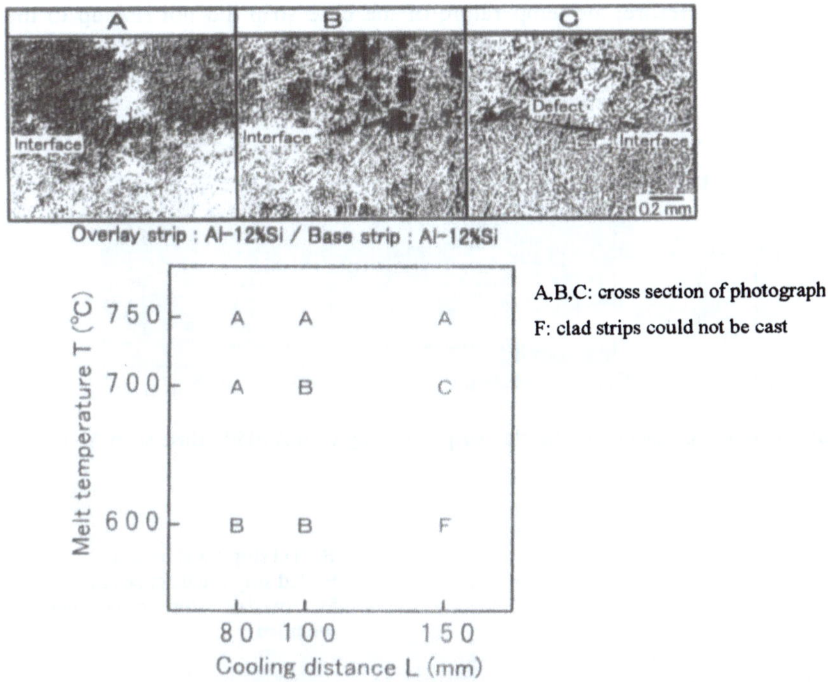

Fig. 2 Relationship among the cooling distance, melt temperature of the overlay strip, ability of connection of the two strips, and condition of interface between the two strips

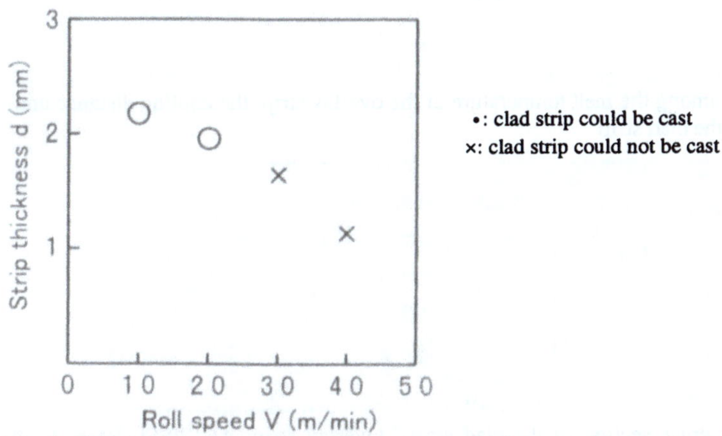

Fig. 3 Relationship among the roll speed, strip thickness and connection of the two strips

Figure 3 shows relationship among the roll speed, strip thickness and connection of the two strips. When the roll speed was 10 and 20 m/min, the clad strip was obtained. However, when the roll speed was 30 and 40 m/min, the overlay strip did not connect with the base strip. The higher the roll speed, the smaller the puddle. The higher the roll speed, the shorter the contact time between the base strip and the puddle. Therefore, the temperature of the base strip did not rise up to the proper temperature for connecting. As the result, the overlay strip could not connect to the base strip. Thickness of the clad strip became thinner as the roll speed became higher like the conventional twin roll caster.

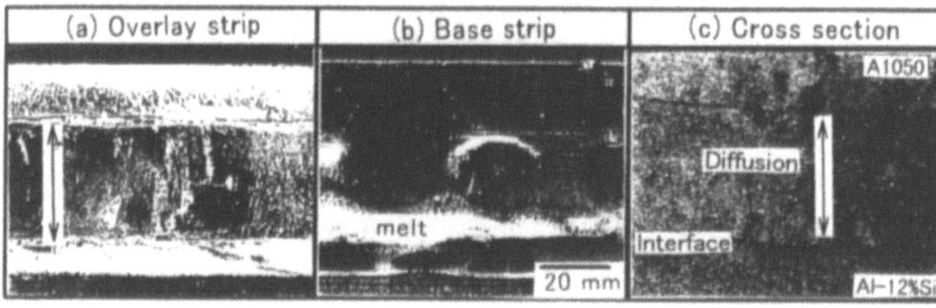

Fig. 4 Surface and cross section of the clad strip. Overlay strip:A1050, Base strip:Al-12%Si

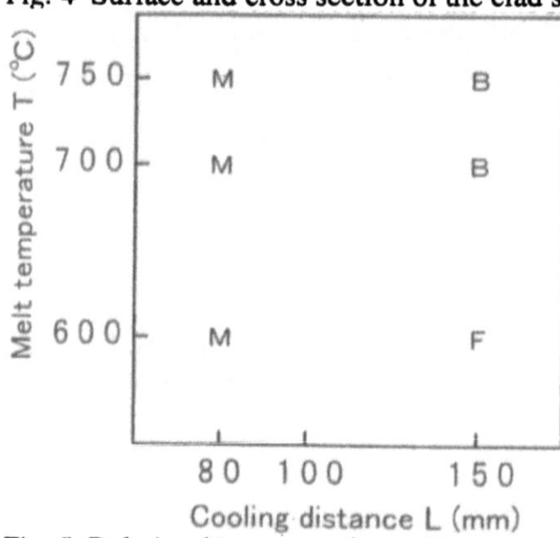

B: clad strip could be cast
F: clad strip could not be cast
M: overlay strip could not be solidified

Fig. 5 Relationship among the melt temperature of the overlay strip, the cooling distance and ability of casting of the clad strip

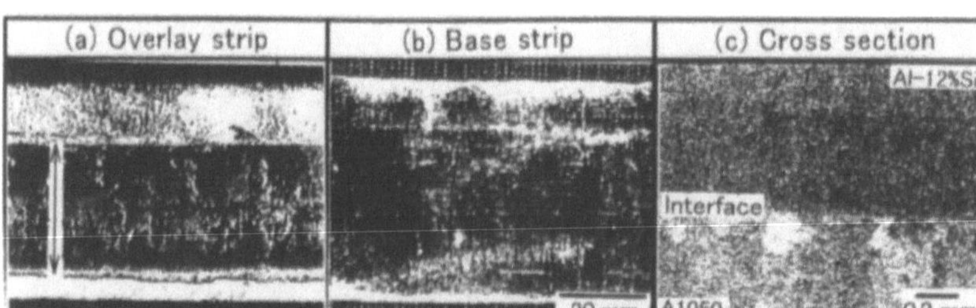

Fig. 6 Surface and cross section of the clad strip. Overlay strip: Al-12%Si, Base strip: A1050

When the overlay strip was A1050 and the base strip was Al-12%Si, the melting point of the overlay strip was higher than that of the base strip. The base strip was melted by the overlay strip as shown in Fig. 4. Si of the base metal diffused into the overlay strip. Figure 5 shows relationship among the melt temperature of the overlay strip (Al-12%Si), the cooling distance and fabrication of the clad strip, when the melting point of the overlay strip (Al-12%Si) was lower than that of the base strip (A1050). When the cooling distance was 80 mm, the overlay strip was not completely solidified (M in Fig. 5). It is thought that the temperature of the base strip was higher than the melting point of the overlay strip at the connecting point. When the melt temperature of the overlay strip was 600 • and the cooling distance was 150 mm, the overlay strip did not connected with the base strip (F in Fig. 5). The melt temperature of the overlay strip was lower than the melting point of the base strip. Therefore, the temperature of the base strip did not rise up to the temperature that was suitable for the connecting. When the temperature of the overlay strip was 700, 750 • and the cooling distance was 150 mm, the clad strip could be fabricated (B in Fig. 5). The base strip was not melt by the overlay strip and the interface existed clearly as shown in Fig. 6. Figure 7 shows SEM image and line analyzing around the interface of the clad strip, which overlay strip was Al-12%Si and the base strip was A1050. The interface was clear and Si of the overlay strip did not diffused into the base strip.

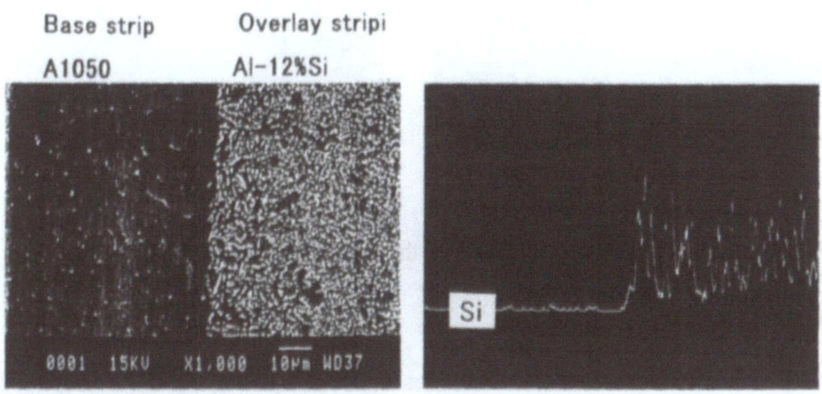

Fig. 7 SEM image and line analyzing around the interface of the clad strip

Fig. 8 cross section near the broken area of the clad strip

Continuous bending of the clad strip was carried out until broken of the clad strip. Figure 8 shows cross section near the broken area of the clad strip. It was apparent that the overlay strip and the base strip were connected firmly at the interface.

Figure 9 shows the relationship between the thickness of the overlay strip and the base strip. Clad ratio (thickness ratio : overlay strip / base strip) was in the range of 1:10 • 2:1. The thickness of each strip could be controlled by contact length between the melt and the roll. The longer the contact length, the longer the solidification time. As a result, the strip becomes thick. The contact length could be controlled by melt head in the nozzle.

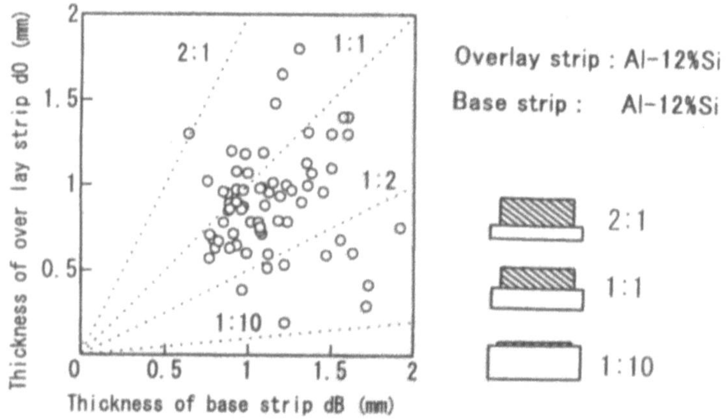

Fig. 9 Relationship between the thickness of the overlay strip and the base strip.

5. Conclusion

A twin roll caster that can cast clad strip was devised. This twin roll was obtained by combining a melt drag twin roll caster and a downward melt drag twin roll caster. Clad strips of aluminum alloys could be cast using this twin roll caster. The clad strip had clear interface between the two strips. Two strips were connected strongly. Strips did not peel off at the interface after continuous bending until broken. Temperature of the base strip, roll speed, melting point and temperature of the melt for overlay strip were important factors to cast sound clad strips. The alloy which melting point was lower than the other was suitable for the overlay strip in order not to melt the base strip. Clad ratio of the thickness between the overlay strip and the base strip was 1:10•2:1.

References

•

[1]T.Haga, N.Inoue, S.Suzuki, et al, Proc. ICAA-6, Vol.1,(1998), 327-332.
[2] T.Haga, N.Inoue, S.Suzuki, et al, Proc. AMPT, Vol.1,(1998),105-112.

Electroforming of a Stereolithography Master Pattern for use as an Electro-Discharge Machine Electrode

Allan E W Rennie[1], Chris Bocking[2], Graham R Bennett[3]

Centre for Rapid Design and Manufacture, Buckinghamshire Chilterns University College, Queen Alexandra Road, High Wycombe, Bucks, HP11 2JZ
Email: [1] allan.rennie@bcuc.ac.uk ,
[2]Chris.Bocking@bcuc.ac.uk,[3]Graham.Bennett@bcuc.ac.uk

ABSTRACT. Electroformed EDM electrodes have been proposed and studied by a number of workers and have been found to be an acceptable method of construction. They have an excellent capability of copying minute surface details (<0.1 microns). However, much of the work has been carried out using thick electroforms (in excess of 3 mm). Unfortunately, such electroforms take many days to produce. This paper examines the feasibility of using thin walled electroforms (<1.5 mm) for use as EDM electrodes that can be produced in copper in time-scales of between 14 and 38 hours. Stereolithography models were used as mandrels on which to electroform the copper. Dimensional changes were monitored at each stage.

1.Introduction

Electrical Discharge Machining (EDM) is frequently used in the production of injection mould cavities where conventional machining methods cannot produce the desired geometry such as internal sharp corners. The electrodes used are generally machined but again, certain geometries of the electrode require multiple electrodes to be produced. This can be an expensive process as up to 50% of the total cavity machining costs may be attributed to the production of the electrodes [1].

Electroforming is an established technology for the production of free standing objects [2]. It is an extension of electroplating, in which metal is deposited from aqueous solutions containing salts of the metal being deposited. By the application of a suitable voltage across two electrodes immersed in the solution, a current flows, leading to deposition of the metal at the cathode. The current density and the time of deposition control the mass of metal deposition.

There are limitations to the magnitude of the current density that can be applied, beyond which the deposits are powdery and incoherent. Therefore, to obtain thick deposits, extended time periods are used. For copper electrodeposition, around 1 micron is deposited in 2 minutes at a current density of $20mA/cm^2$.

Electroforming is achieved by electroplating for extended time periods onto a model or mandrel until sufficient metal has been deposited. The mandrel has the reverse geometry of that of the desired electroform. If the electroform is of an appropriate thickness, it will retain an exact reverse copy of the geometry of the mandrel and will be free standing. Surface detail as small as 0.1 microns can be reproduced [3]. This paper describes a feasibility study of the fabrication and use of thin walled electroformed electrodes for machining injection mould cavities.

2.Experimental

2.1.Electrode design

Conventional methods of producing EDM electrodes such as CNC Vertical Milling cannot produce certain internal features such as sharp corners in internal recesses of the tool. The electrode design chosen for this work contained such features and the final cavity geometry required is shown in **Fig 1**. The mandrel has the same shape as the desired cavity. The electroformed electrode, therefore, has the reverse of this geometry. As these were feasibility trials, the original CAD model did not take into account the additional dimension of the spark gap, which alters the final cavity dimensions by the spark gap setting of the EDM machine. A number of stereolithography (SL) mandrels were produced on 3D Systems SLA 250/50 using Ciba Geigy resin SL5170. No hand finishing of the mandrels was conducted. A total of eight electrodes were produced of various thicknesses, shown in Table 1.

Part No	Plating Time Hours	Average Thickness (mm)	Backing	Performance
B1	26	0.7	PU	Sectioned
B2	118 (reference)	3.0	PU	OK
B3	14	0.4	PU	Sectioned
B4	18	0.5	Metal	Sectioned
B5	38	1.0	PU	Edge split
B6	38	1.0	PU	Edge split
B11	38	1.0	Metal	OK
B12	38	1.0	Metal	OK

Table 1. Electroformed electrode details

2.2.Metallisation of the mandrel

Electroforming requires that the surface of the mandrel be electrically conductive. The mandrels were metallised using a high silver content conducting paint specifically designed for electroforming. This may be brush painted or sprayed onto the mandrel. It has been found in the past that thinner metallisation may be achieved by brush painting, thus preserving a closer dimensional accuracy.

2.3.Electroforming conditions

Electroforming was carried out in the following copper electroplating solution;

Copper sulphate $5H_2O$	75 g/l
Sulphuric acid	100 ml/l
Chloride	60 mg/l
Addition agent	12 ml/l
Temperature	25 - 30°C
Current density	20 mA/cm^2

The addition agent is used to control the crystal structure of the deposit and to ensure that the internal stress is maintained at close to zero. Significant internal stress may lead to distortion of the electroform, especially those with thin walls. The plating thickness was controlled by varying the time of deposition and varied between 14 to 38 hours. This corresponded to an average plating thickness range of between 0.36 to 1.0 mm. It should be noted that the local thickness of deposit at any one point is dependent on the geometry of the part - recessed regions in the mandrel will attract less deposit whilst prominent regions will attract more.

2.4.Backing the electroform

Some initial EDM trials were conducted on samples as they were removed from the mandrel. However, these showed that early failure occurred due to corner weakness, which caused deformation of the electrode (see Para. 3.1). It was clear that the electroforms required some form of additional support. Two methods of back-filling the electrodes were used; Alchemix PU368 polyurethane with Fillite sand mixed at a ratio of 4 to 1, vacuum cast and a tin/bismuth alloy (Fusible 17, MP 138°C). In the case of the metal filled electrode, adhesion was promoted by the use of a flux (Bakers Fluid No 3). The metal was poured into the electroform at a temperature of between 190 to 200°C.

2.5.Removal of the electroform

The electroform was removed from the stereolithography mandrel after filling with the respective material. SL5170 resin has a glass transition temperature of between 65 and 90°C and by immersion in boiling water, the resin becomes flexible and the electroform can be gently eased from the mandrel.

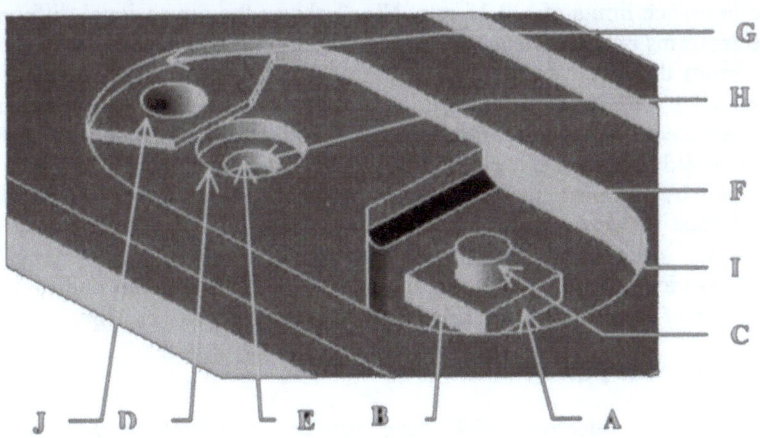

Figure 1. CAD representation of the SL model and positions of measurements taken at each stage of the process.

2.6.EDM settings

The electrodes were attached to a System 3R pallet using conducting epoxy resin. Machining was conducted using a Charmilles ROBOFORM 31 EDM machine. Currents of between 12A and 6A were used for roughing and finishing respectively, dependent upon the stage of the operation. Material removal rates of a maximum of 50mm^3/min were achievable during the roughing stages of machining whilst

5mm^3/min were estimated during the finishing stage. Surface roughness was kept at a constant value of 1.62 microns.

2.7. Dimensional measurements

At each stage of the process, measurements were made in millimetres using calibrated vernier callipers, accurate to two decimal places. Measurements for this study were taken of ten key features on the model, that was created to test the complexity and capabilities of the electroformimg process, and are as follows:
A-length of square; **B**-width of square; **C**-diameter of cylinder; **D**-diameter of base cylinder; **E**-diameter of upper cylinder; **F**- overall width of part; **G**-overall length of part; **H**-height of upper cylinder; **I**-distance from edge to inner edge of square cavity; **J**-diameter of nodule. These features are identified in the image shown in **Fig. 1**.

3. Results

In total, eight parts were built and a selection of these were used in trials as EDM electrodes after suitable backfilling. **Fig. 2** shows the dimensional differences between the original CAD and the resultant mandrels. These results are acceptable within the tolerances quoted by 3D Systems of ±0.15mm.

It should be noted that dimensions A, B and C are internal on the electroform, whereas features D-J are all external. It was found that the electroforms were within the tolerance limits of ± 0.15 mm. **Fig. 3** shows the dimensional differences of the electroforms in comparison to their respective original mandrels.

From the electroforms created, some were used as EDM electrodes and the remainder were retained for sectioning to observe the nature and distribution of the copper electrodeposition. The wear of the electroforms is shown in **Fig. 4** as a normalised percentage wear rate defined as the original electroform dimension divided by the post-sparking dimension and then multiplied by the dimensional difference between the two. In this way, direct comparisons between features of differing sizes could be made. The normal way of representing tool wear (volumetric tool wear) was not suitable with this technique due to the potential charring of the polyurethane filler if exposed during spark erosion.

3.1. Electrode performance

The reference electrode, B2, gave the best performance with no significant breakdown of the electroform. B6 failed at an early stage during use due to separation and curling of the metal on the deepest edge as shown in **Fig.5**. These results are not shown in **Fig. 4.** Electrode B5 exhibited some splitting of the same edge after the EDM process. However, this was not as severe as in the former case. It was decided that electrodes B11 and B 12 would be used sequentially, with B11 acting as a roughing electrode and B12 as the finishing electrode. The resultant cavity from this operation is shown in **Fig. 6.** Using this double electrode regime, it was found that a cavity could be successfully produced with no degradation due to electrode wear and curl.

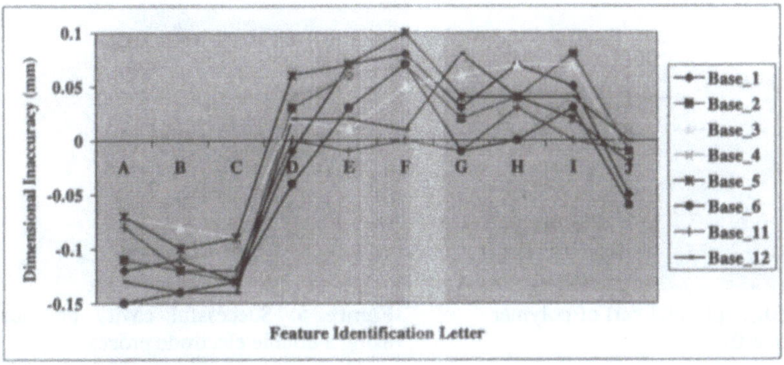

Figure 2. Dimensional inaccuracies of eight SL5170 models in comparison to original CAD dimensions

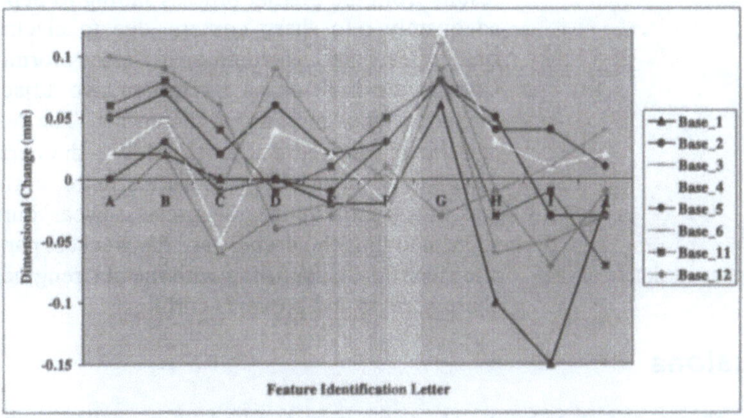

Figure 3. Dimensional changes of eight electroforms after extraction from their respective master pattern

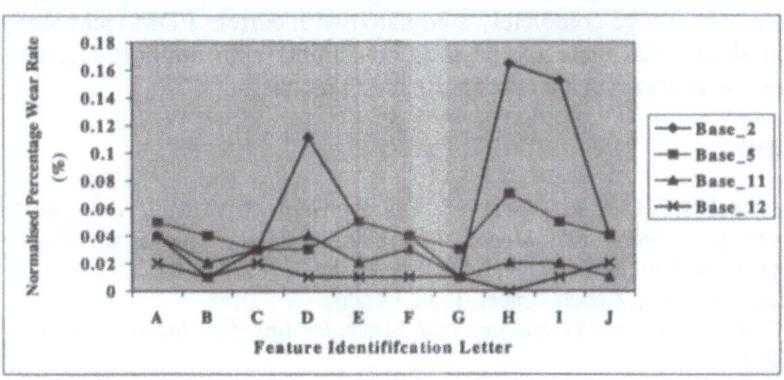

Figure 4. Normalised percentage wear rate of four electroforms created using SL5170 master patterns

Figure 5. Edge split and curl of polymer filled electrode B6

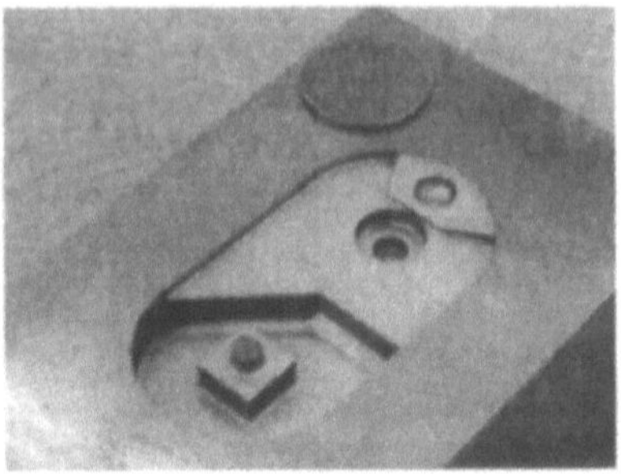

Figure 6. Successful cavity produced using a double electrode process

Examination of the microsections of the electroforms revealed that the sharp internal corners exhibited a weakness, common to electroforming. Corner weakness results from the plating process failing to deposit adequately into sharp corners, due to electrical field effects and is common to electroforming. Corners are built up as a result of two adjacent planar surfaces merging at the corner. The result is a line of weakness at around 45° to the corner. This can be seen in **Fig. 7**. Furthermore, corners and edges experience the greatest wear during EDM and thus the corners are the weakest part of the electrode. Backfilling with metal strengthens these corners and prevents curling.

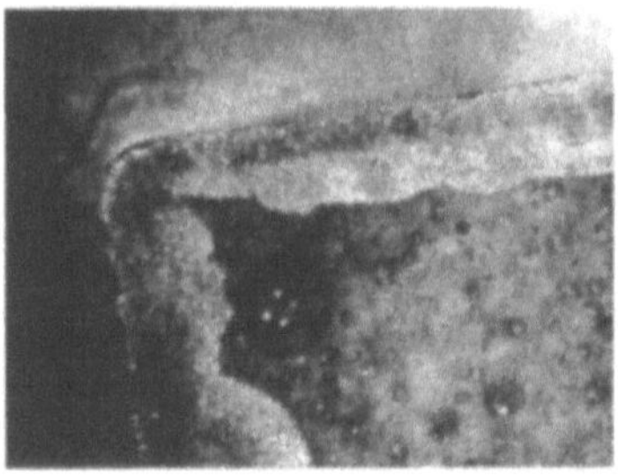

Figure 7. Corner weakness of the electroform

4.Conclusions

This paper shows the feasibility of using metal filled thin walled copper electroforms as EDM electrodes. More suitable would be the employment of two electroforms to act as roughing and finishing electrodes in a sequential EDM process, which can now be completely automated with current EDM technology. It has also been shown that there are no time constraints with multiple electroform production, as electroforms can be produced simultaneously.

References

[1] Arthur A, Cobb RC Dickens PM, 1995, *The Potential for Rapid Prototyped EDM Electrodes Using Stereolithography Models.* Proceedings of First National Conference on Rapid Prototyping and Tooling Research, Ed G Bennett, pp 221- 235, UK

[2] Spiro P, *Electroforming*, Robert Draper, 1970, London

[3] Legierse P E J, *Mastering Technology and Electroforming for Optical Disc Systems,* Trans. IMF, **65**, 13, 1987

Acknowledgements. Thanks to Peter Moreton and David Mantle of BCUC Engineering Workshop for their kind assistance.

Modelling of the Friction Surfacing Process for Turbine Blade Reclamation

Jaworski, B.[1], Voutchkov, I. I.[1], Vitanov V. I.[1], Hughes V.[1]

[1] Department of Mechanical and Manufacturing Engineering, University of Portsmouth, Anglesea building, Anglesea road, Portsmouth PO1 3DJ, United Kingdom
Tel: +44 2392 842520 Fax: +44 2392 842351 Email: val.vitanov@port.ac.uk

ABSTRACT: Friction surfacing/coating provides a valuable mechanism for the repair of steam turbine blades, which can have a very high replacement cost. Optimum process conditions are essential and this paper details a finite element model based analysis that takes into account the dynamics of temperature gradients due to complex geometry. The model can determine appropriate values for traverse speeds required to produce an optimum interface bonding temperature. The research will lead to the inclusion of the process model in a real-time control system.

1. Introduction

The high cost of producing turbine blades makes any consideration of reducing waste, or extending life by repair, of significant interest. The friction surfacing process (fig. 1) has a proven capability for depositing a hot forged coating at specific blade locations, without any fundamental changes to the blade structure or condition. Repair can be made at a lower cost than other techniques available with a higher degree of repeatability due to the much-localised nature of the process. In order to optimise the quality and repeatability of repairs suitable analytical models are needed.

So far existing models - as proposed by Fukakusa and Satoh [1] - can be used for the friction surfacing process, but have limited value because they do not take into account the traverse speed of the substrate nor do they consider the full range of critical process variables. In previous models [2], the mechanism of friction surfacing has been discussed using temperature distributions obtained from heat transfer equations involving heat-generating layers. The current lack of an adequate process model in the area of friction surfacing, as well as a general inability of the existing models to update control variables when complex changes in blade geometry occur has held up commercial exploitation of the friction technology in this area.

Turbine blades have variable thickness and constantly changing conduction conditions as the deposition proceeds. The required process control conditions can be found by experiment, but only after considerable effort and even then the results can only be applied to a specific geometry.

The need has been for an analytical model that can be applied to different forms.

The developed technique involves the establishment of a number of assumptions regarding boundary conditions and other parameters. Following testing and analysis a set of simpler assumptions have been derived and their validity established leading to a viable model for process control purposes.

2. Experimental Procedure

2.1 The friction surfacing process

The use of friction energy to deposit metals in the solid phase on a substrate is an area of considerable industrial importance known as *Friction Surfacing*. The friction surfacing process solid phase has hot forged microstructures and exhibits an excellent metallurgical bond [3].

The process involves rotating a rod (called mechtrode™) of the coating material onto a substrate (such as turbine blade) and developing, through the friction and pressure generated at the rubbing interface, sufficient temperature to cause the end of the coating rod to plastically deform. By moving the substrate across the face of the rotating mechtrode™, a layer of plasticised material is deposited (Fig. 1). The coating is regular and flat, without the familiar meniscus section profile experienced with methods based on fusion welding. The thickness of the coated layer is typically between 0.5 and 4mm depending on mechtrode diameter and material.

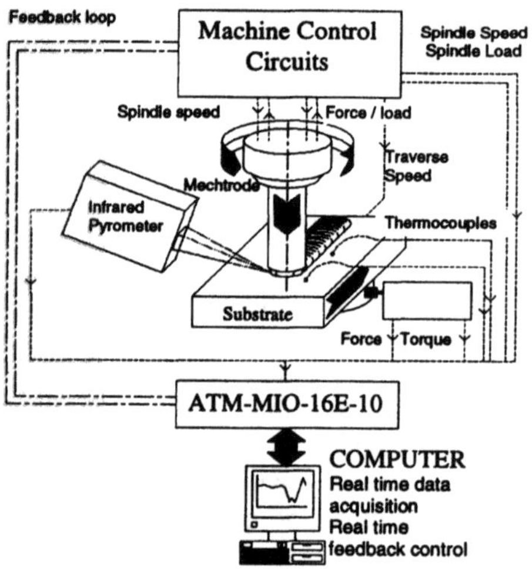

Figure 1: Friction surfacing

Research over the last four years has shown that the quality of the bond between the coating and the substrate is mainly influenced by the applied pressure and the temperature. In this context, it becomes important to develop mathematical models capable of analysing temperature and forces into the process, and especially in specific areas like the rubbing interface and the bond interface. An experimental set-up has been developed as shown in Fig. 1. In which forces in the process are measured online using a pressure sensor mounted below the substrate. The nature of the process itself does not allow the measurement of temperatures inside the deposition area. In developing the Finite Element (FE) model two techniques for measuring the temperature have been used to acquire the temperature at the surface of the substrate and on the surface of the deposit:

- **Thermocouples**: they have a fast response time due to their small size (wire *Ø0.2mm*), and deliver accurate readings (*+/-4degC* at *1000degC*) making thermocouples a very efficient way of measuring the temperature at the surface of the substrate.
- **Infrared pyrometry**: it is used to measure the surface temperature of the deposit and the rotating rod, close to the rubbing interface. The pyrometer currently in use has a spot size of *0.5mm* or *0.3mm*, depending on the lens in use (respectively

250mm and *80mm* focal length). The accuracy of infrared pyrometry is dependant on the surface characteristics of the measured component, in particular its emissivity, that changes with the surface characteristics. For stainless steel a temperature gradient from *1300degC* to *700degC* has been measured, from the rubbing interface to the bond interface with *+/-7degC* accuracy.

2.2 The finite element model

In order to establish the bond temperature, a numerical model has been developed, using the finite element method.

The model replaces the Mechtrode™ by a moving heat flux source having sensibly the same area. That heat flux moves along the generated coating, online created on the substrate (of any shape). The heat flux is considered constant in time, as suggested by the experimental temperature curves shown in Fig. 3. Process temperature, while moving on top of the substrate, remains constant, expressing a constant heat flux in equilibrium with a consistent heat sink i.e. the substrate. The heat flux value used in the model has been determined through experiments.

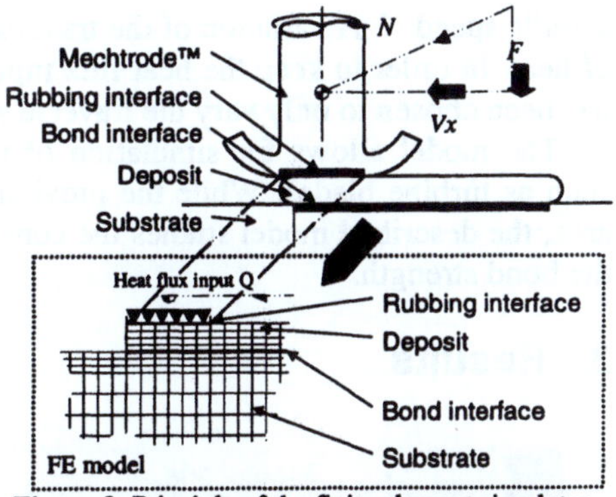

Figure 2: Principle of the finite element simulator

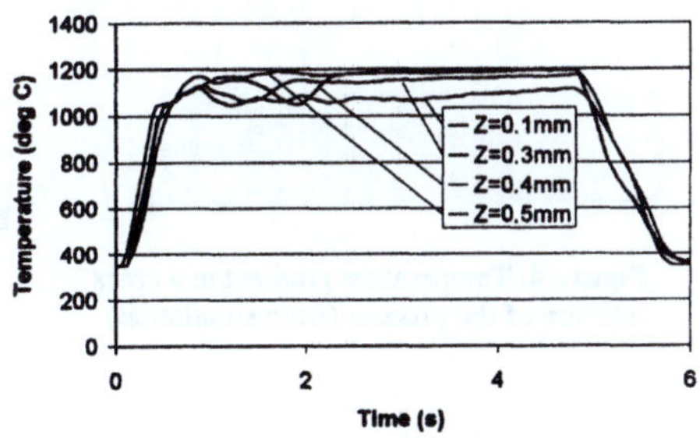

Figure 3: Series of measurement at different elevations onto the coating.

The model takes into account material characteristics and process parameters, and is capable of calculating the temperature gradient in both the coating and the substrate. Once the temperature gradient has been calculated, a second part of the simulator extracts for convenience the bond interface temperature, and temperatures for thermocouples and the infrared pyrometer.

As the simulator can calculate the bond interface temperature and its variation in time, it is possible to use it when changing process parameters. At this stage only traverse speed (*Vx*), applied force (*F*), and spindle speed (*N*) can be varied online. The Bowden and Ridler formula [4] expressing the amount of heat equivalent of the mechanical energy input into a system has been used under the assumption that :

- It is correct up to the region of the rubbing interface
- The rubbing interface is a disc with a radius equal to the radius of the rod
- The pressure is evenly distributed at the bond interface

- The sliding coefficient μ is constant

$$\dot{Q}\,rod = \frac{2.\mu.F.N.R}{3.J} \qquad (1)$$

where F : force on the top of the rod ; μ : sliding friction coefficient ; R : radius of the rod ; N : rotational speed of the rod (rad/s) ; J : mechanical equivalent of the heat

$\dot{Q}\,rod$ is the rate of heat energy generated at the rubbing interface. According to the previous equation $\dot{Q}\,rod$ can be considered linked only to the force F and the spindle speed N. A variation of the traverse speed Vx does not influence the amount of heat. In order to keep the heat flux input value constant during the simulation, it has been chosen to only vary the traverse speed Vx.

The model allows for simulation of the thermal effects in complex geometry such as turbine blades. While the previous models focused on the heat generation area, the described model studies the conditions at the bond interface that influence the bond strength.

3. Results

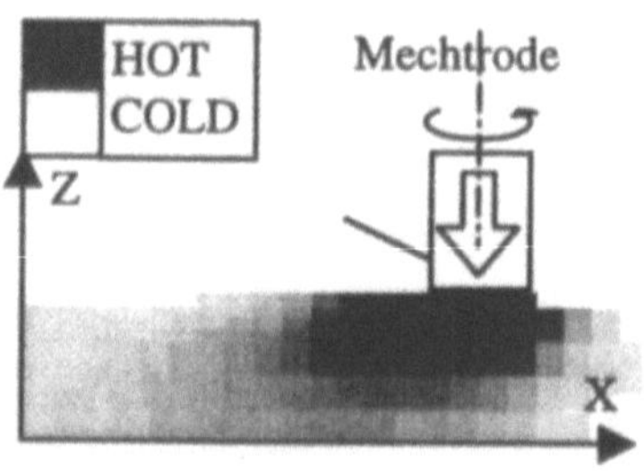

Figure 4: Temperature gradient in a cross section of the process (from simulation)

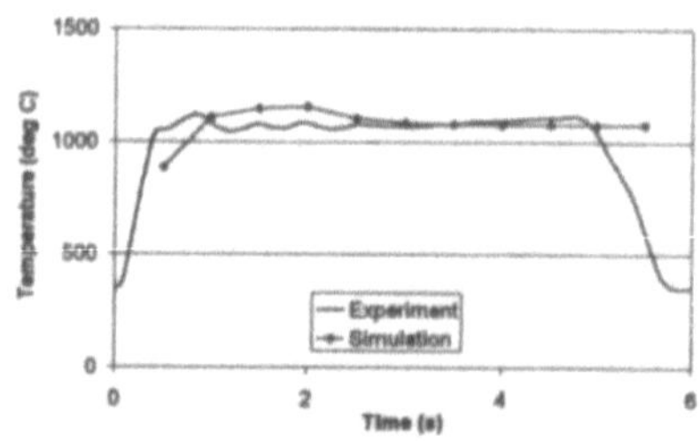

Figure 5: Comparison between experimental infrared measurement and simulated temperatures

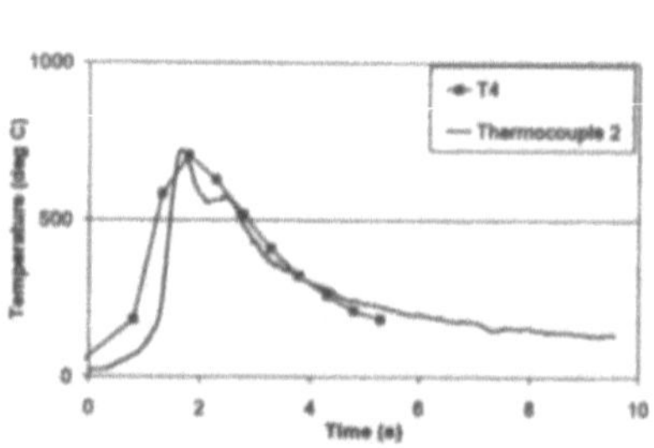

Figure 6: Comparison of temperatures obtained from a thermocouple and those derived by an FE model.

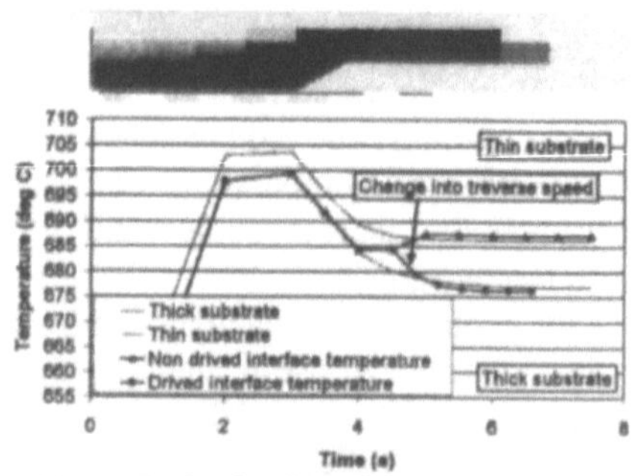

Figure 7: Simulated bond temperature using a changing traverse speed on substrate with a variable thickness

Experiments so far have involved the use of SS304 for the rod, which has been deposited on a mild steel substrate. Readings from thermocouples and infrared pyrometer have then been used in the FE model to establish the value of the heat source at the rubbing interface. A typical process model showing temperature gradients through a cross section is shown in Fig. 4. Comparisons between model temperatures and those from experimental readings are shown in Fig. 5 and Fig. 6. A good correlation can be seen over the time span when temperatures are increasing. A weaker match has been found over the cooling phase.

4. Discussion

The main problem in determining process conditions is due to the variations of the substrate shape. The set of simulation results shown in Fig. 7 models a substrate changing its thickness under the coating. Using a previously determined heat flux value to simulate the rod, the simulation shows that the bond temperature changes to a higher value when the coating reaches the thinner part of the substrate. An increase in the traverse speed at the point of change has been simulated in order to keep the bond temperature constant. The response curve obtained shows that the increase of traverse speed Vx is followed by a decrease in the bond temperature. During this simulation, the applied force F and spindle speed N are kept constant. According to equation 1, the heat input remains the same. The research has shown that changes in the coating thickness are of major importance in the distribution of the temperature gradient. To this end, the driving conditions of the process are currently developed to take into account the variations of the coating thickness.

The proposed finite element model has shown reasonable agreement with the experimental data. The main assumptions - a simple heat source instead of the complex Mechtrode™/Substrate system - allows a greatly simplified model. The experimental procedures are now simplified, using a standard experiment involving a simple shape substrate. From a single set of experiments, bond temperature and heat input are evaluated. Using the defined amount of heat, a simulation of the response of the system onto different shapes of substrate using different traverse speeds allows optimum conditions to be determined in order to achieve the same bond interface temperature. The model allows for a change in the traverse speed at each step of the simulation.

Up to this stage, no expensive substrate has to be produced. Leading to a better knowledge of the thermal gradient inside the process, the finite element model allows a drastic reduction in the time and money spent in experimenting. Replacing the case by case experimentation by a single set on a standard substrate, done once for a couple of materials (rod/substrate), and using the established condition to any new case will drastically minimise the further need for experiments, leading to a "first time first success". In the case of reclamation of components such as turbine blades, where no experimentation can be done on the component, such method is of first importance.

5. Conclusions

1. The work has concentrated on the control of interface conditions to produce 'First-time' bonding. This breakthrough has made it feasible to consider turbine blade reclamation as a new application for the friction surfacing process. Further work will lead to on-line process control with the model used in the feedback loop.
2. The proposed model has achieved reasonable agreement with experimental results and for different substrate shapes.
3. The proposed model is capable of maintaining feasible values for process quality parameters when perturbations in the thermal system occur.

References

[1] Fukakusa K., Satoh T., 1986, Travelling phenomena of Rotational Plane during Friction Welding, OSAKA Institute of Technology

[2] Fukakusa K., 1995, On real rotational plane in friction welding of different diameter materials and dissimilar materials. OSAKA Institute of technology

[3] Bedford G. M., 1990, Friction surfacing for wear applications, Metals and Materials, November, pp. 702-705

[4] Hutchings I., 1992, Tribology, Ian M Hutchings, ISBN: 0-340-56184-X

Modelling, Optimization and Control Integration of the Friction Surfacing Process

Voutchkov, I. I., Vitanov V. I, Jaworski, B., Bennett, N. G.

Department of Mechanical and Manufacturing Engineering, University of Portsmouth, Anglesea building, Anglesea road, Portsmouth PO1 3DJ, United Kingdom; Tel: +44 2392 842520; Fax: +44 2392 842351
e-mail: Val.Vitanov@port.ac.uk

Abstract: Friction surfacing is an advanced manufacturing process that has been successfully developed and commercialised over the past decade. This paper considers a novel approach that will allow the expansion of the range of potential applications in the area of turbine blade repairs and manufacture including manufacture of bladed components, aerofoil replacement repairs, blade tip extensions, surface build up (including erosion/corrosion damage) and shoulder wear. The process is modeled using two types of model - Controlled AutoRegressive Integrated Moving Average (CARIMA) for the dynamics and neurofuzzy relationships for the static. Self tuning extremum control integrates both models and produces optimal control in real time. The algorithm is shown to follow successfully a predefined setup pattern.

1. Introduction

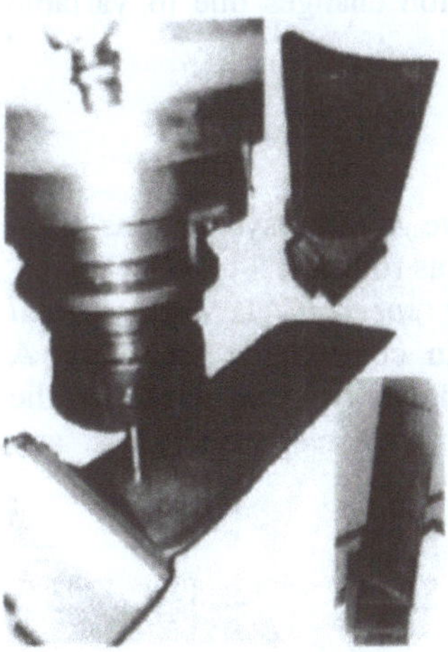

Figure 1 Friction surfacing used to repair turbine blades

The friction surfacing process is a derivative from friction welding that involves rotating a rod of the coating material onto a substrate and developing, through the friction generated at the rubbing interface, sufficient heat to plastically deform the end of the coating rod (mechtrode™). By moving the substrate across the face of the rotating rod, a layer of mechtrode™ material is deposited. Fig. 1 shows the process that is used for turbine blade repairs. Normally such blades are produced using precision casting processes, however small imperfections on the face may be left after the blade is taken out of the casting mold. Such imperfections, as well as service corrosion and other wear effects, usually render the blade unusable. Friction surfacing offers an efficient and inexpensive way to repair various machine parts with complex shapes [1].

2. Variable Process Conditions

The friction surfacing process depends on factors such as the thermal conductivity of the coating material and substrate, as well as their geometry. When the source is moving away from the origin point, assuming that the geometry of the substrate is constant, the process is in a stable state and optimal parameters can assure a good quality bond as shown in Fig 2a.

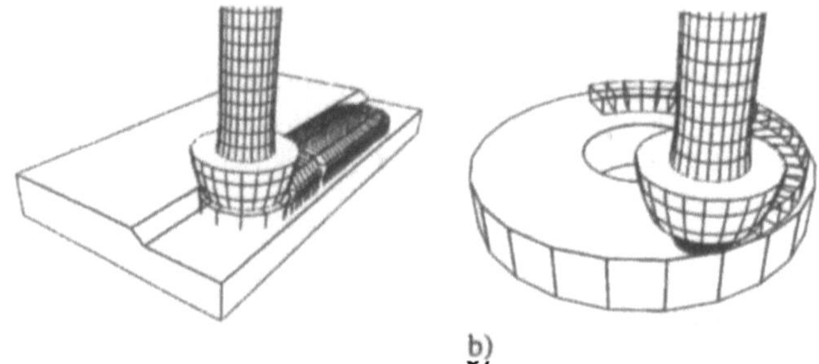

Figure 2 (a and b) Substrate geometries with different thermal systems. a) Heat source moving away from the origin; b) Heat source returning to the origin

However, when the source is returning to its origin point, as shown in fig. 2b, the amount of heat builds up inside the substrate, and its temperature gradually rises, making the coating thinner and the bond quality poorer [2]. This becomes of critical importance in micro friction surfacing, where the internal energy of the thermal system changes rapidly as similar effect is observed when coating components with variable thickness, such as the profile of a turbine blade. In these cases, although the heat flux is constant, its distribution changes due to variable heat sink conditions.

3. Feedback Control

To compensate against such disturbances an adaptive control system has been introduced. The main process parameters are identified as follows:

Mechtrode rotational speed (N), substrate speed (V_x), and vertical mechtrode speed (V_z) are considered to be the main controllable variables. A conventional adaptive approach would produce such inputs ($\mathbf{u}_i$) that minimize the difference between a given setpoint ($\mathbf{w}_i$) and the object's response ($\mathbf{y}_i$) at any moment *i*, as shown in equation (1).

$$(\mathbf{y}_i - \mathbf{w}_i)^2 \xrightarrow{u_i} \min \tag{1}$$

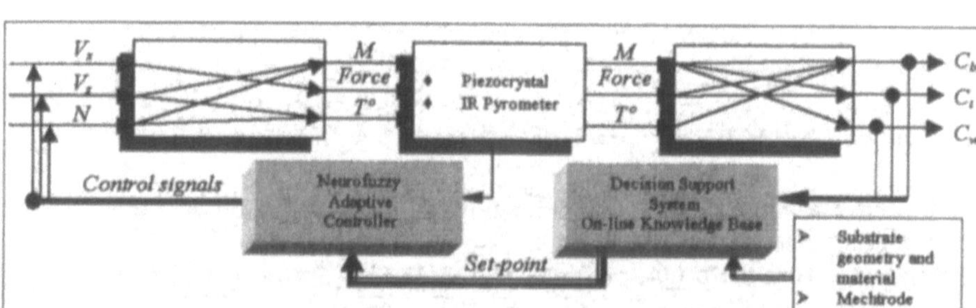

Figure 3. Feedback control of friction surfacing process

Friction surfacing responses are the coating characteristics such as coating thickness C_t, coating width C_w and bond strength C_{bs}. These three parameters, and the bond strength in particular, cannot be measured in real time. Instead three other parameters can be readily measured; *temperature* at the substrate/coating interface (*T*), *friction torque* (*M*) and *resultant force (F)* - and used further as process indicators. Figure 3 shows a schematic of a developed real time process monitoring and feedback control system. It enables monitoring of indicator parameters and to calculate the output state variables of the object by using neurofuzzy models. These models have been derived on the basis of 1800 experiments with different mechtrode and substrate materials, environmental conditions and geometry.

4. Modeling

As described above two different types of models were needed, one to forecast the process output as the indicator parameters are measured, and the other, to account for the dynamic relationship between the process inputs and the indicators.

4.1 Indicator-output Models

Since the indicator/output model can be represented as a steady state relationship, a non-linear, one hidden layer neural network was found to fit the data best as shown in Fig. 4. The model consists of 30 neurons with a sigmoidal transfer function. The output layer contains two neurons with a linear transfer function.

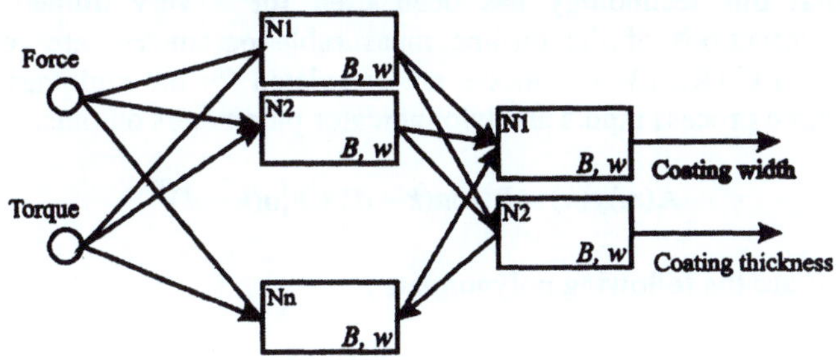

Figure 4. Neural network based model for the indicator/output parameters

The temperature at the substrate/coating interface is an important indicator parameter, but its influence can be registered by measuring the other two indicators, force and torque. That is why temperature is omitted from the indicator group for this model. A complete model should include the bond strength as the third output, as identified earlier however due to the lack of a reliable test procedure capable of accurately measuring bond strength, neural networks or statistical models alone are unable to produce an adequate forecast. To overcome the existing data uncertainty ill-defined problems are tackled well by appropriate set of functions using fuzzy logic set, which is used to complete the model [3]. Figure 5 shows response surfaces for the three outputs with respect to the indicator parameters. The graphs were obtained, using the proposed neurofuzzy model, and can be used for optimisation and real time control methods.

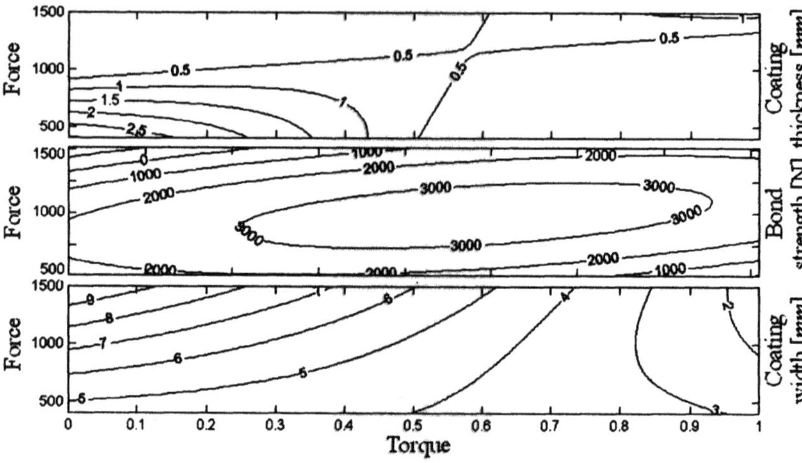

Figure 5. Response surface for the indicator/output parameters

4.2 Input-indicator Models

A review of published literature post-1946, confirms a complete lack of publication on the use of feedback control techniques for friction surfacing, indicating that this technology has been used for a very limited number of applications. Reactions of the on-line measurable parameters are recorded and coefficients of a CARIMA type model are calculated for the multivariable models which have three process inputs and two indicator parameters outputs.

$$\mathbf{A}(m)\mathbf{y}(k) = \mathbf{B}(n)\mathbf{u}(k-d) + \mathbf{F}[\mathbf{u}(k-d)]$$

where **A**, **B**, **F** are the following polynomials :

$$\mathbf{A}(m) = 1 + a_1 q^{-1} + a_2 q^{-2} + \ldots + a_m q^{-m}; \quad \mathbf{B}(n) = b_0 + b_1 q^{-1} + b_2 q^{-2} + \ldots + b_n q^{-n};$$

$$\mathbf{F}(r) = 1 + c_1 f\left(q^{-1}\right) + c_2 f\left(q^{-2}\right) + \ldots + c_r f\left(q^{-r}\right)$$ - non-linear functions

where

$\mathbf{u}(k)$ - input at sample k, vector 3x1 $\quad$ $\mathbf{y}(k)$ - output at sample k, vector 2x1

The coefficients of the model are calculated using the Recursive Least Squares Method. A known problem using this method is the decreasing of the trace of the covariance matrix, especially for slow variations of the process parameters, which leads to reduced sensitivity to parameter variations. The problem may occur in when coating large components, where the process is relatively slow. That is why an adaptive forgetting factor is used to keep the trace of the covariance matrix away from zero, as well as to assure convergence of the estimations [4]. Figure 6 illustrates the reaction of the process to a step change of the vertical mechtrode speed.

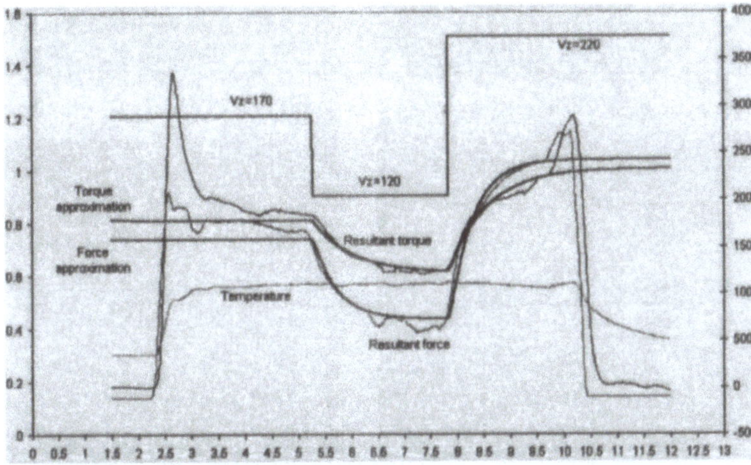

Figure 6. Measurable parameters' reaction to stepwise change of *Vz*

5. Adaptive Control Algorithm

Using the previously derived and validated models, a real time adaptive control algorithm was implemented. A major advantage of using this approach is that only the dynamic part has to be updated to reflect the variation in the thermal system. Such change does not affect the indicator/output relationships, and there is no need to continuously update the neurofuzzy model, which leads to a significant reduction in calculations in real time. This makes the adaptation algorithm fast and reliable, since it is based only on recursive calculations. Outputs in the cost function (1), can now be substituted with the predicted values. Control values are therefore shown in equation (2), where the cost function is minimum, or zero.

$$\mathbf{u}_k = \min\left(\hat{\mathbf{y}}_k(\mathbf{u}_{k-1}) - \mathbf{w}_k\right)^2 \tag{2}$$

Since a real time digital control system is expected to converge to an optimum solution within the time of discretization, a simple method of scanning must be selected. Each input is varied between predefined minimum and maximum values with a certain step. After all combinations are calculated, the one produced minimum output is selected. In order to reduce the time for calculations, the variation range is always kept to 10% of the previous selected parameter.

$$\begin{aligned}&\mathbf{u}_k = \mathbf{u}_k^{\min} : \mathbf{u}_k^{step} : \mathbf{u}_k^{\max};\\ &\mathbf{u}_k^{\min} = 0.9\mathbf{u}_{k-1};\ \mathbf{u}_k^{\max} = 1.1\mathbf{u}_{k-1};\ \mathbf{u}_k^{step} = 0.001\left(\mathbf{u}_k^{\max} - \mathbf{u}_k^{\min}\right)\end{aligned} \tag{3}$$

Such an approach takes 30% of the discretization time, CARIMA update takes 10%, and about 15% is used by scaling and other signal conditioning.

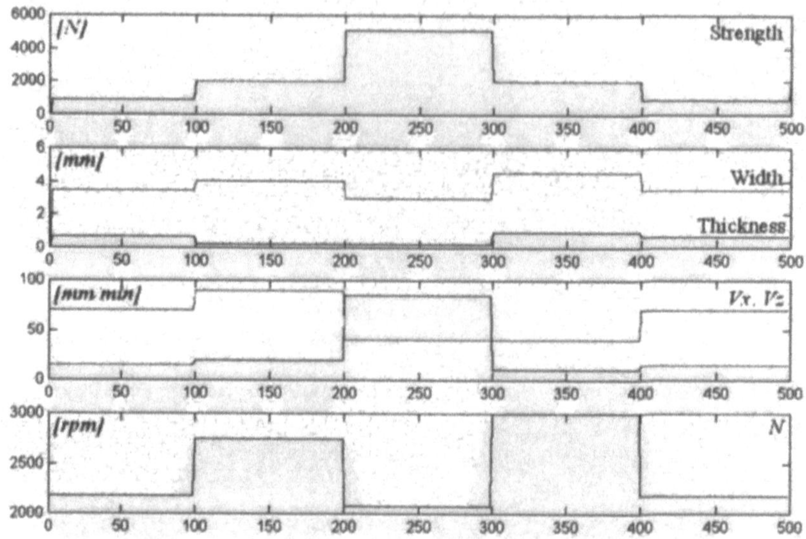

Figure 7 Calculated values for the control signals and object's output

Figure 7, shows the control signals, calculated by the control algorithm, to maintain the coating strength, width and thickness to a predefined setup pattern. Graphs only show the steady state of the process, which is achieved after the process dynamics converges to a steady state, since these are the values of major practical importance.

6. Conclusions

1. Steady state models are derived, using neural networks and fuzzy logic.
2. The dynamics of the friction surfacing process is modelled, using CARIMA models.
3. The continuously updating, dynamic part is separated from the neurofuzzy model, significantly reducing the calculations carried out in real time.

References

[1] Bedford G. M., Davies A., Sharp J. R., 1995, Micro-Friction Surfacing in the Manufacture and Repair of Gas Turbine Blades, *Third International Charles Parsons Turbine Conference, Newcastle*, pp 683-693

[2] Bedford G. M, Vitanov V. I, Voutchkov I. I, 1999 On the thermo-mechanical events during friction surfacing of high speed steels, *Proceedings of the International conference on advances in materials and processing technologies AMPT'99, pp.931-938*, August 3 - 6, Dublin, Ireland

[3] Zengqi S., Zhidong D., 1995, A fuzzy neural network and its application to controls, *Artificial Intelligence in Engineering*, v.10, pp. 311-315, Elsevier Science Ltd.

[4] Parkum, J. E, Poulsen, N. K., Holst, J 1991, Analysis of Forgetting Algorithms, *IFAC Symposium, Identification, Budapest*, pp 134 - 138

Casting of Ingots for Thixo-Forming Using a Downward Melt Drag Process

[1]Toshio HAGA and [2]Shinsuke SUZUKI

[1] Osaka Institute o f Technology, 5-16-1 Omiya Asahi-ku Osaka city, Osaka 535-8585 , JAPAN,
haga@med.oit.ac.jp
[2] Technical University Berlin, suzuki@hmi.de

Abstract: A downward melt drag single roll caster (DMDSRC) was devised to make ingots for thixo-forming. DMDSRC was assembled from a roll and a nozzle. Therefore, process was very simple. A copper roll surface was not coated by lubricant. However, the metal did not stick to roll and semi-solid slurry could be made continuously. Primary crystal became globular, and eutectic Si became fine and globular when Al-Si alloy ingots was heated up to semi-solid condition. Diameter of the globular primary crystal was about 90 •m.

1. Introduction

There are various processes to make ingots for thixo-forming[1-3]. In these processes, electric stirring and SIMA (strain inducing melt activation) are famous[1,2]. The electric stirring is popular[1].•A cooling slope is the simplest process[3]. However, there is a demerit that solidified metal tends to stick to the slope. In the present study, a simple casting process without sticking of the solidified metal to a cooling unit was devised. It was a downward melt drag single roll caster (DMDSRC). DMDSRC was assembled from a single roll and a nozzle. Therefore, DMDSRC was very simple. Fundamental properties of DMDSRC were investigated in the present study.

2. Casting Process and Experimental Conditions

Figure 1 shows processes tested in the present study. (a) is the cooling slope. In the process (b), the melt was poured on a rotating roll coated with lubricant. However, solidified metal stuck to the roll. (c) is DMDSRC. In DMDSRC, the roll was not coated with the lubricant. However, the melt did not solidify at the roll surface. Figure 2 shows an experimental flow, nozzle-roll gap G and melt-roll contact length L. Experimental parameters were roll speed, contact length between the melt and the roll, and nozzle-roll gap. Effect of experimental conditions on morphology of primary crystal and eutectic Si was investigated. Composition of aluminum alloy used in this study is shown in Table 1. Pouring temperature of the melt was 650•. Liquidus temperature of this alloy was 610•. Melt was cast into a metal mold. Ingots cast using DMDSRC was heated up to semi-solid condition (600•) , and water quenched. A copper roll was used, and diameter of the roll was 250 mm. The nozzle was made from insulator, and width of the nozzle was 20 mm.

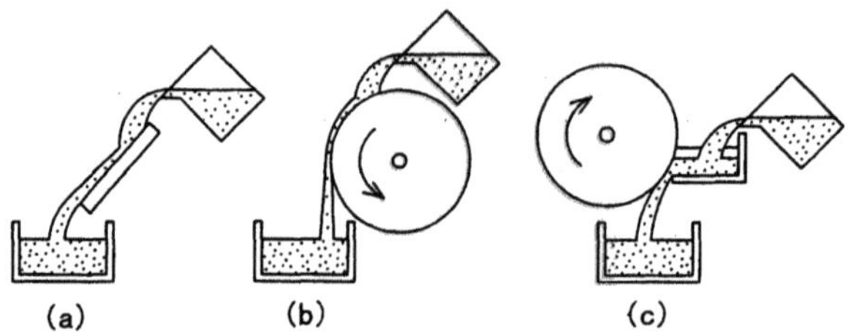

Fig. 1: Processes to cast ingots for thixo-forming investigated in the present study
(a) cooling slope, (b) cooling roll, (b) downward melt drag single roll caster (DMDSRC)

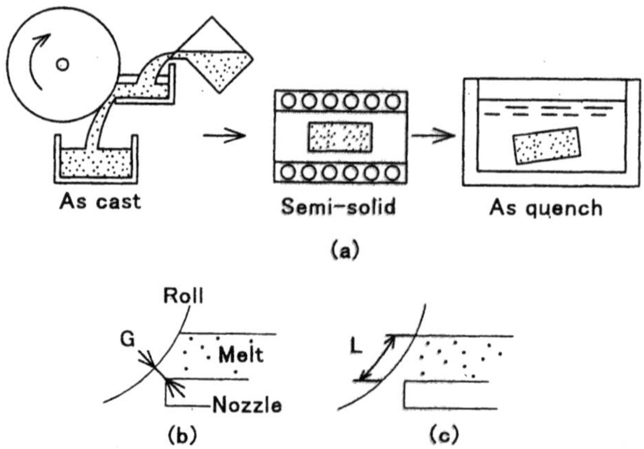

Fig. 2: Schematic illustration of experimental flow of casting using DMDSRC, and nozzle-roll gap G, melt-roll contact length L

Table 1: Compositions of material (mass%)

Si	Cu	Mg	Ni	Mn	Fe	Ti	Al
6.50•7.50	•0.20	0.20•0.40	•0.05	•0.10	•0.20	•0.20	bal.

3. Result

Primary crystal of as cast ingot and as quench ingot was classified into A, B and C as shown in Fig.3. C shows that the primary crystal of as cast ingot was dendrite and the primary crystal of as quench ingot did not become globular. B shows that the primary crystal of as cast ingot was dendrite and the primary crystal of as quench ingot became globular. A shows that the primary crystal of as cast ingot and as quench ingot became globular. Primary crystal of as quench ingot of the A was more globular than that of B. The primary crystal of A was excellent. Table 2 shows relationship between nozzle-roll gap G, melt-roll contact length L, roll speed V and morphology of the primary crystal. A, B and C of Table 2 coincide with A, B and C of Fig. 3. When the roll was stopped or speed was too low, the metal solidified at the tip of the nozzle. When the roll speed was too high, the melt was solidified into strip. There was tendency that the primary crystal became globular at the condition as below. The nozzle –roll gap G was narrow, for example G=2,3 mm. The melt-roll contact length was long. The roll speed was high.

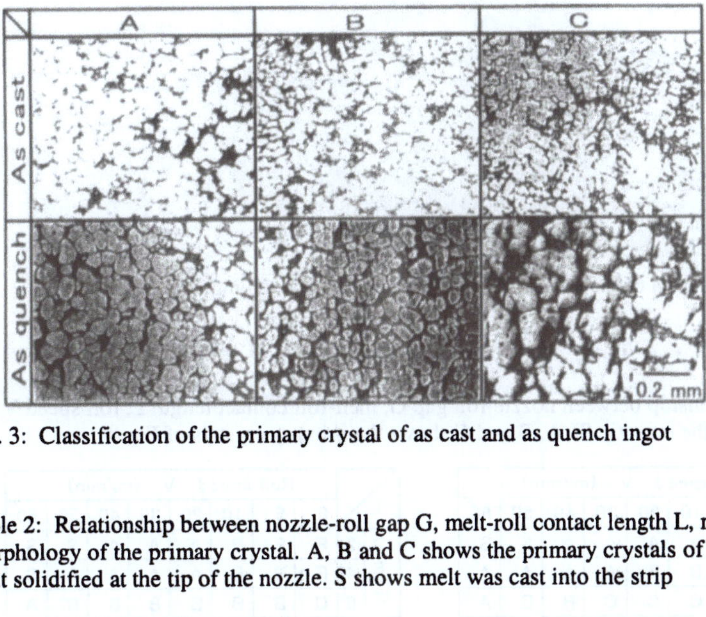

Fig. 3: Classification of the primary crystal of as cast and as quench ingot

Table 2: Relationship between nozzle-roll gap G, melt-roll contact length L, roll speed V and morphology of the primary crystal. A, B and C shows the primary crystals of Fig. 3. F shows melt solidified at the tip of the nozzle. S shows melt was cast into the strip

G (mm)	Roll speed V (m/min) 0	5	10	20	30	40	50	60
2	F	A	A	A	A	A	S	S
4	F	A	A	A	B	B	A	A
6	C	C	C	C	C	B	B	A

Contact length L : 15 mm

G (mm)	Roll speed V (m/min) 0	5	10	20	30	40	50	60
2	F	F	A	A	A	A	S	S
4	F	A	A	A	B	A	A	S
6	B	C	C	C	C	B	B	A

Contact length L : 30 mm

Eutectic Si of as cast ingot and as quench ingot was classified into A, B and C. Eutectic Si of as cast ingot of A, B and C was acicular. C shows that the eutectic Si of as quench ingot was acicular. B shows that the eutectic Si of as quench ingot became small acicular. A shows that eutectic Si of quench ingot became small and globular. The eutectic Si of A was excellent. Table 3 shows relationship between nozzle-roll gap G, melt-roll contact length L, roll speed V and morphology of the eutectic Si. A, B and C of Table 3 coincide with A, B and C of

Fig. 4. The eutectic Si became excellent when the primary crystal became excellent. The effect of the melt-roll contact length L on the eutectic Si was larger than that on the primary crystal. E of Table 4 shows that classification was A in both Table 2 and 3. Excellent ingots for thixo-forming could be cast at the conditions of E in Table 4.

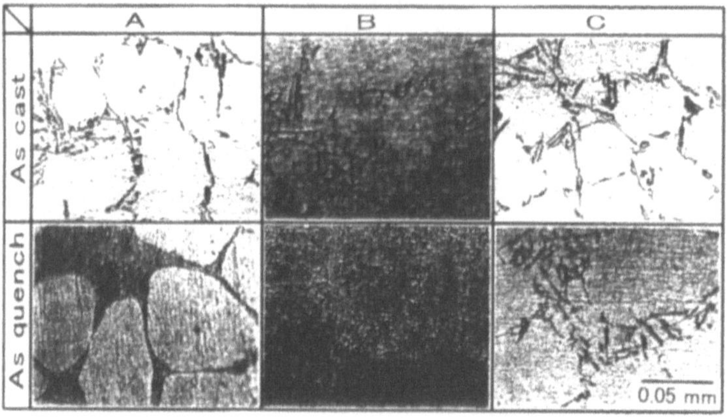

Fig. 4: Classification of the eutectic Si of as cast and as quench ingot

Table 3: Relationship between nozzle-roll gap G, melt-roll contact length L, roll speed V and morphology of the eutectic Si. A, B and C shows the primary crystals of Fig. 4.

		Roll speed V (m/min)							
		0	5	10	20	30	40	50	60
G (mm)	2	F	A	A	A	A	A	S	S
	4	F	B	B	B	A	A	A	A
	6	C	C	C	C	C	B	B	A

Contact length L : 15 mm

		Roll speed V (m/min)							
		0	5	10	20	30	40	50	60
G (mm)	2	F	F	B	A	A	A	S	S
	4	F	B	B	A	A	A	A	S
	6	C	B	B	B	B	B	B	A

Contact length L : 30 mm

Table 4: Excellent conditions both of the primary crystal and the eutectic Si. (E: excellent)

		Roll speed V (m/min)							
		0	5	10	20	30	40	50	60
G (mm)	2	F	E	E	E	E	E	S	S
	4	F						E	E
	6								E

Contact length L : 15 mm

		Roll speed V (m/min)							
		0	5	10	20	30	40	50	60
G (mm)	2	F	F		E	E	E	S	S
	4	F			E		E	E	S
	6								E

Contact length L : 30 mm

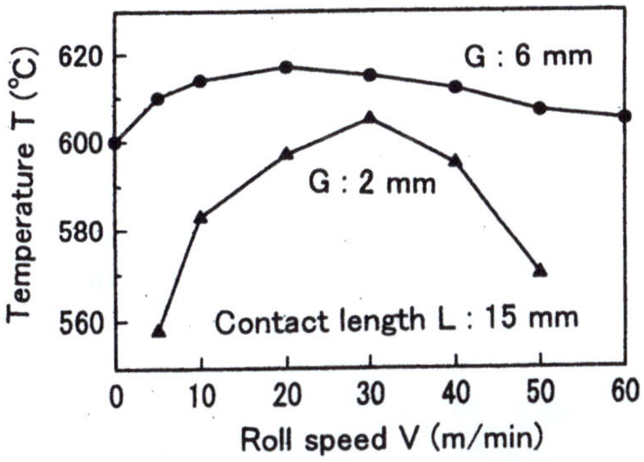

Fig. 5: Temperature of the metal cooled by DMDSRC

Figure 5 shows temperature of the metal when the metal was poured into the mold. Temperature of the metal was lower at the conditions that the primary crystal and eutectic Si were excellent. The primary crystal and eutectic Si became better, as the temperature of the metal poured into mold became lower.

Relationship between melt temperature and primary crystal was investigated. Experimental flow is shown in Fig.6, and microstructures were shown in Fig.7. When temperature was higher than liquidus line, the primary crystal did not become globular. Therefore, the metal must become semi-solid slurry when metal was poured in the mold. The primary crystal of the as quench ingot cast using a metal mold was more globular and smaller than that cast using an insulator mold. Cooling rate of the melt in the mold affected the microstructure of the as quenched ingot. The higher the cooling rate, the more globular the primary crystal of as quench ingot.

Average diameter D and globular ratio R of the primary crystal of the as quench ingot cast using DMDSRC were measured. Average diameter was about 80•90 •m, and globular ratio was 90 %. Average diameter and globular ratio were not affected remarkably by casting conditions.

4. Conclusion

A downward melt drag single roll caster (DMDSRC) was devised to cast ingots for thixo-forming. Semi-solid slurry could be made continuously using DMDSRC. Primary crystal became globular, and eutectic Si became small and fine when ingots was heated up to semi-solid condition. Roll speed, nozzle-roll gap and melt-roll contact length affected the microstructure of ingots heated up to semi-solid condition. There was tendency that primary crystal and eutectic Si became excellent at the conditions as below. The roll speed was high, the nozzle-roll gap was narrow, and the melt-roll contact length was long.

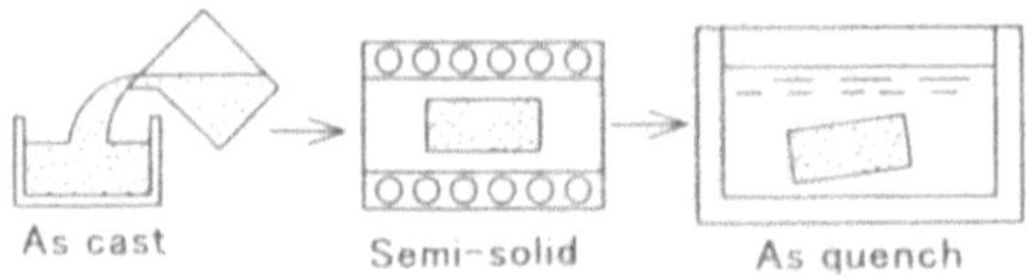

Fig. 6: Experimental flow to investigate the effect of melt temperature on primary crystal

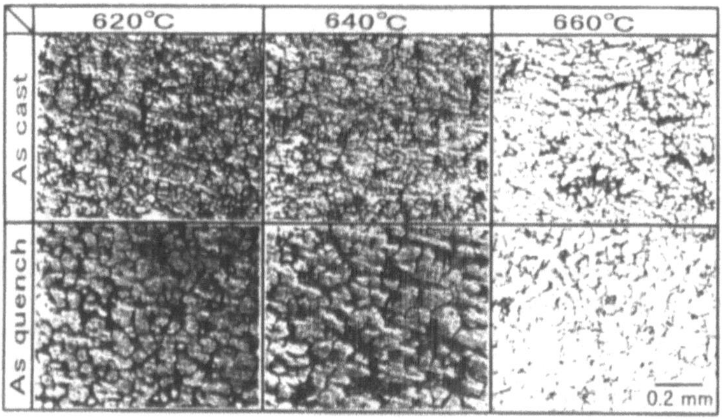

(a) metal mold

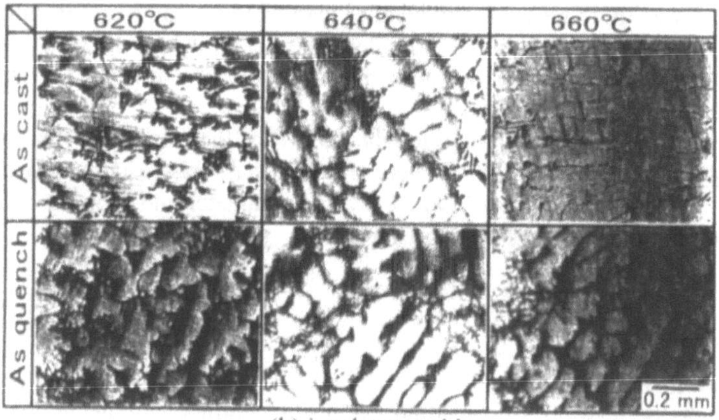

(b) insulator mold

Fig. 7: Primary crystal of as cast and as quench ingot of Fig. 6.

References

[1] C.Pluchon, W.Loue, P.Y.Menet , et al. Proc. Light Metals 95,(1995)1233-1241.
[2] S.P.Midson, N.H.Nicols, R.A.Nichting, et al, Proc. Processing of semi-solid alloys and composites, (1992)140-148.
[3] T.Motegi, N.Ogawa,i, K.Kondo, et al, Proc. the ICAA-6,(1998)297-326.

UNCONVENTIONAL CUTTING

Factors Affecting Electrochemical Honing

H. A. Senbel

Ain Shams University, Faculty of Engineering, Cairo, Egypt, email: hsenbel@yahoo.com

Abstract: Electrochemical honing is one of the non-equilibrium gap processes in ECM and is a new technique, which in spite of being used in some industrial plants especially to smooth surfaces, is still not fully described due to the variety of the factors affecting the process. More information about the process is required especially the effects of the working parameters on the produced surface roughness. A special honing tool was designed by using different tool tip shapes (rectangular, circular, triangle & inclined) to study the ability for improving the surface roughness. This work presents a study for the factors affecting the electrochemical honing process especially the machining time, workpiece material, initial working gap, tool rotational speed, tool tip shape and the inclined tool tip angle. The results are finally furnished with the aim to generalize a useful guideline for the user to enable proper selection of conditions for obtaining good surface quality.

1. Introduction

One of the main industrial applications of electrochemical machining ECM especially in auto industries is the enlargement of pre-drilled holes with very smooth surfaces. Several conventional and non-conventional techniques have been adopted in industry to smooth rough surfaces. The mechanical methods are often based on abrasive cutting such as grinding, lapping and honing, whereas the non-conventional methods utilize different mechanisms such as ultrasonic machining, abrasive flow deburring and nylon abrasive filament [1]. The production engineer should be able to define the suitable working parameters to achieve the specified hole size with better surface roughness and out of roundness. Despite the various investigations carried out to study the enlargement process by ECM [2] and the produced surface roughness [3,4&5], the literature still fails to supply production engineers with sufficient data regarding hole surface quality by electrochemical honing.

To enlarge and finish a circular hole several electrochemical techniques can be applied. Such techniques could be divided into two main groups, equilibrium and non-equilibrium. For equilibrium processes such as boring and broaching, the main efforts are directed towards the metal removal. An analysis of experimental data has been carried out using the response surface methodology [6] in boring pre-drilled holes using bit tools. The results indicated that with electrochemical bit boring tools the process was more stable than with conventional tools. The honing process is similar to stationary cutting with the exception of tool rotation during the process. The combination of both mechanical and electrochemical honing actions has been investigated by many authors, [7-9]. The main advantage of using rotating tools with or without reciprocation is to remove geometrical errors quickly when compared to other techniques.

During the electrochemical polishing action the electrolyte is designed to form a film. In general, two factors control this film formation [10], one is a layer of viscous liquid with a high concentration of dissolved products and the other is a layer of gas which is generated during the passage of electricity through the anode. This film acts similar to a layer of wet paint where it fills the valleys and allows the peaks to be projected through that layer. The resistance to electric current flow is less at these projections than the main surface, so that a higher current density is acting on the projections and hence they are dissolved first. Then a general improved smooth surface occurs.

In the present work different types of honing tools were used under different electrochemical machining conditions to achieve a better surface roughness for different common workpiece materials.

2. Experimental Work

A special test rig was designed as shown in Figure (1) to suit the honing process. The electrolyte was pumped at the front of the tool which has a conical tipped shape made of a non-conductive material to facilitate the electrolyte flow to enter the electrode gap as shown in Figure (1). The test conditions are given in Table (1). Different honing tools with straight tips, along the tool axis, are used which take the form of rectangular, circular and triangle indicating that each honing tool contains the same two tips at **180°** to each other as shown in Figure (2). Also triangular tips with different inclined angles β namely **zero, 10, 20** and **30** degree were used [the inclined angle β is measured in the direction of tool rotation, as shown in Fig. (2).

The workpiece shape takes the form of a cylindrical predrilled hole and the specimens of Aluminum, Steel 37 and Stainless Steel 314 were tested as common engineering materials. In order to avoid the problems associated with the distortion of insulating material in the inter-electrode gap, the insulating material completely covered the tool cylindrical surface in spite of the cutting tip surface. The predrilled holes of the workpiece surface were produced through drilling and internal turning processes with an average surface roughness of [7-8] μm to ensure a common value of roughness for all the electrochemically machined specimens. The metal removal thickness and consequently the removal rate was determined on the basis of weight difference of the annular machined part, and the surface roughness was measured by using a Talysurf SV 414. To minimize the problem of temperature changes during the ECM process, a large electrolyte tank (70 l) was used. The details of the experimental conditions are given in Table (1).

3. Results and Discussions

3.1 Effect of workpiece material

The effects of the workpiece material on surface roughness at different machining times are shown in Figure (3). It is observed that Steel 37 gives the highest value of the average roughness Ra due to the highest value of its metal removal than both of Aluminum and Stainless Steel 314 workpieces. It must be noticed that Steel 37 has the smallest grain size as compared with Aluminum and Stainless Steel 314 i.e. the material having smallest grain size the produced average surface roughness will be greater than those having bigger grain size as reported in [11&12]. It is also observed by increasing the machining time the average roughness decreases for different workpiece materials. This is due to the fact that as machining proceeds and radial gap increases, thus the large scale roughness is successively removed leading to improve the surface quality. This same trend was previously declared by Ebeid and Abdel-mahboud [11].

3.2 Effect of initial working gap

The increase of the initial working gap between the electrode and the workpiece causes the decrease of dissolved metal at the same machining time and hence decreases the amount of metal removal which results in decreasing the average roughness as shown in Fig. (4). Figure (5) also shows that the workpiece material of Steel 37 causes the highest value of average surface roughness compared with both Aluminum and Stainless Steel 314 under a certain predefined initial working gap.

3.3 Effect of tool rotational speed

The effect of tool rotational speed on surface roughness by using a cylindrical tool was previously illustrated by Ebeid & Abdel Mahboud [2&13] in the case of boring processes, they showed that with the increase in tool rotational speed the roughness improved, up to rotational speeds of 11000 rpm. The same results were obtained in the present work but for limited speeds not exceeding 4000 rpm due to the experimental test rig limitations. The authors [2&13] indicated that the flow pattern improves as the tangential speed of the suspended sludge in the electrolyte increases. In the present work the honing tool shape (using a tipped tool) is found to improve the electrolyte flow pattern as compared to the cylindrical one. The amount of the produced average surface roughness is less than those previously obtained by Ebeid and Abdel Mahboud [2&13], the effect of tool rotational speed on surface roughness is illustrated in Figure [6].

3.4 Effect of tool tip shape

Figure (7) shows the effect of tool tip shape on the produced workpiece surface roughness at different machining times. It is observed that the triangular tool tip produces a better surface roughness compared with both circular and rectangular shapes. This is accounted for the concentration of the current lines in the case of the triangular tip which leads to improve the electrochemical machining and consequently improve the surface roughness.

3.5 Effect of tool tip inclined angle [triangular shape]

The effect of the tool tip inclined angles in the case of triangular shapes on the final surface roughness is shown in Figure (8). It is clearly observed that by increasing the inclined angle the roughness decreases. The smallest value of the roughness occurred by using inclined angles ranging between **10°** to **20°**. This is accounted for improving the electrolyte flow pattern through the inter-electrode gap. It is observed also that by increasing the inclined angle above **20°** the roughness will increase due to increasing of the tool tip cutting area which leads to increasing the current density.

Finally, a general empirical relationship between the average workpiece surface roughness and the main working parameters can be assumed as follows:

$$Ra = R_o \cdot e^{-\Sigma \lambda_i t}$$

Where the values of λ_i , dimensionless, are according to the following table:

Workpiece material		St 37	Al	St.St 314	
λ_1		0.0115	0.019	0.034	
Initial working gap, mm		0.3	0.5	0.75	1
λ_2		-0.0015	0.00	0.0004	0.0031
Tool rotational speed, rpm		2000	3000	4000	
λ_3		0.00	0.0005	0.0133	
Tool tip shape	Rectangular	Circular	Triangular	Inclined Tri., β= zero	
λ_4	0.00	0.0001	0.0047	0.01	

4. Conclusions

The present work studies the factors affecting electrochemical honing especially the workpiece material, initial working gap, tool rotational speed, tool tip shape and the inclined tool tip angle. The results show that:

- Increasing the machining time decreases the average surface roughness. Steel 37 shows a greater surface roughness compared with both, in order, Aluminum and Stainless Steel 314.
- Increasing the initial working gap and the tool rotational speed decreases the average surface roughness.
- The triangle tool tip shape produced a better surface roughness compared with both circular and rectangular shape.
- In the case of triangle tipped tool the recommended tipped inclined angle ranged between **10** to **20** degrees to produced the better surface roughness.

References

[1] Mahadev P., 1996 , " Finishing Technology Smooths The Roughest Jobs" American Machinist, V. 140, No. 6, June, USA.

[2] Ebeid S.J. and Abdel Mahboud A.M., 1988, " High Speed Electrochemical Boring" J. Precision Engg. Vol. 10, No. 2, pp. 80-84.

[3] Kozak J., Dabrowski L., Osman H. and Rajurkar K.P., 1991, " Computer Modelling of Electrochemical Machining with Rotating Electrode" J. of Materials Processing Technology, Vol. 28, pp. 159-167.

[4] Osman H.M. and Abdel-Rahman M., 1993, " Integrity of Surface Produced by ECM", Journal of Materials Processing Technology, Vol. 37, pp. 667-677.

[5] Abdel Mohsen Mahdy M., 1999, " Factors Affecting The Roughness of Surfaces Produced by ECD and ECPD", 5 th. Conference of Egyptian Society of Tribology, pp. 215-226, Cairo, Egypt.

[6] Jain V.K., Vinod K. and Pandey P.C., 1982, " Computer Aided Analysis of ECBD Process", Proc. 23 rd Conf. MTDR, pp. 257-264.

[7] Ghabrial S.R. and Ebeid S.J., 1985, " Electrochemical Broaching and Stationary Cutting – A Comparative Study", MDP 3rd Conf., pp.477-483, Cairo, Egypt.

[8] Horgan I.F., 1962, "Electrolytic Boost for Honing", American Mach., V.67, No.3, USA.

[9] Pahlitzsch G., 1972, " Mechanical and Electrochemical Honing with Diamond Sticks", SME, MR 72-181.

[10] Madwed A., 1976, " Some Fresh Approach to Electrochemical Deburring " SME, Technical Paper, MR 76-129, Michigan.

[11] Rajurkar K.P. and Hewidy M.S., 1988, " Effect of Grain Size on ECM Performance", Journal of Mechanical Working Technology, 17, pp. 315-324.

[12] Kops L. and Quack V.B., 1976, "Investigation of the Role of Workpiece Grain Size in Electrochemical Machining Process", ASME, vol. 98, pp. 360-368.

[13] Ebied S.J. and Abdel Mahboud A.M., 1986, " On The Prediction and Control of ECM Surfaces", 2nd AME Con. TP-3, pp. 27-34, Cairo, Egypt.

Table (1), Experimental conditions

Workpiece materials	Aluminum, St37, Stainless Steel 314
Workpiece length, mm	30
Internal hole diameter, mm	30.6, 31, 31.5 & 32
Tool outside diameter, mm	30
Electrolyte	20% w/v NaCl
Electrolyte flow rate, l/min	10
Back pressure, bar	2
Machining time, sec.	20, 40, 60 & 80
Tool Rotation speed, rpm	2000, 3000& 4000
Inclined angle, β, degree	zero,10, 20 & 30
Tool tip shape, Fig.(2)	A: Rectangular B: Circular C: Triangular, [β=-15] D:Inclined, [β=zero,10,20,30]

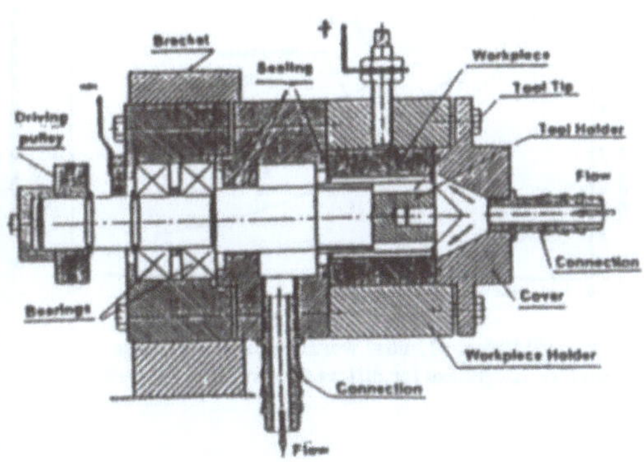

Fig. (1) Test rig set up

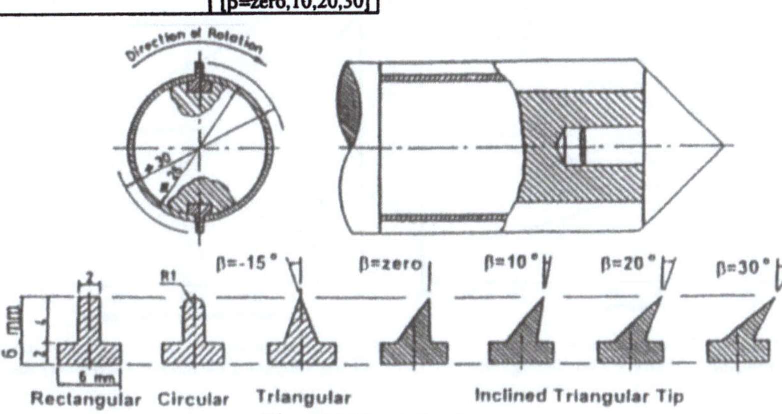

Fig. (2) Tool tip shapes

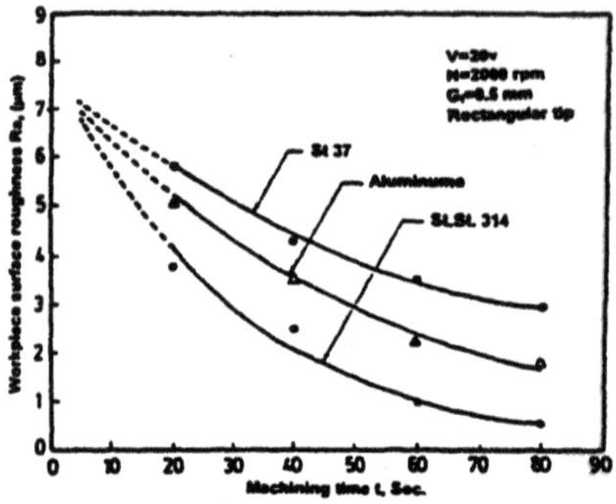

Fig. (3) Effect of workpiece material on average surface roughness at different machining times

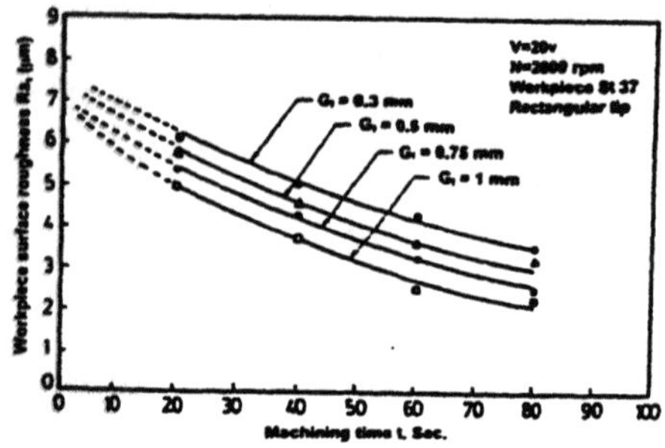

Fig.(4) Effect of initial working gap on average surface roughness at different machining times

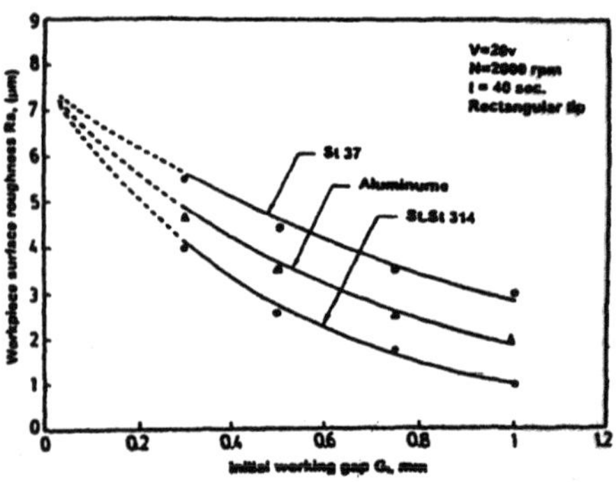

Fig. (5) Effect of initial working gap on average surface roughness for different workpiece materials

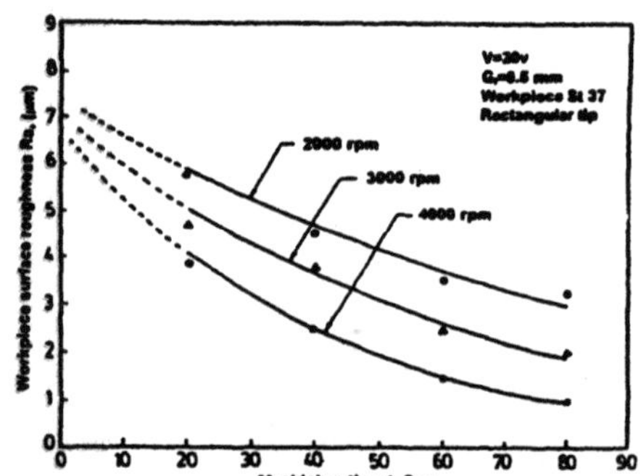

Fig.(6) Effect of tool rotational speed on average surface roughness at different machining times

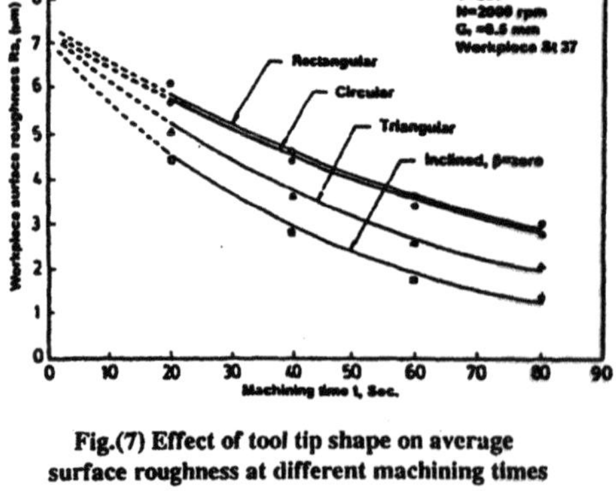

Fig.(7) Effect of tool tip shape on average surface roughness at different machining times

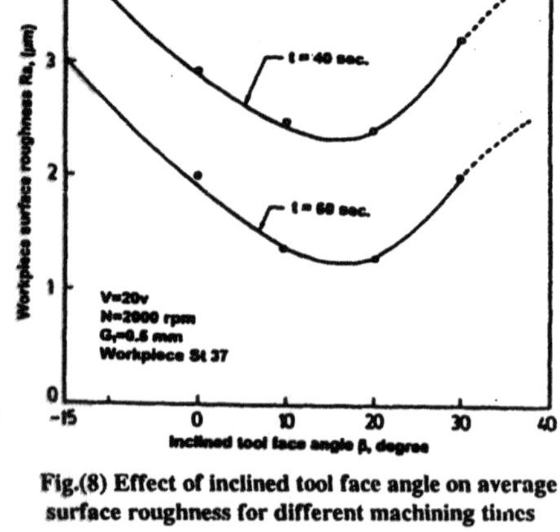

Fig.(8) Effect of inclined tool face angle on average surface roughness for different machining times

Laser and Abrasive Water-jet Cutting: Complementary Technologies

Davorin Kramar[1], Miha Junkar[2]

[1,2] University of Ljubljana, Faculty of Mechanical Engineering,
Askerceva 6, 1000 Ljubljana, Slovenia, davorin.kramar@fs.uni-lj.si

Abstract: Laser cutting and abrasive water-jet (AWJ) cutting can replace conventional mechanical removal methods in many industrial applications, particularly in the processing of difficult-to-machine materials. The selection of which manufacturing process to apply to a particular material is influenced by a number of factors such as part geometry, production volume, part quality, and particularly physical properties of the material. Based on experimental work on several materials and workpiece thickness the comperation of AWJ and laser cutting is presented in the form of technology data base. This base serves as an orientation for technologist when deciding upon appropriate cutting process.

1. Introduction

Non-conventional cutting processes: laser cutting and cutting with AWJ are well introduced in different industrial applications, from aircraft and automobile industry, tool-manufacturing, sheet metal industry to paper and food production. Hardened metals, ceramics, composites such as aluminium embedded in rubber and honeycomb structures, or soft rubber, foam and paper are just a few good examples.

Both technologies are suitable on the fields where the flexible tool is needed. Also the generated surface is in a certain sense similar (striations, kerf taper, burr), but the removal mechanism is completely different. Laser cutting is thermal process; on the other side AWJ cutting is mechanical process. So which manufacturing process to apply to a particular material depends on thermo-physical and mechanical properties of the workpiece material, and also on the performances of both technologies.

2. Processes Removal Mechanism

Laser cutting is a thermal process similarly like flame or plasma cutting. The difference is in a heat source and its energy density. AWJ cutting is mechanical type of process, but it is not tool shape manufacturing like convectional mechanical processes.

2.1 Physical Mechanism for Laser Cutting Material Removal

The incoming laser beam energy is balanced by the conduction heat, energy for melting or vaporization of material, and heat losses to the environment.

Due to the relative beam/workpiece movement, the erosion front is formed in front of the laser beam. Thus the temperature inside the workpiece depends on the distance to the erosion front. The accumulated molten and/or evaporated material is ejected out from the kerf with aid of a coaxial gas jet.

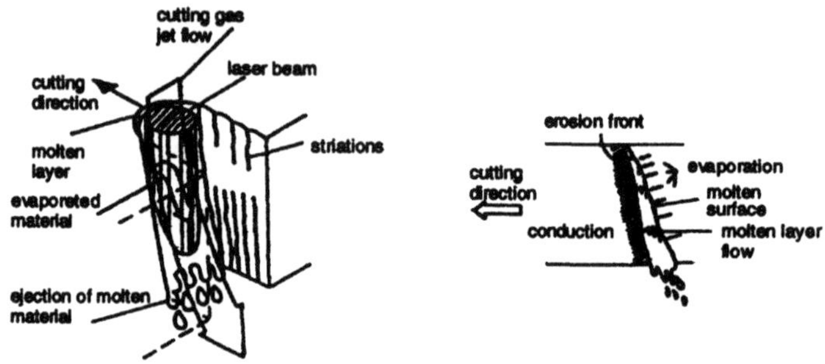

Figure 1: Physical Mechanism for Laser Cutting Material Removal

There are three mechanisms of laser cutting which can be used for processing different materials:

A/ Laser Sublimation Cutting: Most of material is vaporized by high intensity laser beam and blown out by inert gas flow;

B/ Laser Fusion Cutting: Material is heated and molten by laser beam and blown out by an inert gas flow;

C/ Laser Flame Cutting: Reactive working gas is used for an exothermic reaction and burns the laser heated material.

Laser machining causes beside phase changes many secondary phenomena related to surface quality of the cut; such as micro-cracks, striations formation. In metals the conduction heat resulting from the high energy density on the workpiece creates HAZ (heat affected zone) in the vicinity of the erosion front in range of few hundredths for thin materials and few tenths for thicker materials. This is in some cases not allowed. In plastics and polymer-matrix composites, material decomposition may occur as a result of elevated temperatures, which cause the breakdown of the polymer into charred residues and gaseous products.

2.2 Physical Mechanism for AWJ Cutting Material Removal

High-pressure pump raises water pressure to some hundred bar. Water at that pressure is leaded to the water-nozzle, reducing flow cross-section for ten or more times. Nozzle so increases the water pressure and forms water jet with pressure up to 400 MPa.

AWJ is made by mixing abrasive particles with high-velocity waterjet in the mixing tube. When AWJ hits material surface the erosion starts. The impact of single solid-particles is the basic event in the material removal by AWJ. The literature about solid-particle erosion is extensive. Generally four sub-mechanisms by which solid-particles separate material from a target surface are defined. These mechanisms are cutting, fatigue, melting, and brittle fracture. Clearly, these mechanisms generally do not act separately, but in combination. Their importance for the particular erosion process depends on several factors, such as the impact angle, the particle kinetic energy, the particle shape, environmental conditions and target-material properties [1].

Regarding the concept in Hashish's model et.al [2] the total depth of cut is divided into two distinct zones due to different modes of abrasive/material interaction (see figure 2).

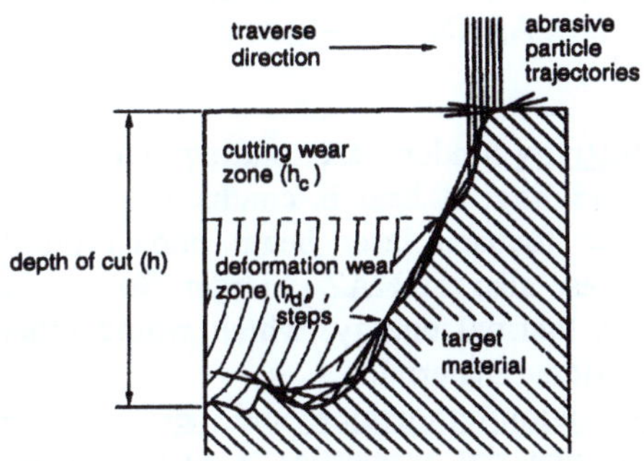

Figure 2: Physical Mechanism for AWJ Cutting Material Removal

In the upper zone, material is removed by particle impact at shallow angels. The material is removed by micro cutting - "cutting wear mode". A geometrical expression of this zone is fine roughness of cut surface. In the lower zone, sequential steps are formed which is believed to lead to large angel impacts. In this stage, the AWJ penetrates the material at a decreasing rate as the depth of cut increases. The material is removed by a different erosion mode that is associated with particles bombardment, surface hardening by plastic deformation and crack formation - "deformation wear mode". The striations and waviness geometrically characterize this zone.

3. Material Properties

3.1 Important material properties in laser processing

The effectiveness of laser machining depends upon the thermal properties and some optical properties of material to be machined.

The absorption of the laser beam energy depends on both the wavelength of the laser radiation and spectral absorptivity characteristics of the material (see Figure 3), and it is most important property of material to be processed. Metals 1 exhibit high reflectivity for CO_2 laser radiation, so Nd-YAG lasers are more effective. Also, for metals and ceramics, the presence of a molten layer changes the absorptivity value. Furthermore, absorptivity of the surface depends on its orientation with respect to beam direction, the surface roughness and also the temperature of material.

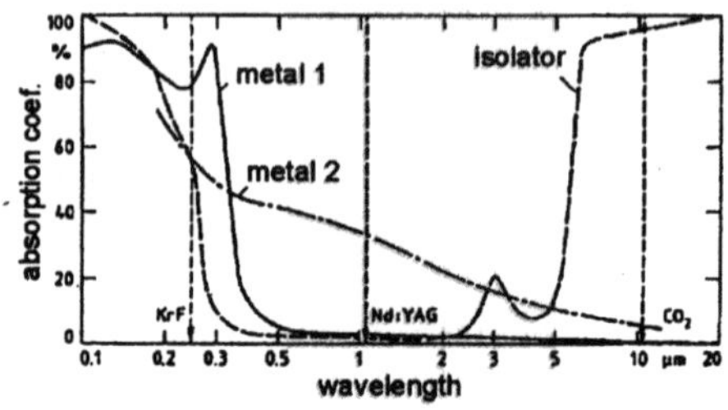

Figure 3: Influence of laser wavelength on absorption coeficient. for different types of materials. Metals 1 (Au, Ag, Cu, Al); Metals 2: Fe, Cr, Ni, Ti; Isolator: plastics, wood, rubber, glass [3]

Thermal conductivity hinders laser cutting and local surface treatment, but it is useful by welding. If heat is conducted into inner material it is difficult to warm the focused laser beam impact point to vaporization temperature. In this case also the HAZ is larger. So the presence of thermal conduction dissipates thermal energy, which would otherwise be used to melt or vaporize additional material.

Melting / vaporization temperature and melting / vaporization heat are also very important material properties. Lower this features of certain material are, easier is laser machining of this material by vaporizing (cutting, drilling) or by melting (welding, alloying, surface treatment).

Material combustibility; as it is said above (laser flame cutting) we can use O_2 as a reactive working gas. By laser flame cutting of mild steel the oxidation is connected with rise of a temperature and so exothermal burning delivers up to 70% of process energy. Iron oxide raises the absorption factor for CO_2 radiation to 90%. The purity of oxygen jet has also a serious effect upon the cutting process. With decreasing O_2 purity the cutting speed is reducted and generated surface quality is also weaker. When cutting of non-metal organic materials oxidation is not desired. Protective atmosphere using inert gas (N_2, Ar or He) is required. In most cases use of air jet is adequate.

Thermal expansion: Cracking can be caused by presence of a hot thermally expanded zone surrounded by the cold brittle substate. The liquid phase of material is cooled too rapidly by conduction to the body of the workpiece. The stresses thus built up and exceed the fracture stress, and cracking occurs. Typical materials where this problem appears are glass and ceramics. Preheating the workpiece and pulse processing can solve these problems.

3.2 Important material properties in AWJ processing

Based on two-zone cutting model, a combination of mechanical properties to describe the material resistance against the 'cutting wear' and the 'deformation wear' are defined [1]. According to this model, the entire depth of cut is (*figure 2*) $h = h_C + h_D$. Some authors, assume that material hardness determines the erosion resistance under shallow angles; whereas the volume removal due to large angles depends on the Young's modulus (E_M).

There are models that include some other material properties to describe machinability of certain material. For example authors in [4] develop the machinability numbers of sever engineering materials.

Table 1: Mechanical material properties that characterize material machinability by AWJ cutting

Material Property	Machinability
↑ hardness (H_V)	↓
↑ Young's modulus (E_M)	↓
↑ grain size (D_M)	↑ (↓ for brittle materials)
↑ surface energy (γ_M)	↓
↑ flow stress (σ_f)	↓
↑ porosity	↑
↑ th. shock resistance	↓ (for brittle materials)

4. Complementarity of Technologies

In an economic context, both technologies provide the following advantages over other processes:

- eliminate the replacement costs associated with tool wear and breakage,
- machine downtime for tool replacement and recalibration are eliminated,
- can be integrated in a computer numerical control (CNC) system to maximize machining rates and minimize production times,
- closer nesting of parts can be accomplished, so less material is wasted.

The drawbacks of laser and AWJ technologies are that they produce a tapered kerf shape, compared to the straight vertical kerf walls achievable by convectional methods. Laser cutting can reduced the kerf taper by adjusting the focal point of laser beam to the interior of the workpiece instead of on the surface. Reducing the kerf taper produced by AWJ cutting is achieved only by reduction in traverse rate.

Table 2 shows technical data for laser and AWJ cutting of different materials. Laser cutting is much faster cutting technology for laser suitable materials and thicknesses. AWJ cutting is useful for materials where laser is inefficient (reflective or thicker materials). There is no problem in selection which cutting technology applies to certain material if they are both available. These two technologies are complementary and not competitive.

5. Conclusion

In general most metals and non-metals have thermal and optical properties favourable for laser machining. However, the economic viability of applying laser machining depends critically on properly matching the material with the laser type and size. The thermal properties of a material determine the power requirement for an incident beam in order to achieve desired depth of cut or material removal rate. Since ceramics, plastics and composites have high absorptivity values for 10.6 μm radiation, they can be machined effectively with CO_2 laser. Ceramics and composites generally have higher melting and vaporization temperatures than plastics; therefore, higher laser power is required. Some metals such as aluminium, copper and brass exhibit low absorptivity values for CO_2 laser radiation, so most of the beam energy will be reflected. These materials are more suitable for machining with an Nd-YAG laser or AWJ.

Any material that we cannot cut with a laser we can with an AWJ. Since AWJ machining relays on selecting appropriate abrasive, virtually any kind of material can be machined. The small force (<10 N) applied on the workpiece by the jet allows machining of fragile or deformable materials or structures. Using a cold-cutting tool, AWJ leaves a surface free of a HAZ. By eliminating HAZ, AWJ technology saves time and money that were usually spent on secondary operations.

Table 2: Representative Data for Laser and AWJ Cutting of different types of material

material	properties	thick. [mm]	LASER			AWJ	
			type	power [W]	t. rate [mm/min]	pressure [MPa]	traverse rate [mm/min]
alumina ceramics	++hard	1	C	150	250	300	170
	- - M_F	4	A/NY	170/180	20/30	300	35
	+t. expansion	10	A/C	280	13/20	300	13
tool steel	+hard	2	B	1500	2300	310	530
		10	B	2100	650	310	100
		15	B	3500	220	310	55
stainless steel	hard	2	B/C	1500	2500/4500	380	600
		6	B/C	2100/1500	1100/700	380	250
		15	B	3500	240	380	85
mild steel	-hard	1	C	1500	7000	380	1800
	+combast.	6	C	1500	2000	380	350
		15	C	3500	1000	380	125
alumin.	+reflective	1	A	2100	6000	380	3000
	+ t.conduct.	2	A/NY	2100/1200	3000/1500	275/380	1400/2000
	+soft, - M_F	6	B	2100	400	380	600
cooper	+ reflective	1	C	1200	3000	310	1200
	++ t.conduct.	3	C/NY	3900/120	3000/50	310	425
	soft, - M_F	5	C	3900	800	310	250
titanium	- density	1	B	1000	3000	275	1300
	hard	3	B	1500	1800	310	450
plexi-glass	+ absorpt.	1	A			310	2200
	++combast.	5	A	520	3000	310	1300
	+soft	15	A			310	450
wood	++ absorpt.	1	A	520	30000	310	2500
	++combast.	6	A	280	2000	310	2200
	++soft	20	A	250	300	310	1200

NY... Nd-YAG

6. References

[1] W. Momber, R. Kovacevic, Principles of Abrasive Water Jet Machining. Springer, 1997

[2] M. Hashish, A Model for Abrasive-Waterjet (AWJ) Machinig, Journal of Engineering Materials and Technology, Vol.111, 154-162, 1989

[3] H. H] gel, Strahlwerkzeug Laser, Teubner Verlag, 1992

[4] J. Zeng and T.J. Kim, Machinability of Engineering Materials in Abrasive Water Jet Machining, International Journal of Water Jet Technology, Vol. 2, No. 2, 103 – 110, 1995

[5] J. Powell, CO_2 Laser Cutting, Springer - Verlag, 1993

High Tolerance Plasma Arc Cutting of Titanium

Michele Monno[1], Barbara Previtali[1]

[1]Politecnico di Milano, Dipartimento di Meccanica, Via Bonardi, 9, 20133 Milano, Italy, previtali@mecc.polimi.it

Abstract: The aim of present work is to analyse the application of the HTPAC process to cutting of titanium. By systematic experimental study, the paper addresses the influence of HTPAC parameters on titanium quality.

1. Introduction

Titanium and its alloys find widespread applications because of their excellent corrosion resistance and high specific strength properties. Due to reactive behaviour of the titanium exposed to high temperature, the thermal cutting of titanium has always been considered a rough application (except the laser cutting). Significant improvements in the plasma arc cutting process have been recently achieved, namely the development of High Tolerance Plasma Arc Cutting (HTPAC). The present work is addresses to the definition of the process parameters suitable to the HTPAC of titanium and attempts to identify the relationship between the process parameters and the quality of the cut.

2. Preliminary Analysis

The preliminary phase is aimed to define the following items:

- *type of gas.* The HTPAC is a double gases torch process. The cut and shield gases selection depends on the workpiece material attributes.
- *torch design.* The selection of the gases affects the selection of the torch design.
- *process parameters range.* After the selection of torch and gases, the feed rate f of the torch and the arc voltage V should be set. In order to perform the selection, the higher and lower limits of variation for both the parameters should be known. However, since the HTAPC of Ti is a new application, the range of variation for f and V should be found, depending on torch and gases.

The material used for this study originated from a 5.2 mm sheet of cp Ti grade 1. The experimental runs have been performed making use of a Hypertherm HD 3070 system recently integrated at the laboratories of the Politecnico di Milano. The selection of the cut gases arises from the following considerations [1]:

- *partial analogy with stainless steels*

Due to great reactivity of the material at high temperature to the more diffuse gases and mixtures, the thermal cutting of Ti is often compared to thermal cutting of stainless steel (SS). Due to its low cost, relatively inert behaviour and great thermal conductivity, nitrogen (N_2) is widely used for HTPAC of SS. Therefore, the same solution has been adopted in this case, even if the reaction of Ti with N_2 at high temperature should be considered. Indeed, N_2 dissolves interstitialy in Ti and forms a hard and brittle chemical compound.

- *HTPAC improvements*

The HTPAC technology has been introduced mainly to allow fine thermal cutting with oxygen (O_2). Since the heat energy in O_2 is greater than in N_2, higher feed rate can be used. Cutting Ti with O_2 at high temperature not only causes oxidation (TiO_2) but also solid solution hardening of the surface as a result of O_2 diffusion.

• *technical information*

Despite very little information is available about the HTPAC of Ti, a considerable number of references dealing with the other thermal cutting processes offer indirect indications [2][3]. Particularly, the references confirm the use of O_2 and N_2 and in addition suggest to use argon (Ar) or mixtures [4]. Indeed, the Ar is the best cut gas to use because it does not react with the Ti. Due to its low thermal conductivity however Ar could not allow edge separation. Therefore, also an argon-hydrogen mixture, named H_{35} (35% of hydrogen) has been tested. Addition of H_2 to Ar increases the heat input, allowing higher feed rate and increasing depth of kerf.

Table 2: Sets of gases

Cut gas	Shield gas	Results
N_2	N_2	OK
N_2	Ar	OK
O_2	N_2	OK
O_2	Ar	OK
Ar	Ar	No cut
Ar	N_2	No cut
H_{35}	Ar	Dross
H_{35}	N_2	Dross

Table 3: HTPAC of SS

SS torch parameters	70 A
Cut gas	N_2
Shield gas	N_2
Cut pressure (bar)	4.4
Shield pressure (bar)	0.9
Feed rate (m/min)	1.8
Arc voltage (V)	150

The aim of the shield gases is to prevent contamination of the cutting zone. Ar and N_2 are the more suitable gases for shielding . Table 2 shows the combined sets of gases which have been tested. The preliminary experimentation has been carried out by a 70 A torch with the conditions for the SS (Table 3). Therefore, all the sets of gases have been preliminary experimented in order to obtain the separation of the kerf walls and the lack of dross on the bottom, and to find out the higher and lower limits for both *f* and *V*. The main conclusions from the preliminary tests are:

• Ar is not suitable as cut gas in the HTAPC of Ti. The edge separation could not be obtained without a considerable diminution of *f* (about 75% of the settled value).

• the HTPAC of Ti with the H_{35} mixture is not suitable. The H_{35} has two negative aspects. Former, the cool presence of Ar changing the surface tension of the liquid Ti increases the dross formation [5]. Latter, the H_2 supplies an excessive amount of energy leading to a coarse and wavy cut surface.

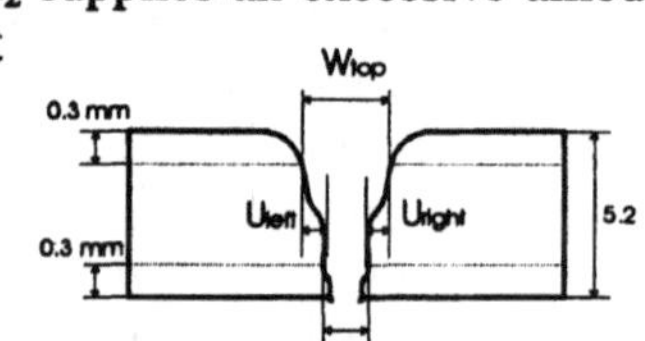

Fig. 1: Quality attributes

Table 4: Threshold limits for *f* and *V*

Set of gases		Voltage		Feed rate	
Cut	Shield	L	H	L	H
N_2	N_2	-10%	+10%	-50%	+50%
N_2	Ar	-10%	+10%	-50%	+50%
O_2	N_2	-10%	+10%	-50%	+300%
O_2	Ar	-10%	+10%	-50%	+250%

At the same time, for the four suitable sets of gases the higher and lower limits of *f* and *V* have been found (results in Table 4, the reference is the HTPAC of SS).

3. Assessment of Cut Quality

Nowadays, several specific standards deal with the qualitative and quantitative attributes of the thermal cutting processes [6][7]. This study has adopted the cut evaluation procedures from these standards even if qualitative evaluations have been substituted by quantitative ones. Indeed, the reactive behaviour of Ti makes qualitative evaluations on the cut surface damage extremely confused. The main quality attributes are (Fig.1):

- *Kerf width*: since the plasma jet loses energy from the top to the bottom of the edge, a top kerf width (W_{top}) and a bottom kerf width (W_{bott}) have been identified.
- *Unevenness:* that can be defined as the greatest perpendicular distance between the actual and the expected surface and is the sum of angular deviation and concavity/convexity (irregularity) of the obtained surface. As in the Fig.1, it could be identified a left and right unevenness (U_{left} and U_{right}).
- *Heat damaged zone (HDZ) width:* can be assumed as the width of any visible change, or alteration in the material. More precisely, the following three different zones characterise the HDZ:
  - the oxidised zone (OZ), the zone covered by a film of TiO_2 and dissolved O_2;
  - the fused zone (FZ), the layer melted and suddenly solidified;
  - the heat affected zone (HAZ), the zone altered by the heat without being fused.

The assessments of quality do not consider the surface roughness since it does not give indicative results for the damaged surfaces. Beside, the threshold limits allow to obtain free of dross cuts, so the dross measurement have not been considered.

4. Experimental Plan Definition

Aiming to the investigation of the quality of HTPAC of Ti, an experimental plan has been designed using the Design of Experiments technique and the Analysis of the Variance procedure [8]. The preliminary analysis has outlined that the process parameters mainly depend on the gases. Therefore, for each set of gases, the following experimental plan has been designed.

- variable factors: *f* and *V*;
- fixed factors: material (cp Ti), specimen thickness (5.2 mm), torch design (70 A);
- levels: the range of the *V* and *f* belongs to the interval above determinate. replications: three. The 27 runs for each set have been performed in random order.
- response variables: top and bottom kerf width, left and right unevenness.

Table 5: Variable factors level

Factor	Level	Gases	Level value	Factor	Level	Gases	Level value
f (m/min)	3	N_2/N_2	1.3-1.8-2.2	V (V)	3	N_2/N_2	143-150-157
		N_2/A_r	1.3-1.8-2.2			N_2/A_r	143-150-157
		O_2/N_2	1.8-3.5-5.3			O_2/N_2	143-150-157
		O_2/Ar	1.8-3.5-4.4			O_2/Ar	143-150-157

5. Results

The fundamental hypotheses of the ANOVA applicability to the experimental data (normality and homogeneity of the variance of the residuals) have been verified. Beside, a test hypothesis has been performed to compare the response variables and the sets of gases. Since the data are paired, the one-sample t test has been performed on the mean of the differences between the paired observations.

Table 6: Range for kerf width and unevenness (in mm).

N_2/N_2		N_2/Ar		O_2/N_2		O_2/Ar	
W_{top}	W_{bott}	W_{top}	W_{bott}	W_{top}	W_{bott}	W_{top}	W_{bott}
1.9-2.5	0.9–1.2	2.1-2.7	1.1–1.6	1.9–3.5	0.5–1.9	2.3–3.7	0.7–1.9
U_{left}	U_{right}	U_{left}	U_{right}	U_{left}	U_{right}	U_{left}	U_{right}
0.3–0.9	0.2–0.7	0.4–0.8	0.3–0.9	0.7–0.8	0.5–0.8	0.8–1.1	0.6–1.1

The Table 6 shows the range of variation for the unevenness and the width of the kerf as well as the Table 7 shows the significant factors of the analysis performed.

5.1 Kerf width

- It should be outlined that the best case (narrowest W_{top} and W_{bott}) is obtained with the set O_2/O_2 and the highest feed rate (5.5 m/min).
- When f belongs to the inert gas, only V influences the W_{top}. Indeed, the W_{top} increases when the V increases. When the cut gas is reactive, f becomes the more significant parameter. If f increases both the widths became tighter.
- The comparison between the sets of gases outlines that the Ar as shield gas produces a wider W_{top} and W_{bott} whatever the cut gas is. This is probably due to characteristics of the cooler Ar arc, that realises a wider plasma jet.

Table 7: Significant factors

Set of gases	W_{top}	W_{bott}	U_{left}	U_{right}
N_2/N_2	V, V*f		V,f,V*f	V*f
N_2/Ar	V		V	
O_2/N_2	f	f		f
O_2/Ar	f	f		

Table 8: Distance from the border (in µm)

	Point				
Set of gases	1	2	3	4	5
N_2/N_2 N_2/Ar	30	60	80	80	80
O_2/N_2 O_2/Ar	20	40	60	80	80

5.2 Unevenness

- With the exception of the N_2/N_2 set on the U_{left}, f and V do not seem to significantly influence the unevenness of both the sides of the cut . This is probably due to the behaviour of cut surface. The interaction of the Ti with the plasma jet leaves an irregular and rough surface, that does not allow easy measurements. This is particularly true in the case of Ar as shield gas.
- Nevertheless, the comparison between the U_{left} and U_{righ} for each set offers some considerations. Observing the sets O_2/O_2 and N_2/N_2 it should be noticed that the torch cutting direction and the swirling direction of the plasma gas determine a "good" (the right) and a "bad" (the left) side of the cut in terms of unevenness.

Probably the main reasons of asymmetric unevenness lay on the swirling of the gas that flattens the circular arc on the good side of the. The asymmetry is not evident in the case of Ar, where the edge has a bad and irregular aspect.

- The comparison between both the unevenness (U_{left}, U_{righ}) confirms that the bad properties of Ar in terms of irregularity and inclination of the edge. Indeed, the worse unevenness is obtained when the shield gas is Ar whatever the cut gas is.

5.3 Heat damaged zone width

Usually, the influence of process parameters on the thermal alteration of the border edge is evaluated by means of the micro hardness profile (Table 8). However, the preliminary analysis of the microstructure and some micro hardness measurements on the HDZ have led to not consider the micro hardness measurements as response variables. Indeed, due to average dimension of the diagonal (ranging from 10 to 40 micron) the size of the HV micro hardness indentations has been too extent to point out significant differences as a function of f and V. Indeed, about only one indentation could be obtained in each of the three zones independently by the set of f and V. Therefore, the influence of the process parameters on the single specimen could not be measured. However, an average value in the three characteristic zones, depending on the set of gases, could be measured (Table 9 and Table 10).

Table 9: Average HV micro hardness

	1	2	3	4	5
N_2/N_2	463	309	217	217	208
N_2/Ar	601	353	328	275	230
O_2/N_2	1648	699	373	325	218
O_2/Ar	1734	795	280	219	214

Table 10: HDZ width (in mm)

Set of gases	OZ	FZ	HAZ
N_2/N_2	no	no	0,05-0,1
N_2/Ar	no	0,03-0,05	0,1-0,15
O_2/N_2	0,01	no	0.01
O_2/Ar	0,03-0,1	0,05-0,1	0,05-1

The reference value of HV micro hardness of cp Ti sheets used ranges from about 190 to 230 HV. The results in Table 10 are consistent with the micro structural analysis. Indeed, at the 50 or 100 magnification the following microstructures could be identified [8]:

- *Not heat damage zone.* The cp Ti at ambient temperature has the equiaxed close packed hexagonal structure (alpha phase). The grains have been elongated in a preferential direction due to previous cold working (average HV: 190-230).
- *Oxidised zone (OZ).* Cutting Ti in O_2 results not only in oxidation but also in solid solution hardening of the surface as a result of the diffusion of O_2. The O_2, in fact, stabilises the alpha phase and forms a brittle and hard layer named "alpha case" with a white and homogeneous coloration (average HV: 1100 -1800).
- *Fused zone (FZ).* The fused zone has a black structure of acicular alpha phase, the transformation product deriving from the beta phase during cooling. This phase indicates that the material has been heated to a temperature above the beta transition temperature. The acicular alpha microstructure could be observed primarily in the FZ and sometimes in HAZ (average HV: 400-800).

- *Heat affected zone (HAZ).*With the exception of few acicular alpha phase, the HAZ consist only in a grain grown (average HV: 250-350).

The Table 11 indicates the average value of the three different zones width measured by the means of the optical microscope (where they are present). The analysis of the HDZ microstructure (Fig. 2) and the HV hardness measurements put in evidence that:

- cutting by means of O_2 produces a narrow film of TiO_2 and an "alpha case" layer, whatever the shield gas is (N_2 or Ar). This white layer is very brittle and hard.
- shielding by means of Ar produces an irregular FZ, which is absent when the shield gas is N_2. Moreover, the Ar shield realises the widest total HDZ.
- cutting by means of the set N_2/N_2 does not produce the OZ and the FZ, and allows to obtain a regular border profile.

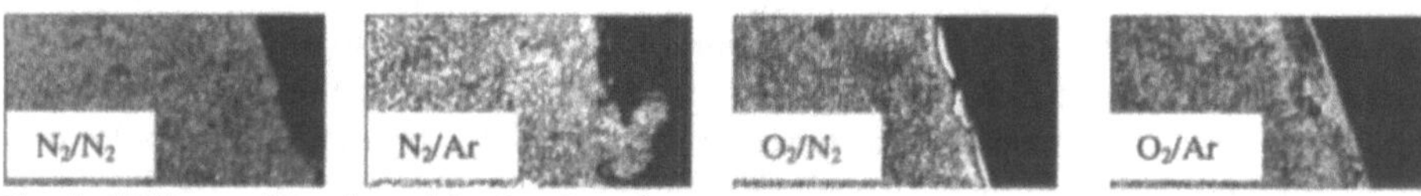

Fig. 2: 50X magnification of the border edge

6. Conclusions

The present work deals with the application of a recent technology, the HTPAC, to the cp Ti. The objective is aimed to identify the influence of the process parameters on the quality of the cut. The analysis has been carried out in two steps: Former, the cutting conditions have been defined. Four sets of gases have been identified as well as the suitable range of *f* and *V*. Later, the relationship between the cutting conditions and the attributes of the quality has been found. This first contribute of the authors to the HTPAC of Ti, has outlined the following future developments:

- the extension to the Ti alloys, first of all to the most used Ti6Al4Va;
- the extension to another gases and mixtures (Ar with He, CH_4, H_2).

Acknowledgements

The authors are grateful to the Italian Ministry of University and Technological Research (MURST) for partly funding this study in the frame of the "CO.TE.STA" project; to Hypertherm and CRM for the helpful cooperation in the integration of the experiment facility; to the Titania SpA for supplying the specimen material and to CNR-ITIA Milano for the availability of metallurgic laboratory, the technical competence and the scientific support.

References

[1] 1994, *Recommended Practices for Shielding Gases,* ANSI/AWS C5.10-94, AWS.
[2] Schutz J., 1988, *Thermal Cutting of Thick TA6V4 Products,* 6th World Conference on Ti.
[3] Maher A. et al., 1998, *Laser Cutting of Titanium with Pulsed and Modulated Pulsed Nd:YAG lasers,* Proceedings of the SPIE, Vol. 3265.
[4] Decker I. et al., 1985, *Laser Cutting of Ti Materials,* Schweissen Schneiden, 37, (8).
[5] Stenbacka N. et al., 1989, *Shielding Gases for GMAW,* Welding Journal, 68 (11).
[6] *Lasers and Laser-related Equipment,* ISO/TR 11552: 1997(E).
[7] *Thermal Cutting,* DIN 2310, teil 1-5, 1987.
[8] Montgomery D., 1991, *Design and Analysis of Experiment,* Wiley, NY.
[9] 1991, *Metals Handbook,* Ninth Edition, American Society for Metals.

METAL FORMING I

Ironing with Superimposed Force

Zlatko Kampuš

University of Ljubljana, Faculty of Mechanical Engineering,
Ašker• eva 6, SI-1000 Ljubljana, Slovenia,
zlatko.kampus@fs.uni-lj.si

Abstract: The paper presents equations for the calculation of stresses in ironing with superimposed force which were derived on the basis of the elementary theory of plasticity. The process of ironing with superimposed force was simulated using the finite element method (FEM). Simulation showed that cups can be made either with internal or external wall thickening. Experiments were conducted to verify the predictions obtained by the model using the elementary theory of plasticity and the finite element method.

1. Introduction

The first patent which contained the principles of ironing was taken out in the USA as early as 1904 [1], however, intensive research into this procedure around the world took place only after 1950. Most papers have dealt with the calculation of stresses and the forming force. Most are based on the elementary theory of plasticity [2, 3], but also on the deformation work method [4], the upper boundary method [5, 6], slip-line theory [7], multifactorial planning method [8, 9] and the finite element method [10, 11, 12, 13]. Some authors have researched the possibility of increasing the formability of ironing, either by ultrasonic vibration [14] or by the action of a superimposed force on the cup edge. The published results of ironing with superimposed force are given only on the basis of experiments [15, 16], but no in-depth theoretical analysis has been performed. Our research therefore focused on the derivation of stresses and forces in ironing with superimposed force using the elementary theory of plasticity, and the theoretical model was verified by experiment, while certain possibilities were also checked using the finite element method (FEM).

2. Derivation of Stresses

Formability is the ability of materials to withstand plastic deformation without fracture or other defects. Due to the complexity of influential parameters, formability cannot be defined in a single manner. For identical materials and constant forming conditions, the forming limit also depends on the stress state [17] (Figure 1). Formability in ironing can be improved by increasing the percentage of compressive stresses in the forming zone, which can be achieved with an additional compressive superimposed force on the cup edge (Figure 2).

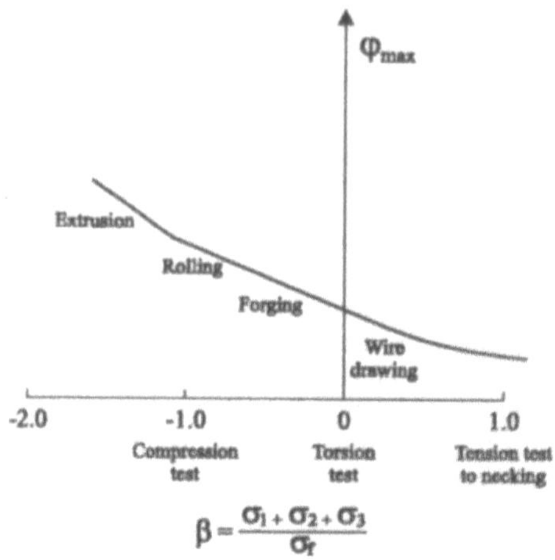

Figure 1: Influence of stress state on strain to fracture [17]

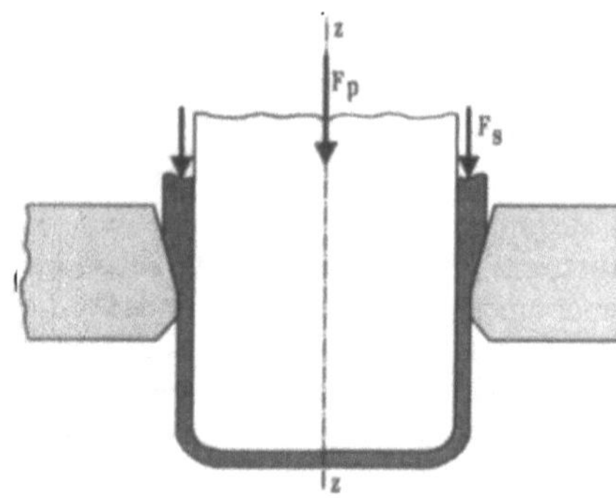

Figure 2: Principle of ironing using superimposed force

The elementary theory of plasticity was used to derive stresses. A differential equation for equilibrium was set in the direction of the z-axis:

$$\sum dFz = 0 .$$

It was assumed that the superimposed force varies linearly in the forming zone. Tresco's plasticity condition was applied:

$$\sigma_1 - \sigma_3 = \sigma_z - \sigma_r = \frac{2}{\sqrt{3}} \sigma_f .$$

The longitudinal stress at die outlet σ_z is

$$\sigma_z = 1.15\sigma_f - (1.15\sigma_f - \sigma_{z1}) e^{(-2\pi R_2 \mu_d h)/A_2} ,$$

and the radial stress σ_r is

$$\sigma_r = -(1.15\sigma_f - \sigma_{z1}) e^{(-2\pi R_2 \mu_d h)/A_2} .$$

Figure 3 shows the distribution of longitudinal and radial stresses in the forming zone. Due to compressive stresses, both stresses are displaced to the left – the radial stress σ_r increases, while the longitudinal stress σ_z decreases. The maximum strain, which depends on the longitudinal stress σ_z, therefore also increases.

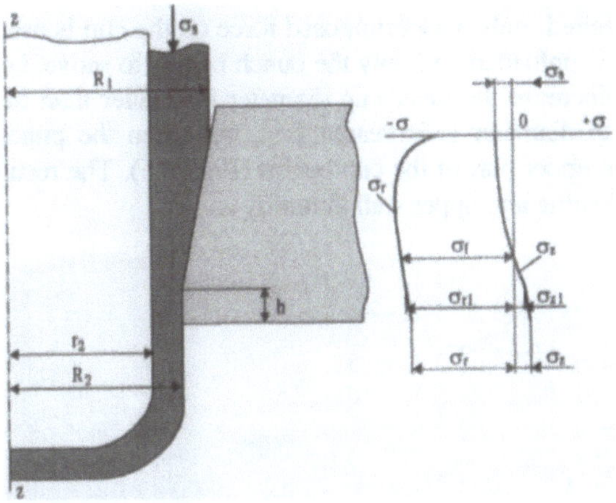

Figure 3: Distribution of longitudinal σ_z and radial σ_r stresses in the forming zone.

3. Numerical Simulation

The ABAQUS software package was used for numerical simulation. The objective was to research the possibilities of manufacturing various products which can be obtained by adding a superimposed force on the cup edge.

Since we were not interested in product accuracy here, rigid elements were used for the punch and die and axi-symmetric quadratic elements for the cups. In order to obtain a suitable preform, deep drawing without a blank holder was first simulated. Unhardened forming properties were taken as the properties of the material in ironing ($\sigma_f = 690^{0.25}$ N/mm^2). The influence of heating was not taken into account in the model.

In some cases (products with grooves or splines), a thickened cup wall is desired. With respect to different time variations of the superimposed force and force on the punch, thickening can be made on the outside or inside of a cup.

Figure 4 shows a simulation of the production of a cup with internal wall thickening. In the first phase of drawing, only the superimposed force acts on the cup edge. The lower part of the cup is pressed through the die. The outer diameter decreases, while cup wall remains unchanged. Then the punch force begins to act on the cup in addition to the superimposed force. The process of ironing begins. In this manner, a cup is produced with internal wall thickening on the lower end and a thinned wall at the upper end.

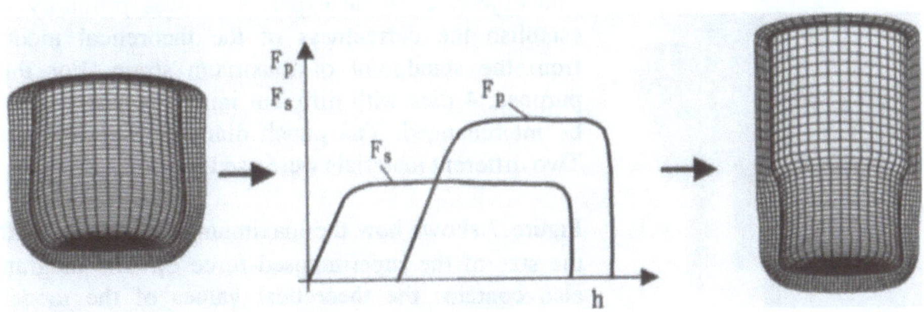

Figure 4: Production of a cup with internal wall thickening

Even if external wall thickening is desired, only a superimposed force on the cup is used in the first phase. Then the cup edge is unloaded and only the punch begins to move, i.e. ironing. Since the force required to increase the inner cup diameter is smaller than the required ironing force, the lower cup diameter is increased first, but when the punch reaches the cup bottom ironing of the upper part of the cup begins (Figure 5). The result is a cup with lower external wall thickening and upper wall thinning.

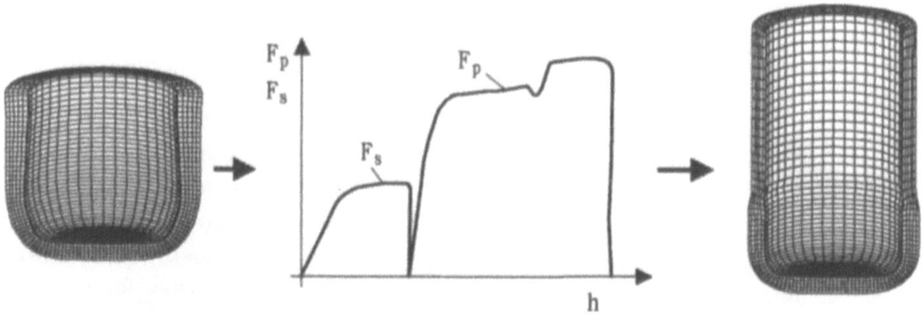

Figure 5: Production of a cup with external wall thickening

Figure 6: Tool for ironing with superimposed force

4. Experiments

A double-action hydraulic press was used for experiments. An experimental tool was made. The upper part of the tool, the die, was clamped into the ram. The superimposed force could be changed as desired during drawing via four spacers, which were connected with the hydraulic cushion of the press (Figure 6). Preforms for ironing were cups made by deep drawing. The punch diameter was 40 mm and that of the inlet die opening with a tractrix shape was 55 mm. After deep drawing, the cups were annealed for 1 hour at 650 ^{0}C, phosphated and soaped.

The objective of the experiments was primarily to establish the correctness of the theoretical model from the standpoint of maximum strain. For this purpose, 4 dies with different inner diameters could be interchanged. The punch diameter was 40 mm. Two different materials were used.

Figure 7 shows how the maximum strain varies with the size of the superimposed force σ_s. The diagram also contains the theoretical values of the model, which were calculated from the condition that the maximum tensile stresses in the cup wall can only be equal to the tensile stress of the material:

$\sigma_z \leq R_m$.

It can be seen that theoretical predictions agree well with the results of experiments, especially for material 20MnCr5. When material C10 is used, the maximum strain increases by 40% with the action of a superimposed force, and for 20MnCr5 by about 30%.

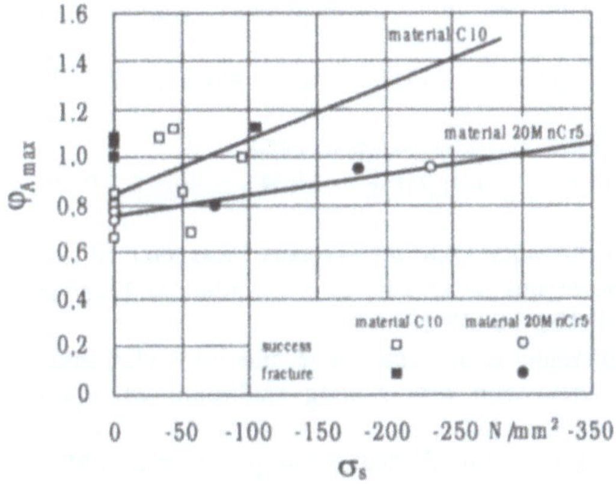

Figure 7: Comparison of theoretical and actual maximum strain φ_{max} vs. the magnitude of superimposed stress σ_s

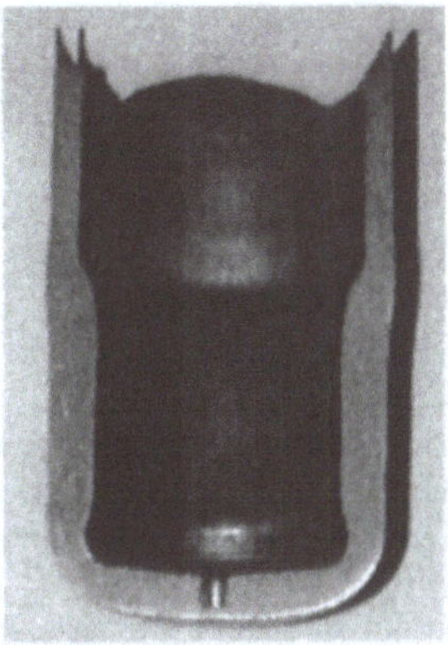

Figure 8: Photograph of a cup with internal wall thickening

When cups with nonuniform wall thickening were produced, there were problems due to the insufficient size of the hole in the die. The production of cups with external wall thickening therefore failed. Figure 8 shows a photograph of a cup with internal wall thickening. This was made from AlMg3.

5. Conclusion

The derived equations of stresses in ironing with a superimposed force are similar to those for normal ironing. If the superimposed force equals zero, these equations become the equations of classical ironing [3].

The theoretical model and experiments showed that the maximum strain can be increased by up to 40% with the use of a superimposed force. Even a small superimposed force yields a straight cup edge, which is unusual in the ironing of cups previously made by deep drawing.

A superimposed force increases the total forming force and energy, therefore it should be only as large as is required for the reliability of the process.
Numerical simulation showed (and experiments confirmed) that it is possible to produce cups with partially thickened walls. However, due to technical difficulties, we succeeded in making only one cup with internal wall thickening.

References

[1] Campion D.: Deep drawing & ironing – theory and practise. Sheet Metal Industries. February 1980, 111-119.

[2] Busch R. K.: Abstreckringen von zylindrischen Hohlkörpern mit einem oder mehreren Abstreckringen. Industrie-Anzeiger 92 (19709 73, 1741-1742.

[3] Nagel H. G.: Die Fertigung von langen zylinndrischen Werkstücken durch die Kombination der Verfahren Kaltfliespressen und Abstreckziehen. Umformtechnik (1972) 3, 33-39

[4] Popov E. A.: Osnovy teorii listovoi štampovki. Mašinostroenie, Moskva 1977, 199-212.

[5] Musafia B.: Primjena teorija plasti• nosti. Svjetlost, Sarajevo 1974.

[6] Fukui S.: Analytical Study of Wall Ironing, Considering Work Hardening. C.I.R.P. 18 (1970), 593-599.

[7] Hill R.: The Mathematical Theory of Plasticity. Oxford University Press, Oxford 1950.

[8] Djuki• H., Popovi• P.: Metoda prora• una izvla• enja sa redukcijom debljine stene zida. Obrada deformisanjem metala 9 (1984) 2, 45-62.

[9] Djuki• H.: Eksperimentalno odredjivanje zakona promjene sile pri izvla• enju sa redukcijom zida omota• a. 16. Savjetovanje proizvodnog mašinstva Jugoslavije. Mostar 1982.

[10] Odell E. I.: A study of wall ironing by the finite-element technique. Trans. ASME, J. Eng. Ind., 100 (1978) 1, 31-36.

[11] Huang You-Min, Yuung-Hwa Lu, Jia-Wine Chan: An elasto-plastic finite element and experimental study of the ironing process. Journal of Materials Processing Technology, 26 (1991), 53-80.

[12] Baillet L., Brunet M., Berthier Y.: Experimental and numerical dynamic modelling of ironing process. Journal of Materials Processing Technology, 60 (1996) 677-684.

[13] Yanran Zhan, Wang Z. R., Chein Weimin: Numerical simulations for extrusion and ironing and die-angle optimisation. Journal of Materials Processing Technology, 55 (1995) 48-52.

[14] Jin M., Kaewtatip P., Murakawa M.: Utility of ultrasonic vibration applied to metal-forming processes. Advanced Technology of Plasticity, Vol. III, Nüremberg 1999

[15] Brückner L., Pöhlandt K., Lange K.: Gebrauchseigenschaften von durch Tiefziehen und nach-folgendes Abstreckgleitziehen gefertigten Näpfen. Blech Rohre Profile 41 (1994) 9.

[16] • vik I., M., Šarov A. S.: Erhöhung der Genaugkeit und des Umformgrades beim Abstreckziehen. Umformtechnik (1976) 2

[17] ASM International: Metals Handbook; Ninth Edition; Volume 14; Forming and Forging. USA 1989

The Continuous Extrusion Process CONFORM: a full 3D Thermal and Mechanical Analysis

A. Barcellona[1], L. Cannizzaro[1], R. Riccobono[1]
[1] Dipartimento di Tecnologia e Produzione Meccanica, Università degli studi di Palermo, viale delle Scienze, 90128 Palermo, e-mail: barce@dtpm.unipa.it

Abstract: The principle of the Conform extrusion process is based on the action of a rotating wheel that drags and push the feed material by means of the frictional force existing between them. The Conform process has several advantages over the traditional extrusion processes. The preheating phase may be neglected, so that furnaces for heating are not necessary; furthermore, the product is continuous and welding phase of the billets is not required. The influence of three operative parameters, die geometry, velocity and die cooling rate, has been analysed in this paper by means of an "outranking approach" able to give precious indications for the assessment of the process.

1. Introduction

The principle of the Conform extrusion, invented in UK, is shown in the figure 1. An extrusion rotating wheel drags and push the feed material by means of the frictional force existing between them. The material is pushed into dies of various shapes.
The frictional force between the rod feed and the wheel must overcome not only the force required for the deformation but also the frictional force existing between the rod feed and the stationary dies. The extrusion occurs when the material meets the abutment and the die cavity. The product is continuous and welding phase of the billets is not required. A wide range of products may be manufactured, such as wire, profiles, multi-void tubes and strips. During the process, heat is generated due to the friction between material and shoe, and due to the deformation. Some experimental and numerical results of the process, are reported in literature [1, 2, 3, 4, 5].
One of the most serious problem in the process is due to the eventually excessive heat generation in the material; for this reason in a previous paper [6] a numerical analysis has been performed in order to evaluate some limits of the process and to study the quantity of heat generated by friction and by the imposed high strains. In this paper the choice of some operative parameters, such us die geometry, velocity and cooling rate of the dies, has been performed by means of an outranking approach [7, 8] in order to select the best combination of them, by comparing the numerical results in the different cases.

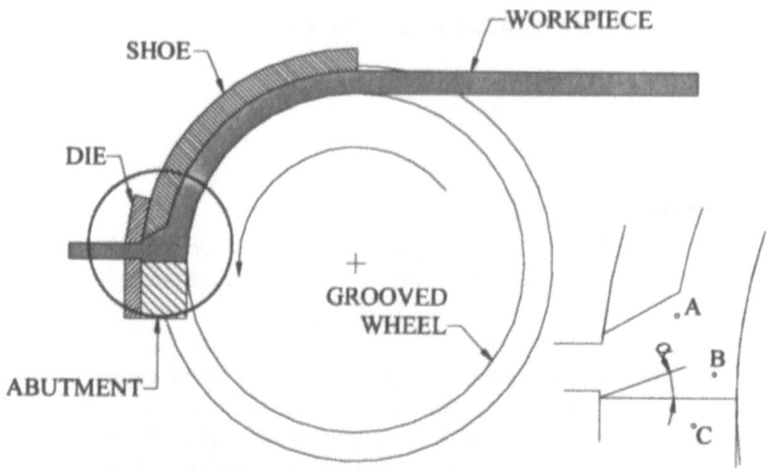

Fig. 1: The Conform process, the analysed points and die geometry

2. The 3D Numerical Simulations

2.1 The model

Simulations have been carried out by means of the finite element DEFORM3™ code [9], that has shown interesting results in 3D numerical analysis. The feed aluminium rod is 16 x 38 mm, the die diameter considered is 14 mm; the rotating extrusion wheel has a diameter equal to 300 mm; all the dies are made of H13 steel; as far as regard the operative parameters the analysed velocity are 10, 90 and 500 mm/sec, the die angle is equal to 0, 20 and 40 degree and cooling rate was considered by circulating cold water existing or absent by means of opportune convection coefficient.

Dies were modelled as rigid surfaces, thus leading to a thermal-mechanical analysis of the workpiece and to a thermal analysis of the dies; in fact, in order to study also the variation of the temperature in the tooling, die, wheel, abutment and shoe were meshed. The numerical simulations were executed on a high performances IBM workstation with a mean CPU time equal to 40 hours. Due to the die geometry, several automatic remeshings were required by the simulation. The three velocities have been obtained by assigning to the extrusion wheel three different rotations per minute.

The workpice is the 6063 Aluminium alloy; the plastic flow curve of the material employed for the simulation has been considered as a function of temperature, deformation and deformation rates; flow data have been obtained in [10] and introduced in the code by points. Frictional effects at the interface workpiece-dies have been taken into account by means of the constant shear friction model. The friction factor was assumed equal to 0.8 between the material and the rotating

wheel and equal to 0.1 between the material and the others dies. The simulations have been carried out according to the values of the thermal characteristics of the materials reported in Table 1.
It should be noted that, before the beginning of the grooved wheel rotation, an upsetting phase of the material inside the die is necessary to generate a pressure inside the dies. For this reason, firstly, a movement of the shoe in the direction of the wheel, while the workpiece is located between them, has been numerically simulated.

Table 1: Thermal properties of Aluminium 6063 and H13 steel.

Workpiece material and tooling	Melting point °C	Heat capacity $N/mm^2/sec$	Thermal conductivity W/m °K	Linear expansion coefficient (20-100°C)
6063	650	1.80	218	23 * 10E-6 / °C
H13	-	3.77	30	12 * 10E-6 / °C

2.2 The simulation plan

In order to find the best combination of process parameters it has been decided to simulate the full factorial plan of experiments. The simulations have been carried out according to the values of the operating parameters reported in Table 2.

Table 2: Operative parameters and the factorial plan of experiments.

Case	Velocity [mm/sec]	α [degree]	Coolant	Case	Velocity [mm/sec]	α [degree]	Coolant
1	10	0	no	10	10	0	yes
2	10	20	no	11	10	20	yes
3	10	40	no	12	10	40	yes
4	90	0	no	13	90	0	yes
5	90	20	no	14	90	20	yes
6	90	40	no	15	90	40	yes
7	500	0	no	16	500	0	yes
8	500	20	no	17	500	20	yes
9	500	40	no	18	500	40	yes

The final mesh, for the case 7, is show as an example in Figure 2a. The simulation have been validated in the previous research [6]. As results from the simulations, the values in Table 3 have been considered: in the point A and B (see fig. 1), the maximum value of the strain (ε_{max}), the equivalent stress (σ_{eq}); in the point C of the abutment, the peak contact pressure (p_{max}); the maximum temperature (T_{max}) of the part. The minimization of the contact pressure to increase the die life and the avoiding of an excessive workpiece heating are the greatest targets in the design of the process. As an example, with respect to the pressure (p_{max}) criterion, the case 3 is better then the case 1; in this way the alternative 3, only with regard to this criterion, will have a priority on the alternative 1.

2.3 Numerical results

In the figures 2a and 2b it is possible to see the extruded rod for the case 7; in fig 2b, wheel and shoe are intentionally omitted to simplify the observation.
The illustrated situation is characterized by the following geometric and operative parameters: die angle α is equal to 0 degree, velocity is equal to 500 mm/sec, coolant is considered not acting.

Fig 2a, b: The workpiece entering and exiting the dies.

The increasing of temperature of the feed material inside the dies due to deformation and due to friction between itself and the stationary tools reduces the heating costs, and could be employed to avoid at all the necessity of a furnace.
The difficulty to compare all the different production alternatives is related to the large number of criteria to be used in the comparison.
A positive fact is that all the criteria have a "quantitative" characteristic and that it is possible to quantify a weight for the relative importance of each criterion.
The ELECTRE method [7] appears well suited to build an outranking relation for this application. This method enables to assemble *pre-orders* among the alternatives.
The method consists in applying both a concordance test and a discordance test to every couple of alternatives.
A weight is chosen and assigned to each criterion. Calling k_j the weight associated to the j-th criterion g_j, J^+, $J^=$ and J^- the subsets of criteria for which the first alternative is preferred, indifferent or non preferred to the second, the following concordance tests have been applied to every couple of alternatives:

$$\frac{\sum_{j\in J^+} k_j + \sum_{j\in J^=} k_j}{\sum_j k_j} \geq C \,, \frac{\sum_{j\in J^+} k_j}{\sum_{j\in J^-} k_j} \geq 1 \,,\ g_{\max}(a) < 1.2 * g_{\max}(a')$$

The threshold C employed to build the *pre-order* has been selected equal to 3/4. The non concordance test has been applied, chosen a 20% max exceeding, to the maximum temperature criterion, where a and a' indicate two generic alternatives; the non concordance condition imposes a limit to the amount of temperature with respect to the other criteria.

3. Results and Discussion

The criteria and the chosen pertinent weights are shown in the Table 3. For each criterion, the result from numerical computation is reported for several cases.

Table 3: Weights, criteria and numerical results for some cases.

Criterion	ε_{max} in A [%]	ε_{max} in B [%]	σ_{eq} in B [MPa]	p_{max} in C [MPa]	T_{max} in B [°C]
(Weight)	(0.2)	(0.2)	(0.1)	(0.25)	(0.25)
Case 1	1.37	2.1	110	600	121
" 2	2.4	1.8	95	570	201
" 3	3.4	1.5	85	550	281
" 4	1.4	2.0	100	830	217
" 14	2.4	2.75	110	700	97
" 15	3.4	3.55	100	690	176
" 16	1.45	2.0	100	500	509
" 17	2.4	2.65	95	470	559
" 18	3.5	3.5	85	450	669

These results were elaborated by means of the above mentioned algorithm and the pre-order was found. The obtained "best" sequence, going from the best to the worst alternative, is the following:

P : (17) (16) (6) (5) (3) (4) (2) (15) (1) (14) (12) (9) (8) (7) (18) (13) (11) (10)

The number in parenthesis are the alternatives (cases). Cases 17, 16, and 6 are respectively the first, the second and the third of the sequence: this means the highest velocity may be employed even with a square abutment, but only in presence of a coolant of the dies. Case 6, still the third of the sequence, is characterized by absence of coolant, medium velocity and highest angle of the abutment. If coolant circuit is considered too complex to realize this should be the best combination of the operative parameters. The cases 5 and 3, still among the

first positions of the obtained sequence, indicate that die geometry angle brings greater relevant benefits then the presence of the coolant.
Furthermore, by means of the built approach it is possible to add, to the knowledge database, new numerical results, for example varying just one parameter, with the aim to observe the single effect of it on the obtained sequence.

4. Conclusions

An analysis of an extrusion forging process has been realized by means of 3D numerical simulations. Avoiding expensive trial and error experimental methods, the improvement in the design of a non conventional extrusion process has been obtained. The results have been reached by means of an outranking approach, founded on the ELECTRE algorithm able to perform a sensitivity analysis for the assessment of the process parameters.

References

[1] Reinikainen, T., Korhonen, A. S., Andersson and K., kivivuori, S., 1993, Computer-Aiede Modelling of a NEW Copper Extrusion Process, Annals of the CIRP, Vol. 42/1/1993.
[2] Reinikainen, T., Andersson, K., kivivuori and S., Korhonen, A. S., 1992, Finite-element Analysis of Copper Extrusion Process, J. Mat. Proc. Technology.
[3] Newman, R., 1995, Precision Extrusion with Conform Technology by Holton Machinery Limited, UK, Int. Conference on Aluminium Extruded Products, Brescia, Italy.
[4] Green, D., 1972, Continuous Extrusion-Forming of Wire Sections, J. Institute of Metals.
[5] Tomiatu S., Fukuoka S., Ozaki m. et al., 1999, Thermal Management in Conform Etrusion of Multi-Voids Tubes, Proc. Of the 6th ICTP, Vol. III. Germany.
[6] Barcellona A., 1999, On The Influence of some Operative Parameters in the Conform Extrusion Process, Proc. of the IV Convegno AITEM, Brescia, Italy.
[7] Roy, B., Bertier, P., 1973, La Méthode ELECTRE II - Une Application au Média-Planning, OR '72, M.Ross, North-Holland Pub. Company.
[8] Antonelli, D., Barcellona, A. and Cannizzaro, L., 1999, Application of a Decision Making Method to Improve an Industrial Hot Extrusion Forging Process, accepted for the publication on the Annals of the CIRP 1999.
[9] Oh, S.I., Wu, W.T., Tang, J.P. and Vedhanayagam, A., 1991, Capabilities and Applications of FEM code DEFORM: the Perspective of the Developer, J. of Mat, Processing Technology, 27:25-42
[10] Clode, M. P., 1992, Hot Torsion Testing to Model the Extrusion Process, J. Aluminium Industry, 11:24-39.

Acknowledgments

This work has been made using 40% funds. Authors wish to thank Mr. Filippo Ciofalo for his helpful contribution in the solid modelling.

Experimental Verification of Dieless-Drawing Theories

Oskar Pawelski, Wolfgang Rasp, and Walter Wengenroth

Max-Planck-Institut für Eisenforschung GmbH, Düsseldorf, Germany

Abstract. The dieless-drawing method is described briefly. Because of the strong thermo-mechanical coupling it is difficult to solve the underlying differential equations. Therefore several assumptions have been made to obtain simplified but nevertheless realistic models of the process. For an axi-symmetric cylindrical model the heat-conduction equation can be solved analytically and the radial dependency can be studied. For thin specimens a one-dimensional model is capable of predicting the most important parameters like axial temperature distribution and the final shape of the deformation zone. In contrast to conventional drawing the dieless-drawing method provides the possibility of producing variable cross-sections. The necessary control function is given for an example of a linearly varying final shape.

1 Principle of Dieless Drawing

Dieless Drawing is a flexible forming process to draw wires, bars, and tubes. The great flexibility is gained from the replacement of the outer die. Instead of the rigid drawing die an induction coil combined with a cooling device provides the plastic deformation of the workpiece. Fig. 1 shows the principle of the dieless-drawing process. The induction heating raises the temperature in

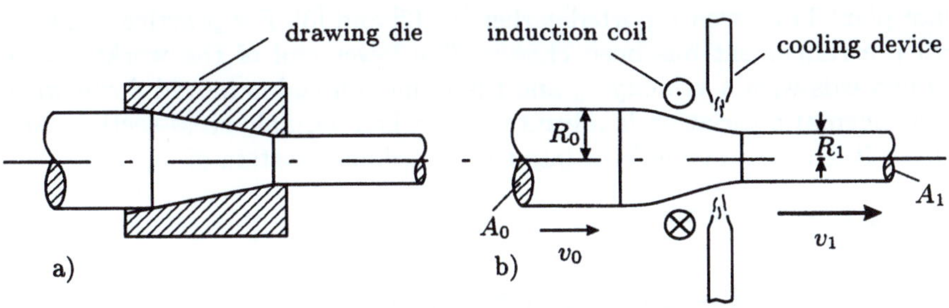

Fig. 1. Conventional drawing (a) and dieless drawing (b)

the bar which leads to the lowering of the yield stress until the applied drawing stress is matched. At this point the plastic deformation starts. The subsequent cooling device inhibits necking of the workpiece and thus stabilizes the process. Fig. 1 depicts the generic kinematic variation with fixed induction coil and ingoing (with v_0 at R_0) and outgoing (with v_1 at R_1) material. Fig. 2 shows the laboratory pilot plant at the MPI in Düsseldorf. The characteristic features of

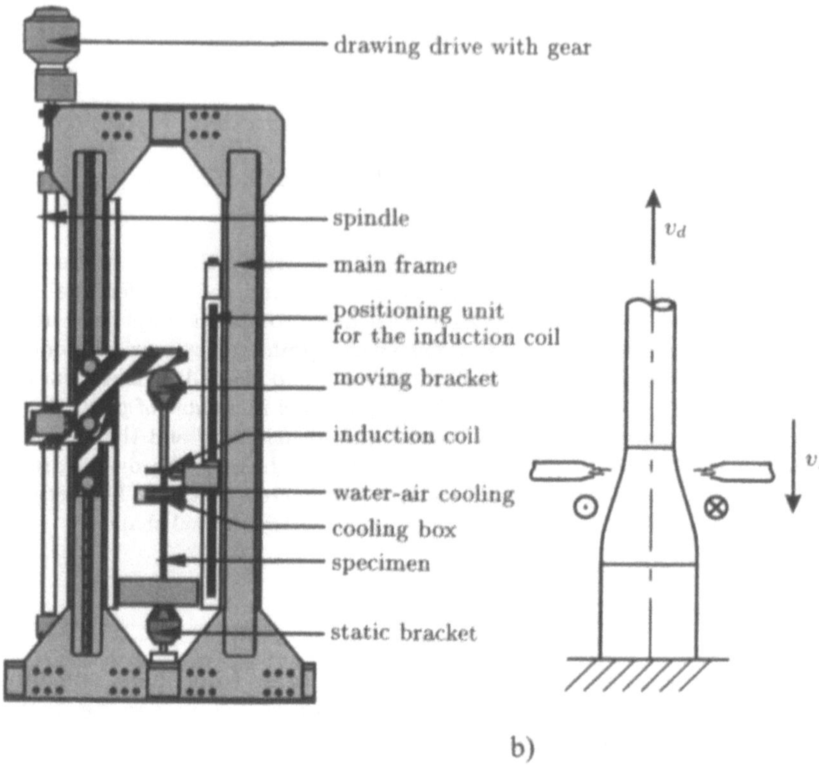

Fig. 2. Laboratory pilot plant (a) and kinematic realisation (b)

the pilot plant have been reported earlier [1], [2] and [3]. For practical reasons a vertical arrangement has been chosen. The lower end of the workpiece is moved upwards with a velocity v_d and the induction coil is moved down with v_I. This kinematic variation is advantageous when variable cross-sections are produced. It is easy to transform into the co-ordinate system of fig. 1

$$v_0 = v_d + v_I, \quad v_1 = v_I. \tag{1}$$

2 Simplified Mathematical Models

To investigate the dieless drawing process theoretically several models have been developed. This paper concentrates on the temperature distribution in the workpiece which is of major importance for the plastic deformation.

2.1 Axi-symmetric cylindrical model

The stationary heat-conduction equation in cylindrical co-ordinates is

$$v_x \frac{\partial T}{\partial x} + v_r \frac{\partial T}{\partial r} = a \left(\frac{1}{r} \frac{\partial T}{\partial r} + \frac{\partial^2 T}{\partial r^2} + \frac{\partial^2 T}{\partial x^2} \right) + \frac{p}{\varrho c} \tag{2}$$

with $a = \lambda/\varrho c$, thermal conductivity λ, density ϱ, and specific heat capacity c. Heating and cooling are applied as boundary conditions. If the deformation is small, the reduction in diameter may be neglected ($R \neq R(x)$, $v_r \approx 0$) and the solution domain is cylindrical. In this case eq. (2) can be solved analytically with an exponential-bessel-series expansion

$$T(x,r) \propto J_0(\beta r) \cdot exp\left(\frac{v_x}{2a} \pm \sqrt{\frac{v_x^2}{4a^2} + \beta^2}\right) \tag{3}$$

with constants β, which are determined by the boundary conditions. For details of the exact input parameters and the boundary conditions, see [4] and [5]. This analytical solution also suits well to verify FEM simulations (points), as can be seen in fig. 3. In the case of thin rods the radial dependency can be

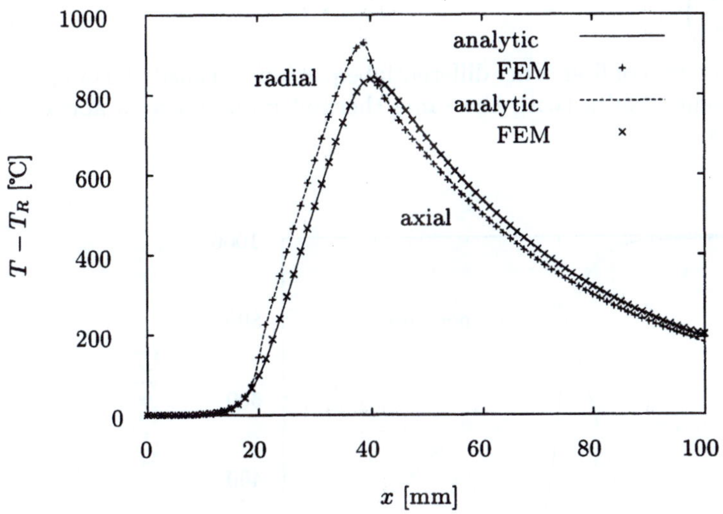

Fig. 3. Axial and radial temperature distribution $T(x)$ in cylindrical domain

neglected. In this case only the length co-ordinate x has to be dealt with. In a one-dimensional model it is easy to incorporate the plastic deformation (i.e. $R = R(x)$). A second differential equation is needed for the determination of the two unknown functions: temperature $T(x)$ and shape of the deformation zone $R(x)$. For this purpose serves the equilibrium of the applied drawing stress and inner stress.

2.2 One-dimensional coupled model

The simplified one-dimensional heat-conduction equation reads

$$\frac{dT}{dx} + \frac{2\alpha R}{c\rho v_0 R_0^2} T - \frac{2PR}{c\rho v_0 R_0^2} = 0 \tag{4}$$

with heating power P and heat transfer-coefficient α. In the above two-dimensional model appropiate boundary conditions correspond to these additional terms. The diffusive part $(\partial^2 T/\partial x^2)$ has been neglected, which is justified in some cases [6]. The detailed derivation of the second differential equation is given in [6]. The result is

$$\frac{dR}{dx} = -\frac{R^3}{2v_0 R_0^2} \left\{ \frac{\sigma \exp(m_1 T)}{C(\varphi + \varphi_0)^{m_2} \exp\left[m_4/(\varphi + \varphi_0)\right]} \right\}^{1/m_3} \tag{5}$$

with material dependant constants C, $m_1 \ldots m_4$, φ_0, strain

$$\varphi = 2 \ln \left(\frac{R_0}{R} \right) \tag{6}$$

and stress

$$\sigma = \sigma_0 \left(\frac{R_0}{R} \right)^2 . \tag{7}$$

The above system of first order differential equation can easily be integrated numerically. Some results (with the same thermal input data as above) are plotted in fig. 4.

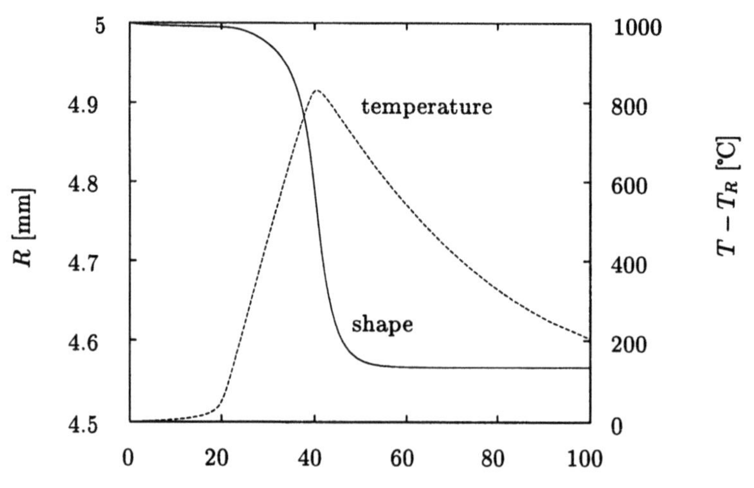

Fig. 4. Distribution of shape $R(x)$ and temperature $T(x)$

3 Variable Cross-Sections

One of the most attractive features of dieless drawing is the production of bars with variable cross-sections $R(x)$ over the length. This can be achieved by varying the entry (v_0) or exit (v_1) velocity (or both) with time in the continuity equation (8)

$$vR^2 = v_0 R_0^2 = v_1 R_1^2. \tag{8}$$

Since the variation of the entry speed v_0 would lead to an inhomogenous heating of the inflowing material only the exit speed v_1 has been varied as a function in time, (i.e., $v_1 = v_I + v_d(t), v_0 = v_I = const.$). In the following sections the control function for $v_1(t)$ is calculated to obtain a desired end-profile $R_1(x)$.

3.1 Control function

The exit speed is

$$v_1(x) = \frac{dx}{dt} = \frac{v_0 R_0^2}{R_1^2(x)}. \tag{9}$$

Solving for dt and integrating yields

$$t(x) = \frac{1}{v_0 R_0^2} \int_0^x R_1^2(\xi)\, d\xi. \tag{10}$$

By eliminating x from eqs. (9) and (10) $v_1(t)$ can be established. Generally eq. (10) can only be integrated and inverted numerically. Because of the great practical relevance (e.g. for the production of springs with a progressive characteristic) the case of a cone contour has been studied more closely.

3.2 Cone contour

The desired cone-contour can be expressed as

$$R_1(x) = a + b \cdot x \tag{11}$$

with constants a and b. Integration yields

$$t(x) = \frac{1}{v_0 R_0^2}\left(a^2 x + abx^2 + \frac{b^2 x^3}{3}\right). \tag{12}$$

Solving for x gives

$$x(t) = \frac{-a + (a^3 + 3btv_0 R_0^2)^{1/3}}{b}. \tag{13}$$

Substituting x into the continuity equation (8)

$$v_1(t) = \frac{R_0^2 v_0}{(a^3 + 3bv_0 R_0^2 t)^{2/3}} = v_0 \left(\frac{R_0}{a}\right)^2 \left(1 + \frac{3bv_0 R_0^2}{a^3} \cdot t\right)^{-2/3} \tag{14}$$

or, with $a = R_0$ and $b = -\Delta R/l_R$,

$$v_1(t) = v_0 \left(1 - \frac{3\Delta R v_0}{R_0 l_R} \cdot t\right)^{-2/3}, \tag{15}$$

with a reduction of ΔR over a length of l_R. Fig. 5 shows the control function $v_z(t)$ and the final shape $2R_1(x)$. In [6] the control function for a sine contour is given.

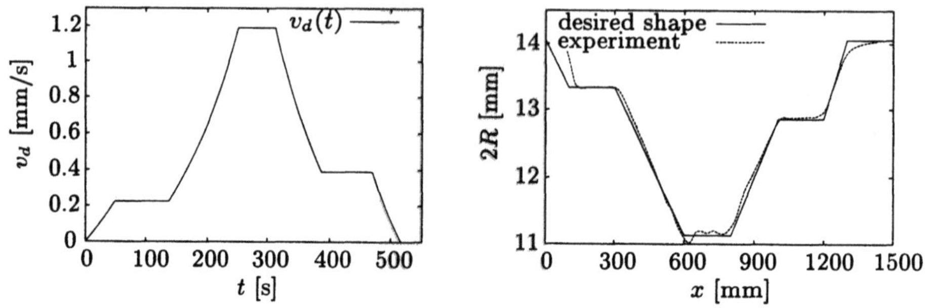

Fig. 5. Drawing velocity $v_d(t)$and resulting contour $2R_1(x)$

4 Conclusion

It has been demonstrated that it is possible to produce variable cross-sections with the dieless-drawing method if the control-velocity function is chosen appropiately. Different models for the process have been compared. Meanwhile the one-dimensional model has been extended to simulate time-dependant phenomena.

References

1. Pawelski, O., Rasp, W., Kolling A., Schmeißer, K. (1993) Dieless Drawing, An Example for a flexible Metal Forming Process. Proceedings of the 4th ICTP, 969–979
2. Pawelski, O., Rasp, W., Kolling A., Schmeißer, K. (1995) Dieless Drawing. Proceedings of the 31st International MATADOR Conference, 401–406
3. Schmeißer, K. (1998) Laboruntersuchungen zum freien Längen von Stäben (Dieless Drawing), Umformtechnische Schriften, Vol.80, Shaker-Verlag
4. Pawelski O., Kolling, A. (1995) Calculation of the temperature distribution in dieless drawing. steel research**66**, 50–54
5. Kolling, A. (1997) Entwicklung eines Prozeßmodells für das freie Längen von Stäben. Umformtechnische Schriften, Vol.70, Shaker-Verlag
6. Pawelski, O., Rasp, W., Wengenroth, W. (1999) Investigation into processing of bars with variable cross-sections by dieless drawing. Proceedings of the 6th ICTP, 1753–1762

METAL FORMING II

Ceramic Inserts For Forging Tools – Optimizing The Design

Hyuk-Hong Kwon [1], Chao-Cheng Chang [2] and Alan N Bramley [2]

[1] Dept. of Mechanical Design Engineering, Daejin University, Kyonggido 487- 711, Korea, hhkwon@road.daejin.ac.kr

[2] Dept. of Mechanical Engineering, University of Bath, BA2 7AY, U.K., A.N.Bramley@bath.ac.uk

Abstract: The use of ceramic inserts in steel forging tools offers significant technical and economic advantages over other materials of manufacture. These potential benefits can however only be realised by optimal design of the tools so that the ceramic inserts are not subjected to stresses that lead to their premature failure. Process simulation and stress analysis are thus combined during the design, and a data exchange program has been developed that enables optimal design of the dies taking into account the elastic deflections generated in shrink fitting the die inserts and that caused by the stresses generated in the forging process. The stress analysis of the dies is used to determine the stress conditions on the ceramic insert by considering contact and interference effects under both mechanical and thermal loads. Simulation results have been validated as a result of experimental investigations.

1. Introduction

As the demands made upon hot and warm forging and extrusion tooling to perform to new levels of productivity continues to grow, the challenge to engineers, and material scientists alike calls for a new approach. Until now, hot working tool steels have been the only cost effective die material option. Recently, the use of ceramic materials as functional materials has increased rapidly in a wide field of industry, since they have many excellent technical properties such as high hardness, high corrosive resistance, high elastic modulus, etc. Clearly they are going to be used increasingly in many engineering fields [1-2].

Near net shape forging technologies offer the possibilities to produce parts with good mechanical and technological properties combined with economical benefits. To realise precision forging processes of parts, with their demands on the accuracy, some important requirements must be fulfilled. These are: computer aided design of the tools, especially of the forming dies; FEM simulation of the forging process to determine shrinkage effects and mechanical stresses in the tool system; an exact adjustment of the main influence parameters; and finally, the high accuracy of the forging tools [3]. The use of ceramic inserts in steel forging tools offers significant technical and economic advantages over other materials of manufacture. This arises largely due to the very high wear resistance of certain ceramic materials. These potential benefits can however only be realised by optimal design of the tools so that the ceramic inserts are not subjected to stresses which lead to their premature failure. This is now possible since reliable information on the loading conditions in forging tools can be obtained from process simulation. In some cases die stress calculations can be coupled with the plastic flow analysis. Currently this approach can lead to excessive computing times and has not therefore been adopted in the work described here.

In this paper the data on the loading of the tools is determined from a commercial

forging simulation package as the contact stress distribution on the die-workpiece interface and as temperature distributions in the die. This data can be processed as load input data for a finite element die-stress analysis. Process simulation and stress analysis are thus combined during the design, and a data exchange program has been developed that enables optimal design of the dies taking into account the elastic deflections generated in shrink fitting the die inserts and those caused by the stresses generated in the forging process. The stress analysis of the dies is carried out to determine the stress conditions on the ceramic insert by considering contact and interference effects under both mechanical and thermal loads.
An example is given of the application of this methodology and subsequent experiment trials for the optimal design of a ceramic punch insert in a simple forging configuration.

2. Forging die design

Process simulation is now being used routinely in many parts of the forging industry [4-5]. It not only provides information on material flow, die fill and defect formation, but also very detailed information on the contact stress distribution at the die-workpiece interface. The contact stress distributions are far from being uniform in the tooling, an assumption often made in tool stress analysis. But, by combining process simulation and tool stress analysis, more exact stress data can be obtained, and possible failure due to high localised loads in the tooling can be identified and avoided through modifications in the design before the tooling is produced. The general methodology for optimising the design ceramic inserts for forging tools which was taken is shown in Figure 1 and is as follows:

- Process simulation to determine the loads during a forging cycle and the appropriate contact stress distribution at the die workpiece interface.
- The exchange of data between the different programs is necessary. Reliable information on the loading conditions of a die can only be obtained from process simulation and is given as the contact stress distribution at the die-workpiece interface and as temperature distribution in the die. This data can then be processed as load input data for a die stress analysis. The post-processing of DEFORM-HT results and use of data exchange program of die temperature and

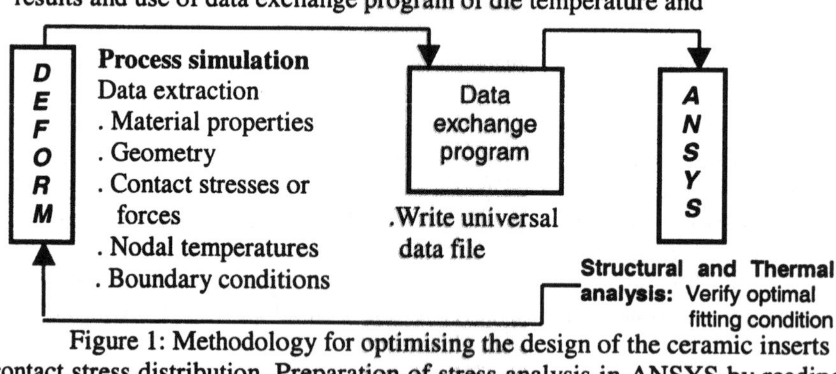

Figure 1: Methodology for optimising the design of the ceramic inserts

contact stress distribution. Preparation of stress analysis in ANSYS by reading in the universal files created by the data exchange program and adding boundary conditions and material data to the models.

• Stress analysis of the tooling using the previously determined contact stress distribution as load input data. ANSYS was used for this task.

2.1 Process simulation of forging tools

To demonstrate the applicability of this technique it was applied to the optimal design of a ceramic punch insert in a simple forging configuration as shown in Figure 2. Experimental trials were also performed.

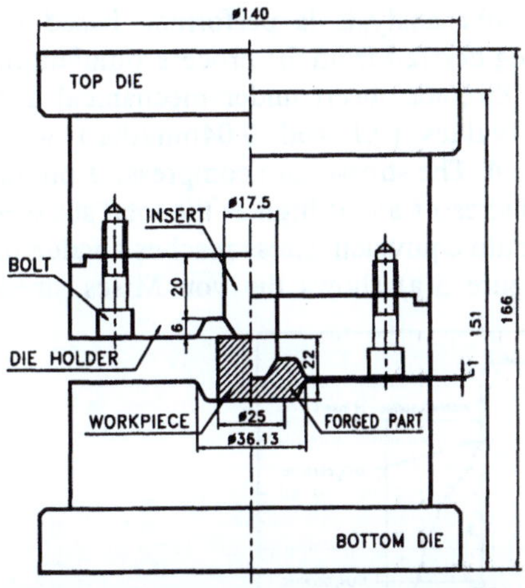

Figure 2: Schematic of the forging process and tooling system used in the laboratory tests (units: mm)

The process conditions of the non-isothermal forging simulation with rigid dies and a rigid-viscoplastic workpiece were as follows: The velocity of the upper die was taken to be 21 mm / s, friction factor was taken to be 0.3, the initial temperature of the workpiece and the die used in non-isothermal forging simulation were taken to be 1073K and 473K, respectively. The size of workpiece in this model was 25 mm in diameter and 22 mm in height. The mechanical and thermophysical properties used in the analysis are summarized as follows.

Material	H13	ZrO_2	Si_3N_4
Young modulus (GPa)	190	205	290
Poisson ratio	0.3	0.31	0.24
Thermal expansion coefficient ($K^{-1} \times 10^{-6}$)	11.3	10	3
Thermal conductivity ($N \cdot s^{-1} K^{-1}$)	28.6	2.5	25
Heat capacity ($N \cdot mm^{-2} K^{-1}$)	3.588	4	8
Emissivity ($N \cdot s^{-1} mm^{-1} K^{-4}$)	0.7	0.7	0.7
Interface heat transfer coefficient ($N \cdot s^{-1} mm^{-1} K^{-1}$)	27.5	7.6	7.6

Temperature gradients of the dies and workpiece are shown in Figure 3. Thermal expansion due to local temperature gradients introduces some stresses. The plastic deformation which is localised near the nose of the insert is transformed into heat. The maximum temperature of the die is located near the insert nose because in this region the contact time is long and the temperature of the workpiece is very high.

2.2 Structural and thermal analysis

In forging die design, pre-stressing is useful in preventing excessive tensile hoop stresses which may cause die failure during metal forming [6-7]. It is therefore necessary to determine accurately the interference value between the die insert and die holder for pre-stressing. Also, the thermal stress analysis is focused on the insert. The mechanical and thermal stresses are calculated with ANSYS software by modelling the interference using contact elements between insert and die holder. For a thermal analysis, transient heat transfer analysis is performed based on temperature-time profiles during forging being derived from the process simulation. The von Mises stress distributions on the Zirconia insert under mechanical and thermal loads with different interference values (0.1 and 0.04mm/dia.) were analysed by ANSYS and are shown in Figure 4. The stresses are compressive on the inclined surface of the insert and the thermal stresses are as high as the total stresses in the inclined areas of the insert. The maximum equivalent stress reaches a value of 977MPa at the point A (see Fig 4(b)). Figure 5(a) shows the von Mises stress

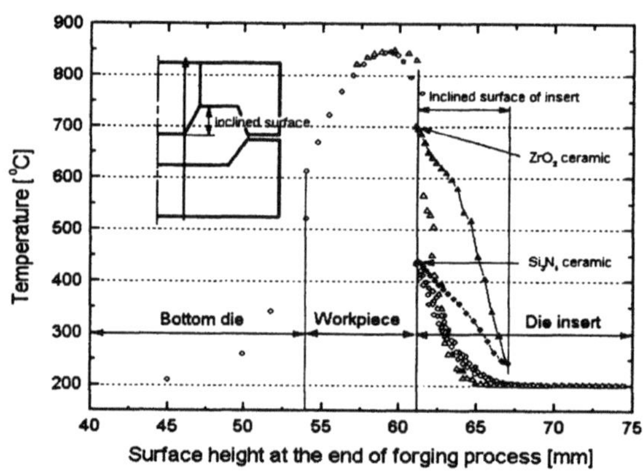

Figure 3: Temperature gradients in the die and workpiece at the end of forging process

distributions on the Silicon nitride insert, the peak of von Mises stress is smaller than the flexural strength in the insert case showing that the interference value is 0.1mm. Figure 5(b) shows the hoop stresses under the mechanical and thermal loads on the Silicon nitride insert. The hoop stresses are largely compressive, with a small tensile value in the central region of the insert. This does not however exceed the flexural stress 450MPa (at 1073K) or the maximum compressive stress 2000MPa of the Silicon nitride. Ceramic inserts have very high compressive stress capabilities, but low tensile stress and low thermal shock capabilities. These aspects have to be accommodated in the design of the insert.

3. Experimental investigations

The forging dies were made from H13 tool steel, hardened and tempered to HRC55-56, the tool was pre-heated up to 723K and pre-stressed by shrink fitting with two different interference values. Forging tests were conducted on a Rhodes 1000kN hydraulic press.

Ceramic inserts accept a high polish and provide excellent surface finish to the forged part as shown in Figure 6(a). Figure 6(b) shows the fractured pieces and scanning electron microscope of the Zirconia insert with interference 0.04mm after 50 forging strokes. The location of the fracture is consistent with the point of maximum equivalent stress, point A in Figure 4(b). This exceeds the flexural stress 354MPa (at 1073K), but does not exceed the maximum compressive stress 1700MPa of Zirconia. Silicon nitride insert dies survived more than 30 forging cycles per die without showing excessive wear or improper finish to the forged product. Figure 7 represents the load-stroke curves obtained from the simulation and the experiments. There is good agreement between the experimental and simulation results up to a die stroke of about 7 mm. At this point the workpiece makes contact with the inter part of the upper die. Thereafter the simulation load is significantly greater than the experimental load. This could be a result of the variation in friction between the workpiece and the tooling depending on the location of the ceramic material.

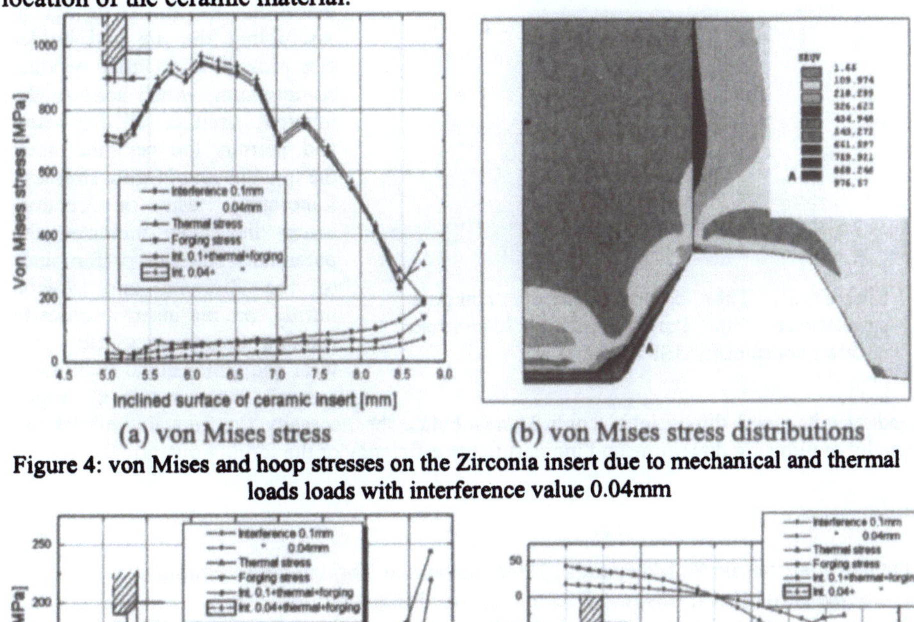

(a) von Mises stress (b) von Mises stress distributions

Figure 4: von Mises and hoop stresses on the Zirconia insert due to mechanical and thermal loads loads with interference value 0.04mm

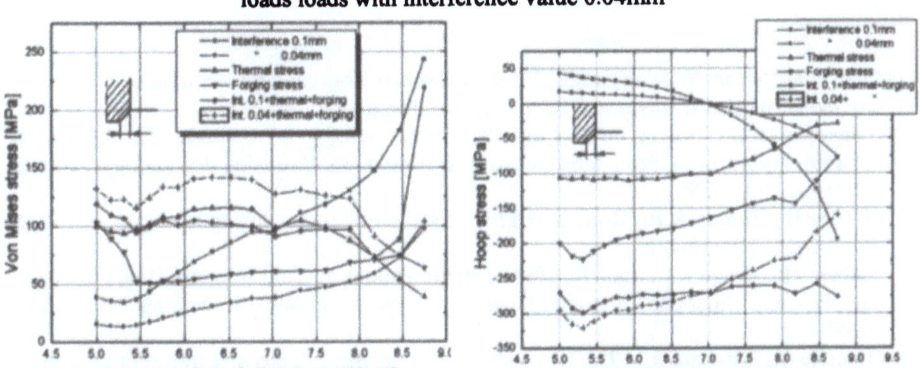

(a) von Mises stress (b) Hoop stress

Figure 5: von Mises and hoop stresses on the Silicon nitride insert due to mechanical and thermal loads

(a) Forged part (b) Fractured pieces and SEM of Zr insert

Figure 6: Photos of the forged part, fractured pieces and SEM of Zirconia insert

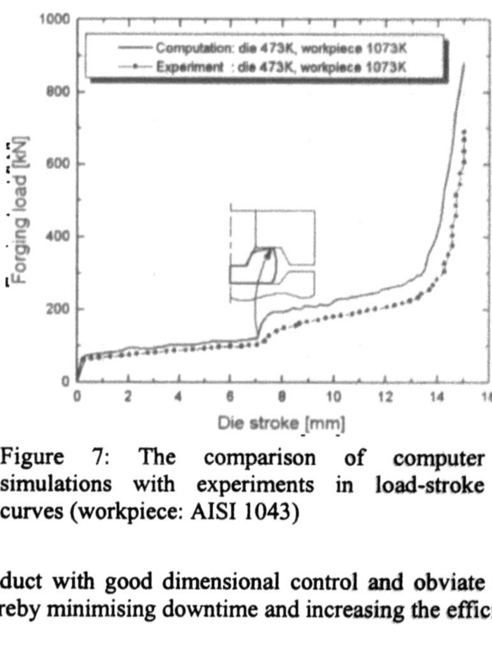

Figure 7: The comparison of computer simulations with experiments in load-stroke curves (workpiece: AISI 1043)

4. Conclusions

This paper describes a computer aided die design procedure for incorporating ceramic inserts in hot forging tools. The analysis also incorporated the mechanical loads due to shrink fitting, in assembling the die and holder can provide substantial residual compression, which adds to the inherent strength of the insert and permits the ceramic insert die to withstand forging stresses. Laboratory tests on ceramic insert dies have indicated the potentially superior performance of the Zirconia and Silicon nitride ceramic insert in order to prolong maintenance life. Low wear and dimensional stability of ceramic insert result in forged product with good dimensional control and obviate the necessity for frequent refinishing, thereby minimising downtime and increasing the efficiency of the forging operation.

References

[1] Ohuchi K., Sasaki S., Matsyno K., 1990, Isothermal Forging with Ceramic Die on Industrial Basis, Advanced Technology of Plasticity, 1:271-278.

[2] Guyoncourt D., Huang H.C., Piller R., 1996, Housing of Ceramic Die Inserts for Aluminum Extrusion, Proc. 6th International Aluminium Extrusion Technology Seminar, Chicago, 2:133-139.

[3] Doege E., Bohnsack R., Romanowski C., 1998, Tool Technologies for Near Net Shape Forgings, ICFT:311-322.

[4] Knoerr M., Lange K., Altan T., 1994, Fatigue Failure of Cold Forging Tooling: Causes and Possible Solutions through Fatigue Analysis, J. of Mat. Processing Technology, 46:57-71.

[5] Brucelle O., Bernhart G., 1999, Methodology for service life increase of hot forging tools, J. of Mat. Processing Technology, 87:237-246.

[6] Lange K., Cser L., Geiger M., Kals J.A.G., 1992, Tool Life and Tool Quality in Bulk Metal Forming, Annals of the CIRP, 41:667-675.

[7] Doege E., Andreis G., Romanowski C., 1997, Ceramics for Forging Tools, ICTMP 97, Japan.

A Simplified Procedure for the Computation of Temperature Changes in the Tools in Cold Forging Production

Y. Qin, R. Balendra and K. Chodnikiewicz

Design, Manufacturing and Engineering Management
University of Strathclyde, Glasgow G1 1XJ, UK

Abstract: Research in high-precision forging required to define changes of temperature in tools during forging production, since the changes could result in variations of component forms. The definition may be effected by FE simulation. Existing methods were, however, not applicable to the cases when a general-purpose FE code was employed for analysis tasks. A simplified procedure, which was implemented using a commercial FE code, was used to effect the analysis. Applications of the method showed high computational efficiency of the analysis.

1. Introduction

During forging, temperature in the tools could rise due to being subjected to the repeated heat-loads (plastic-work-generated heat and friction-generated heat). At same time, heat could also transfer from the tools into the workpiece when temperature in the tools rises to a value higher than that in the workpiece. This, again, would influence temperature distributions in the workpiece and tools. Temperature in the tools is, obviously, a function of manufacturing cycles. Changes of the temperature in the tools would influence forming-pressure-requirement, tool-deflection, material expansion and contraction, and hence, dimensions of the component. Before variations of component-errors could be qualified for forging production, temperature changes in the tools with manufacturing cycles would have to be defined.

Due to extremely large computational-requirements, repeating a full-scale, coupled thermo-mechanical FE simulation to simulate multi-cycle loading characteristics of a forging production process is unrealistic. A procedure, in which a "step-wise move mode" was used, was proposed for the prediction of temperature-rise in cold extrusion of can for repeating operations [1]. In another analysis for hot-forming processes [2], material flow was assumed to have similar patterns for different manufacturing cycles, therefore, calculation of the velocity for each cycle was eliminated. As a result, computational requirements for the analysis of temperature changes during multi-cycle forging were reduced.

In cold forging production, it could take thousands of cycles to establish a steady temperature distribution in the tools. Therefore, a more efficient method was proposed [3] for the computation of temperature changes in the tools. The method was to define individual heat-flux density functions. These functions were, then, applied to the die-surfaces for initiating a repeated heat-loading analysis of the die.

Since heat transfer analysis was only conducted for the die, no plastic-deformation of work-material, and no material-/tool-movement, were involved, high computational-efficiency of the analysis was achieved.

Based on the theory established [3], further efforts were made to develop a simplified procedure, with a view to enabling implementation of the method, using general-purpose, commercial FE codes. This is required since these codes are, now, widely used for analysis of forging processes.

2. FE Simulation

In order to develop a procedure to simplify the computation, at same time, maintain computational accuracy, firstly, FE simulations were conducted, using FE code ABAQUS [4], to examine the mechanism of temperature changes in the tools for a single forging-cycle. The simulations were conducted for several processes, such as closed-die forging, forward-extrusion and back-can extrusion. The analyses considered each stage of the forging cycle – loading, unloading, ejection of the component, as well as idle time of the process.

Fig. 1 shows an example of temperature changes during a forming cycle, at the die-surface for a forward extrusion. Through the analysis, influences of several factors on temperature distribution in the material and tools were qualified; the major factors include deformation-rate of the work-material, interfacial friction, thermal contact conductivity at the workpiece/die interface, and heat convection from the material and tools to the environment. Further analyses were also conducted by repeating FE simulation of a forging process for a number of cycles, with a view to initially examining temperature changes with manufacturing cycles. In these analyses, temperature computed for each tool-model was recorded after each cycle. The recorded data was then imported, as an initial condition, to initiate the simulation of the subsequent cycle.

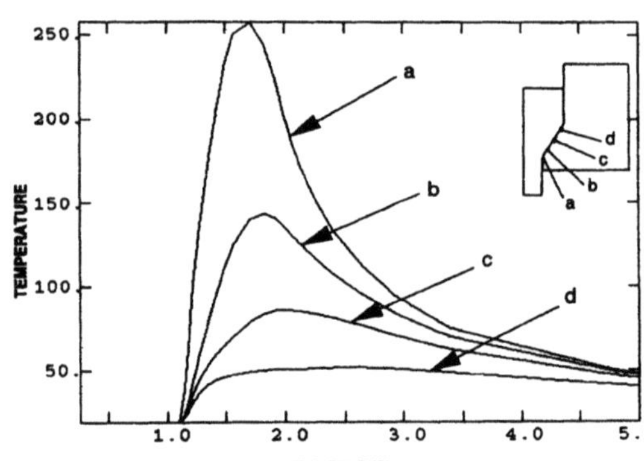

Fig. 1 An example of temperature changes during a forming cycle

3. Examining of Variations of Heat Flux

Accurate prediction of temperature changes in the tools may require to examine the whole forging-system which should include several elements such as machine, tools and work-material. Alternatively, the prediction may be effected by examining individual components of the system. Taking an individual die as an example, changes of thermal conditions at die-surfaces during forging would result in the changes of temperature in the die. Further, thermal boundary-conditions are prescribed by heat-flux densities across the boundaries of the die, which are functions of forging cycles. The value of a heat flux density across a die-surface could be positive, which may suggest that there is heat transferring into the die (Fig. 2(a)), due to plastic-deformation of the work-material and friction at the workpiece/die interface. It could also be negative (Fig. 2(b)) after a number of cycles, which may indicate that the die lost heat (heat transfer to either the workpiece, neighbouring components, or the air). If the changes of heat flux densities across the die-surfaces can be determined, temperature changes in the die can then be derived. However, the fact that the heat-flux densities are functions of the manufacturing cycles adds difficult to the analysis.

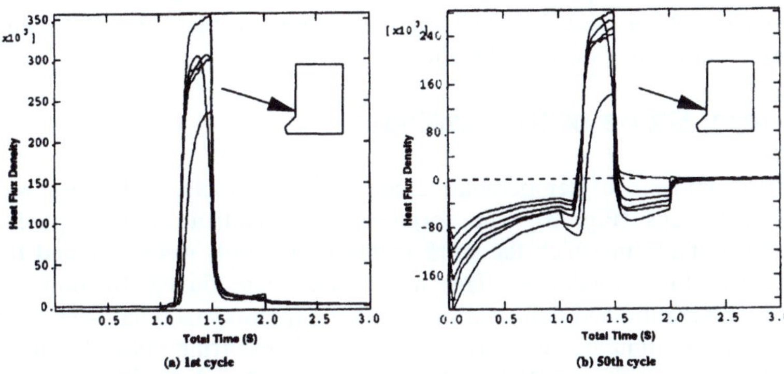

Fig. 2 Variations of heat flux densities with time for two manufacturing cycles for a forward extrusion

4. Definition of Heat Flux Density Function

Strategy for developing a procedure for the prediction of temperature changes in the tools with manufacturing cycles was to separate the tools from the forging system; then, heat-flux density functions were defined for each working-surface of the tools. Using these functions, tools can be analysed individually, by conducting heat transfer analysis.

Initially, heat fluxes across tool-surfaces were decomposed as several components -

- Heat flux from the workpiece to the tool.
- Heat flux from the tool to neighbouring parts.
- Heat convection from the tool to the air.
- Heat flux from the tool to the workpiece.

The methods for defining each of these components were discussed in the literature [3]. The decomposition is necessary, particularly, for the understanding of changes of thermal, interfacial conditions of the tool. Through further trails on the methods, it was determined to further simplify the methods to create a simplified analysis procedure - a procedure which would better suit the use of general-purpose FE codes. Such an issue was necessarily addressed, since commercial FE codes are, now, widely used for the analysis of forging processes.

Heat transfer from the tool to the workpiece is a major component which varies with manufacturing cycles; to define heat-flux density of this component, several iterations were required. The function, which was defined using these iterations, was still not sufficiently accurate, because some assumptions were necessarily introduced. Further modifications were, normally, required to correct the function. New development of the method was to reduce the number of simulations and iterations at early stages of the analysis, while replacing these by defining an assumed, unified heat-flux-density function for each finite element at the die-working-surface. These functions represent total heat-flux densities across the die-working-surface (sum of several heat-flux components). A unified function consists of several linear segments which respectively, define the change of heat-flux density across the die-working-surface, within a particular time-phase during forging. The functions have similar patterns as illustrated in Fig. 2, and have distinguishing differences at different stages of the forging, such as loading, unloading, ejection of component, and cooling down of the die.

5. General Steps of the Analysis

A complete analysis of temperature changes in the tools in forging production included five steps (Fig. 3). The forging was modelled as a coupled thermo-mechanical problem in which the work-material and tools were included for a full-scale, plastic flow simulation. The model was conditioned by comparing the computed temperature and deformation with the experimental data. A single forming cycle was, firstly, simulated to determine the heat transferring to the die, due to heat-generation of plastic deformation of the work-material and the friction between the work-material and the die. A unified heat-flux-density function at the die-working-surface was then defined for each element. Subsequently, the die-sets were modelled for heat transfer analysis. The assumed functions were applied, as distributed heat loads, to the die-working-surfaces, for cyclical loading. A criterion was also prescribed to control whether or not the temperature distribution has reached a steady-sate. Cyclical heat-loading analysis was then conducted. The modelling for the heat transfer analysis only included the die-sets, no material flow and tool movement was considered. Therefore, high computational efficiency was achieved. After the heat transfer analysis was conducted for certain cycles, a full-scale, coupled thermo-mechanical plastic simulation (material flow analysis) was conducted for comparing the assumed functions with FE-calculated values; the assumed heat-flux-density functions were then corrected. Further heat transfer analysis of the die-sets was then initiated, until the temperature distribution reached a steady-state.

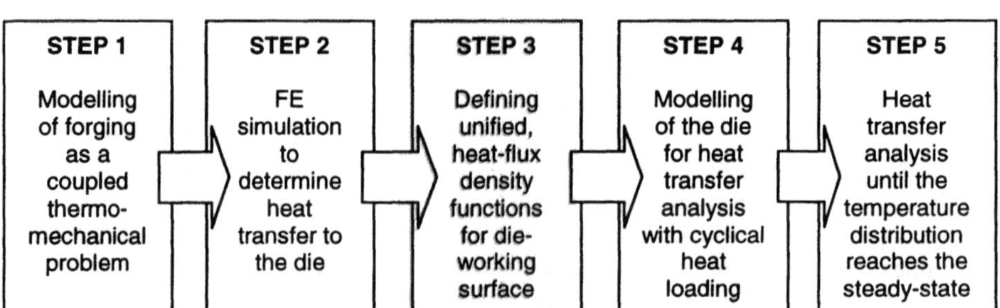

Fig. 3 Steps of the analysis of temperature changes in tools

6. The Applications of the Procedure

Several forging processes have been analysed using the procedure described above; an example of forward extrusion is shown in the paper (Fig. 4) and discussed in the following texts. With consideration of that heat could transfer to neighbouring components during forging cycles, both, die-set and supporting-set, were modelled. Temperature in the air was assumed to be constant (T=20 ^{0}C), and reference film coefficient, a rate of the heat across the die/air interface [4], 10 (JT^{-1} L^{-2} θ^{-1}). Work-material used was C15 steel and die-material common tool steel. Coefficient of friction at workpiece/die interface was assumed to be 0.08, punch velocity 60.0 mm/s, thermal contact conductivity 10 kw/m^{2}K at the die/workpiece interface. A manufacturing cycle was assumed to be completed within 5 seconds, which included material delivery, loading (extrusion), unloading (punch retraction), ejection of component and cooling down of the component and the tools.

Fig. 5 shows the computed temperature-changes at the die-exit – the envelope represents the global rise of the temperature at the end of each cycle, left-side window temperature changes during first fifty cycles, and right-side window the temperature changes during last fifty cycles. The results suggest that the temperature changes in the die were in a cyclical form. In each cycle, temperature sharply increased during the extrusion but gradually decreased after the ejection of component. Although the temperature change (difference between maximum and minimum temperature) for each cycle was very significant, it being about 200 ^{0}C, global increase of the temperature in the die was a slow procedure.

Heat transfer analysis was conducted for a total of 5050 manufacturing cycles. The trend of the envelope suggests that there was a sharp increase of the temperature at the die-surface during the first 1000 cycles. After this, temperature gradually tended to stabilise; maximum temperature was 135 ^{0}C. The curves shown in the right-side window may indicate that the temperature in the die had stabilised after the 5000 cycles. Although there was still a cyclical change of the temperature for each cycle at this stage, temperature at the end of each cycle did not increase significantly. Temperature saturation in the die indicates that thermal equilibrium had established between the heat-components which transferred from the workpiece to the die, and that which convected to the air and transferred to the workpiece from the die.

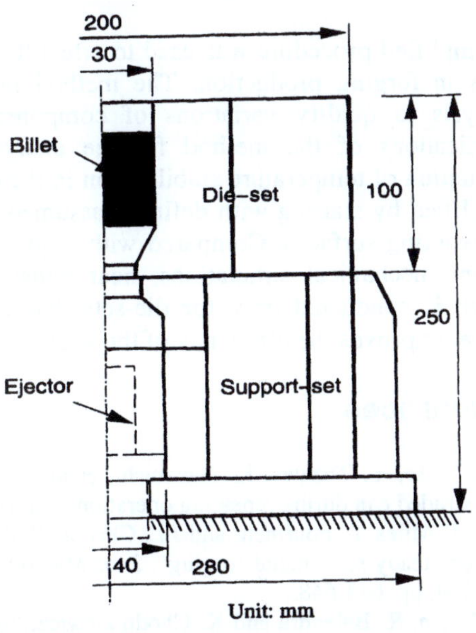

Fig. 4 Process model and die dimensions of a forward extrusion

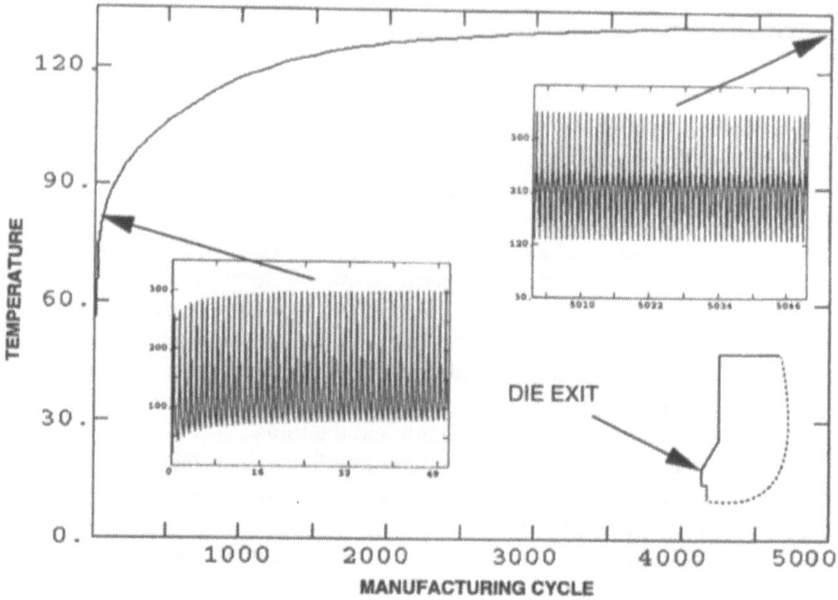

Fig. 5 Temperature changes with manufacturing cycle during the forward extrusion

7. Conclusions

A simplified procedure was used to effect the analysis of temperature changes in the tools in forging production. The method renders a practicable means for further analysis to qualify variations of components-errors with manufacturing cycles. Applications of the method for the analysis of cold forging confirm that the simulation of temperature stabilisation in the tools during multi-cycle forging can be simplified by starting with defining assumed, unified heat-flux-density functions for die-working-surfaces. Compared with requirements for repeating full-scale, coupled thermo-mechanical simulations, heat transfer analysis with cyclical heat-loading, which is conducted only for die-sets, has small computational requirements, and hence, improves the efficiency of the analysis significantly.

References

[1] T. Kato, Y. Tozawa, K. Nakanishi, et al, "Prediction of dimensional accuracy of cold extruded can during repeating operation", Annals of the CIRP, 35(1), 1986.

[2] M.P. Miles, L. Fourment and J.L. Chenot, "Calculation of tool temperature during periodic non-steady state metal forging", J. of Material Processing Technology, Vol. 45, No. 1-4, 1994, pp. 643-648.

[3] Y. Qin, R. Balendra and K. Chodnikiewicz, "A method for the simulation of temperature stabilisation during multi-cycle cold forging operations", Proc. of the 15th Int. Conf. on Computer-aided Production Engineering, April, Durham, UK, 1999, pp. 499-506.

[4] Hibbitt, Karlsson and Sorensen, Inc. ABAQUS Manual, 1994.

Advanced Design of Shrink Ring Interface for Metal Forming Dies using Optimum Analysis Technique

Atsushi. YOKOYAMA[1], Tuneo.HIRAI[2]

1 Kyoto Institute of Technology, Matsugasaki, Sakyo-ku, Kyoto, JAPAN
yokoyama@ipc.kit.ac.jp

2 Doshisha University, Kyo-Tanabe, Kyoto, JAPAN

Abstract: The form of the normal pressure distribution imposed on the die by the shrink ring influences the tensile stresses developed at the inner surface of a metal forming die. In order to maximize die life, it is necessary to optimize the shrink ring interference configuration.
This paper deals with design methodology of die dimensions to optimize stress distribution in die by numerical analysis method. This design methodology is the combination of the finite element method and optimization algorithm. The aim of optimum design is to achieve a reduction in stress and consequent improvement in die life.

1. Introduction

The study considers the optimization of the shrink ring configuration for minimizing the stress concentration at the fillet radius in the bore of a reinforced metal forming die. It was shown in precious study that an effective fitting pressure distribution is achieved when the shrink ring extend no further than the bottom of the die cavity. However, in such a design it is necessary also to consider the stress concentration, which occurs at the interface between the die and shrink ring, due to the difference in length. In this respect it is preferable that the shrink ring should extend over the full length of the die insert.
This study deals with the effect of the geometry of the die-shrink ring interface on the stress concentration, for two types of configuration. The aim of course is to achieve a reduction in stress and consequent improvement in die life. The work is mainly analytical, using finite element methods to determine the stresses, but there is a final remark on a practical application, which is quite encouraging.

2. Modelling of Metal Forming Dies

The Metal forming die system is composed of several components. Figure 1 is a schematic diagram of a typical metal forming die. In the previous study, it is shown that a minimum value of tangential stress in the bore is obtained when the length of the shrink ring, applying uniform interface pressure, is less than the impression depth of the die as shown in Fig.2. It is also evident that the circumferential stress decreases with increasing length of stress ring. As a practical approach to the design of the shrink ring, aimed at minimizing stresses both in the die bore and the die-shrink ring interface, two types of interface configuration are investigated in this study.

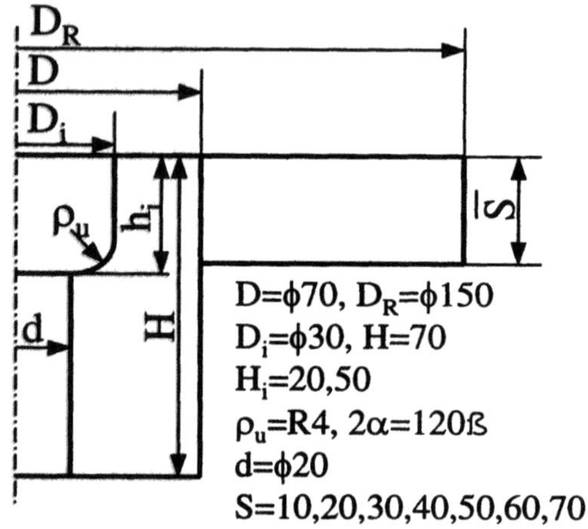

Fig. 1 Target metal forming die

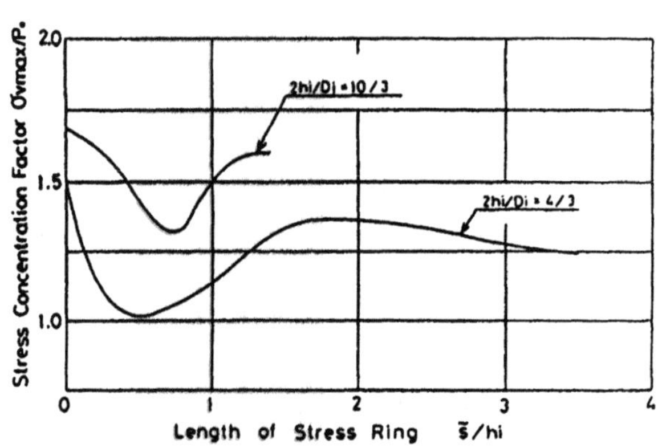

Fig.2 The effect of the length of stress ring on the tangential stress concentration factor

3. Numerical Analysis Approach

In this approach, the finite element method is adopted as a numerical technique. The die system is assumed to be in axisymmetry with contact stresses. The system is constructed of separated blocks with contact faces. Evaluation of the stress concentrations requires a small finite element size and non-linear stress analysis. These requirements lead to difficulties in calculating the stress distribution. The numerical approach is performed using several assumptions to reduce the difficulties of calculation. These assumptions are as follows [1].

(1) Plastic deformation is ignored.
(2) An estimation of contact area is made before calculation which does not change during deformation.
(3) The interface region between contacted dies is represented by an element with a low material constant.

These assumptions allow a linear analysis to be performed to give the stress distribution. Figure 3 shows finite element divisions of the die. The analytical model is limited to half of the die by considering symmetry due to the axisymmetry assumption. From preliminary analysis, the interfaces between die and ring do not open during deformation. This interface region is represented by one layer of elements which have a low material constant as compared with that of the die. The slippage at the interface can be controlled be changing the material constant of interface elements. The working forces vary during the forging process due to deformation of the work piece.

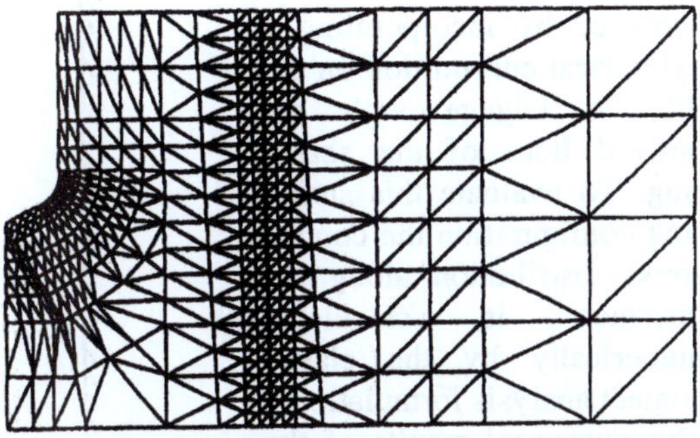

Fig. 3 Finite element division of die

4. Assessment of Overhang

In considering the maximum stress in the tangential and circumferential directions, it has been shown that fracturing is most effectively avoided by making the length of the shrink ring equal to or slightly less than the depth of die cavity. But, because of the difference in length, a stress concentration may be produced at the outer surface of the die, where the die projects beyond the shrink ring.

A practical method of avoiding this stress concentration is by providing a progressive transition in interference at the end of the shrink ring, as shown in Fig.1. The transition is achieved by incorporating a cylindrical end portion on the die, to engage with the tapered bore of the shrink ring. To evaluate this shrink ring configuration the contact stress distribution along the interface is calculated numerically by the elastic contact analysis formulation. The numerical results of the radial stress distribution along the interface are shown in Fig 4, for three values of overhang length of the shrink ring. The value of specific interference fit is $2\delta/D=2.86\times10^{-3}$. The curves shows the stress concentration produced at the end of the ring, the lowest value being obtained when the overhang length is 1mm, corresponding to the ratio s/h = 1.05. Fig. 5 shows the influence of specific overhang length on the contact stress concentration factor, for various values of specific interference.

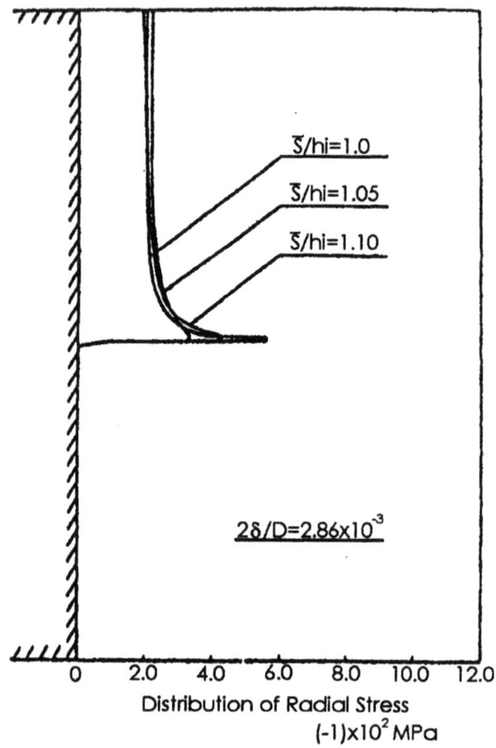

Fig. 4 Stress distribution on interface

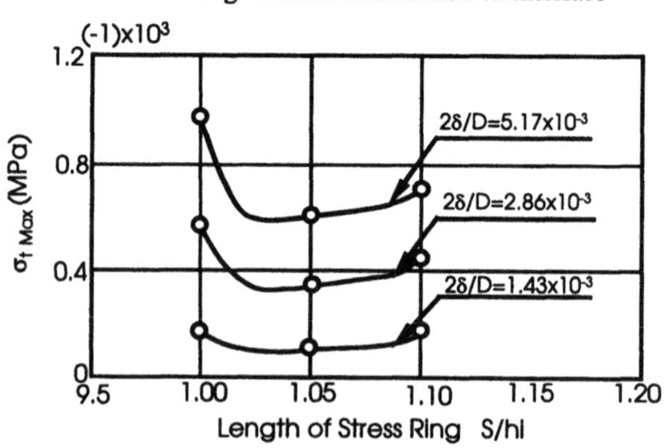

Fig.5 The effect of length of stress ring on maximum stress value

It shows that the contact stress concentration increases somewhat, because the compressive pre-stress energy increases towards the end of the shrink ring, as the overhang length becomes longer than necessary for a progressive transition in interference. The contact stress concentration can be reduced by providing the appropriate transition in interference at the end of the shrink ring.

5. Optimization of Die Dimensions for Uniform Stress Distribution

In the previous section, it is clear that the length of overhang greatly affects the stress distribution in the die and there is an optimum value of overhang length to realize a high durability die system. Further, recent investigations have shown that it is possible to obtain an optimum design. This die design aims to produce a uniform stress distribution in the die system. The die enhances a compressive stress at the corner of a shrink ring by elastic deformation and Poisson's effect as shown in the previous section. Suitable dimensions of the die are required to achieve uniform stress distribution. Therefore, the dimensions of the proposed die system are calculated by the optimum analysis method considering elastic deformation and stress transmissibility at interfaces. In this optimization method, the die dimensions are set as design parameters, because these parameters can be changed easily as actual die dimension. The optimum condition is to realize the uniform stress distribution in the die system during the working process. Therefore, the objective function is set as follows.

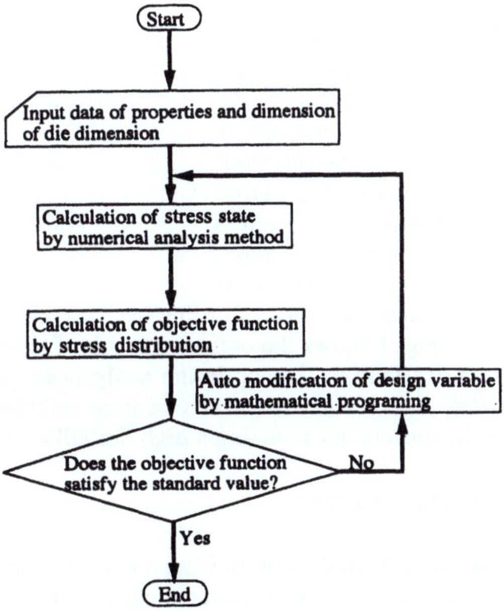

Fig.6 Flowchart of optimum analysis

$$\textit{Objective function} = \sum \left| \sigma_i - \sigma_j \right|$$

where σ is radial stress in die. The steepest descent method is used as an optimization algorithm. Fig 6 shows the flowchart of the optimum analysis.

For the optimization, the length (S) of the shrink ring will be the design parameter in this study. To investigate the effectiveness of the shrink rings, numerical methods are used in applying elastic contact analysis and stress analysis. In this procedure, finite element division is changed automatically according to change the length of theshrink ring.

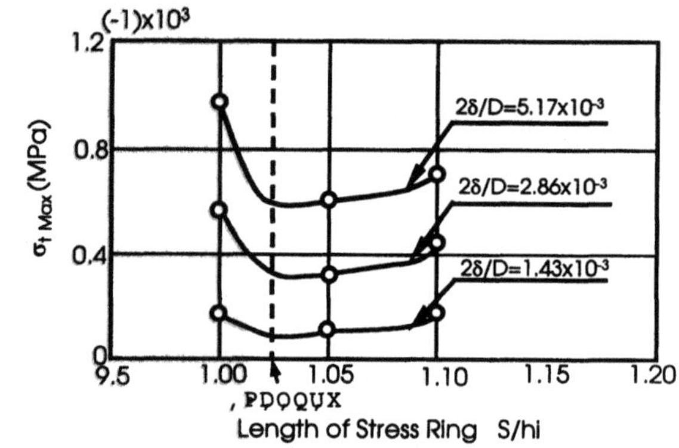

Fig.7 Optimum result of overhang length

Fig. 7 shows the optimum result of length of stress ring to achieve uniform stress distribution. The optimum analysis suggests the best length of stress ring is 1.0269. This result shows the optimum technique is effective to estimate adequate die dimensions for realizing a high durability die system.

6. Conclusions

A numerical method is developed to estimate the die dimensions of the metal forming dies using conventional elastic analysis. This numerical method consists of adding a genetic algorithm of an optimum analysis to calculate the optimum conditions. This modifies the conventional elastic analysis for the stress analysis. The proposed optimum operation is a great candidate for an optimum design method of this design problem such as the stress concentration.

Reference

[1] A.YOKOYAMA, K.AKAMATSU, J.B.HAWKYARD, T.HITAI(1995), Proceedings of the Thirty-first International MATADOR Conference, p.389

Surface Topography and Lubrication in Ring Compression Tests

Z.M.Hu and T.A.Dean

School of Manufacturing and Mechanical Engineering
The University of Birmingham, Birmingham B15 2TT, UK
Email: Z.M.Hu@bham.ac.uk, T.A.Dean@bham.ac.uk

Abstract: Lubricants play an important role in metal forming operations by reducing friction, increasing tool life and affecting the surface finish of formed components. Friction and surface conditions are affected by the surface topography of the workpiece and tools, but sources of information on this topic are relatively sparse. This paper investigates the friction behaviour and surface topography using various lubricants and different workpiece surface textures in ring compression tests.

1. Introduction

Lubricants are widely used in metalforming operations to; separate workpiece and tool surfaces, reduce interface friction, ease metal flow in order to produce sound components and increase tool life [1]. The emphasis of this paper is on the effects of surface topography on friction and surface finish when lubricant is present during forming.

Over the years, lubricants have been conveniently tested using ring compression to determine friction [2], but not much attention has been given to the conditions of surfaces that lubricants are applied to and the influence of the surface topography is not fully understood. Schey [3] demonstrated the effect of different machined surfaces but did not examine in detail the effect of surface topography as such. To date most studies of surface topography and lubricants have been concerned with aspects ranging from lubricant trapping to real area of contact. Early theoretical work by Kudo [4] suggested that friction was affected by lubricants trapped between asperities. This is an indication of the significance of surface topography. Later Nelleman [5] investigated the effect of different surface topography geometries by varying asperity angles and concluded that asperity angle had little effect on frictional stress. More recently, Geiger *et al* [6] identified two types of lubricant pockets, closed and connected, and were able to determine the proportion of surface area at the interface by analysis of the surface topography. Post forged surface finish is also a subject of increasing importance, particularly in net-shape forming. Although previous work has shown that friction is most effectively minimised when the die-workpiece interface is kept completely separated by a lubricant, the resulting post forged surface finish can be relatively poor. Two types of surfaces; one with a random texture and the other with a directional texture, were used by the present authors [7] to study surface topography, friction and lubrication using ring compression and rod extrusion with mineral oil as lubricant. The results suggested the amount of oil lubricant retained increases with decreasing initial roughness on random surfaces (Polished and EDMd), which in turn led to decreasing values of friction in ring compression. The turned surface initially retained more lubricant than other textures, which resulted in great reduction in friction in ring compression, but the resultant surface roughness was high. This paper continues the investigation of the friction behaviour and surface topography using various lubricants and different workpiece surface roughness in ring compression tests. Workpieces are made of commercially pure aluminium and the surfaces are prepared by turning to different levels of roughness or polishing.

2. Experimentation

2.1 Test specimens and tools

The test specimens were rings made from commercially pure aluminium bars with an external diameter of 30 mm, an internal diameter of 15 mm and height of 10 mm. Forging dies were made from AISI L6 and were hardened to approximately 55 HRC and polished to a sub-micro surface finish and flatness of one lightband. The end faces of the rings were manufactured to different degrees of roughness using polishing and turning. The initial parameters of the rings are listed in Table 1.

Table 1 Initial surface parameters of ring specimen

Machining method	Polish	Smooth turning	Medium turning	Rough turning
Roughness R_a (μm)	0.167	0.48	1.00	3.18
Skewness S_{sk}	-0.188	0.524	0.243	0.476
Kurtosis S_{ku}	4.444	2.809	1.842	1.962

2.2 Lubricating and Upsetting

2.2.1 Lubricating

Effort was made to apply lubricants on the surfaces with similar amount using the following methods.

Oil (liquid lubricant)

Mineral oil Shell Tellus 23 was applied on both end faces of ring. The ring was immersed in the lubricant and removed immediately. Then the ring was hung for one minute before it was put on a cone to allow extra oil to drip off. The ring was then turned over and allowed to rest for another minute [7].

Lanolin (viscous lubricant)

Lanolin was used only for the turned surfaces. The lubricant was spread on a clean piece of Perspex to form a thin even film. Then the ring was placed carefully down onto the film of lanolin and twisted until the surface was covered with lubricant. A flat blade was used to scrape the excess lubricant from the face.

Gardolube (solid luricant)

Gardolube refers to an organic polymer based lubricant Gardolube VP 10047/2. It is an aqueous dispersion used for cold forming. It represents an alternative to molybdenum disulphide graphite dispersions and forms a transparent, touch resistant film with similar separating and sliding properties to such materials. In application, the lubricant was diluted to 1 part lubricant: 2 parts water. The rings were then dipped entirely in the diluted lubricant solution and threaded onto a piece of clean steel bar. The bars of steel containing the rings were placed over a beaker and left to dry for 24 hours.

2.2.2 Upsetting

The lubricated ring specimens with various textured surfaces and roughness were compressed between two flat polished dies to one of three height reduction 20, 40 or 60%, at a strain rate of 0.1 s^{-1}. The die surfaces were cleaned with trichloroethane between forging operations.

2.3 Friction and Surface Measurements

The internal bores of specimens were filled with plasticine and average internal diameters were calculated according to the volume of plasticine and the forged height. The coefficient of friction was calculated using a standard calibration chart for each case [2]. The surface of the contacting faces was measured using a 3-D Form-Talysurf stylus machine after cleaning the ring. Rings forged with mineral oil/lanolin were cleaned in an alcohol based solvent in the ultra-sonic cleaner. Those forged with Gardolube were cleaned using ammonia solution, as recommended by the lubricant manufacturer.

3. Results

3.1 Friction

Fig.1 shows the changes of friction coefficient with increasing reduction in height for different surfaces and lubricants. In all cases, the coefficient of friction increases with reduction in height. When oil was used, there was no large dissimilarity between polished surface, smooth, medium and rough turned surfaces. At low to medium degree of deformation, the friction of the rings forged with lanolin is higher than that forged with oil. But they were close for higher deformation. Rings forged with Gardolube produced far higher friction than all of the others.

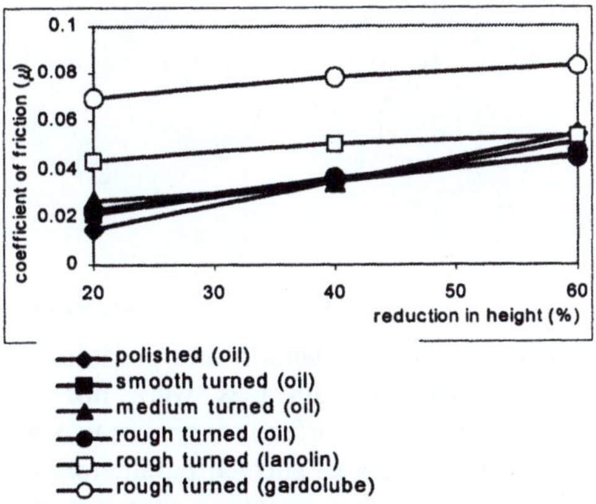

Fig.1 Coefficient of friction from various conditions

3.2 Surface topography

The axonometric plots of the surfaces of various textures before and after forging using oil as lubricant are shown in Fig.2. Physical changes on the surfaces are easy to identify by visual examination. During forging, the polished surface experienced severe roughening, especially at early stages of deformation, 20% and 40% reduction. The surface roughness increased during deformation for all three types of turned surfaces also when mineral oil was used as the lubricant. The values of starting roughness were different, but they reach similar values at 60% deformation. But the surface changes were different when other lubricants were used in the process. Fig.3 shows the axonometric plots of the changes of rough turned surface

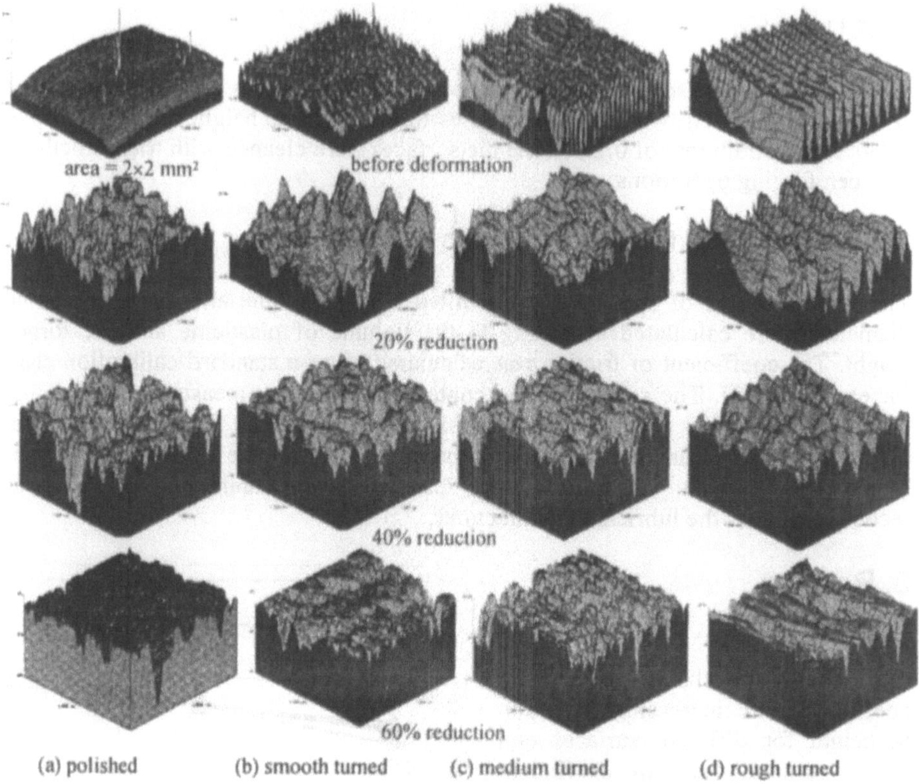

Fig. 2 Changes of various surfaces during upsetting (oil as lubricant)

with mineral oil, lanolin and gardolube as lubricant. When lanolin and gardolube were used, the surfaces were not roughed by forging and a very obvious improvement of surface roughness was achieved through the use of Gardolube.

The changes of surface parameters of average roughness R_a , skewness S_{sk} and kurtosis S_{ku} for various surfaces and lubricants are shown in Fig.4. Wherever mineral oil was used, there was an increase of R_a value for all surfaces, with the maximum of about 10 times increase for the smooth surfaces (polished and smooth turned). R_a values decreased from 3 to 1 μm when gardolube was used while the changes were small in the case of lanolin.

Changes in skewness, S_{sk}, a measure of the asymmetry of surface deviations about the mean plane, can be also seen in Fig.4 (b). The skewness decreased with deformation for all types of surfaces and lubricants. That is, the surfaces initially dominated by peaks were transferred to ones dominated by toughs after deformation. As reported previously [7], the turned surfaces show a near linear decline from positive to negative skewness.

Kurtosis, a measure of the peakedness or sharpness, appears to have an almost exponential relationship with deformation.

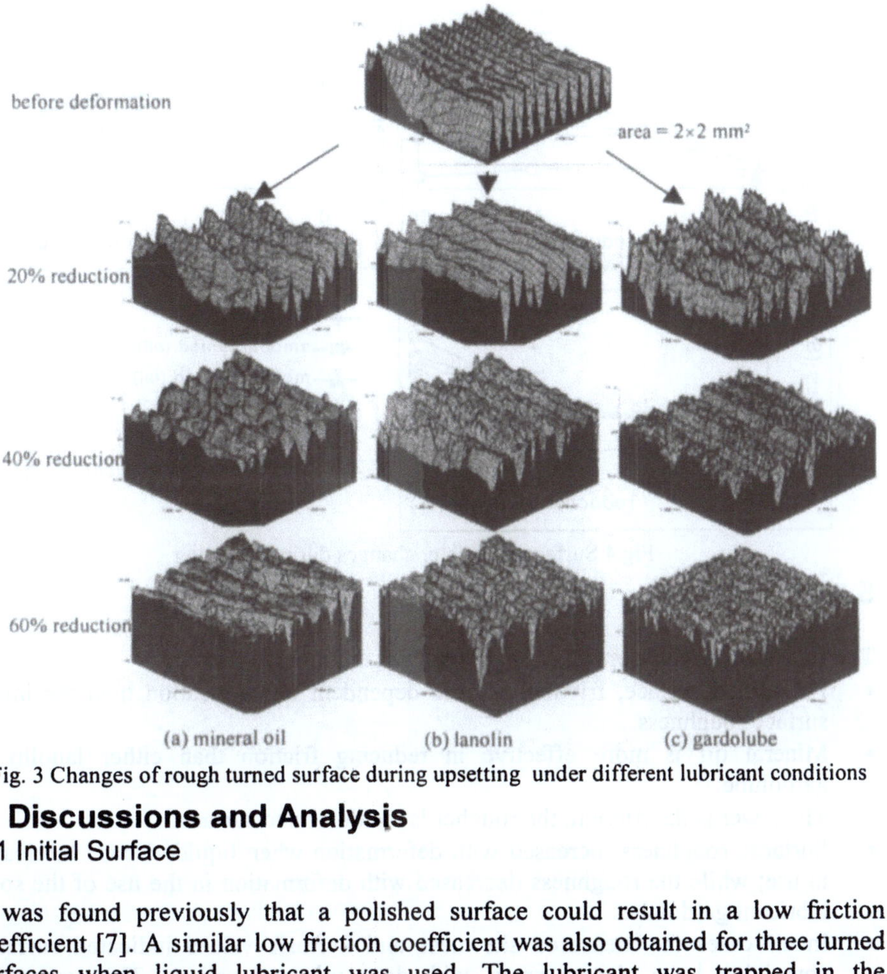

Fig. 3 Changes of rough turned surface during upsetting under different lubricant conditions

4. Discussions and Analysis

4.1 Initial Surface

It was found previously that a polished surface could result in a low friction coefficient [7]. A similar low friction coefficient was also obtained for three turned surfaces when liquid lubricant was used. The lubricant was trapped in the concentric turned grooves to form static pockets [6], which are extremely effective in minimising friction [4]. The values of the coefficient of friction range from approximately 0.015 to 0.085 for polished and turned surfaces. The differences would have very little effect on loads and stresses during most industrial forming operations, but the influences on the surface topography and final surface finish are large. Both the polished and turned surfaces became roughed after compression with the use of oil. But the polished surface was better than the turned surfaces at the largest deformation of 60%.

4.2 Different types of lubricant

Despite the tendency of increasing friction when gardolube and lanolin were used, the benefit was from the surface finish. When lanolin was used, surface was roughened only at the later stage of deformation while the use of gardolube greatly improved the roughness, resulting in a reduction of R_a value from 3.18 μm to about 1 μm.

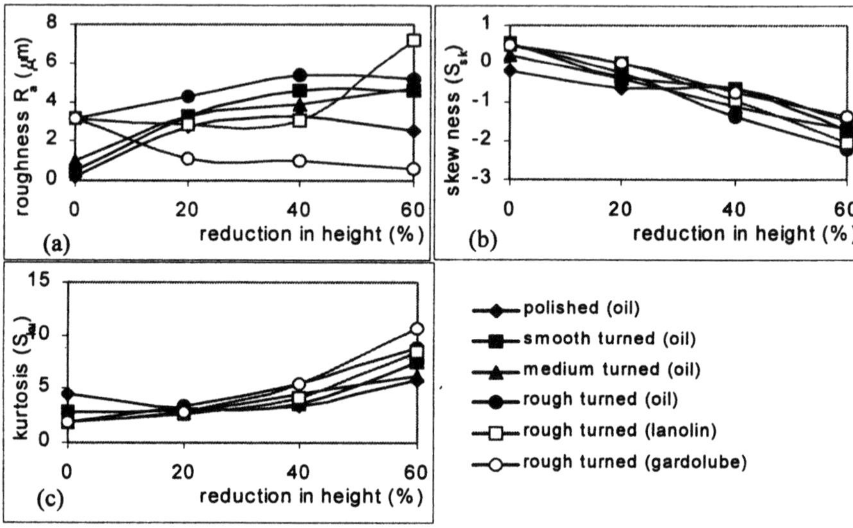

Fig.4 Surface parameter changes during upsetting

5. Conclusions

The following conclusions may be drawn from the investigation.

- For turned surface, friction is more dependent on lubrication than the initial surface roughness.
- Mineral oil is more effective in reducing friction than either lanolin or gardolube.
- The lower is the friction, the rougher is the deformed surface.
- Surfaces roughness increased with deformation when liquid/viscous lubricants in use; while the roughness decreased with deformation in the use of the solid lubricant gardolube.
- In all cases the measured skewness and kurtosis values indicated surfaces containing large plateau areas, with deep pits and valleys. This was more pronounced in the surfaces with higher initial roughness.

Acknowledgements

The authors would like to thank Dr. K.S. See and Mr. J. Smith for their efforts in providing the experimental results.

References

1. Schey, J.A., Tribology in Metalworking, friction, lubricant and wear, American Society for Metals, 1983
2. Male, T. and Cockcroft, M.G., J. of Inst. of Metals, 93, 1964, p38
3. Schey, J.A. et al, Eoro Trib'77, 2nd European Tribology Congress, Paper 67
4. Kudo, H, Int. J. Mech. Sci., 7, 1965, p383
5. Nelleman, T., Bay, N. and Wanheim, T., Wear, 1977, p45
6. Geiger, M. et al, ICFG plenary meeting, Sept. 1995, Denmark
7. Hu, Z.M. and Dean, to be published in Int. J. Mach. Tools Manufact.

METAL FORMING III

Spring Back and Secondary Yielding Analyses Using A Physically Based Constitutive Model

X. Peng[1] and R. Balendra[2]

[1] DMEM, University of Strathclyde, G1 1XJ, xianghe.peng@strath.ac.uk
[2] DMEM, University of Strathclyde, G1 1XJ, r.balendra@strath.ac.uk

Abstract: A material user subroutine, UMAT, was programmed for embedding in the commercial software, ABAQUS, together with a new constitutive model for large elastoplastic deformation and the corresponding numerical algorithm. The spring back and secondary yielding of cylindrical billets in elastic tools during upsetting and closed-die upsetting were analysed. The comparison between the computed and experimental results shows satisfactory agreement.

1. Introduction

Spring back and secondary yielding phenomena are significant in precision forming. They not only occur during unloading, but may also occur during loading while local unloading take place. These phenomena are related to the material properties and the deformation history of the workpiece and the deflection of tools. The basic challenge to the constitutive model for precision forming is that it should be able to describe realistically not only the triaxial nonproportional thermomechanical behaviour caused by complex histories of stress and temperature in a wide range, but also the material response during unloading and reloading processes. In the past decades, great progress has been achieved in the development of new constitutive relations [1-5], but few of the achievement were extended to the case of large elastoplastic deformation and applied to metal forming analysis.

A material user subroutine UMAT was programmed for embedding in the commercial software ABAQUS together with a new thermomechanically consistent 3-D constitutive model for large elastoplastic deformation and the corresponding numerical algorithm. It satisfies the fundamental constraints of irreversible thermodynamics and takes into account the energy stored in the residual microstress fields. Strain-rate and thermal effects and the temperature dependence of material property can also be taken into account. Spring back and secondary yielding phenomena during upsetting and closed-die upsetting were analysed. Compared with the result obtained by assuming rigid tools, it is shown that the rigid-die analysis may be inadequate in the design of precision components.

2. Constitutive Model and Numerical Algorithm

Based on previous work[6-8], a set of constitutive equations were obtained for finite elastoplastic deformation by using the mechanical model shown in Fig.1, in which **e** and **s** denote rthe deviatoric components of logarithmic strain ε and the conjugated stress **T**, espectively. **e** is further partitioned into an elastic component $\mathbf{e}^e$ representing the response of the spring G (with elastic shear modulus G) and a plastic component $\mathbf{e}^p$ representing the response of the parallel branches. In the rth branch, the dashpot-like block B_r (with plastic "damping" coefficient b_r) and the spring E_r (with stiffness E_r) are used to describe the rth dissipative mechanism. It should be noted that the spring E_r (r=0, 2,..., n) is related to residual microstress fields and does not contribute to the macroscopic elastic property of material. The energy stored in E_r represents that stored in the residual microstress fields in plastically deformed materials due to its non-homogeneous nature. Assuming the material is initially isotropic and plastically incompressible, the following relations [6-8] would apply,

$$\mathbf{s} = \sum_{r=0}^{n} \mathbf{Q}^{(r)}, \qquad \mathbf{Q}^{(r)} = E_r(\theta)(\mathbf{e}^p - \mathbf{p}^{(r)}) \tag{1}$$

with

$$\mathbf{e}^p = \mathbf{e} - \mathbf{e}^e, \qquad \mathbf{e}^e = \frac{\mathbf{s}}{2G(\theta)}, \tag{2}$$

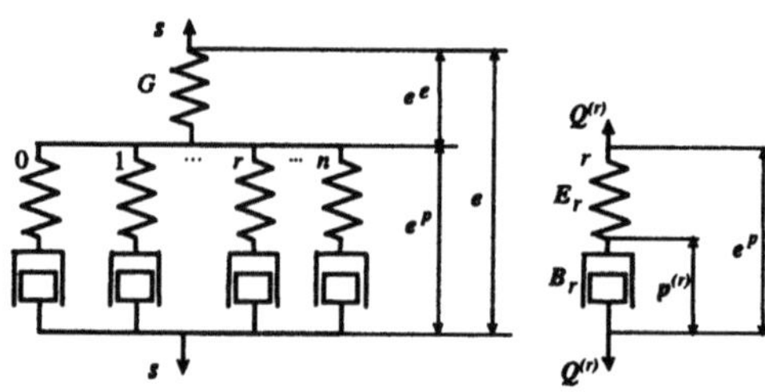

Fig.1 Mechanical model of constitutive equation for large elastoplastic deformation

where θ denotes temperature, $\mathbf{p}^{(r)}$ and $\mathbf{Q}^{(r)}$ represent respectively the rth internal variable and the corresponding generalized force satisfying the dissipation inequality $\mathbf{Q}^{(r)}:d\mathbf{p}^{(r)} \geq 0$ (r=0, 2 ..., n). $\mathbf{Q}^{(r)}$ can phenomenologically be assumed to relate $\mathbf{p}^{(r)}$ by the following relationship [8]:

$$\mathbf{Q}^{(r)} = b_r(\theta, \varsigma, \dot{\varsigma})\frac{d\mathbf{p}^{(r)}}{d\zeta}, \qquad (r = 0,1,2,\ldots,n) \tag{3}$$

where $d\zeta = \sqrt{d\mathbf{e}^p : d\mathbf{e}^p}$. Assuming $b_r(\theta, \varsigma, \dot{\varsigma}) = b_r^0(\theta) f(\varsigma) k(\dot{\varsigma})$, equation (3) can be rewritten as

$$\mathbf{Q}^{(r)} = b_r^0(\theta)\frac{d\mathbf{p}^{(r)}}{dz}, \qquad \text{where} \quad dz = \frac{d\varsigma}{f(\varsigma)k(\dot{\varsigma})}, \tag{4}$$

where z is non-negative and increases monotonically and can be used as generalized time in plastic deformation analysis, $f(\zeta)$ is the hardening function and $b_r^0(\theta)$ defines the initial state of the material which varies with temperature. $k(\dot{\varsigma})$ describes strain-rate effect. In Fig.1, C_r and b_r should retain positive values since stored energy and dissipation are non-negative by nature. The combination of equations (1) and (4) provides

$$d\mathbf{Q}^{(r)} = E_r d\mathbf{e}^p + (-\alpha_r dz + \frac{1}{E_r}\frac{dE_r}{d\theta}d\theta)\mathbf{Q}^{(r)} \tag{5}$$

and

$$d\mathbf{s} = \sum_{r=0}^{n} d\mathbf{Q}^{(r)} = C d\mathbf{e}^p + d\mathbf{H} \tag{6}$$

where $$C=\sum_{r=0}^{n} E_r, \qquad d\mathbf{H}=\sum_{r=0}^{n}\left(-\alpha_r dz+\frac{1}{E_r}\frac{dE_r}{d\theta}d\theta\right)\mathbf{Q}^{(r)} \tag{7}$$

and $\alpha_r = E_r / b_r^0$. The differential form of deviatoric elastic response can be obtained from equation (4)

$$d\mathbf{s} = 2G(\theta)(d\mathbf{e} - d\mathbf{e}^p) + \frac{G'(\theta)}{G(\theta)}\mathbf{s}d\theta \tag{8}$$

and the volumetric response is assumed elastic and can be expressed as

$$T_k^k = 3K(\theta)[\varepsilon_k^k - 3\alpha(\theta)(\theta-\theta_0)] \tag{9}$$

$$dT_k^k = 3K(\theta)d\varepsilon_k^k + \left[\frac{K'(\theta)}{K(\theta)}T_k^k - 9\alpha'(\theta)K(\theta)(\theta-\theta_0) - 9\alpha(\theta)K(\theta)\right]d\theta \tag{10}$$

in which K and α denote elastic bulk modulus and the coefficient of thermal expansion, θ and θ_0 the current and reference temperatures. In order to reduce the numerical error[9], we define

$$d\bar{z} = \alpha_r dz - \frac{1}{E_r}\frac{dE_r}{d\theta}d\theta \tag{11}$$

and derive the incremental form of equation (6) as

$$\Delta\mathbf{s} = \sum_{r=1}^{n}\Delta\mathbf{Q}^{(r)} = A\Delta\mathbf{e}^p + \mathbf{B}\Delta z + \mathbf{R}\Delta\theta \tag{12}$$

where $$A=\sum_{r=1}^{n}k_r E_r, \quad \mathbf{B}=-\sum_{r=1}^{n}k_r\alpha_r\mathbf{Q}^{(r)}(\bar{z}_n), \quad \mathbf{R}=\sum_{r=1}^{n}k_r\frac{E_r'(\theta)}{E_r}\mathbf{Q}^{(r)}(\bar{z}_n), \quad k_r=\frac{1-e^{-\Delta\bar{z}}}{\Delta\bar{z}}. \tag{13}$$

Combining equation (12) with equation (8) yields the following deviatoric elastoplastic constitutive equation

$$\Delta\mathbf{s} = 2G_p\Delta\mathbf{e} + \lambda\mathbf{B}\Delta z + \varphi\Delta\theta, \tag{14}$$

where $$\lambda=\left[1+\frac{A}{2G}\right]^{-1}, \qquad 2G_p=\lambda A, \qquad \varphi=\frac{G_p}{G^2}G'\mathbf{s}+\lambda\sum_{r=1}^{n}k_r\frac{E_r'}{E_r}\mathbf{Q}^{(r)}. \tag{15}$$

The incremental constitutive equation can be derived as

$$\Delta T_j^i = \mathrm{E}_{jl}^{ik} : \Delta\varepsilon_k^l + \Theta_j^i\Delta\theta \tag{16}$$

where

$$E_{jl}^{ik} = 2G_p\delta_l^i\delta_j^k + (K-\frac{2}{3}G_p)\delta_j^i\delta_l^k + \frac{2}{a}(G-G_p)HB_j^i\frac{(\Delta\mathbf{e}^p)_l^k}{f^2k^2\Delta z} \tag{17}$$

and

$$\Theta_j^i = [\mathbf{M}^{-1}]_{jl}^{ik}\left[\left(\frac{\lambda G}{2G^2 f^2 k^2}B_k^l\frac{[\Delta\mathbf{e}^p]_m^n}{\Delta z}T_n^m + \varphi_k^l + \left(\frac{K'}{K}T_m^m - 3\alpha' K(\theta-\theta_0) - 3\alpha K\right)\delta_k^l\right]\right. \tag{18}$$

$$H=\frac{\lambda}{2Gf^2k^2},\quad a=1+HB_j^i\frac{(\Delta e^p)_i^j}{\Delta z},\quad M_{jl}^{ik}=\delta_l^i\delta_j^k+\frac{\lambda}{2Gf^2k^2}B_j^i\frac{(\Delta e^p)_l^k}{\Delta z} \tag{19}$$

In the updated Lagrangian approach, the increment of Cauchy stress increment can be derived by assuming that the current configuration coincides with the reference configuration

$$\Delta\sigma=\Delta\mathbf{T}+\Delta\Omega\cdot\sigma-\sigma\cdot\Delta\Omega-\Delta\varepsilon_k^k\sigma \tag{20}$$

where $\Delta\Omega=\omega\Delta z$, ω denoting spin. In isothermal and axisymmetrical case, equation (16) can be expressed as

$$\{\Delta T\}=[D]\{\Delta\varepsilon\},\qquad \text{where}\qquad [D]=[D_e]+\frac{2(G-G_p)}{a}[D_p] \tag{21}$$

$$\{\Delta T\}=(\Delta T_r,\Delta T_\theta,\Delta T_z,\Delta T_{rz})^T,\qquad \{\Delta\varepsilon\}=(\Delta\varepsilon_r,\Delta\varepsilon_\theta,\Delta\varepsilon_z,\Delta\gamma_{rz})^T \tag{22}$$

$$[D_e]=\begin{bmatrix} C_1 & C_2 & C_2 & 0\\ & C_1 & C_2 & 0\\ & & C_1 & 0\\ sym. & & & G_p\end{bmatrix} \tag{23}$$

$$[D_p]=H(B_r,B_\theta,B_z,B_{rz})^T(\frac{\Delta e_r^p}{\Delta z},\frac{\Delta e_\theta^p}{\Delta z},\frac{\Delta e_z^p}{\Delta z},\frac{\Delta e_{rz}^p}{\Delta z}) \tag{24}$$

3. Spring Back and Secondary Yielding Analysis

Based on the proposed constitutive equation and the corresponding numerical algorithm, a material user subroutine, UMAT, was programmed and embedded in the commercial software, ABAQUS, with which the spring back and secondary yielding during upsetting and closed-die upsetting were analyzed.

In metal forming analysis, tools were often regarded as rigid bodies, which is in general inadequate if very high pressure and severe friction are involved, especially in the case of precision forming. An upsetting process with elastic tools was investigated. The specimen with initial diameter D_0=13mm and height H_0=13mm was compressed between two elastic parallel platens with diameter 18mm and height 10mm (see Fig.2). The die was made of tungsten carbide with Young's modulus 618 GPa and Poisson's ratio 0.21. The specimen was made of mild steel, the material constants of which were determined as follows by choosing n=3, and $f(z)=d-(d-1)\exp(-\beta z)$[6] and ignoring strain-rate effect (see equations (1) and (4))

E=192 GPa, ν=0.3, $E_{1,2,3}$=1050, 18.4, 3.1 GPa, $\alpha_{1,2,3}$= 5200, 360, 35, d= 2.51, β=2.0.

Coulomb frictional coefficient, μ=0.15, was prescribed at the die-billet interface [10]. The quarter-specimen analysis was conducted because of symmetry; elements of type CAX4R were used for both billet and die. The computation is stroke-controlled and the displacement of the upper surface of platen was 2.08mm (32% of the original semi-height of billet). Comparison of the load-displacement relationship between the computed and experimental measurement [10] is shown in Fig.3. The deformed mesh at reduction of 32% height is shown in Fig.4, where the deformation of the die is amplified to 50 times its original value for ease of observation. Fig.5 shows, respectively, the computed profiles of the upper surface of billet at full loading and unloading, in which it can be seen that there is a marked difference between the results corresponding respectively to elastic and rigid dies.

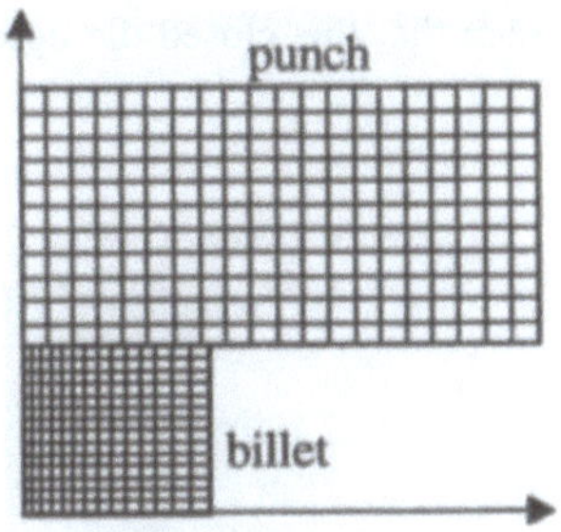

Fig.2 Upsetting set-up

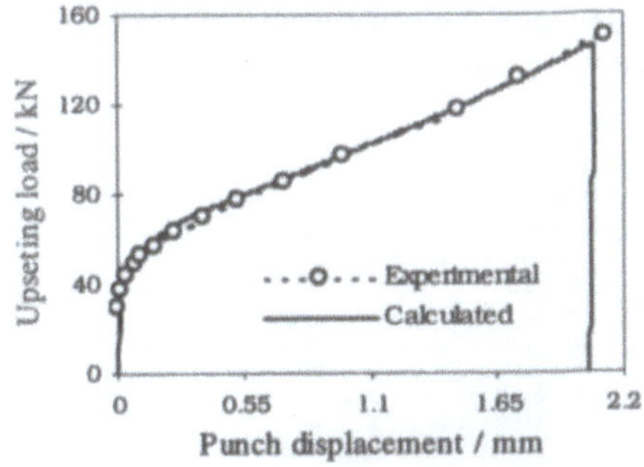

Fig.3 Load-displacement relationship

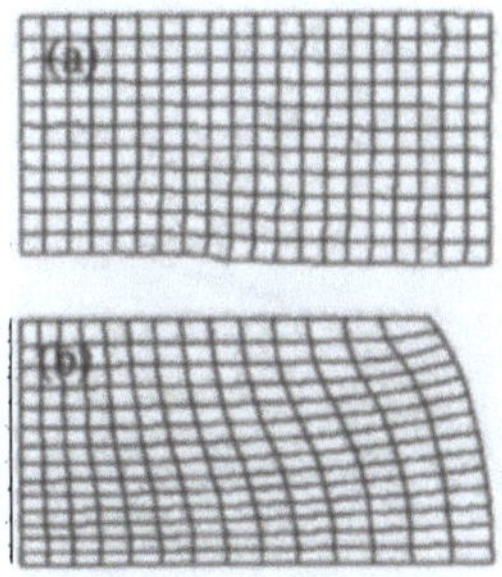

Fig.4 Deformation of die (a) and billet (b)

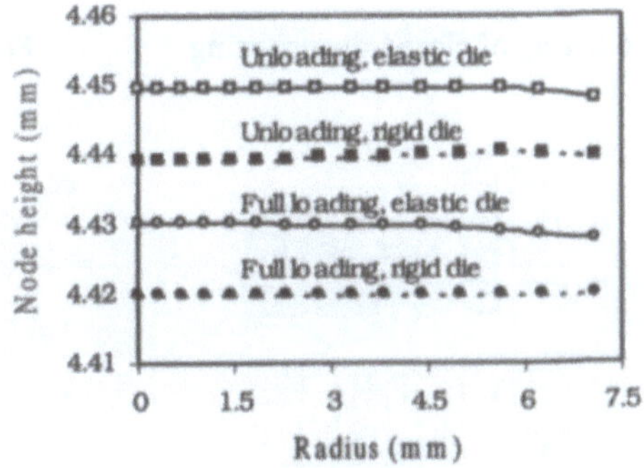

Fig.5 Spring back of billet

In closed-die upsetting, the billet undergoes complicated plastic deformation history including secondary yielding due to spring-back of tools. The set-up of the closed-die upsetting simulated is shown in Fig.6, in which the initial diameter and height of billet were 20.029mm and 20mm, respectively, and those of die were 120mm[11]. The frictional coefficient was prescribed as 0.09[11] between contact interfaces. The material of the billet was aluminium, the experimental [11] and fitting σ–ε^p curves of which are shown in Fig.7, where the corresponding material constants were identified as follows

$$E=70.6\text{ GPa},\quad \nu=0.345,\quad E_{1,2,3}=141,3.7,1.3\text{ GPa},\quad \alpha_{1,2,3}=2000,150,85,\quad f=1.0$$

Both the die and the punch were made of steel of Young's modulus 210GPa and Poisson's ratio 0.3. The quarter-specimen analysis was conducted and the elements of type CAX4R were used for billet, punch and die (see Fig.6) in the simulation. The upsetting process consisted of three phases: loading (punch moving downward), unloading (punch moving upward) and removing contact between billet and die. The original profile of billet and the profiles of billet at full loading, unloading and after removing contact are shown in Fig.8(a), (b), (c) and (d), respectively, in which the deformation was amplified to 30 times its original value for ease of observation. It can be seen that the billet underwent axial deformation confined by the elastic deformation of die in the first phase; in the second phase the spring back of the die makes the billet deform back, which can be observed by comparing Fig.8(c) with Fig.8(b); during the third phase the removal of confined pressure results in re-distribution of stress in billet and may result in new plastic deformation. Fig.9 shows the relationship between the final diameter and the corresponding maximum punch load, Compared with the experimental measurement, it can be seen that this closed-die upsetting process was satisfactorily replicated.

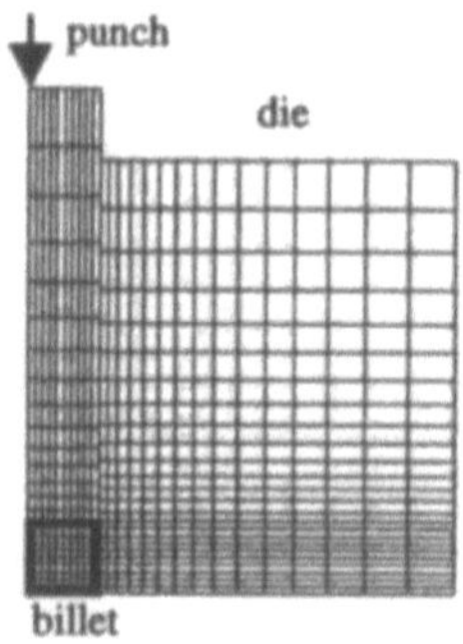

Fig.6 Set-up of closed-die upsetting

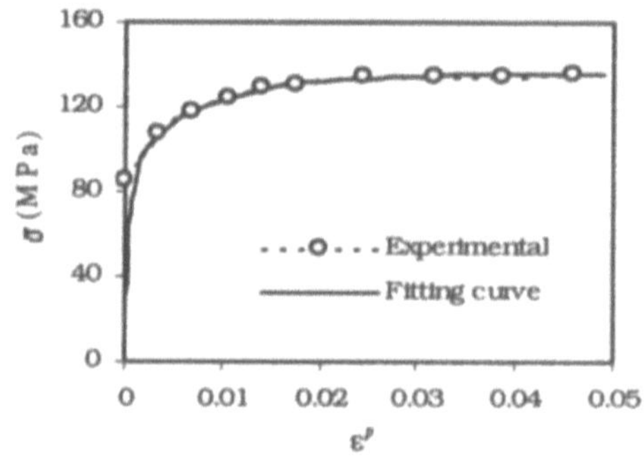

Fig.7 Tensile stress-plastic strain curves

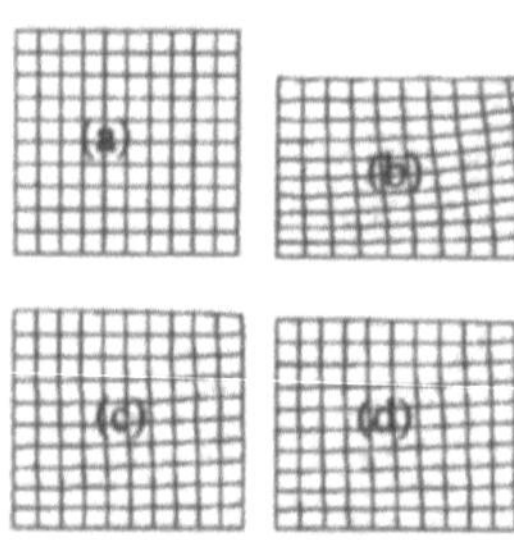

Fig.8 Profiles of original and deformed billet

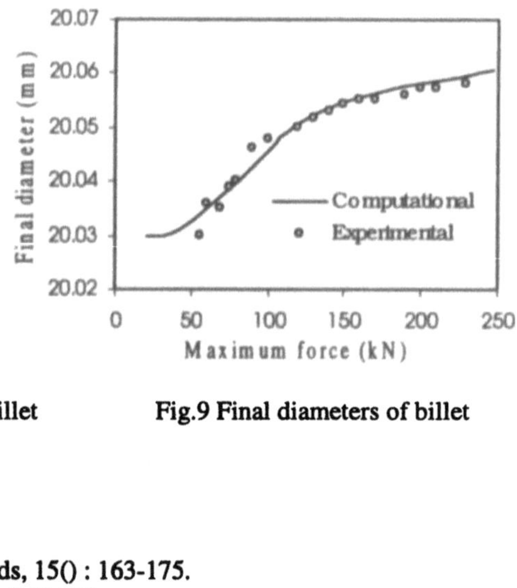

Fig.9 Final diameters of billet

References

[1] Mroz Z., 1967, *J. Mech., Phys.*, Solids, 15() : 163-175.

[2] Valanis K.C., 1970, *Arch. Mech.*, 23: 517-551.

[3] Dafalias Y. F. and Popov E. P., 1975, *Acta Mechanica*, 21: 173-192.

[4] Chaboche J.L. and Rousselier G., *J. Pressure Vessel Technology*, 105(1983): 153-164.

[5] Mcdowell D. L., 1985, *J. Applied Mechanics*, 52: 298-308.

[6] Jinghong F. and Xianghe P., 1991, *J. Engng. Mat. Tech.*, 113: 254-262.

[7] Peng X. and Ponter A. R. S., 1994, *Int. J. Solids Struct.*, 31: 1099-1111.

[8] Peng X., Zeng X. and Fan J., 1998, *Int. J. Solids Struct.*, 35: 2733-2747.

[9] Peng X. and Fan J., 1993, *Computers & Structures*, 47: 313-320.

[10] Lin S.Y., 1990, *Journal of Material Processing Technology*, 86: 73-80.

[11] Rosochowski A. and Balendra R., Secondary yielding of forged components due to unloading, submitted to *Journal of Material Processing Technology*.

Springback Evaluation In Basic Sheet Metal Stamping Processes

L. D'Acquisto*, L. Fratini**
University of Palermo - viale delle Scienze 90128 Palermo - ITALY
* DIMA - *e-mail*: dacquisto@dima.unipa.it
** DTPM - *e-mail*: abaqus@dtpm.unipa.it

Abstract: In the present paper a measurement technique based on the shadow moiré method has been set up in order to evaluate the springback phenomenon in deep drawing operations. Two proper coefficients are introduced to quantitatively describe the effect considered. In particular the proposed technique is able to acquire the final profile of the drawn parts along one or more generatrices, once extracted out from the dies and to compare such profile with the ideal one derived from the dies CAD geometry. The proposed technique has been applied to an axysimmetrical deep drawing operation and moreover the influence of the blankholder force level and of the lubricating conditions have been taken into account.

1. Introduction

The optimization of forming processes has a fundamental role aimed to obtain sound components free from both shape defects and fractures. As far as the former is concerned, the springback phenomenon should be taken into account too, since it determines a discrepancy between the die geometry and the final part shape.
At the end of the forming process, as far as the dies are close together, the stress state characterizing the component is in equilibrium with the external actions determined by the die surfaces. Subsequently - as the dies are opened - the internal stress state is not balanced any more and the component undergoes an elastic deformation up to a new equilibrium configuration only characterized by the residual stress state. Such elastic deformation is indicated as "elastic springback".
In literature heuristic models aimed to the quantitative determination of the phenomenon have been often presented [1] referring anyway to simple stamping processes. Actually, they seem to be effective just for limited ranges of operative parameters considered. Furthermore a few researches have applied numerical techniques based on the finite element method for the determination of the final shape of the stamped parts [2]. A deep investigation of these researches shows that, no definitively powerful tools have been presented yet.
For the measurement of out-of-plane surfaces several techniques present useful characteristics [3][4], showing different sensitivity levels and, what is more, utilizing different setups which are very complex and troublesome to be implemented within an industrial environment.
The proposed application requires a measurement technique allowing a simple acquisition, processing and output procedure for the measurement signal, by the use of customary and low cost instruments. Starting from the former considerations, the geometric moiré method seems to be the most effective measurement technique even if it presents a few shortcomings, connected with the characteristics of the measurement to be developed. These shortcomings have been overcome through the proposed image acquisition and processing method. In the present paper the

authors utilize the proposed measurement technique to determine the profile of an axisymmetric deep drawn part along one or more generatrices.

2. Measurement technique

In the present paper the shadow moiré technique is employed. This enables one to obtain the topography of the specimen surface is employed. Its use in the industrial environment requires the full automation of this technique to make it easy to use and economically effective. In recent years, it has been widely employed to investigate strain in electronic components [4], to characterize both initial strain and temperature induced strain.
The intrinsic accuracy of the above mentioned technique is well suited for the required measurements. However, it presents some difficulties in resolving micrometer displacements on a measurement fields of some mm. Classical shadow moiré applications have overcome this problem by acquiring multiple images with a known phase shift between the fringe patterns in the different images. To distinguish concavity or convexity on the surface to be measured, at least two images, i.e. two fringe patterns showing light intensity signals mutually shifted by an amount less than π, have to be compared. Therefore it requires a long time to run while it doesn't improve the resolution and can introduce further uncertainty elements .
To overcome the above mentioned difficulties of the traditional shadow moiré technique, an optical technique to increase the number of fringes in the measurement signal is employed. A carrier signal is then introduced by frequency modulating the optical measurement signal. It enables to improve the sensitivity of the method and fulfills Fourier Transform based signal processing techniques. In particular, it draws significant advantage from the capabilities of filtering the useful frequency bandwidth from those affected from noise. Also when the original optical signal presents less than one fringe, this new technique enables to employ the shadow moiré method, measuring the object profile with no ambiguity about its direction.
A sharp prismatic steel bar, illustrated in fig. 1, is employed. It has the double function to hold up horizontally the specimen to measure and to visualize the carrier fringe pattern on its frontal white surface. Through the image acquisition of the carrier fringe pattern, it allows to directly calibrate the measurement signal and at the same time to obtain a calibration of the dimensional measurement scale of the acquired image. The proposed method then relates distance z between the surface on the prismatic bar and the points on the specimen profile (see fig.1-a) to the phase of light intensity signal visualized on the specimen caused by the mechanical interference between the moiré grid and its shadow projected from the light source.
The following experimental setup has been utilized:

- a white light parallel source
- a moiré grid on glass plate (Graticules SAG4) parallel fringes, pitch 0,127 mm
- a sharp prismatic bar to hold the specimen having measurement surface white opaque painted to visualize the carrier optical signal;
- a polarizing filter (to remove the unwanted reflex on the camera lens);

- a CCD B/W camera (Video Standard CCIR 795x596) with s/n ratio >48 dB
- a camera lens Nikkor AF Micro 60mm, 1:2,8 D;
- a personal computer with 8 bit digitizing board Matrox Comet 768x576 pixel.

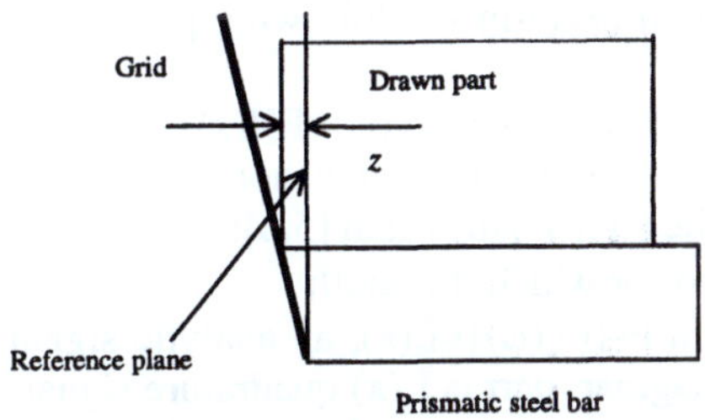

Figure 1: a) specimen–grid lay-out

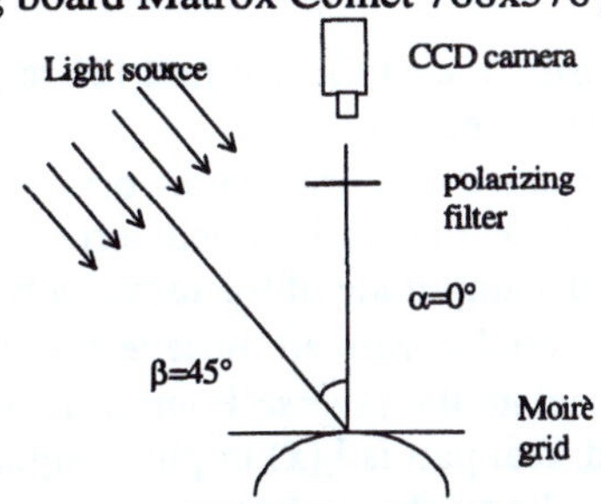

b) image acquisition scheme

3. Digital processing of the light intensity experimental signal

The profile to measure together with a fringe carrier signal gives rise, in presence of noise sources, to a light intensity signal expressed by the following equation:

$$I(x) = I_0 + I_1 \cos[\phi(x) + \omega_c x] + I_n \qquad (1)$$

when $\omega_c >> \phi(x)/2\pi$, being $I_0(x)$ the background intensity of the signal, $I_1(x)$ the intensity of the modulation term, $I_n(x)$ the intensity of the high frequency added noise and $\omega_c=2\pi f_c$ the carrier frequency (rad/s). The unknown phase $\phi(x)$ contains information related to the profile's geometry.

To ensure an effective use of the Fourier Transform method to a carrier modulated measurement signal, phase $\phi(x)$ should vary slowly with respect to the carrier frequency f_c. Some authors [4] suggest the use of a carrier with frequency at least twice the fastest term contained in the signal to measure. To carry the evaluation of $\phi(x)$ it is also necessary to remove the terms $I_0(x)$ and $I_n(x)$, by filtering, to obtaining the in-phase signal $I_f=I_1 \cos[\phi(x)+\omega_c x]$. Introducing the use of a quadrature signal I_q, the wrapped phase $\phi'(x)$, variable in the range $\pm\pi$ described by the following relation:

$$\phi'(x) = \tan^{-1}(I_q(x)/I_f(x)) \qquad (2)$$

Continuous phase $\phi(x)$ is obtained by the unwrapping process described by equation: $\phi(x) = \phi'(x) + 2k\pi$, with k integer number. This process requires the former knowledge of the signum of k.

In the unwrapping procedure, when an initial fringe pattern with constant spatial frequency is introduced by the use of a carrier signal, the signum of each 2π jump is obtained by evaluating the phase and frequency change in the final fringe pattern with respect to those of the carrier fringe one. The Fourier transform method enables to obtain the wrapped phase along a profile or a surface in a semi-automated manner subtracting $\phi_c(x) = \omega_C x$ from the phase obtained by the application of the Fourier Transform Method to the experimental fringes signal.

4. Measurement signal processing by the Fourier Transform method

The Fourier Transform method [3] requires to perform the following procedure to the measurement signal:
1. to evaluate the Fourier Transform of the acquired signal $Y(\omega)=F[I(x)]$;
2. to remove the background term I_0 and the high frequency noise term I_n, by zeroing the amplitude of the terms outside the signal bandwidth [5,6];
3. to put equal to zero all those terms in $Y(\omega)$ for which it is $\omega<0$;
4. to calculate the inverse Fourier Transform $F\text{-}1[Y(\omega)]=I_a(x)$, an analytic signal is obtained: real part is $I_f(x)$ in-phase signal imaginar part is $I_q(x)$ quadrature signal;
5. to evaluate the unknown wrapped phase $\phi'(x)$ as $\tan^{-1}[Im(I_a(x))/Re(I_a(x))]$, assuming values in the range from $-\pi$ to π.

Applying unwrapping procedure to $\phi'(x)$, the phase signal $\phi''(x)= \phi(x)+ \phi_c(x)$, that represent the unwrapped phase of the carrier modulated measurement signal, is obtained. Then $\phi(x)= \phi''(x)- \phi_c(x)$ is calculated, being $\phi''(x)$ and $\phi_c(x)=\omega_c x$ already known. Therefore, $\phi(x)$ has been obtained without any signum uncertainty. This is the most meaningful advantage obtained using the Fourier Transform technique.

5. Application

The proposed measurement technique has been applied to the determination of the elastic springback which occurs after an axisymmetric deep drawing process. The basic process mechanics is determined by the restraining actions applied on the drawing part as it is pushed inside the die by the punch. Such actions start from the combined effect of two operative parameters: the blankholder force and the lubricating conditions. A too heavy combination of the two effects would determine an excessive action on the blank with subsequent necking phenomena and tearings while a too low combination these effects would lead to insurgence of wrinkles. The deep drawing process considered is characterized by a drawing ratio, (ratio between the diameter of the undeformed blank and the punch diameter) R=1.78. The obtained drawn parts showed an average height equal to 25 mm. The punch and the die radii (Rp and Rd respectively) are equal to 3 mm. An EN 10130-Fe P01 steel has been utilized for the developed tests with an initial thickness equal to 0.5 mm. Different levels of blankhoder force have been taken into account referring to the initial blankholder force on the blank expressed as percentage of the material yield stress (196 N/mm^2). Lubrication between the blank and the dies has been obtained utilizing Teflon, mineral oil and EP grease to highlight the influence of such parameter on the elastic springback. Such phenomenon determines an opening of the cup walls which at the end of the process presented a greater tapering with respect to the closed dies geometry.
In order to develop the measurement of the elastic springback a moiré grid with 200 lines per inch has been utilized (step equal to 0.127 mm). The drawn part has been placed on a proper prismatic steel basement (fig. 1) in order to observe in the same image (fig.2) both the measurement fringes on the specimen and the carrier fringes on the basement. In figure 3 the signal behavior (a) corresponding to the profile

shown in figure 2, is reported together with the average component (b) to be removed and with the signal (c) already filtered from the average component and from the high frequency noise. The fringes on the specimen surface represent the sum of two effects: a) the distance along the z direction between the point of the profile and the visualization plane of the carrier fringes, b) the modulation of the carrier signal obtained sloping the moiré grid with respect to the referring plane (fig. 1a). The following parameters (see fig.1) have been chosen in order to acquire the images utilized in the measurement procedure: lighting angle $\beta=45°$, vision angle $\alpha=0°$, distance between the grid plane and the camera 2000 mm, stop f=32. In the described conditions the assumption of infinity observing point is acceptable and the distorting geometric effect due to the utilized lens can be neglected.
These camera shooting conditions have determined an image resolution equal to 0.18 mm, which is the dimension on the specimen corresponding to a pixel on the token image. on the ground of the equation $z(x) = \phi(x)\, p/2\pi(tg\alpha + tg\beta)$ the proposed measurement technique provides a combined uncertainty [7] $u_C < 4*10^{-3}$ mm, with p = pitch of the grid, z α and β illustrated in fig. 1 a-b.

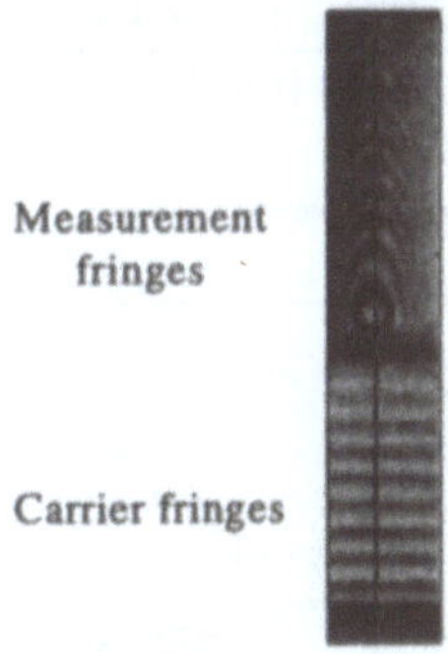

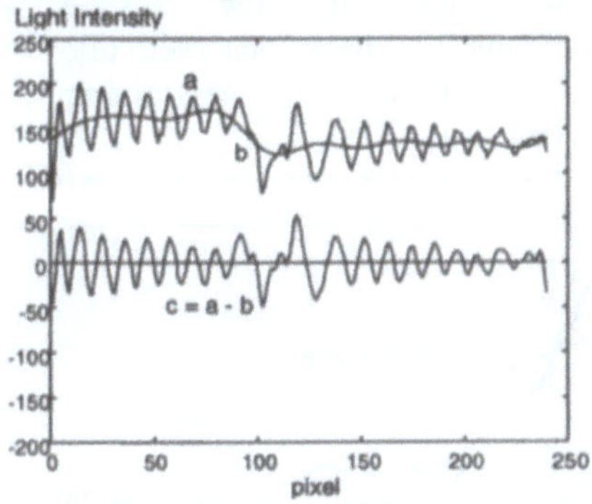

fig.2: image of the measured profile fig.3: measurement signal and its processing

6. Analysis of the results

In the present paper in order to quantify the elastic springback of the drawn part two numerical parameters are proposed, namely S_a and D_a which allow to quickly describe the different shape of the measured profile compared to the ideal one.

$$Sa = \frac{1}{L}\int_0^L \left|z_m(x) - z_i(x)\right| dx \qquad Da = \sqrt{\frac{1}{L}\int_0^L \left(z_m(x) - z_i(x)\right)^2 dx}$$

The parameter S_a represents the mean absolute error between the two profiles, being L the height of the measured profile, z_m the ordinate of the measured profile and z_i the ordinate of the ideal profile. It provides quantitative information regarding the global discrepancy between the two profiles apart from the shape of the final one.
The parameter D_a is introduced with the aim to quantify the shape difference between the two taken into account profiles. The proposed measurement technique has been utilized to investigate the influence of the blankholder force (bhf) on the elastic springback phenomenon. Constant levels corresponding to an initial blankholder pressure on the blank equal to the 0.5%, 1%, 2% and 3% of the

material yield stress respectively, have been tested, an EP grease has been utilized as lubricant. As figure 4 shows, for increasing levels of the blankholder force the obtained profile tends to became closer to the ideal profile reducing the angular distance which macroscopically characterize the elastic springback in the drawn part, due to the higher restraining actions on the part.

Such results is quantitatively confirmed by the S_a and D_a values reported in table 1. Finally the influence of the lubricating condition on the elastic springback has been taken into account, utilizing a constant blankholder force corresponding to an initial pressure on the blank equal to the 1% of the material yield stress. In particular EP grease, mineral oil, and Teflon have been used as lubricants. An increment in the elastic phenomenon has been observed due to the reduction in the restraining forces (fig. 5, tab. 2).

	S_a	D_a
3%	0,238	0.007
2%	0,315	0,310
1%	0,384	0,292
0.5%	0,863	0,266

Table 1 - S_a and D_a values [mm]

	S_a	D_a
Grease	0,303	0,288
Oil	0,384	0,292
Teflon	0,793	0,414

Table 2 S_a and D_a values [mm]

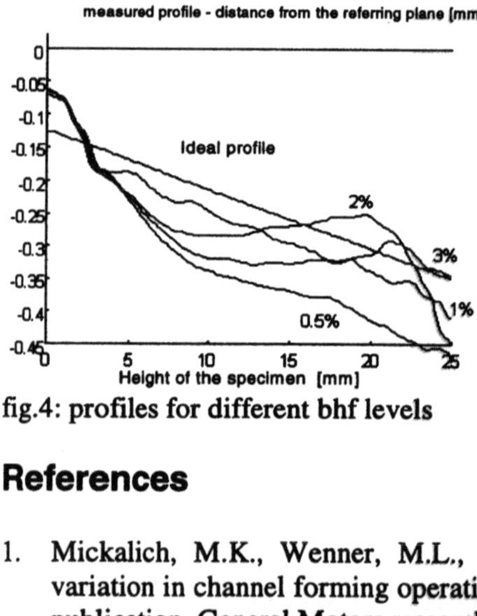

fig.4: profiles for different bhf levels

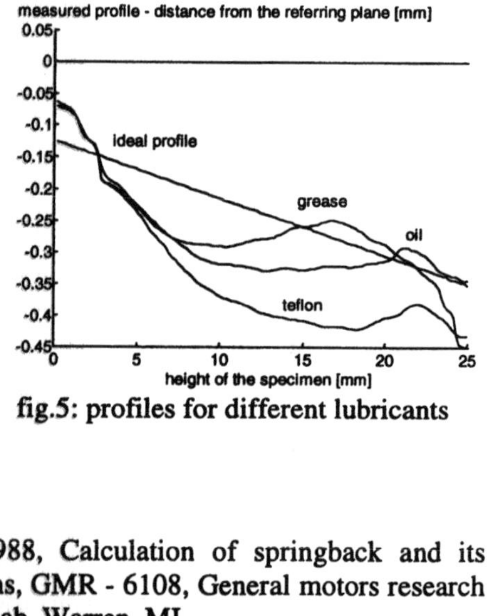

fig.5: profiles for different lubricants

References

1. Mickalich, M.K., Wenner, M.L., 1988, Calculation of springback and its variation in channel forming operations, GMR - 6108, General motors research publication, General Motors research lab, Warren, MI
2. Makinouchi, A., 1996, Sheet forming simulation in industry, J. of Materials Proc. Technology, 60:19-26
3. Sciammarella, C.A., Ahnadshahi M.A.: Detection of fringe pattern information using a computer based method (1986), Proc. of the VIIIth Int. Conf. on Experimental Stress Analysis,pp.359-368.
4. Post, D., Han, B., Ifjiu, P., High sensitivity Moiré, (1994) Springer Verlag Ney York, p.119-124, p. 419-426

5. Takeda M., Ina H., Kobayashi S.: Fourier transform method of fringe pattern analysis for computer based topography and interferometry; J. Opt. Soc. of Am. (1982), vol.72, pp.156-160.
6. Papoulis, A.: Signal Analysis, (1984), McGraw-Hill
7. ISO – Guide to the Expression of Uncertainty in Measurement (1993) Switzerland, first ed.

Application of Kohonen Neural Network to Tool Monitoring in Blanking

L. Moszczynski[1], A. Rosochowski[2] and R. Balendra[2]

[1]Institute of Materials Processing, Warsaw University of Technology, Warsaw, Poland, l.moszczynski@wip.pw.edu.pl
[2]Department of Design, Manufacture and Engineering Management, University of Strathclyde, Glasgow, UK, a.rosochowski@strath.ac.uk

Abstract: A dynamic growth of artificial neural network applications in engineering is mainly confined to multi-layered feed-forward back-propagation neural networks which are based on supervised learning. In this research an alternative approach based on unsupervised learning was investigated. It was implemented in the form of a modified self-organizing Kohonen network and used for the classification of tool wear in blanking. The modification, based on Ahalt's implementation of a conscience mechanism, enabled very close tool states to be distinguished. The results achieved suggest that the method could be successfully used in monitoring and diagnostic tasks in all kinds of industrial processes.

1. Introduction

Artificial neural networks (ANNs) are increasingly used in engineering as the alternative or complementary method to traditional problem solving techniques. One of the earliest applications of ANNs in metal forming was in a cold forging expert system [1] and included such functions as the determination of suitable forming methods, the number of forming steps and critical conditions leading to die or component failure. The concept of employing ANNs in preform design was formulated in [2] and tested in [3] on FEM generated results. A popular application proved to be the evaluation of flow stress at various strains, strain rates and temperatures [4-6]. Other mechanical properties such as the yield strength, tensile strength and elongation resulting from rolling were also evaluated [7]. ANNs seem to be well suited for the prediction of forces required for metal forming operations. Most attention in this context was again given to rolling [5, 8].

Relatively few applications of ANNs refer to controlling and/or monitoring of metal forming operations. For example, an ANN was used to identify and compensate roll eccentricity in rolling mills by measuring the strip exit thickness [9]. A concept of monitoring of blanking operation was proposed in [10] to account for tool wear. The idea was based on a process autograph (blanking force vs. displacement) representing a particular state of blanking tools but its implementation using a multi-layered feed-forward back-propagation neural network did not produce good results. An attempt was made to characterize the blanking process better by including acoustic emission, but still using the back-propagation learning algorithm [11].

In fact, all the above examples employed the feed-forward back-propagation network. It represents the so-called supervised learning which requires both the inputs and the outputs to be known. An alternative approach, mainly used for classification tasks, requires only inputs while outputs are arranged according to patterns or classes revealed in the inputs (unsupervised learning).

The latter approach, and its particular representation known as the self-organizing Kohonen network, is used in this research to investigate a possibility of obtaining better results in the above mentioned blanking monitoring problem. Since this approach is not so widely known some basic information on data classification and the Kohonen network is given first. This is followed by the presentation of a blanking tools' classifier based on the classical as well as the modified Kohonen network. The results of numerical experiments indicate that this modification substantially improves the network diagnostic capability.

2. Data Classification

Machine or tool monitoring based on ANNs would normally follow a standard layout of data recognition and classification system (Fig. 1).

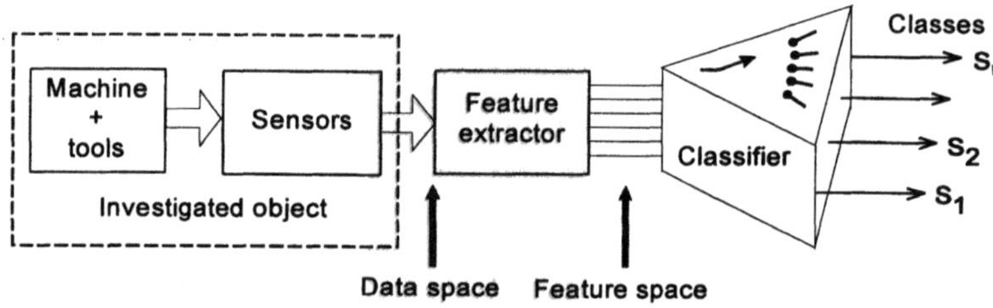

Fig. 1. Data recognition and classification system.

The measured data can be represented as a set of data vectors in a multi-dimensional data space. The features most representative for the machine operation can be extracted from that space to create a feature space of reduced dimensionality. The classifier in Fig. 1 implements a certain decision function which associates feature vectors with a certain number of classes. This mapping must be unambiguous so that each feature vector belongs to only one class. On the other hand one class can be associated with a group of vectors, each group represented by a vector called prototype. Prototypes, therefore, reflect the distribution of classes in a feature space. If the mapping function is a scalar function $G(\mathbf{x},S)$, the association of a feature vector $\mathbf{x}$ with a proper class S_{opt} from the set S can be described as

$$G(\mathbf{x}, S_{opt}) = \min_{S} G(\mathbf{x}, S) \qquad (1)$$

3. Kohonen Neural Network

A convenient method of resolving large classification problems is to assume an initial classification and correct it using an iterative optimization. This is the working principle of self-organizing ANNs. Classification based on self-organization is best achieved by neuron competition. In the classical version of the Kohonen algorithm, known as a self-organizing map (SOM) [12], competition is

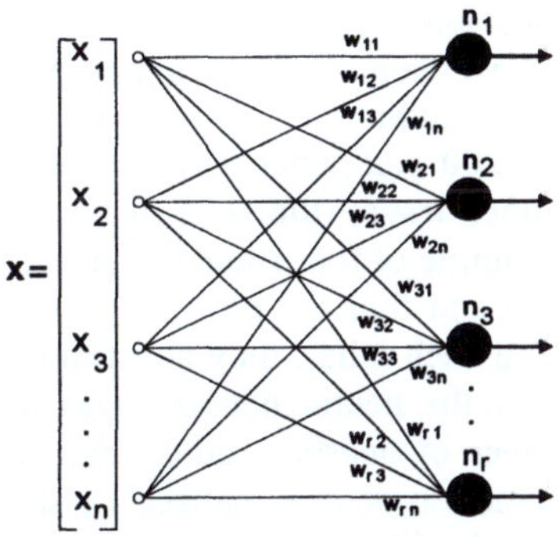

Fig. 2. Structure of the network.

based on the Euclidean distance between the input vector $\mathbf{x}(x_1, x_2, \dots x_n)$ and the weight vector $\mathbf{w}_j(w_{j1}, w_{j2}, \dots w_{jn})$ of a given neuron j (Fig. 2). The sequence of work with this kind of network can be described as follows:

Step 1: The number of classes (neurons) is assumed.

Step 2: Random weight vectors are specified.

Step 3: For a consecutive input vector $\mathbf{x}(t)$, the nearest neuron (a winner) s is identified according to the following norm:

$$\|\mathbf{x}(t) - \mathbf{w}_s(t)\| = \min_j \{\|\mathbf{x}(t) - \mathbf{w}_j(t)\|\} \tag{2}$$

Step 4: The winning neuron s, and other neurons k from its neighbourhood $h(s,k)$, is modified by changing its weights w_{si}. This is realized by a competitive learning rule:

$$w_{si}(t+1) = w_{si}(t) + \alpha(t)h(s,k)[x_i(t) - w_{si}(t)] \tag{3}$$

where $\alpha(t)$ is a learning coefficient from the range $\alpha \in (0, 1)$.

The steps 3÷4 are repeated until all input vectors are presented to the network. This creates one learning epoch. The whole process is then repeated many times for different orders of input vectors.

Step 5: After the learning process has been completed, the classification capability of the network is tested using the input vectors not belonging to the learning set.

4. Neural Classifier of Blanking Tool Conditions

Due to only limited success of earlier attempts of using a feed-forward back-propagation network as a monitoring aid in blanking, the Kohonen network described above was chosen as the alternative approach. The training of this network should be carried out using a large set of process autographs gathered over the whole life of the blanking tools and representing different modes of tool wear and failure.

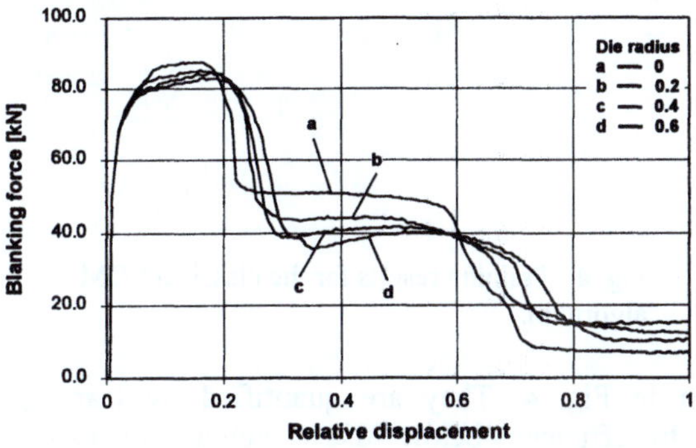

Fig. 3. Force autographs obtained in blanking.

In this paper, however, just to test the feasibility of the proposed approach, the experimental data from controlled blanking experiments was used. It was extracted from [13], where blanking was carried out on a 3 mm thick low carbon steel sheet and using a 20 mm diameter die. To simulate various types of tool wear, the die and the punch edges were radiused and the punch diameter reduced. The 48 force-displacement autographs used in this paper represented the case of a 0.25 mm punch-die clearance and all possible combinations of 3 material states, 4 punch radii and 4 die radii.

Each autograph constituted an input vector whose 200 components represented consecutive force samples taken over the relative punch travel (the ratio of punch displacement and sheet thickness). Fig. 3 shows an example of four such autographs obtained for the cold-rolled material, the sharp punch and four die radii of 0, 0.2, 0.4 and 0.6 mm. Despite the autographs being close to each other, they constituted separate classes already. Therefore, the task faced by the neural network was to associate 48 autographs with exactly the same number of neurons rather than to group the autographs. As explained in the previous section, it was realized by the weights of the winning neuron and its neighbourhood adjusted in a way which made them move closer to the learning set.

The challenge for the learning algorithm was to achieve unambiguous assignment of all neurons to closely distributed multi-dimensional input vectors. A successful learning process should avoid repetitions, that is neurons representing more than one input vector. This could be assessed by learning efficiency, defined as the percentage of winning neurons, and/or the average mapping error over all input vectors j and their components i.

$$Q_{error} = \frac{1}{p} \sum_{j}^{p} \sum_{i}^{n} \left\| \mathrm{x}_i - \mathrm{w}_{ji} \right\|^2 \tag{4}$$

Thus, in the ideal case, the learning efficiency is 100% (no idle neurons left) and the mapping error is 0 (weight vectors equal to input vectors).

5. Numerical Experiments

The first neural classifier for blanking tools was built using the classical SOM algorithm. In the first phase of the learning process, the initial neighbourhood $h(s,k)$ was set wide to enable fast global organization of the network, while in the second phase it was narrowed to enable fine tuning. This was accompanied by the initial learning coefficient α being reduced from 0.5 in the first phase down to 0.1 in the second phase.

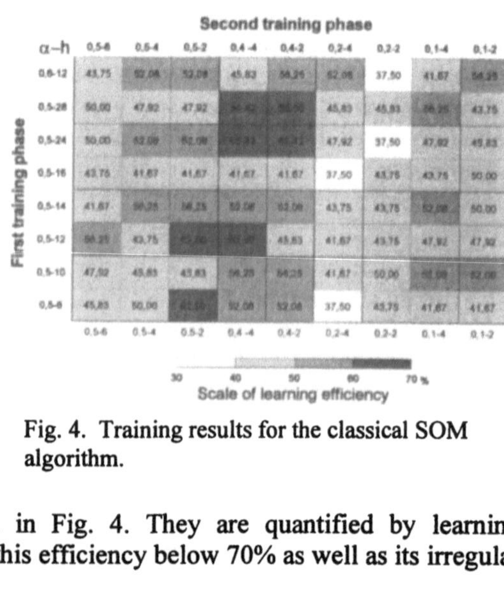

Fig. 4. Training results for the classical SOM algorithm.

The training results, based on the above parameters, are shown in Fig. 4. They are quantified by learning efficiency. The maximum value of this efficiency below 70% as well as its irregular distribution with regard to the learning parameters indicate poor performance of the classical SOM algorithm.

To improve this performance, the learning algorithm was modified by incorporating into it the so-called conscience mechanism. Its particular form used here was proposed by Ahalt [14]. Compared to the classical SOM algorithm, the winning neuron is nominated on the basis of the minimum "distance"

$$\min_j \left\{ \left\| \mathbf{x}(t) - \mathbf{w}_j(t) \right\| \mathrm{u}_j(t) \right\} \tag{5}$$

where $u_j(t)$ is the number of wins of j neuron after time t. The conscience mechanism equalizes the number of wins by all neurons and, therefore, improves the chance of finding the global minimum. This was confirmed by the training results in Fig. 5 where, in addition to learning efficiency, the Q_{error} is also presented. For the initial value of α=0.6 and only 200 learning epochs, this error vanishes (learning efficiency 100%) which indicates the fully trained network.

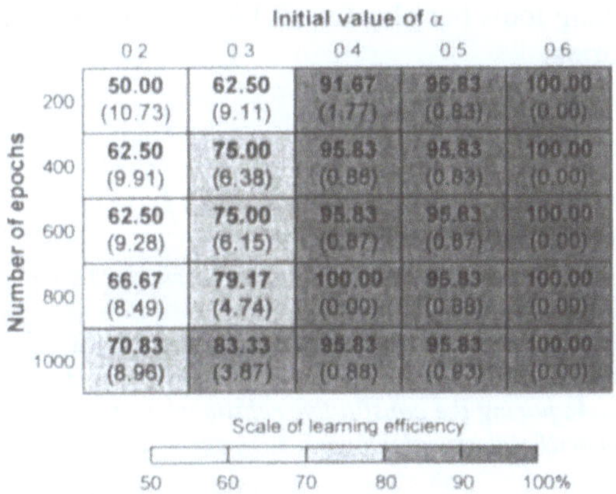

Fig. 5. Training results for Ahalt's modification.

After the completion of training, the network should unmistakably classify a new input vector. Nevertheless, in order to test the network in this mode, an input vector was generated exactly between two adjacent training autographs; it was then shifted slightly toward one of them and then toward another. Each time, when presented to the network, it was correctly classified as belonging to the class represented by the closer autograph. Another test was carried out for an input vector generated by taking two training autographs with at least two common points and exchanging portions of these autographs which lay between these points. This created a "noisy" input which, however, was easily classified by the network as belonging to the nearest class.

6. Conclusions

Earlier experience with the most widely used neural network, based on the feed-forward back-propagation learning algorithm, was that it did not perform very well as a classifier of different conditions of blanking tools. In an attempt to find a better approach, two neural networks based on unsupervised learning were tested. The first case was that of self-organizing Kohonen network which employs a classical competitive learning algorithm combined with the concept of neighbourhood. In this concept, the winning neuron shares its success with other neurons from a predetermined neighbourhood to make sure that all neurons participate in the learning process. To achieve the same result in the second network considered, its learning algorithm was equipped with a conscience mechanism. The Ahalt version of this mechanism uses the winning frequency of neurons as an additional means of promoting the winner.

To test the efficiency of these two approaches, the force-displacement autographs from controlled blanking experiments were used for training. They represented a difficult case of close input vectors in the multi-dimensional space. The task was to map all the input vectors to the same number of classes/neurons.

The classical Kohonen network did not perform well despite dividing the learning process into two phases and using various combinations of learning parameters and neighbourhoods. The Ahalt modification, on the other hand, enabled fast and complete organization of the network which manifested itself as the lack of unutilized neurons and the final weight vectors equal to the training vectors.

This encouraging result allows us to believe that unsupervised learning based on Ahalt's modification of the Kohonen competitive learning algorithm is not only a viable option for monitoring of blanking tools but also for all kinds of monitoring and diagnostic activities in manufacturing.

References

[1] Osakada K., Yang G., *Application of neural networks to an expert system for cold forging*. Int. J. Mach. Tools Manufact., 31 (1991) 577-587.

[2] Roy R., Chodnikiewicz K., Balendra R., *Interpolation of forging preform shapes using neural networks*. J. Mater. Process. Technol., 45 (1994) 695-702.

[3] Kim D. J., Kim B. M., Choi J. C., *Determination of the initial billet geometry for a forged product using neural networks*. J. Mater. Process. Technol., 72 (1997) 86-93.

[4] Qingbin L., Zhong J., Mabao L., et al, *Acquiring the constitutive relationship for a thermal viscoplastic material using an artificial neural network*. J. Mater. Process. Technol., 62 (1996) 206-210.

[5] Chun M. S., Biglou J., Lenard J. G., et al, *Using neural networks to predict parameters in the hot working of aluminium alloys*. J. Mater. Process. Technol., 86 (1999) 245-251.

[6] Rao K. P., Prasad Y. K. D. V., *Neural network approach to flow stress evaluation in hot deformation*. J. Mater. Process. Technol., 53 (1995) 552-566.

[7] Liu Z. Y., Wang W. -D., Gao W., *Prediction of mechanical properties of hot-rolled C-Mn steels using artificial neural networks*. J. Mater. Process. Technol., 57 (1996) 332-336.

[8] Larkiola J., Myllykoski P., Nylander J., et al, *Prediction of rolling force in cold rolling by using physical models and neural computing*. J. Mater. Process. Technol., 60 (1996) 381-386.

[9] Aistleitner K., Mattersdorfer L. G., *Neural network for identification of roll eccentricity in rolling mills*. J. Mater. Process. Technol., 60 (1996) 387-392.

[10] Choy C. M., Balendra R., *Prospects for the application of neural networks to the control of metal forming operations*. Proc. of the 11th National Conference on Manufacturing Research, De Montfort University, Leicester, September 1995.

[11] Wadi I., Balendra R., *Using neural networks to model the blanking process*. J. Mater. Process. Technol., 91 (1999) 52-65.

[12] Kohonen T., *Self-organization and associative memory*. Berlin: Springer Verlag, 1984.

[13] Choy C. M., *An investigation of parameters influencing constrained shearing (blanking)*. PhD Thesis, University of Strathclyde, Glasgow, May 1997.

[14] Ahalt S. C., Krishnamurthy A. K., Chen P., et al, *Competitive learning for vector quantization*, Neural Networks, 3 (1990) 277-290

Flexible Method for Punching a Thin Metal Sheet Using a Urethane Sheet

[1]Hisaki Watari, Department of Mechanical Engineering, Oyama National College of Technology, 771 Nakakuki, Oyama, Tochigi, 323-0806 Japan, watari@oyama-ct.ac.jp
[2]Hiroshi Ona, Department of Mechanical Engineering, Takushoku University, 815-1 Tate-machi, Hachioji,Tokyo, 193-0944 Japan
[3]Yu Yoshida, Department of Mechanical Engineering, Nagaoka University of Technology 1603-1 Kami-Tomioka, Nagaoka, Niigata, 940-0285 Japan
[4]Mitsunari Ohkusa, Department of Mechanical Engineering, Oyama National College of Technology, 771 Nakakuki, Oyama, Tochigi, 323-0806 Japan

Abstract: An experimental approach has been conducted to ascertain the effectiveness of a flexible punching method using a urethane elastomer for a relatively thin metal sheet. In this punching process, an elastic tool that consists of a urethane sheet and a metal die is used for punching instead of a conventional metal punch and die. By applying the urethane sheet as a flexible punch that is compatible with many shapes and sizes of holes, a more economical and simpler way of punching could be performed. The effects of the hardness and thickness in addition to the contact conditions between the metal sheet and the urethane sheet on the cutout are examined in detail. The mechanism of the punching process using urethane as a punch is investigated and a suitable punching condition for making a simple circular hole is also proposed.

1. Introduction

In recent years, micro piercing technology has been considered to be increasingly important, especially in the field of electronics. Due to the growing needs of product manufacturers to consider the impact on the environment, it is now widely recognized that attempting to lighten the total weight of the product is a desirable goal. Utilizing technology to lighten product weight is necessary to manufacture various electronics parts. For example, piercing method with high accuracy is required to make holes less than 1.0 mm in diameter in sheet with a thickness of less than 0.2 mm. Generally, in the micro piercing process, a metal punch and die, drilling machine, electron discharge machine or etching has been used. However, it has been said that problems in micro piercing are mainly due to the increasing costs of manufacturing special tools, in the case of using drills or a conventional metal punch and die. In addition, the buckling of punch owing to the insufficiency of the tool's strength is a further problem, even if the proper tools could be made. Also, in using a drilling process, the shape of the hole is limited to only a simple circular shape.

The authors, therefore, have proposed a flexible punching method using an elastic tool instead of a metal punch and die, which is compatible with many varieties of products as well as small productivity [1-2]. This process draws upon Guerin Process, which was used for manufacturing aircraft parts in the 1960's in the USA [3-4]. Despite the low production cost contributed by the Guerin Process, the accuracy of products is not regarded very highly. In this paper, an experimental investigation for a relatively thin metal sheet has been conducted to ascertain the adaptability of this application using urethane elastomer as a punch. In order to improve the accuracy of the products, a simple hole can be cut by subjecting a sheet plate on a metal die to an impact load by the urethane elastomer. By applying the urethane sheet as a flexible punch that is compatible with many shapes and sizes of holes, a more economical way of punching can be performed.

For this reason, this paper firstly describes the effects of the hardness and thickness of the urethane as a punch on the results of cutout. Secondly, the contact condition in terms of the radius of a metal punch, which subjects an impact load to the urethane elastomer, is investigated. Also, the mechanism of the punching process using urethane as a punch is discussed. Finally, the relationship between diameter-thickness ratio d/t and the tensile strength of the sheet metal is established with regard to the possibility of manufacturing a simple circular hole.

2. Experimental procedures

2.1 Punching machine

Figure 1 shows an illustration of punching machine. An impact force to the urethane sheet can be given by hitting a metal punch that is taken advantage of the spring force. Subjecting a sheet plate on a metal die to an impact load by the urethane elastomer can cut a simple circular hole. In this process, a urethane sheet can act as an elastic punch instead of using a metal punch. Table 1 shows the principle dimensions and properties of the punching machine that was used in this experiment.

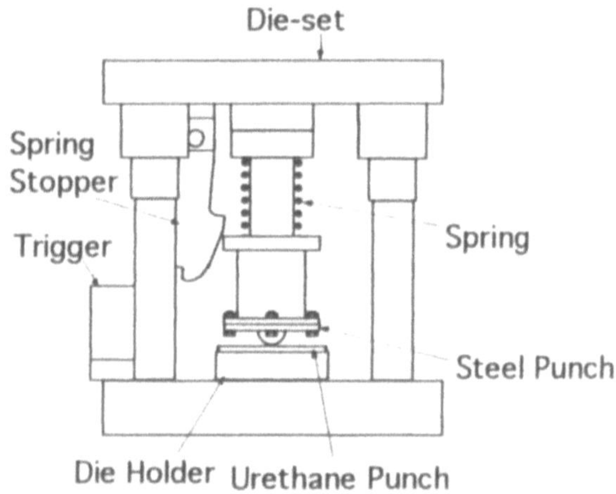

Fig. 1 Schematic illustration of punching machine.

2.2 Punching process

Table 1: Properties of the punching machine.

Spring constant /N · mm^{-1}	94.1
Initial deflection /mm	30
Mass of steel punch /kg	0.96
Impact velocity /m · s^{-1}	9.4

Figure 2 shows the schematic illustrations of the punching process. As shown in Fig.2 (b), the distribution of the pressure, in which shape is a circular arc, is obtained by hitting a metal punch with high velocity as shown in Fig2 (a). Therefore, the urethane elastomer is deformed as shown in Fig. 2 (c). As a result, the deformed urethane sheet flows into the metal die that is fitted under the sheet plate. Thus, a desired hole could be cut in a sheet plate.

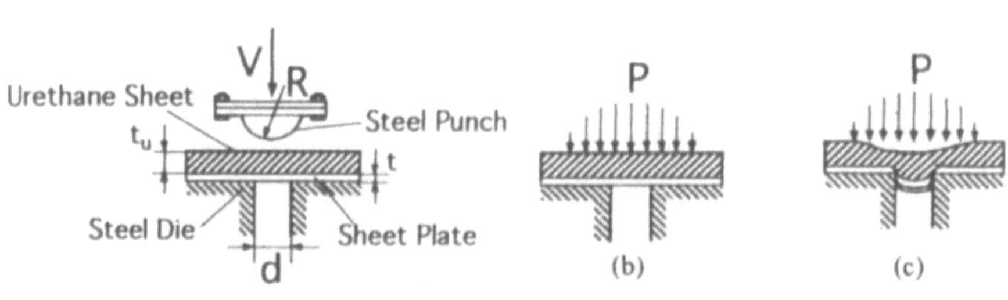

Fig. 2 Punching process.

2.3 Specimen and experimental conditions

The mechanical properties of the materials adopted are shown in Table 2. Table 3 shows the experimental conditions that were performed. In this experiment, a simple circular hole with d mm diameter is made in a sheet plate that is t mm thick. Four kinds of urethane hardness from 70 degrees to 95 degrees were used to examine the best hardness as an elastic punch. The sheet thickness of urethane was varied from 3.0 mm to 6.0 mm in three ways. Three kinds of dimensions of holes were taken to investigate the limit of punching. Also, three kinds of punch radius were used in order to examine the best contact condition between a metal punch and a urethane sheet.

Table 2: Material properties and sheet thickness.

	Yield stress or 0.2% Proof stress σ_Y /MPa	Tensile strength σ_T /MPa	Thickness t /mm
SUS304	1285	1445	0.05, 0.1
PBP2-1/2H	551	582	0.1
A1050-H18	170	196	0.2

Table 3: Experimental conditions.

Hardness of urethane H_u /deg	70, 80, 90, 95
Thickness of urethane t_u /mm	2, 3, 5, 6
Hole diameter d/ mm	1.6, 2.4, 4.8
Radius of metal punch R/mm	5.0, 7.5, 10.0

3. Results and discussion

3.1 Effects of hardness of urethane punch

Some examples of the results of possible cutout in this experiment are shown in from Fig. 3 to Fig. 5. In these figures, white and black round marks indicate possible cutout. White and black triangle marks indicate primary fracture [1] and cross marks indicate impossible cutout. From Fig. 3, it can be seen that the punching is impossible when the hardness of urethane is more than 90 deg. and d/t is less than around 40 in the case of SUS304. This result is obtained in the condition that the thickness of urethane t_u is 6.0 mm, the radius of metal punch R is 7.5 mm. It is also seen that the results of punching are primary fracture in other cases. As shown in Fig. 4, the experiment shows the cutout in the case of PBP2 resulted in almost primary fracture. In the case of A1050, the result is shown in Fig. 5. Punching can be possible when the range of hardness of urethane H_u is 80 or 90 deg., the ratio of d/t is 24 or 48. From these results, it is considered that the punching can be possible in the case of the value of H_u is 80 or 90 deg., when the sheet plate has not very high strength. This is explained by the following mechanism. When the hardness of urethane is decreased, the urethane sheet is easily deformed and comes into the metal die, so that primary fracture tends to occur. However, the punching is not completed owing to the insufficiency of the tool's rigidity as an elastic punch. On the contrary, when the hardness of urethane is increased, the tool's rigidity as an elastic punch is sufficient, but the urethane sheet is difficult to be deformed. As a result, the punching cannot be completed. In other words, an appropriate urethane hardness is necessary to perform punching by the use of urethane

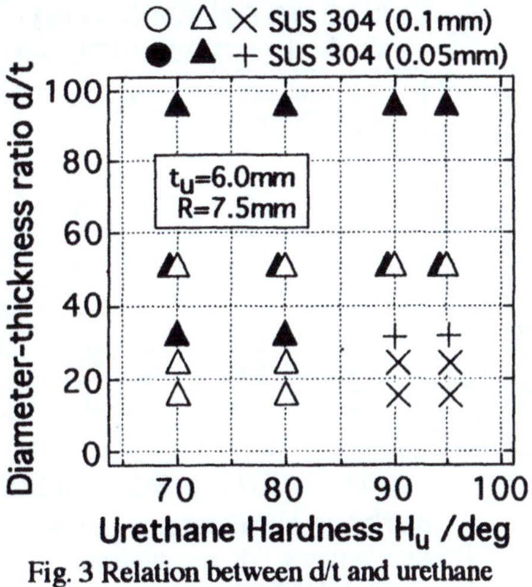

Fig. 3 Relation between d/t and urethane hardness H_u (SUS304).

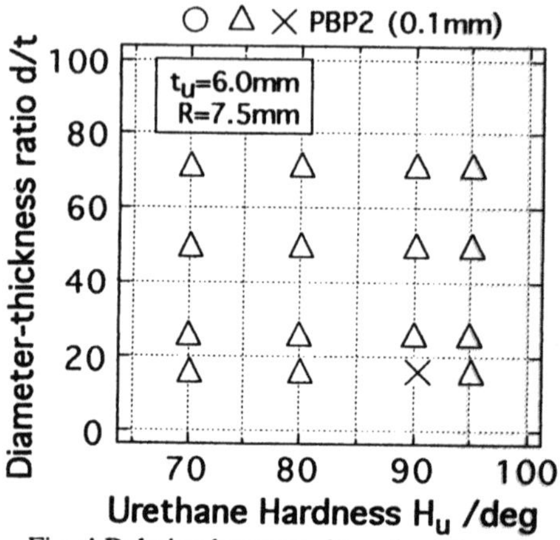

Fig. 4 Relation between d/t and urethane hardness H_u (PBP2).

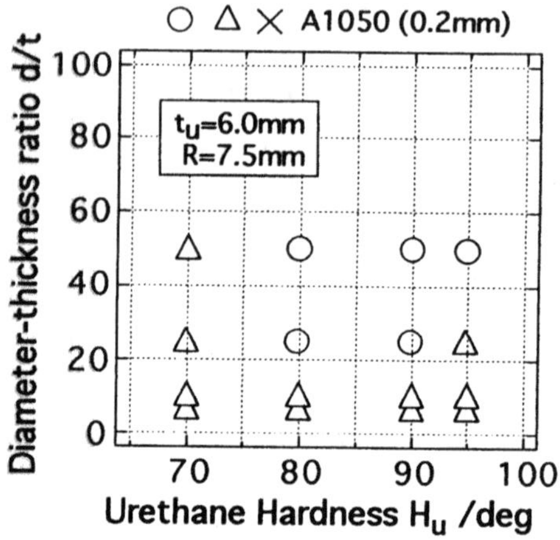

Fig. 5 Relation between d/t and urethane hardness H_u (A1050).

sheet as an elastic punch. In this case, it is considered that the Fig. 5 results from the appropriate ranges of urethane hardness that is suitable for the flexible punching as an elastic tool as well as a deformed body. In this experiment, it is demonstrated that the appropriate range of the hardness of urethane sheet is between 80 deg. and 90 deg. to complete the flexible punching.

3.2 Effects of radius of metal punch

The radius of the metal punch R affects the pressure distribution on the urethane sheet when the metal punch hits the urethane sheet. On the whole, it is considered that as the radius of the metal punch increases, lager magnitude of pressure can be obtained. However, when the radius of the metal punch becomes too small, the area of pressure distribution becomes narrower, so that the punching cannot be possuible for relatively large holes. In order to find the best contact condition between the urethane sheet and the radius of the metal punch, experimental results are denoted on Fig. 6, Fig. 7 and Fig.8. The data are summarized with the axis, namely, the vertical axis is shown with the urethane hardness H_u, the horizontal axis is shown with the ratio between the hole diam-

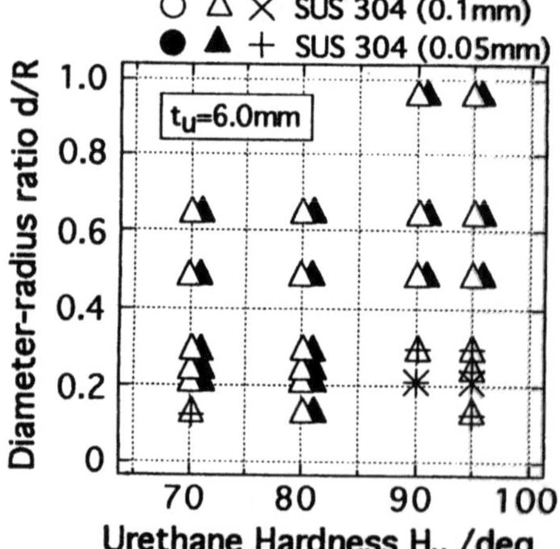

Fig. 6 Relation between d/R and urethane hardnessH_u (SUS304).

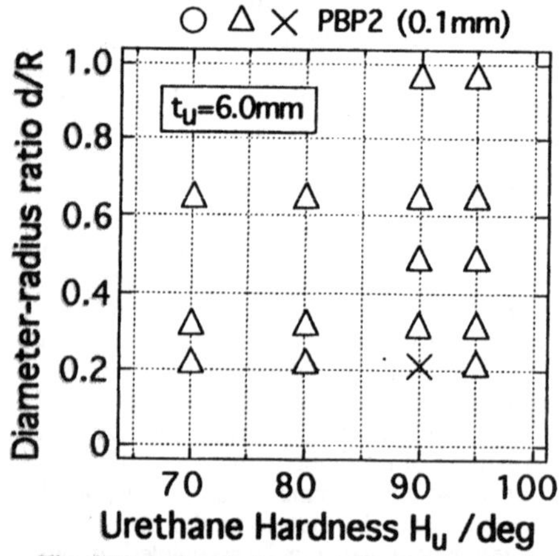

Fig. 7 Relation between d/R and urethane hardnessH_u (PBP2).

eter and the radius of the metal punch as shown in Fig. 6 , Fig. 7 and Fig. 8. From Fig. 6, it is seen that the punching is impossible in the case that the diameter-radius ratio d/R is less than around 4.0 and when the hardness of the urethane is 90 and 95 deg. in the case of SUS304. In the case of PBP2, the relation between the ratio d/R and urethane the urethane hardness H_u is shown in Fig. 7. It is seen that the punching results are almost primary fracture. It also can be seen that this trend is the same as that in Fig. 4. Furthermore, Fig. 8 shows the result in the case of A1050. When the strength of the sheet metal becomes lower, it can be seen that primary fracture can be possible. However, in the case of the value d/R is 0.48 to 0.64 as well as H_u is 80 and 90 deg., the formations in Fig. 8 result in possible cutout. In order to ensure successful punching in these experimental conditions, it is suggested that the appropriate range of the diameter-thickness ratio is 0.48 to 0.64.

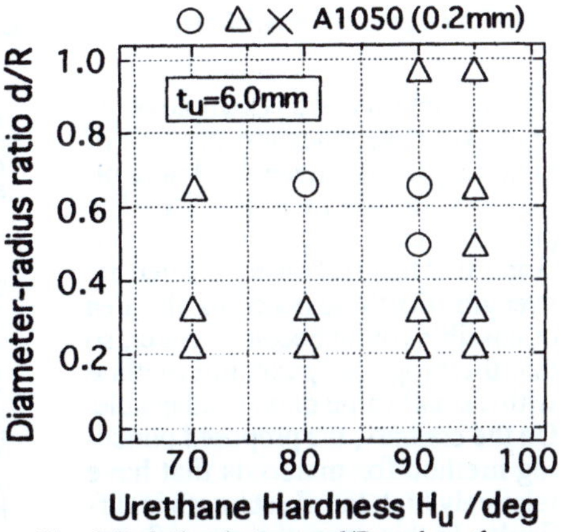

Fig. 8 Relation between d/R and urethane hardnessH_u (A1050).

3.3 Effects of thickness of urethane sheet

Table 4 and 5 show examples of results possible cutout in the case of urethane thickness t_u is 6.0 mm, 5.0 mm, respectively. It is clearly found that more formations of successful punching can be performed as shown in Table 5 compared to the result in Table 4. It is considered that a larger magnitude of punching force can act on the sheet plate through the urethane sheet by hitting the metal punch, when the urethane becomes thinner. In other words, punching is easily performed when an appropriate thickness of the urethane sheet is taken. However, this experiment showed that the usage of the urethane sheet that was too thin caused a damage of the metal punch and the urethane sheet, which made punching impossible. In order to complete successful punching by the use of urethane sheet, the thickness of urethane should be carefully decided so that it can ensure the rigidity of an elastic tool as well as the delivery of punching force. Comparing Table 4 with Table 5, it is suggested that only a difference of 1.0 mm between the thickness of the urethane in the both cases considerably affects the result of possible cutout. In these experimental conditions, it is obtained that an appropriate thickness of the urethane is within the range from 3.0 mm to 5.0 mm.

Table 4: Result of possible cutout (t_u=6.0mm).

Specimen	Hole diameter d/mm 1.6	2.4	4.8
SUS304 (0.05mm)	×	△	△
SUS304 (0.1mm)	×	×	△
A1050-H18 (0.2mm)	△	△	△
PBP2-1/2H (0.1mm)	△	△	△

(R=7.5mm,H_u=95deg)

Table 5: Result of possible cutout (t_u=5.0mm).

Specimen	Hole diameter d/mm 1.6	2.4	4.8
SUS304 (0.05mm)	△	△	△
SUS304 (0.1mm)	×	△	△
A1050-H18 (0.2mm)	○	○	○
PBP2-1/2H (0.1mm)	△	△	△

(R=7.5mm,H_u=95deg)

3.4 Effects of mechanical properties of materials

Figure 9 shows that an example of the relation between tensile strength of materials and diameter-thickness ratio. As shown in Fig.9, flexible punching can be completed for materials that have relatively low tensile strength. For these materials, the area of possible cutout becomes wider, so that the proposed punching method with a small value of d/t is adaptable. On the contrary, the proposed punching method for materials that have relatively high tensile strength is difficult. In this case, the area of possible cutout becomes narrower so that the proposed punching method with a small value of d/t is difficult to adapt by this machine. This can be explained by the insufficiency of rigidity of urethane sheet as an elastic tool, and thus punching could not be completed. On the other hand, punching can be completed for a low tensile strength sheet because the urethane has desirable rigidity as an elastic tool. In order to complete the proposed punching method, it is necessary to use harder urethane sheet to adapt larger force with high speed.

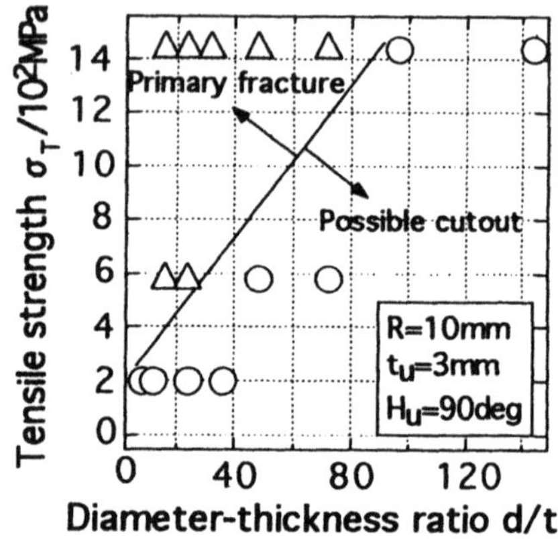

Fig. 9 Effect of material properties on possible cutout.

4. Conclusions

In the present investigation, an experiment was conducted to ascertain the effectiveness of a flexible punching method using a urethane elastomer for a relatively thin metal sheet. The results obtained are as follows:

(1) The suitable hardness of the urethane as a punch is between 80 and 90 deg. The suitable thickness of the urethane is within the range of 3.0 mm to 5.0 mm.
(2) The desirable ratio d/R between the radius of the metal punch R and hole diameter d is within the range from 0.48 to 0.64.
(3) The relationship between diameter-thickness ratio d/t and the tensile strength of the sheet metal is established, and the mechanism of punching using an elastic tool has also been clarified.

References

[1] H. Watari, H. Ona, Application of Urethane Rolls for Producing Section with Holes at Regular Intervals, Proc. 32nd Int. MATADOR Conf. ,UK, (1997) 463-467.
[2] H. Watari and H. Ona, Characteristic features of shape defects occurring in the cold roll forming of pre-notched products, J. of Mater. Process. Technol.,. 80-81 (1998) 225-231.
[3] K. Sato, R. Miya, K. Ozaki, H. Nishimura, Y. Ishikawa, O. Ishishita, Punching aluminum sheets with urethane elastomer tools, J. Jpn. Inst. Light Metals (in Japanese), 31-3 (1981) 164-168.
[4] K. Sato, R. Miya, K. Ozaki, H. Nishimura, Y. Ashikawa, Punching square holes in aluminum sheets with urethane elastomer tools, J. Jpn. Inst. Light Metals (in Japanese), 31-9 (1982) 608-613.

METAL CUTTING

Monitoring of Chatter in High Speed Endmilling Using Audio Signals Method

Yuan Ning[1], M.Rahman[2] and Y.S.Wong[3]

[1]National University of Singapore, MPE, Singapore 119260, engp8755@nus.edu.sg
[2]National University of Singapore, MPE, Singapore 119260, mpemusta@nus.edu.sg
[3]National University of Singapore, MPE, Singapore 119260, mpewys@nus.edu.sg

Abstract: This paper presents an experimental investigation of chatter phenomenon in high speed milling (HSM) of hardened steel using a sound level meter (SLM). A comparison of various sensors (dynamometer, accelerometer, SLM) is carried out. It is found that the signal from the SLM provides a distinct, and in many instances the best signal to identify chatter. A method to judge machining stability by FFT analysis of audio signal is proposed based on the experiment.

1. Introduction.

Chatter is a kind of self-excited vibration that generates chatter marks. It is harmful to metal cutting and may be observed on any chip producing machine tool. The effects of chatter are all adverse on surface finish, dimensional accuracy, tool life and machine life.

Sensing audio signals from the cutting process is one of the most promising methods for the purpose of detection and control of chatter. A great amount of research work has been done in this area [1-6]. This paper made an attempt on using audio signals to monitor chatter in HSM of hardened steel, on which it seems that not much research work was reported.

Sound emitted during HSM of steels mainly comes from two sources. The first is the metal cutting process, to which the primary deformation zone and the secondary deformation zone have principal contributions. The second is from the machine tool, which includes the rotating cutter, spindle motor, transmission system, bearing, chip conveyor, etc. The vibration of the machine tool structure should also be counted in. The latter sound source always has several constant frequencies and it is relatively regular. We refer to this as the machine's background noise.

Based on the principle of superposition, the two sound sources that have pressure fluctuations that vary both as a function of the spatial co-ordinate and time can simply be added at each position and time to yield the net pressure fluctuation produced. In this case, component of the audio signal generated by the cutting process will not be corrupted by the machine's background noise and vice versa.

2. Experiment.

All experiments are conducted on a Makino V55 Machine Center. Machining conditions used in the survey are: spindle speed: 10K-30K rpm and relative cutting speed: 377m/min to 1130m/min. The test material is AISI H13 hardened to HRC 55. Axial cutting depth varies from 0.1mm to 0.8mm and feed-rate is in the range of 0.025mm/tooth-0.05mm/tooth. The cutter is a two-flute solid carbide ball-nose endmill with TiAlN coating. It has a standard helix angle of 30^0. A Kistler dynamometer is used to measure the cutting force. The signal is recorded by a multi-channel DAT (digital audio tape) recorder at the sampling rate of 12000 samples/sec/channel. An optical microscope observes tool wear.

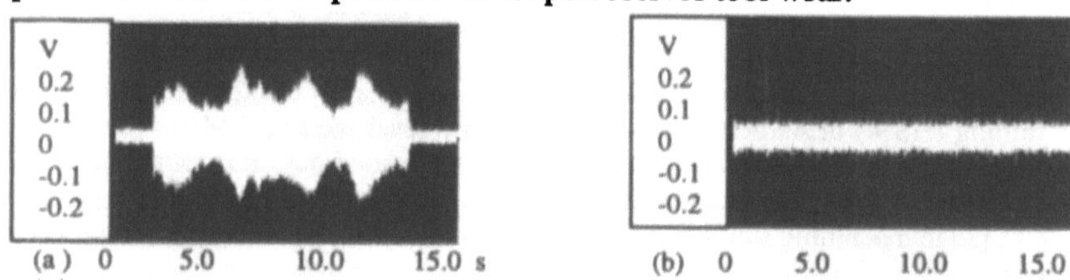

Fig.1 Typical real-time signal from SLM. (a) Chatter, (b) Stable cutting.

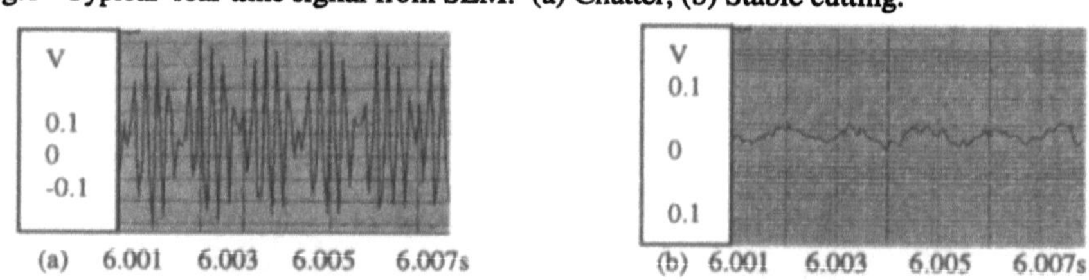

Fig.2 Zoomed real-time signal from SLM. (a) Chatter, (b) Stable cutting.

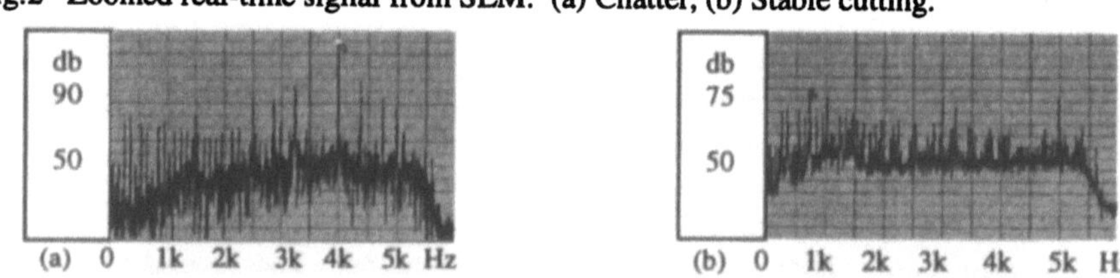

Fig.3 FFT Spectrum of SLM signal. (a) Chatter, (b) Stable cutting.

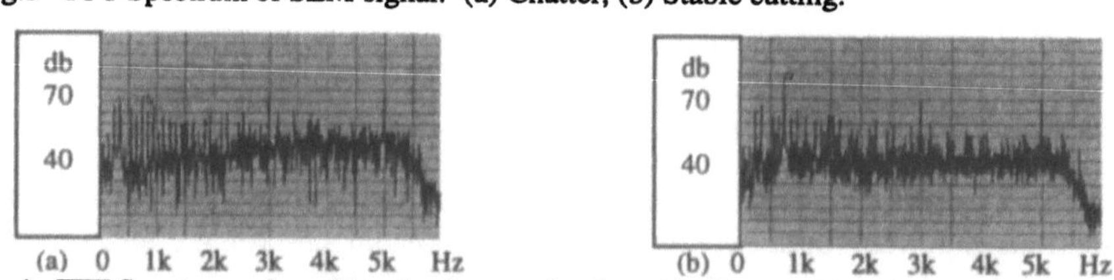

Fig.4 FFT Spectrum of machine background noise. (a) Chatter, (b) Stable cutting.
(Chatter cutting condition of all 'a' figures: RPM-22K, feed-0.03mm/z, cutting depth-0.7mm, Stable cutting condition of all 'b' figures: RPM-22K, feed-0.03mm/z, cutting depth-0.3mm)

When chatter occurs, the effective cutting depth and cutting angles are changing randomly. Usually the effective cutting depth is much larger than designated, thus the total sound intensity is quite larger compared with stable cutting. Some drastic slopes and valleys are found in the waveform as shown in Fig.1(a). In FFT spectrum, the peak frequency in Fig.3(a) is 3999.024hz(peak with arrow), which is much higher than the peak of background noise as shown in Fig.4(a). In stable cutting, the peak frequency (737.047hz) of cutting process is the same as that of the background noise as shown in Fig.3(b) and Fig.4(b). From these FFT spectrums it is clear that in HSM the modes of high frequency dominate in chatter ,which range from 2Khz to 5 Khz. While in stable cutting, modes of low frequency take over.

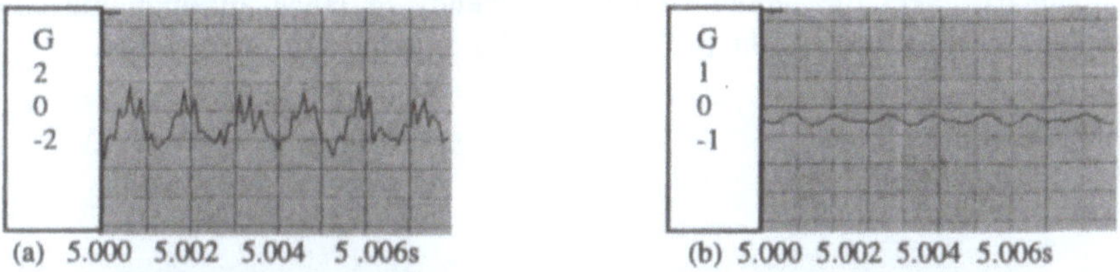

Fig.5 Real time signal from accelerometer. (a) Chatter, (b) Stable cutting.

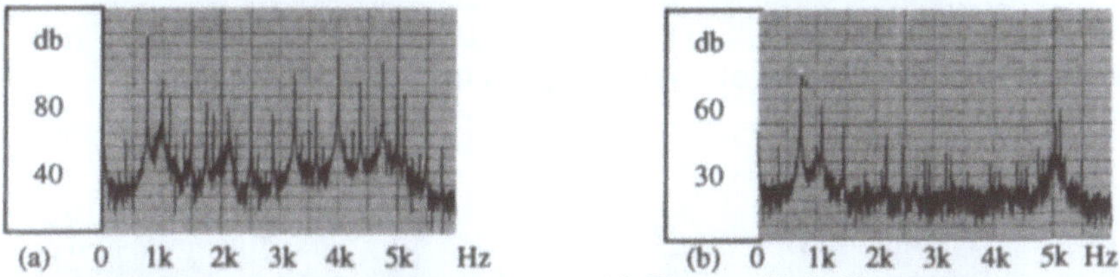

Fig.6 FFT Spectrum of accelerometer signal. (a) Chatter, (b) Stable cutting.

The accelerometer is mounted on the workpiece to measure the vibration of Y-axis. It has a high natural frequency of 76Khz. The real time signal of chatter differs drastically from stable cutting as shown in Fig 5. Accelerometer is in direct contact with the target. So the peak frequency is always the tooth frequency, although the chatter frequency 3999.024hz appears as one of the peaks in FFT spectrum of Fig.6(a). It is also observed that the modes of high frequency dominate in chatter of HSM, as what has been found in the SLM signal as in Fig 3(a).

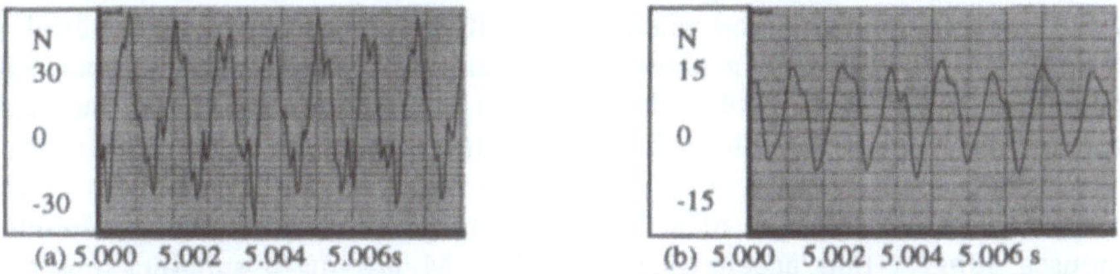

Fig.7 Real time force signal of Y-axis. (a) Chatter, (b) Stable cutting.

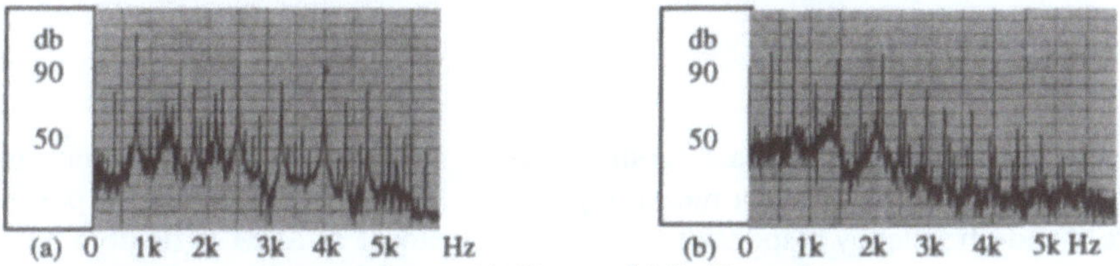

Fig.8 FFT Spectrum of Y-axis force. (a) Chatter, (b) Stable cutting.

In HSM, force of Y-axis is most sensitive to chatter because the damping ratio of the machining process is the smallest compared with other axes. There is a noticeable difference in the real-time signal between chatter and stable cutting from Fig.7, while not much difference is found in the X-axis force signal from Fig.9. X-axis is the feed direction, so the process damping ratio is much higher than Y-axis. As mentioned before, chatter observed in the experiment is of the range of 2k-5Khz. The natural frequencies of dynamometer are: X-axis 2Khz, Y-axis-3 Khz, and Z-Axis-1.8Khz. It is obvious that the response of the dynamometer is inadequate to detect the modes of high frequency in HSM although the chatter frequency appears as one of the peaks in the FFT spectrum of force signal.

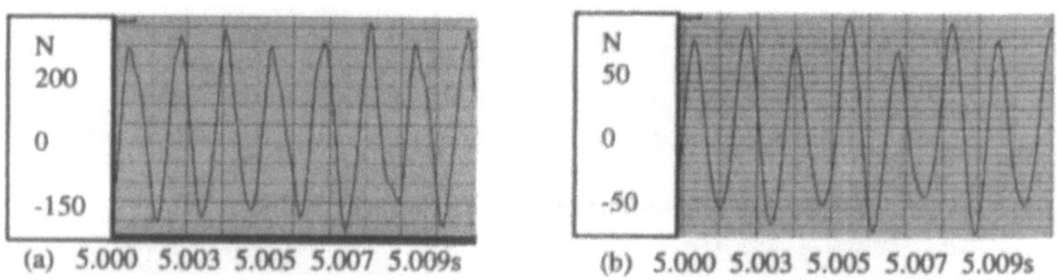

Fig.9 Real time force signal of X-axis. (a) Chatter, (b) Stable cutting.

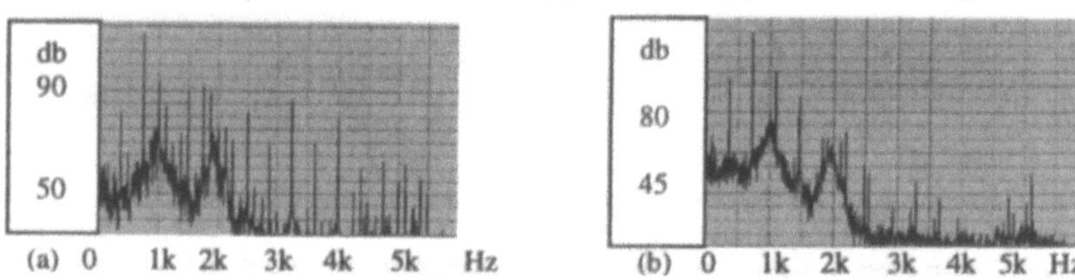

Fig.10 FFT Spectrum of X-axis force signal. (a) Chatter, (b) Stable cutting.

Dynamometers are also contact type sensors. So it is no wonder that in its FFT spectrum, the dominant peak is always the tooth frequency. The chatter frequency 3999.024hz is one of the peak frequencies in the FFT spectrum as shown in Fig.8(a) and 10(a), but it could not be identified without the help of SLM. Additionally, it is doubtful whether the sensitivity of the dynamometer is enough to detect the relative low cutting forces encountered when finish milling certain easy to cut materials like aluminum. The difficulty in identifying chatter of HSM lies in the short cutter engagement time that results from the high rotating speed of the cutter. Consequently it does not allow the chatter vibration to distinctly imprint on the force signal, since force is only generated during cutter engagement. Those limitations and the fact that most force sensors generally have narrow frequency responses suggest that accelerometers and SLM are more suitable sensors for chatter detection and control.

3. Discussion.

Accelerometer and SLM are both suitable sensors, while SLM has many advantages. Most SLM has a measuring range of 200-13K HZ so that it possesses quite a wide frequency response range. The placement of SLM is flexible because it is a non-contact type of sensor, as long as the sound from the process could reach the SLM directly and the distance is set properly. It is found that the best location for accelerometer is to put it on the workpiece. The result is not satisfactory when it is placed on the worktable and the spindle housing because of poor transmissibility. SLM could reflect exactly the characteristic frequencies of chatter and the chatter frequency could be directly read from its FFT spectrum. Besides, one accelerometer can only measure the response of one axis, but usually at least two accelerometers are needed to properly measure the vibration unless the machine is highly coupled. While a single SLM only is enough to monitor the whole machining process.

Real-time and spectrum methods are used frequently to analyze audio signals. RMS is one of the most common ways. It is found in this survey that the RMS value is much higher in chatter than that of stable cutting and it also changes drastically, which is theoretically explained by R.Teti and D.A.Dornfeld [1].

The limitation is that some real cutting must be conducted in advance to obtain certain basic information and the threshold. RMS value is dependent on cutting condition. For different cutting depth, feed-rate and cutting velocity, the critical value varies. Considering various tool and material combination, the situation is made much more complicated. So based on the data obtained from the experiment, a simple power spectrum method is proposed to judge chatter by audio signals. It is described as following:

If $Fdom \neq Fbac$ and $Fdom \neq R \times F_{tooth}$,
Then there is chatter.
Where: F_{dom}- dominant frequency of the power spectrum of SLM.
F_{tooth}- tooth Contact Frequency=NZ/60Hz, N-Spindle RPM.
Z-Number of cutting tooth, R-integer (1,2...).
F_{bac}- dominant frequency of background noise.

Table1: Experimental proof of the frequency method

F_{dom} (Hz)	F_{bac}(Hz)	Chatter	F_{tooth} (Hz)	RPM
3249.024	4998	yes	750	22,500
3184.57	345.703	yes	700	21,000
3137.695	4992.188	yes	666.67	20,000
3999.024	737.047	yes	733.33	22,000
3111.328	345.703	yes	600	18,000
2830.078	546.875	yes	266.66	8,000
2434.57	2997.07	yes	500	15,000

The machine's background noise signal is obtained by recording the signals when the machine does the air cutting, i. e. the machine spindle rotates at a certain rpm and travels at a certain feed-rate, but it does not touch the workpiece. As discussed before, the sound mainly comes from two sources when the cutting is stable. If the dominant frequency in machining process is neither from the machine's background noise nor from the stable cutting process that is noted for the multiple of tooth frequency, that we could expect that chatter has occurred. With regard to the effect of tool wear, as long as it is within the allowable range, it does not have much effect on cutting stability as well as on the pattern of audio signal.

BUE (built up edge) and chip that sticks to cutter were not observed in the experiment. The vibration of the machine tool itself is always of a very low frequency, so we can safely ignore its effect in the analysis.

The SLM must be put close to the cutting process, so the effects of the sounds emanated from other machines in the workshop could be minimized. There should not be any barrier between SLM and the cutting process. Thus it imposes a danger of being spoiled by the flying chip or the coolant. It should also be put away by a enough distance from the cutting process to ensure that the source can be regarded as a point source, normally more than three times the characteristic wavelength [3]. If the SLM is used in a very noisy environment, to improve sensitivity and eliminate unwanted noise, sound isolation should be employed according to J.Tlusty [3]. Thus it complicated the operation. SLM doesn't have enough response for low frequencies. Fortunately chatter always occurs at a high frequency in HSM. So far in the experiment, SLM never fails to detect the occurrence of chatter. SLM reflects chatter from both real-time and FFT spectrum analysis and it is found to be the best sensor in monitoring chatter in HSM.

4. Conclusion.

It is found that a single bare sound level meter, which is placed near to the chamber of the machine center without any sound isolation, provides a satisfactory signal for the use in chatter detection. It is proved to be the best sensor to monitor chatter in HSM. Chatter is always at very high frequencies in HSM ranging from 2Khz to 5kHz. Common dynamometers are not good to detect chatter under HSM. Accelerometer seems capable to give promising results.

Further development in use of audio signals for chatter detection and control is required to obtain better direction sensitivity. A single microphone may not be powerful enough to be used in a highly noisy environment. Techniques as adaptive filtering, isolation, noise cancellation may be required under such situation. The test for chatter control in processing some easy to cut material like brass or graphite using SLM should be conducted to examine its feasibility. Because of the special pattern of chatter signal, some advanced methods like pattern recognition and count rate might be developed, which will be the topic of future research.

References

[1] R.Teti, D.A.Dornfeld., Modeling and Experimental Analysis of Acoustic Emission From Metal Cutting ,Journal of Engineering for Industry, August 1989,Vol111,pp229-237.

[2] T.Blum., I.Inasaki., A Study on Acoustic Emission from the Orthogonal Cutting Process Journal of Engineering for Industry, August 1990,Vol112,pp203-211.

[3] T.Delio., J.Tlusty., S.Smith., Use of Audio Signals for Chatter Detection and Control. Journal of Engineering for Industry, May1992, Vol 112,pp146-157.

[4] R.Teti., U.La., Compare Cutting conditions and Work Material State Identification through Acoustic Emission. CIRP Annals, Vol41/1/1992, pp89-96.

[5] S.Y.Liang., D.A.Dornfeld., Tool Wear Detection Using Time Series Analysis of Acoustic Emission Journal of Engineering for Industry August 1989,Vol111,pp199-205.

[6]Weck.M., Melder.W., Problems in Assessing the Noise Behavior of Machine Tools.CIRP Annals,Vol.26-2,pp.397-402.

High Speed Milling of Hardened Steel Moulds and Press Tools

R. Singh[1,2], H.L. Coldwell[1,3], D.K. Aspinwall[1,4] and R.C. Dewes[1]

[1] School of Manufacturing and Mechanical Engineering, The University of Birmingham
[2] GTI Corporation Ltd, Houghton Regis, Bedfordshire
[3] Portway Tool and Gauge Ltd, Lye, Stourbridge, West Midlands
[4] IRC in Materials for High Performance Applications, The University of Birmingham

Abstract: The paper outlines the use of high speed ball nose end milling for the manufacture of hardened steel moulds, dies and press tools. Information on suitable machine tools and cutting tools is given, together with relevant machinability data. The industry is characterised by large numbers of small companies (typically <50 employees) and details are given of market trends and current working practice. A case study is presented in which the traditional manufacture of a rubber mould is compared with high speed machining practice, using a spindle speed increaser, state-of-the-art coated carbide cutting tools and suitable cutter paths. Machining times were reduced from over 14 hours to less than 2 hours. Chatter occurred during high speed milling, however, component surface roughness of 1-2 μm Ra and minimal cutter wear were achieved.

1. Introduction to High Speed Milling

High speed milling (HSM) is usually associated with end milling at high rotational speeds between 10,000 and 100,000rpm. It has been applied to a range of applications in the aerospace industry for the machining of aluminium alloys, titanium alloys and nickel-based superalloys [1],[2]. Over the last decade, HSM has been employed for the machining of fully hardened tool steels such as AISI H13, P20 and D2 (typically 25 – 62 HRC), for moulds, dies and press tools, with a greater or lesser degree of success.

1.1 Machine tools, controllers and ancillary equipment

A prerequisite of HSM is a machine tool capable of high spindle speeds and feed rates, coupled with rapid acceleration and deceleration. This may be provided by a specialist machine tool or the use of ancillary equipment fitted to a conventional unit. Modern conventional machines typically operate at spindle speeds of up to 10,000rpm. A recent survey of 17 machine tool manufacturers carried out by the authors, showed that a growing number of units are appearing that can operate with spindle speeds of up to 40,000rpm. In addition, machine tools are increasingly available with feed rates of up to 60m/min and accelerations of $10m/s^2$ (~1g). A partial solution to HSM, in the absence of a purpose built machine, is to purchase a spindle speed increaser. Such units however, do not provide a comprehensive solution due to performance issues such as vibration, short life span (approximately 2000 hours or 12 months) and the lack of adequate feed rate / acceleration performance with the existing machine tool.

In parallel to machine tool developments, work on CNC cutter path control and optimisation is underway. Numerical control data for HSM need to be supplied at higher rates and in larger volumes than for conventional machine tools. Some HSM applications require block-processing times down to 1ms. Controllers commonly employ a 'look ahead' facility to maintain accuracy.

1.2 Cutting tools and research / application data

Generally, cutting tools for the majority of HSM applications are solid carbide, although indexable insert cutters are being developed and used successfully [3]. The advantages of solid products include higher accuracy and tool life (in some cases, up to several thousand metres cut), while indexable tools offer lower capital cost. Tool materials that have been successfully used for the HSM of hardened tool steels include: micrograin cemented tungsten carbide [3]-[6], cermets [7] and polycrystalline cubic boron nitride (PCBN) [8]. Table 1 gives a selection of workpiece materials, tooling and typical cutting parameters.

Table 1. Typical cutting parameters for HSM of hardened tool steel [3]-[5].

Workpiece material & hardness	Tooling	Cutting mode	Typical cutting parameters
AISI P20 32HRC	6mm diameter Ti(CN) coated carbide ball nose end mill	Dry, down milling	v_c=200m/min, f_z=0.1mm/tooth, a_p=0.5mm, a_e=0.5mm
AISI P20 30HRC	25.4mm diameter TiN coated carbide ball nose end mill	Dry, down milling	v_c=300m/min, f_z=0.5mm/tooth, a_p=0.5mm, a_e=0.5mm
AISI D2 62HRC	12.7mm uncoated carbide ball nose end mill	-	v_c=220m/min, f_z=0.05mm/tooth, a_p=0.76mm
AISI D2 59HRC	25.4mm diameter TiN coated carbide ball nose end mill	Dry, down milling	v_c=60m/min, f_z=0.7mm/tooth, a_p=0.25mm, a_e=0.75mm

Nomenclature: v_c=cutting speed, f_z=feed per tooth, a_p=axial depth of cut, a_e=radial depth of cut

2. The Mould, Die and Press Tool Industry

The UK mould, die and press tool industry is characterised by large numbers of small companies (generally <50 employees) with highly skilled designers and toolmakers. In the past, the industry has maintained its market share by producing products of high quality, however, with increasing imports from countries such as Japan (81% of UK press tool imports in 1999 [9]), the industry is under increased pressure to also compete on price and lead time. One way in which the UK industry is addressing this is by adopting HSM. Benefits include significant cost / lead time reduction (average lead time in 1999 for tools with a value of £100,000 was >20 weeks [10]), by the elimination of multiple machining operations / hand finishing, allied with quality improvements. These stem from improved component dimensional accuracy / surface roughness / surface integrity. Distortion problems due to heat treatment are also avoided.

2.1 Market statistics

In recent years, the industry has struggled to maintain sales in the UK and global markets. A period of growth occurred over the period 1993 to 1997, with injection / compression moulding tools for rubber and plastics achieving sales approaching £380M. Over the same period, pressing and stamping tool sales also increased to ~£300M, however, a sharp decline (20-25%) in all areas occurred in 1998 [11]. It is unclear whether this is part of an underlying economic cycle or the start of a general decline. Over a similar period, UK rubber and plastic moulding exports and imports have remained fairly constant at ~£20M and ~£65M, respectively [9]. For pressing / punching / stamping tools, the import and export markets have been more volatile; between 1998 and 1999 exports rose slightly from £34M to £39M but imports increased substantially from £28M to £68M [9].

2.2 Current practice

The following section details a profile of mould, die and press tool manufacture, together with components suitable for HSM. Information has been obtained from a programme of data collection carried out by the authors, combined with survey results provided by the Gauge and Toolmakers Association (GTMA) [10].

A survey of 14 components from a rubber moulding manufacturer showed that AISI P20 tool steel with a hardness of 30-32HRC was used for the majority of moulds. Component sizes (width×breadth×depth) ranged from 156×146×18mm to 500×250×25mm. Critical features included internal radii down to 1mm. Similarly, a survey of 20 components from a press-tool maker showed that, AISI D2 tool steel with a hardness of 58-60HRC was typically used for dies and cutting punches. Component sizes ranged from 60×85×36mm to 95×150×160mm. Critical features included internal radii down to 3mm, external radii down to 4mm and M6 to M12 tapped holes. Typically, component surface roughness and dimensional accuracy for a cross-section of moulds / dies ranged from 1-5μm Ra and ±5-50μm, respectively [12].

For rubber moulds, the survey highlighted the use of processes such as drilling, reaming, milling (edge profiling, cavity production & graphite electrode production) and electrical discharge machining (EDM) – die sink & wire. A typical component required 17 hours of CAD and CNC cutter path programming and 45 hours of machining, which together resulted in a manufacturing cost of £1500. For press tool components, processes used included: drilling, tapping, milling, grinding, EDM (small hole drilling & wire), hand polishing and vacuum hardening. A typical individual component required at least 6 hours of CAD and CNC cutter path programming and 63 hours of machining. This resulted in a manufacturing cost of £200-800 depending on component complexity.

A survey of current machine tools within the industry revealed that the majority of machining centres / NC milling machines were between seven and ten years old with a spindle speeds of 2000-6000rpm. One company had in the previous five years, also invested in state-of-the-art wire EDM machines, with facilities such as automatic threading. On average, GTMA members were committed to investing approximately £200,000 per year in new equipment [10]. Examples of capital investment included: high speed machining centres, CAD/CAM hardware and cutter path optimisation software.

3. Case Study

The following case study details the production of a rubber moulding tool using both traditional machining and HSM techniques, specifically ball nose end milling. Figure 1 shows a typical moulding tool (1 of a set of 14) used to produce rubber seals for the automotive industry which has a semi-circular cavity that forms a section of the rubber seal used in a doorframe assembly. Overall dimensions are 196×170×36 mm. The workpiece material is AISI P20 tool steel

Figure 1. Finishing of mould cavity using a speed increaser.

(30-32HRC), pre-ground in a rectangular block. Manufacture involved three stages: drilling & reaming of locating / mounting holes, profiling the mould periphery and machining the mould cavity. The component is currently manufactured using CNC machining centres operating at conventional cutting speeds with high speed steel (HSS) and uncoated carbide tooling. Cutting fluid is applied in all operations. Comparative HSM tests were undertaken using a Bridgeport VMC1000 (10,000rpm) machining centre with a Heidenhain TNC426 controller. A 30,000rpm gear driven Bristol Tool & Gauge speed increaser with a 6:1 speed ratio was used for the finishing operation of the mould cavity, see Figure 1. Initial tests were carried out using existing tool paths, however, alternative cutting tools and machining parameters were used. This entailed Sandvik Coromant drills, end mills and ball nose end mills, together with Titex Plus reamers. Drilling and reaming operations utilised a soluble oil cutting fluid at a concentration of 5% by volume, supplied externally at a flood flow rate of 70 l/min and a pressure of 3 bar. All milling operations were performed dry. Table 2 details tooling and machining parameters for finishing the mould cavity.

Table 2. Tooling and machining parameters.

Operation		Traditional method	HSM method
Finishing mould cavity	Tooling	12mm diameter uncoated solid carbide ball nose end mill	6mm diameter TiCN coated solid carbide ball nose end mill
	Machining parameters	v_c=38m/min, f_z=0.025mm/tooth, n=1000rpm, v_f=200mm/min, a_p=0.3mm, a_e=0.1mm	v_c=250m/min, f_z=0.05mm/tooth, n=13263rpm, v_f=2653mm/min, a_p=0.3mm, a_e=0.1mm

Nomenclature: v_c=cutting speed, f_z=feed per tooth, n=spindle speed, v_f=feed rate, a_p=axial depth of cut, a_e=radial depth of cut

Subsequent tests involved alternative tool paths for the mould cavity which were more suited to HSM. Figure 2 shows both cutter path strategies when finishing. The initial roughing path was modified so that the depth of cut was increased, thereby eliminating the need for a semi-finishing operation. The finishing tool path underwent the most significant change. The traditional method involved constant, rapid changes in cutter direction and

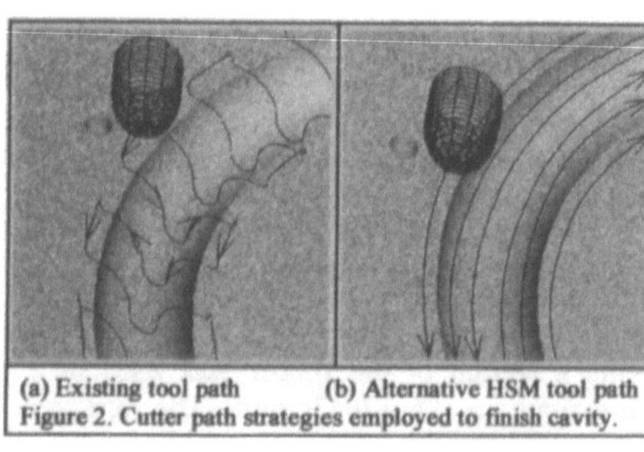

(a) Existing tool path (b) Alternative HSM tool path

Figure 2. Cutter path strategies employed to finish cavity.

therefore significant acceleration / deceleration of feed rate, together with mixed down / up milling. Such practice is likely to result in relatively low tool life and high workpiece surface roughness [13],[14]. The alternative HSM path employed fewer direction changes and solely down milling.

Production time was, as anticipated, greatly reduced using HSM. Table 3 shows the machining times for the various operations when using the alternative machining methods. The total time was reduced by approximately 8 and 12 times, respectively, over the time taken using the traditional technique.

Table 3. Machining times using conventional and HSM methods.

Times in minutes	Method		
Operation	Traditional	HSM with existing tool paths	HSM with alternative tool paths
Drilling	86.5	6	6
Reaming	30	8.5	8.5
Profiling	146	20	20
Cavity roughing	105	24.5	10.5
Cavity semi-finishing	147	19	-
Cavity finishing	360	30.5	24
Total	**874.5**	**108.5**	**69**

Surface roughness, Ra, of the mould cavity was measured using replica casts and a Taylor Hobson Pneumo Form Talysurf with cut-off and evaluation lengths of 0.8 and 5mm, respectively. Average surface roughness, Ra, results obtained after using HSM with the traditional tool paths were 1.66µm (transverse) and 1.07µm (longitudinal). This compared with 1.14µm (transverse) and 0.96µm (longitudinal) when using HSM with alternative tool paths. Roughness, however varied substantially around the circumference of the replica, with the lowest values corresponding with the bottom of the cavity. Chatter marks were clearly visible on the cavity surface after using both types of tool path. These were more prominent on the side walls (hence the higher roughness values) and were thought to be caused by radial deflection of the tool. Another factor was vibration caused by the speed increaser. Tool flank wear was minimal throughout all machining operations, with maximum values of 50-80µm after the machining of a cavity. The wear profile was uniform with no chipping.

4. Conclusions

- In addition to milling, other operations such as drilling, tapping and reaming of hardened tool steels up to 60HRC need to be addressed in order to fully benefit from HSM.
- HSM of a typical rubber moulding tool reduced machining times from over 14 hours with traditional methods to less than 2 hours.
- HSM tool paths provided overall productivity gains of over 30% compared with traditional tool paths.
- Vibration of the spindle speed increaser caused chatter. Despite this, Ra values of the finish machined mould cavity were typically 1-2µm.

Acknowledgements

The authors would like to thank Prof. A.A. Ball (Head of the School of Mechanical and Manufacturing Engineering), the Teaching Company Directorate / DTI / EPSRC, GTI Corporation Ltd, Portway Tool & Gauge Ltd and TRT Manufacturing Ltd for provision of facilities and funding. Thanks also go to the GTMA, Bridgeport Machines Ltd and Sandvik Coromant UK for the supply of equipment / consumables / technical support, together with Dr. Philip Koshy (Research Fellow) who was involved in the case study.

References

[1] Kakino Y., Matsubara A., Kohno Y. et al, "High speed, high productive machining of automobile parts by a machining centre with high speed and high acceleration rate", *Proc. 3rd Int. Conf. on Progress of Cutting & Grinding*, Osaka, Japan, pp. 192-197, 1996

[2] Derrien S. and Vigneau J., "High speed milling of difficult to machine alloys", *Proc. 1st French/German Conf. on High Speed Machining*, Metz, France, pp.284-294, 1997

[3] Fallbohmer P., Rodriguez C.A., Ozel T. et al, "High speed machining of cast iron and alloy steels for die and mold manufacturing", *Journal of Materials Processing Technology (98)*, pp.104-115, 2000

[4] Dewes R.C. and Aspinwall D.K., "High speed machining of hardened steels using tungsten carbide ballnose endmills", *Proc. 2nd Int. German/French Conf. on High Speed Machining*, Darmstadt, Germany, pp.165-174, 1999

[5] Boehner J., Dumitrescu M., Elbestawi M.A. et al, "Effect of carbide tool grades and cutting edge geometry on tool life during high speed machining of hardened tool steel", *Proc. 2nd Int. German/French Conf. on High Speed Machining*, Darmstadt, Germany, pp.37-61, 1999

[6] Munz W.D., Smith I.J., Donohue L.A. et al, "PVD coated cemented carbide tools designed for dry high speed machining applications", *Proc. 1st French/German Conf. on High Speed Machining*, Metz, France, pp.173-183, 1997

[7] Hock S., "High speed cutting in die and mould manufacture", *Proc. 1st French/German Conf. on High Speed Machining*, Metz, France, pp.274-283, 1997

[8] Elbestawi M.A., Chen L., Becze C.E. et al, "High speed milling of dies and molds in their hardened state", *Annals of the CIRP Volume 46/1*, pp.57-62, 1997

[9] Business and Trade Statistics, Surrey, UK, "UK trade statistics", 1999

[10] GTMA, Princes Risborough, UK, "Business Survey-3rd quarter 1999" (commercially confidential), 1999

[11] Office of National Statistics, Newport, S. Wales, UK, "Product commodity report", 1999

[12] Dewes R.C., "High speed machining of hardened ferrous alloys", PhD Thesis (commercially confidential), University of Birmingham, 1997

[13] Tonshoff H.K. and Hernandez-Camacho J., "Die manufacturing by 5- and 3- axis milling (influence of surface shape on cutting conditions)" *Journal of Mechanical Working Technology, 20*, pp.105-119, 1989

[14] Gaida W.R., Rodriguez C.A., Altan T. et al, "Preliminary experiments for adaptive finish milling of die and mold surfaces with ball-nose end mills", *Trans. NAMRI/SME, Volume XXIII*, pp. 193-198, 1995

An Investigation of Temperatures in Form Grinding.

M. N. Morgan[1] and W. B. Rowe[2]
[1,2] Liverpool John Moores University, School of Engineering, Liverpool
E-mail[1]: m.n.morgan@livjm.ac.uk, E-mail[2]: b.rowe@livjm.ac.uk

Abstract: The temperature rise experienced by the workpiece during grinding may be calculated with reasonable confidence for simple shapes: cylindrical and plane surfaces. In the case of more complex forms, the thermal analysis must account for non-uniform temperature distributions that result from variations in contact conformity and material removal rates across the entire profile of the form. In this paper a general approach to the thermal analysis of form grinding is presented. Temperatures measured from experiment are compared with temperatures predicted from theory and shown to exhibit good correlation. The predictive accuracy of the theoretical model is further verified using data obtained from on-site grinding tests.

1. Introduction

The basis of most techniques for the calculation of surface temperatures in grinding lies in the moving heat source model of Jaeger [1]. The Jaeger based models [2-5] concentrated on the analysis of basic geometries such as the flat semi-infinite plane and cylinders. The analysis of temperature distributions in more complex geometries however, has largely been neglected. This paper aims to address that deficiency. As a particular example an investigation of the temperatures in the vee form is described. The vee form is common to many geometries encountered in grinding and is also an important elemental shape because many other forms can be derived from the vee shape.

2. Thermal Modelling

The thermal performance of grinding can be simplified as a band of heat passing a semi-infinite workpiece body at a workspeed v_w. For conventional grinding operations it has been shown [6] that the maximum surface temperature $\theta_{\max}$, in the grinding contact zone can be presented as

$$\theta_{\max} = \frac{C.R_w.q}{\beta_w}\sqrt{\frac{l_c}{v_w}} \qquad [1]$$

where R_w is the partition ratio, q is the average total heat flux in the grinding zone, l_c is the real contact length, β_w is the thermal contact coefficient for the workpiece and C is a constant depending on the pattern of heat flux distribution. In shallow cut grinding operations the surface temperature profile and the position of the maximum temperature has been shown [7] to be best described by a triangular source.

In conventional shallow cut grinding, the real contact length l_c, may be 1.5 to 3 times the geometric contact length l_g, and the use of the geometric contact length can therefore lead to significant error in the temperature predicted. The error can be greater in form grinding where there is a varying contact conformity across the profile of the form. For reliable and accurate temperature prediction it is necessary to use the real contact length determined from a model such as that of Rowe and Qi [8] which takes into account the effect of elastic deflection.

2.1 Power distribution

The approach for the distribution of power is based on the division of the contact zone into elemental strips across the profile of the form [9,10]. The total grinding power P, is first distributed to each surface of the workpiece based on the material removal rate z_c, along each surface. The grinding power is then further distributed to each strip of the surface. The power at each strip is obtained by assuming a proportional relationship with the removal rate at the strips such that: $z_n = ab_n v_w$ where b_n is the width of a strip, and n is the number of each strip, n = 1,2...,N. The power distributed to a strip is

$$P_n = P_c z_n / z_c \qquad [2]$$

The flux on each strip is

$$q_n = R_w P_n / A_n \qquad [3]$$

where $A_n = b_n l_{cn}$ the area of a strip, P_n is the power on a strip and l_{cn} is the contact length of a strip. R_w is the proportion of the total heat flux q_t, entering the workpiece and is termed the partition ratio.

2.2 Partition ratio

The temperature rise experienced by the workpiece depends in part on the proportion of the total grinding energy to which it is subjected. The greater the proportion, the higher the temperature rise. An expression for the partition ratio developed by Rowe, Morgan and Black [11], that includes terms for convective cooling and conduction to the chips was given by

$$R_w = \left(1 + \frac{k_{ge}}{\sqrt{r_o . v_s . (\kappa . \rho . c)_w}}\right)^{-1} . (1 - R_{ch} - R_f) \qquad [4]$$

κ_{ge} is the effective grain thermal conductivity, r_o is the grain contact radius [6].

The partitioning to the chips R_{ch}, is calculated based on the workpiece material melting temperature. A typical value of e_{ch} is approximately 6 J/mm^3 for ferrous materials [12]. An iterative procedure for the determination of convection to the fluid has been developed [9]. At the beginning of the iteration, the initial value of R_f is assumed to be zero and the workpiece partition ratio and surface temperatures are calculated. The flux convected by the fluid is then calculated and the workpiece partition ratio re-calculated. The initial value of workpiece partition ratio is compared with the re-calculated value and the iterative procedure continues until the convergence criterion is satisfied.

2.3 Temperature distribution at a discontinuity

The temperature at a discontinuity is calculated by incorporating a concentration factor, n, at this point. The factor was found to be a function of six non-dimensional groups corresponding to the ratios of applied flux, contact length and included angle for the surfaces adjacent to the junction:

$$n = \varphi\left(\frac{q_1}{q} = p_1, \frac{q_2}{q} = p_2, \frac{\alpha_1}{\alpha} = p_3, \frac{\alpha_2}{\alpha} = p_4,, \frac{l_1}{l} = p_5, \frac{l_2}{l} = p6\right) \qquad [5]$$

The function φ of equation 5 was evaluated using finite element analysis and a linear regression technique. The temperature at the discontinuity was obtained by factoring the plane temperature with the concentration factor for the point [13].

3. Thermal Model Validation

The correlation between temperatures measured from experiments and temperatures predicted was tested by comparing the ratio of the two values to unity. The resulting ratios are shown for the flanks of an inverted vee form workpiece in Figure 1.

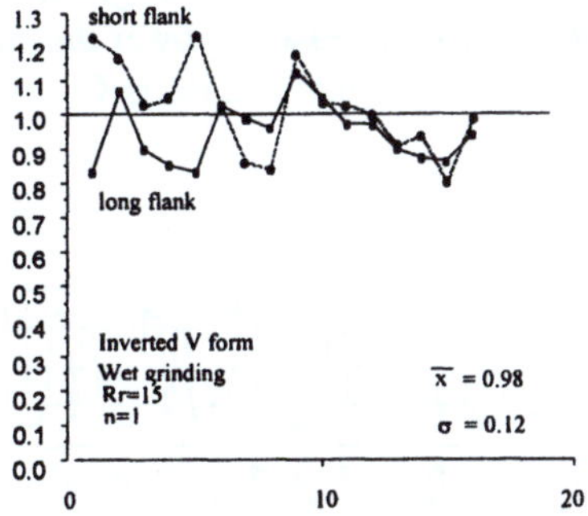

Machine	: Abwood 5025
Material	: En31
Wheel	: 77A601H8VLNAA
Wheel dims	: OD 200 mm, max
Coolant	: Hysol X, 25:1
Wheelspeed	: 4405 rpm
Workspeed	: 0.1 - 0.3 m/s
d.o.c	: various

Figure 1 Temperature ratios for the flanks of the inverted vee form, wet grinding.

The spread of experimental results about the mean value indicates that the theoretical model is reasonably accurate in predicting temperatures when compared with experimental scatter. The scatter evident in the experimental results is considered to be partly a consequence of the subjectivity in depth of cut and power data interpretation and partly the temperature measurements.

Further validation of the theoretical model was achieved through industrial on-site grinding tests. A batch of fifty components with the required geometrical features were selected for the grinding tests.

Each component was measured for size prior to the grinding test. The grinding cycle consisted of two separate operations: (1) angle grinding of the external diameters and (2) angle grinding of the faces. It was reported that the feature subject to thermal damage was the face. The cycle for the face grinding operation involved firstly a z-axis infeed followed with an x-axis infeed.

Process data for industrial tests:

Machine	Curtis D120	*Dressing tool*	Fliece
Wheel	25A100J63VPMCNN	*Dressing Speed*	X – 500 mms^{-1}
Wheel dims	450 x 40 x 203.2mm		Z – 150 mms^{-1}
Wheelhead angle	20 degree	*Dressing frequency*	1 in 14 or 15 (usual)
Wheelspeed	1700 rpm	*Depth of cut*	10 μm
Workspeed	800 rpm	*Rollscan limits*	< 60 – 65
Material	High alloy steel	*Coolant application*	Slot, flood nozzle
Coolant	HL/D1046, 3-4% conc.		

To establish a working set-up the wheel was dressed and one component was ground. The spindle power was recorded via a pc. Consecutive components were then ground until an abrupt change in the power signal was evident. The grinding wheel was gradually loaded by further grinding. On the 38th grind the power on the face grind operation was observed to markedly increase and it was assumed that this phenomenon was indicative of the onset of thermal damage. The power measurements for the set-up grind and first five components are given in Figure 2a, and for grinds 35 – 40 in Figure 2b. The marked increase in power during the 38th grind can be clearly identified on the Figure. The power scale is 0 – 10 volts and the calibration factor was 0.8kW/Volt.

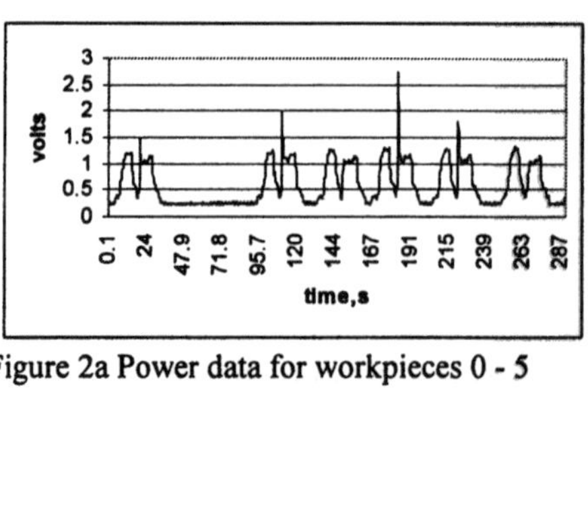

Figure 2a Power data for workpieces 0 - 5

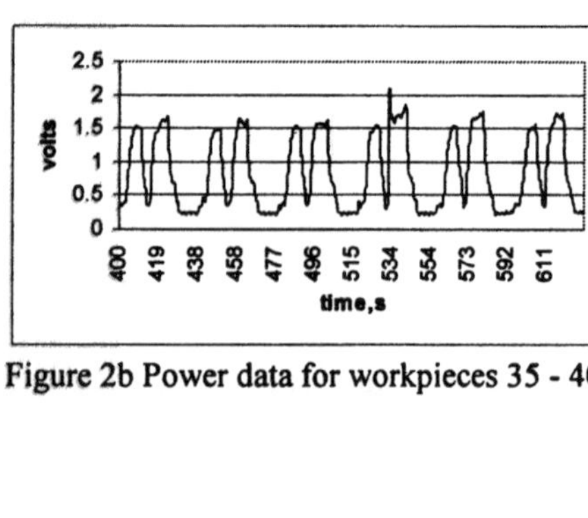

Figure 2b Power data for workpieces 35 - 40

4. Grinding Test and Theoretical Results

The validity of the theoretical model can be established by comparing predicted values of flux q, for the onset of thermal damage with measured values. The fact that the cycle is completed in two operations is helpful in that the power applied to the faces and to the external diameters is already separated. In the case of combined face and shoulder grinding however, the power distribution would be determined using the methodology described earlier. To simplify the comparison l_c was assumed to be $l_c = 1.5l_g$ and the effect of coolant was assumed negligible [11]. The critical workpiece temperature $\theta_{\max}$, was determined [9] at $\theta_{\max}$ =350 deg.C. Values of workpiece speed, v_w and effective wheel diameter, d_e were computed at positions across the face by dividing the diameter of the face into annular strips.

4.1 Measured values

The specific energy was computed from the total power on the faces divided by the total material removal and a proportional relationship between the power and material removal was reasonably assumed. The flux at a point on each annulus was computed and compared with theoretical values. The results are shown in Figure 3. The theory is observed to correctly predict safe grinding, threshold conditions and the occurrence of damage. Early in the tests the process was very efficient with values of e_c typically in the range 15 – 20 J/mm^3. The effect of conduction to the chips was therefore greater and overall partition ratios R_w, in the range 0.55 – 0.6 were computed. Later in the test, process efficiency reduced and values of e_c were typically 25 J/mm^3. R_w increased correspondingly up to a maximum of 0.7.

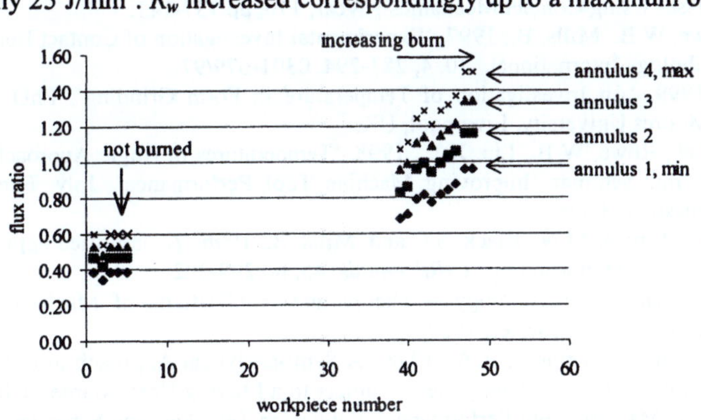

Figure 3 Grinding tests results: ratio of measured flux and transitional damage flux predicted from theory.

5. Concluding Discussion

A theoretical approach has been outlined that can be used to predict the onset of thermal damage in angle approach form grinding operations. An iterative procedure for calculating the energy partitioning to the fluid is included. Grinding temperatures predicted from theory have been shown to correlate well with temperatures measured from grinding experiments. The theoretical approach has been further validated with grinding data recorded from an industrial application. It has been shown that in vee form angle approach grinding operations the geometric discontinuity at the junction of two planes promotes a temperature concentration at that point and as a consequence subjects the workpiece to a higher than normal risk of thermal damage. A small radius imparted to the corner of the wheel significantly reduces the temperature concentration. It has also been demonstrated that power monitoring coupled with a predictive temperature model provides the basis for the development of a thermal damage monitor.

References

[1] Jaeger, J C., 1942 Proceedings Royal Society, New South Wales, 76, 203.

[2] Hahn, R, 1962, 'On the nature of the grinding process', 3rd MATADOR Conf. pp129-155

[3] Shaw, M C.,1990, 'A simplified approach to workpiece temperatures in fine grinding', *CIRP Annals* 39, pp 345-347.

[4] Rowe, W B., Black, S C., Morgan, M N., et al, 1995, 'Experimental investigation of heat transfer in grinding', *CIRP Annals* 44, pp 329-332.

[5] Lavine, A S., Jen, T C., 1991, 'Coupled Heat Transfer to Workpiece, Wheel and Fluid in Grinding and the Occurrence of Workpiece Burn', Int. Journal of Heat and Mass Transfer, 34n 4/5, pp 983-992.

[6] Rowe, W B., Black, S C., Morgan, M N., et al, 1997, 'Grinding temperatures and energy partitioning', Proc. R. Soc. London. A, 453, pp 1083-1104.

[7] Qi, H S., Rowe, W B., Mills, B., 1994, 'An Analysis of Real Contact Length in Abrasive Machining Processes Using Contact Mechanics', Wear, 176, pp 137-141.

[8] Qi, H S., Rowe, W B., Mills, B., 1997, 'Experimental Investigation of Contact Behaviour in Grinding', Tribology International. 30, 4, 283-294. 0301-679/97.

[9] Lin, Z X., 1999, 'An Investigation of Temperature in Form Grinding', PhD Thesis, Liverpool John Moores University, Liverpool, UK.

[10] Morgan, M N., Rowe, W B., Lin, X L., 1998, 'Temperatures in Angle Approach Form Grinding', CIRP Int. Seminar 'Improving Machine Tool Performance', July, Fundacion Tekniker, San Sebastian, Spain.

[11] Rowe, W B., Morgan, M N., Black, S C and Mills, B., 1996, 'A simplified approach to control of thermal damage in grinding', *CIRP Annals* 45, pp 299-302.

[12] Malkin, S., Grinding Technology – Theory and Applications of Machining with Abrasives, Ellis Horwood, Chichester, 1989.

[13] Walsh, D., Izadi, H., Torrance, A A., 1998, 'A Semi-analytical Approach to Evaluating Residual Stress on the Surface of Components Subject to a Moving Heat Source', CIRP Int. Seminar 'Improving Machine Tool Performance', Fundacion Tekniker, San Sebastian, Spain.

Acknowledgements

This work was completed as part of the FormGrind project, sponsored by Brite-Euram under DG XII of the EU as part of the Fourth Framework programme, in collaboration with industrial partners: LucasVarity(UK); Dormer Tools(UK); Unicorn Int.(UK); Castrol(UK); Erasteel(Sweden) Lidköping(Sweden), Trinity College, Dublin(Eire) and SIMR(Sweden).

Creep Feed Grinding of γ-TiAl

S.A.Bentley [1] and D.K.Aspinwall [1,2]

[1] IRC in Materials for High Performance Applications, The University of Birmingham, s.a.bentley@bham.ac.uk

[2] School of Manufacturing and Mechanical Engineering, The University of Birmingham, d.k.aspinwall@bham.ac.uk

Abstract: The paper initially reviews TiAl development in relation to aerospace applications. Following this, experimental data is presented which examines the effect of dresser infeed rate and springback passes at reduced depths of cut, on cutting forces and workpiece surface integrity, specifically residual stress and subsurface microstructure. A 3^2 full factorial experiment with replication was carried out. Plots of cutting forces against dresser in feed rate produced near linear trends, with the normal force rising by up to 75% when the feed rate was reduced from 1.5 to 0.5 μm/rev. In contrast, the tangential force increased by up to 25%. Workpiece residual stress measurements were made using the blind hole drilling method. Large tensile stresses of up to 700 MPa were evident in the direction of cut corresponding to the direction of plastic flow, caused by the moving heat source. In contrast, compressive stresses were measured in the transverse direction with values up to 600 MPa.

1. Introduction

Although titanium was first identified in 1791, it is a relatively new engineering material, which only became commercially available in a usable form in the 1950's. The combined high strength and low density of current α/β titanium alloys makes them attractive for jet engine components, although operating temperature restrictions (< 600°C), effectively confine their use to the cooler intake/compressor sections. This is likely to change however, in the near future, with the use of gamma titanium aluminide (γ-TiAl) intermetallic alloys, based on a stoichiometric combination of titanium and aluminium.

With the potential to operate at up to 800°C and at only half the density of nickel based superalloys, γ-TiAl offers the possibility for significant weight savings in both military and civil engines for compressor stator vanes and blades. Much is also made of its improved specific stiffness over conventional titanium alloys (some 50% higher) and greater burn resistance, however, less attractive, at least from a machinability standpoint, is its room temperature ductility, which is less than 2% at temperatures below ~ 600°C. This compares with 10-15% for Ti-6Al-4V [1].

Research on all aspects of TiAl development (alloy formulation, processing, lifing and machinability) has taken centre stage over the past 15 years, yet currently none of the worlds' leading aero-engine manufacturers (Rolls Royce, Pratt & Whitney and GE), have TiAl components in production. The great potential of these alloys is not disputed, the reasons for delay in commercialisation being the same as for any new high performance, safety critical alloy, namely: cost, supply/availability, risk and development timescales [2]. Progress on the automotive front, where γ-TiAl is seen as a potential replacement for existing valves and turbo impeller materials, has understandably progressed more quickly. As yet however, the authors are unaware of any mainstream production vehicles employing γ-TiAl, despite the development and testing of a wide range of alloys [3].

First generation aerospace γ alloys, such as Ti-48Al-1V, produced in the late 1970's, were found to lack strength and creep resistance. The use of structure / property analysis in the following decade led to the development of alloys such as Ti-48Al-2Cr-2Nb, which in turn gave way in the mid 1990's to third generation grain refined alloys such as Howmets' Ti-45Al-2Mn-2Nb+0.8vol%TiB_2 XD™ (45-2-2+0.8)[4]. Of recent note is the production of higher strength, more highly alloyed compositions such as Ti-45Al-4Nb-4Zr-1B which is capable at operating at significantly higher temperatures than previous alloys [5]. Rolls Royce suggest that there is a high probability of using γ-TiAl in the near future, the 45-2-2+0.8 grain refined alloy being considered for components operating at up to 650°C, "where established casting and machining processes can be relied upon" [6].

2. Experimental Work

Despite progress on alloy production, the machining of γ-TiAl remains "difficult" with regard to a range of key conventional and non-conventional cutting processes. As previously outlined, the materials' low room temperature ductility and its chemical reactivity are chiefly to blame. When turning and drilling, productivity is low, tool wear is high and workpiece surface integrity is generally poor [7-9]. The following experimental work, was undertaken to assess the effect of dresser infeed rate and spark out passes at reduced depth of cut on workpiece surface integrity, when creep feed grinding 45-2-2+0.8.

2.1. Workpiece material, tools and equipment

Blocks of 45-2-2+0.8 were used measuring 200x20x50mm. These were produced by induction skull melting, investment casting and HIPping at 1260°C/170MPa for 4 hours, followed by heat treatment at 1010°C for 50 hours. Test pieces were electrical discharge wire-machined to 20x20x50mm on a Charmille Robofil 200. Grinding trials were performed on an ELB creep feed unit, located at the Institute for Grinding Technology (IGT) Bristol, using continuously dressed vitrified bonded SiC wheels, see Table 1 (specific parameters have been omitted due to the commercial sensitivity of the work). Grinding force measurements were made using a Kistler three component piezoelectric platform dynamometer (type 9257A), fitted with a custom-built test piece holder, connected to a 2-channel oscilloscope, via Kistler 5011A charge amplifiers.

A soluble oil cutting fluid was used supplied at high volume via nozzles either side of the grinding wheel. Photomicrographs of the ground workpiece surface were produced using an Olympus OM-2 35mm camera system attached to a Lietz metallographic microscope.

Residual Stress Measurements were undertaken using the hole drilling method, on a RS-200 Milling Guide, at Rolls Royce Plc. The system, uses a strain gauge rosette array which is bonded to the workpiece and through which a 0.9mm diameter hole is drilled at incremental depths of ~60μm up to a maximum of ~ 0.7mm. Output stress values (calculated from resulting strains), which includes principle stress data, are by convention plotted at half the drilled depth, in the case of the first reading at 30μm. Compared with x-ray diffraction (XRD), the hole drilling method is more problematic and provides less discrimination of the residual stress state in the near surface region of the workpiece. Unfortunately, conventional XRD measurement employing traditional sample tilting, within the x-ray diffractometer, gives unreliable results with γ-TiAl due to significant line broadening [10]

2.2. Experimental procedure

Previous work by the authors, indicated that continuous dressing was necessary in order to prevent workpiece burn [11]. These tests, however, had primarily involved bulk material removal and did not address the effect of dresser infeed rate or a springback pass on the workpiece surface integrity. To accommodate this, a 3^2 full factorial experiment with replication was designed. Table 1 details the test factors and levels. Figure 1 details the experimental arrangement together with the force and stress nomenclature/orientations used.

Table 1: Test factors and levels

Fixed factors	Level *
Wheel	Soft to Medium grade / Medium grain, vitrified bonded SiC
Wheel speed (ms^{-1})	Conventional
Depth of cut (mm)	Low
Workpiece feed rate (mm min^{-1})	Slow
Dressing methods	Continuous
Coolant	Soluble oil
Coolant flow rate (ls^{-1})	High

Variable factors	Levels
Dresser infeed rate (μm rev^{-1})	0.5 / 1.0 / 1.5
Depth of springback pass (μm)	0 / 50 / 100

* Specific parameters omitted due to the proprietary nature of the work.

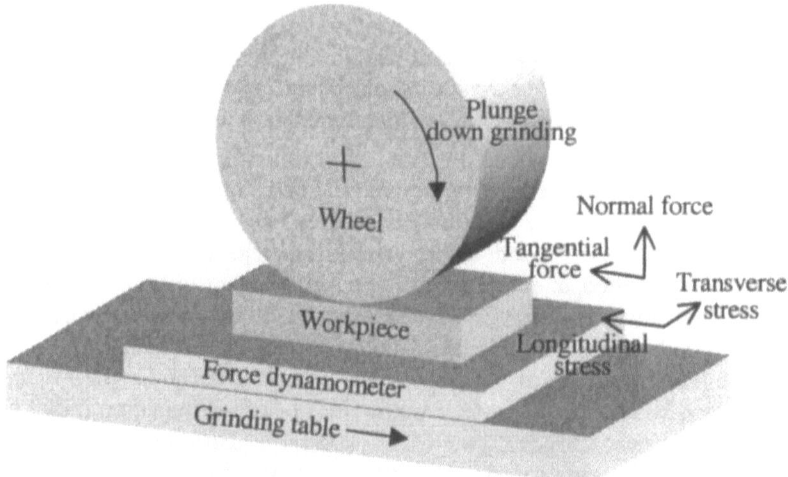

Figure 1: Experimental arrangement

2.3 Results and discussion.

Figures 2 and 3 detail force measurements against dresser infeed rate, measured during the rough cut before the application of a springback pass. The forces imparted by the springback pass were not measured. Figure 4 details the residual stresses in the near surface (~30μm depth) of the testpieces after grinding with and without a 100 μm springback pass.

The force data shows near linear trends with the normal forces rising by up to 75% when the dresser infeed rate was cut from 1.5 to 0.5 μm rev^{-1}. The tangential forces increased by up to 25% over the same range.

The residual stress results detail a clear distinction between transverse and longitudinal measurements. Large tensile stresses are evident in the direction of cut (longitudinal) corresponding to the direction of plastic flow caused by the moving heat source. Conversely, the transverse direction produced large compressive stresses due to the applied normal forces.

Figure 5 details a typical cross–sectional micrograph taken from a sample not subjected to a springback pass. The micrograph clearly shows the deformation in the lamellae to a depth of <10μm, but no surface/subsurface cracking is evident, suggesting a limited degree of plasticity in the material. The lighter phases are the TiB_2 particles which act as a grain.

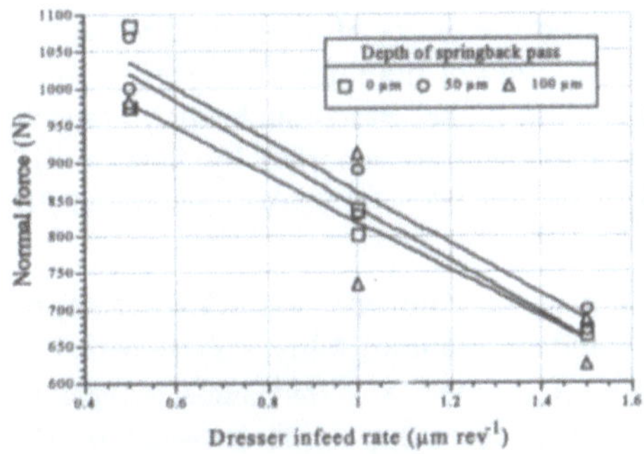

Figure 2: A plot of peak normal force against dresser infeed rate

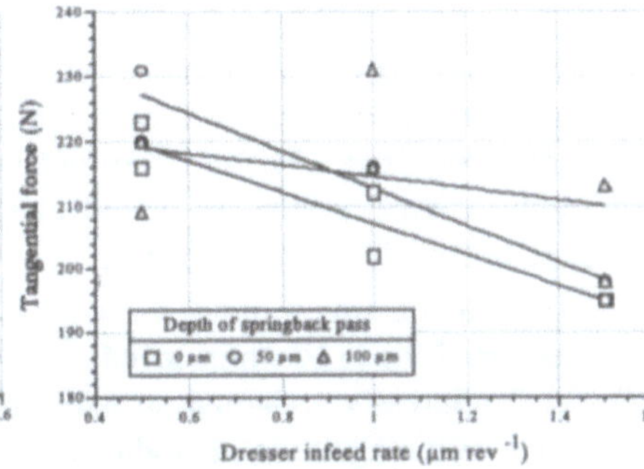

Figure 3: A plot of peak tangential force against dresser infeed rate.

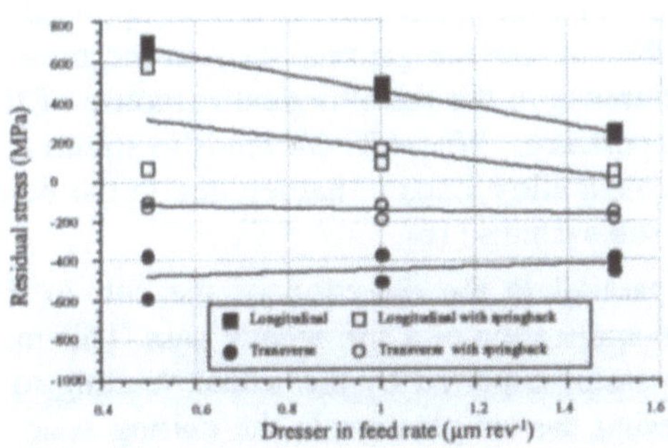

Figure 4: Residual stress data, at 30 µm depth.

Figure 5: A cross sectional micrograph of a workpiece surface

By increasing the area of contact between the wheel and the workpiece, as is the case when the dresser infeed rate is reduced, forces are increased. This leads to an increase in the wear flat area. The tangential and normal components of force vary with the product of the contact stress and wear flat area, according to Equations 1 and 2 respectively.

$$\mathbf{F_t = F_{t,c} + F_{t,sl} = F_{t,c} + \mu\bar{p}A_a} \qquad 1$$

$$\mathbf{F_n = F_{n,c} + F_{n,sl} = F_{n,c} + \bar{p}A_a} \qquad 2$$

where $\mathbf{A_a}$ is the actual area of contact, μ is the coefficient of friction, $\bar{p}$ is the average contact pressure and the subscripts c and sl represent cutting force and sliding force components respectively [12].

As can be seen from Equation 1, the tangential force is also dependent on the coefficient of friction so that as the sliding and ploughing components of the cutting force increase, so overall cutting force must rise to remove the same volume of material. This emphasises the importance of wheel dressing during the grinding process to reduce forces (and therefore power) and prevent workpiece burn.

No reference could be found that addressed the effect of dresser infeed rate on the forces generated during the creep feed grinding of titanium. However, a similar trend to that shown here, is seen when grinding nickel based superalloys (another material recognised as being 'difficult to machine'), as reported by Salmon [13], where an increase in dresser infeed rate, from 0.5 to 1.5μm rev^{-1}, produced 50% decrease in specific grinding energy.

The residual stress data, shown in Figure 4, compare favourably with that reported by Silin [14] when grinding a standard titanium alloy [OT4] using SiC wheels. Compressive residual stresses were also noted in a thin surface layer and were of lower magnitude at higher cut depths, indicating a link between residual stresses and temperature. However, the dressing method was not reported and therefore, a direct comparison with the present gamma alloy cannot be made. Campbell [15] also presents results of residual stress magnitude and direction when creep feed grinding Ti 6/4. His results detailed greater values of compressive residual stress parallel to the cut (longitudinal direction) for Ti 6/4 as opposed to perpendicular (transverse direction) to the cut, for the gamma alloy in the present work. This may be indicative of the difference in the metallurgical response of the two alloys to the induced strains and temperatures. Where Ti 6/4 tends to soften and flow plastically due to overaging, the gamma alloy tends to harden due to the build up of dislocations caused by the limited slip systems [16].

The tensile nature of the stresses parallel to the direction of the cut, in the surface layer, was greatly reduced by the application of a springback pass. This may be due to the reduction in compressive strain imparted by the wheel, leading to a reduction in the chip thickness, so allowing the temperature in the cutting zone to fall below the brittle / ductile transition temperature (600-700°C). Such a reduction in cutting temperature would be crucial as this is a major contributor to the formation of tensile residual stresses in machining [17]. This can be further extended to account for the drop in tensile stress with increasing dresser in feed rate. Here the sharper wheel would be expected to reduce $\mathbf{A_a}$ and therefore, the frictional component of force, which in turn should significantly reduce the cutting temperature. The reduction in compressive stresses perpendicular to the cut, after a springback pass, reflects the influence of compressive strain due to the normal force, on the workpiece surface.

The micrograph in Figure 5 clearly shows the limited deformation occurring in the near surface region of the workpiece (5-10μm depth). The level of "damage" compares favourably with that seen when turning the same alloy, where severe bending of the lamellae (in some cases 90°) is typically 20 to 30μm and is accompanied by extensive cracking [7].

3. Conclusions.

- As the dresser infeed rate increased and, therefore the effective sharpness of the wheel, both normal and tangential forces fell by as much as 75 and 25% respectively

- Residual stresses were tensile in the direction of cut (longitudinal) and compressive in the transverse direction. However, both these stresses were reduced in magnitude by the application of a springback pass leaving a surface that was crack free.
- No major damage to the workpiece surface integrity in the form of cracks, pits, laps, smears of phase changes was noted, in contrast to when single point turning.

Acknowledgements.

The authors would like to thank Professor M.H.Loretto Director of the Interdisciplinary Research Centre (IRC) in Materials for High Performance Applications and Professor A.A.Ball Head of the School of Manufacturing and Mechanical Engineering, both at the University of Birmingham, for the provision of laboratory facilities and workpiece materials. Thanks also go Dr. T.R.A. Pearce of the IGT for access to the Creep Feed Grinding unit and Mr. A.Day of Rolls Royce Plc for carrying out the residual stress measurements. Finally, we would like to express our gratitude to the IRC, Rolls Royce Plc, Unicorn Abrasives Ltd and the Engineering and Physical Sciences Research Council (EPSRC) for financial support.

References

[1] Perry's Chemical Engineers Handbook, 6th Ed., 1984, McGraw Hill, ISBN 0-070-049479, Ch.23

[2] Rugg D, 1999, "Titanium Aluminides, Great Potential – But Not Yet Flying", Gamma Titanium Aluminides, p11-14

[3] Mantle A.L., Aspinwall D.K., 1997, "Machining of Titanium Intermetallic", Proceedings of the 30th International Symposium on Automotive Technology and Automation, p619-626

[4] Austin CM, Kelly TJ, 1993, "Development and Implementation Status of Cast Gamma Titanium Aluminide", Structural Intermetallics, Proceedings of the First International Symposium on Structural Intermetallics, TMS

[5] Cheng TT, Jones I, "TiAl Alloys", European Patent Publication number 0889143IA

[6] Voice W, 1999, " The Future Use of Gamma Titanium Aluminides by Rolls Royce", Gamma Titanium Aluminides 1999, p397-400

[7] Mantle AL, Bentley SA, Aspinwall DK, 1998, "A Comparison of Tool Wear When Turning Ti-45Al-2Mn-2Nb+0.8Vol%TiB2 and Ti-44Al-8Nb-1B titanium aluminide intermetallics", Proceedings of the 4th International Conference on the Behaviour of Materials in Machining: Opportunities and Prospects for Improved Operations, p300-308

[8] Mantle AL, Aspinwall DK, Wollenhöfer O, 1995, "Twist Drilling of Gamma Titanium Aluminide Intermeatllics", Proceedings of the 12th Conference of the Irish Manufacturing Committee – IMC12, p229-236

[9] Aust E and Niemann H-R, 1999, "Machining of γ-TiAl", Advanced Engineering Materials, Vol.1, No.1, ISSN 1438-1656, p53-57
[10] Blackwell P, 2000, Personal communications with authors on γ-TiAl residual stress measurement by XRD
[11] Bentley SA and Aspinwall DK, 1997, "Identification of Significant Operating Factors when Conventional Abrasive Creep Feed Grinding of a Gamma Titanium Aluminide Intermetallic", Proceedings of the 4th International Charles Parsons Turbine Conference 'Advances in Turbine Materials, Design and Manufacturing' p474-488
[12] Malkin S, 1989, Grinding Technology, Theory and Applications of Machining with Abrasives, Ellis Horwood series in Mechanical Engineering
[13] Salmon SC, 1979, "Creep Feed Grinding", PhD Thesis, University of Bristol
[14] Silin SS, Leonov BN, Khrul'kov VA, et.al, 1989, "Special Features of Deep Grinding of Titanium Alloys", Vestmik Mashinostroeniya, Vol.69, No.1, p43-45
[15] Campbell JD, 1991, "A Comparison of Reciprocating and Creep Feed Superabrasive Machining of Titanium", SME, MR910215
[16] Kim YW, Froes FH, 1990, "Physical Metallurgy of Titanium Aluminides", Proceedings of the 1989 Symposium on High Temperature Aluminides & Intermetallics, p1-28.
[17] Wu DW, Matsumoto Y, "The Effect Of Hardness On Residual Stresses In Orthogonal Machining Of AISI 4340 Steel", Trans. of the ASME, Journal of Engineering for Industry, Vol.112, (1990), p245-252

Grinding of Inconel 718 and Udimet 720 Using Superabrasive Grinding Points Mounted on a High Speed Machining Centre

J.M. Burrows [1], R.C. Dewes[2] and D.K. Aspinwall [1,2]

[1] IRC in Materials for High Performance Applications, The University of Birmingham
[2] School of Manufacturing and Mechanical Engineering, The University of Birmingham

Abstract: Following a brief literature review, the paper presents initial feasibility and Taguchi data on the use of CBN and diamond mounted grinding points for the grinding of Inconel 718 and Udimet 720 nickel-based superalloys. Cutting speeds of up to 45 m/s were achieved through the use of a 60,000 rpm retrofit spindle, mounted on a high speed prismatic machining centre. With Inconel 718, tangential and normal forces were typically <30 and <80N respectively, when employing diamond abrasive at a feed rate of 2000 mm/min, depth of cut of 20μm/pass with a 10mm face width. Workpiece residual stresses were compressive, with values of up to 400 MPa in the direction of cut and microstructural damage was minimal. Subsequent L12 Taguchi experiments involving 8 factors at 2 levels, were used to identify key operating variables via main effects plots.

1. Introduction

Nickel-based superalloys are a diverse group of materials, commonly used for elevated temperature applications, in which high strength (Inconel 718 has a tensile strength of 1250 MPa at 650°C), excellent corrosion resistance and good creep / fatigue resistance are required. Their major application is in the aerospace industry for critical jet engine components such as turbine discs, blades, nozzles and vanes. They are also used in land and marine-based gas turbines [1].

Superalloy development and material processing has been the focus of considerable research effort over the last 40 years [2]. A great deal of progress has been made in producing alloys capable of meeting specific operating requirements, however, machinability can still present significant problems dependant on alloy composition/condition. When turning, milling, drilling etc, rapid tool wear is a common feature, irrespective of the tool material, not least because of workpiece susceptibility to strain hardening. The presence of hard abrasive particles such as carbides, oxides and silicates in the microstructure, adds to machining problems.

Relatively little has been published on the grinding of superalloys [3], due in part to commercial restrictions. Tso [4], evaluated the performance of silicon carbide, aluminium oxide and cubic boron nitride (CBN) abrasives when reciprocating surface grinding Inconel 718. From a purely technical standpoint, CBN performed best with grinding speeds of ~30 m/s, depths of cut of around 5-15μm and feed rates of up to 5000mm/min. Gardinier [5], came to similar conclusions. Much of the data that is available in the public domain, concentrates on creep feed grinding (CFG). Here, high removal rates are achieved with conventional abrasive wheels used at grinding speeds of ~30 m/s, cut depths of up to ~5 mm and relatively low feed rates of ~100mm/min [6,7]. Applications centre mainly on the production of blade root forms. Kumar *et.al.* [8] compared resin bonded CBN and diamond abrasives for the CFG of Inconel 718. They observed that CBN outperformed diamond, however in subsequent surface grinding tests they found that diamond required lower specific energy. They also highlighted the fact that electroplated CBN wheels used at higher speeds ~60 m/s, gave higher removal rates and lower wheel wear when compared with resin bonded products. In later work, Kumar [9] compared microcrystalline and monocrystalline electroplated CBN wheels for the CFG of a range of superalloys. The depth of cut used was 1mm and the wheel speed was 30m/s. He concluded that the tougher monocrystalline product required less power, although a higher wear rate was observed. Similar workpiece Ra values were produced irrespective of the particular superalloy, without the use of sparkout passes.

It is widely recognised that heat generated during the grinding process, can lead to thermal damage of the surface layer of the workpiece, resulting in tensile residual stress and poor fatigue life. Zhang *et.al.* [10] studied the CFG of metallic and ceramic surfaces. They showed that high heat input resulted in tensile residual stresses, however, use of an appropriate cooling medium produced compressive stress. Brinksmeier *et.al.* [11], in a comprehensive literature review, discussed the effect of cooling and lubrication when grinding a wide range of alloys, including nickel-based superalloys. Ye and Pearce [12] examined the use of water-based and neat oil fluids for the CFG of superalloys. They concluded that grinding with an oil resulted in less wheel wear and improved workpiece surface roughness, although normal grinding forces increased and the system was susceptible to workpiece burn. Consequently they recommended the use of water based coolant. Webster [13] proposed that optimisation of fluid application to the grinding zone, will in turn optimise the performance of the process. His work detailed considerations for jet velocity, thickness, attitude and flow rate. He recommended that coolant velocity be matched to wheel speed (for optimum fluid inclusion in the grinding zone), except in high pressure applications.

A recent development in grinding technology and coolant application of relevance to the present work is the 'Viper' grinding process [14,15]. The approach uses an ultra-high porosity conventional abrasive grinding wheel, for the CFG of superalloy turbine blade root forms. Typically wheels are ~150mm diameter and periodically, as opposed to continuously dressed. The process utilises a wheel speed of ~40m/s and a high pressure coolant jet ~70bar, which is kept at a fixed distance from the wheel ahead of the grinding zone via a novel mechanical system. A major advantage of the Viper grinding approach is the use of a flexible CNC machining centre.

2. Experimental Work

The following work details initial feasibility / Taguchi experiments undertaken as part of a large research programme to establish machinability / productivity and workpiece surface integrity effects, when utilising small diameter superabrasive wheels on a high speed machining centre [16].

2.1. Workpiece materials, grinding wheels and equipment

Workpiece materials were Inconel 718 (composition 50-55% Ni, 17–20% Cr, 2.8-3.3% Mo, 0.65-1.15% Ti, plus trace elements) and Udimet 720 (composition 55% Ni, 15% Co, 15% Cr and 0.5% Fe plus trace elements). Both were supplied in the solution treated and aged condition with bulk hardnesses of ~43 HRC. Rectangular blocks 50 x 50 x 25mm were electrical discharge wire machined from larger billets for subsequent mounting on a force dynamometer. Specimens were prepared to provide a grinding surface measuring 50mm x 10mm. Diamond and CBN grinding points ~15mm diameter x 25mm long, were collet mounted initially in a Faemat FA80, 45,000 rpm, 3.5 kW retrofit spindle attached to the spindle face of a Matsuura FX-5 high speed machining centre (feed rates up to 15,000 mm/min). In subsequent Taguchi tests a higher speed 60000rpm Faemat spindle was employed. A high pressure manifold ring provided the necessary coupling between the various units and provided the facility for cutting fluid application. This was supplied to the grinding zone via two nozzles connected to a 300 bar / 26 l/min supply system from Pumps and Equipment (Warwick) Ltd. Figure 1 shows a general view of one of the high speed spindles, together with a close-up of the grinding point / workpiece arrangement. The cutting fluid was a 5% soluble oil.

Forces were measured using a Kistler 9257A piezoelectric platform dynamometer and 5011A charge amplifiers, connected to a 4 channel oscilloscope. Residual stress measurements were made using a TEC 1610 X-ray diffraction residual stress analyser at Rolls-Royce. This used a Fe kβ source and 2θ angles of 0°, 15°, 24°, 35° & 45°. Workpiece surface roughness analysis (mean of 5 measurements) employed a Taylor Hobson - Form Talysurf Series 120 stylus profilometer with cut-off and evaluation lengths of 0.8 and 4 mm, respectively.

2.2. Experimental procedure

Initial feasibility tests were confined to Inconel 718 and employed a number of electroplated diamond and CBN wheels together with several vitrified CBN wheels. Grinding speed was 35m/s (as a consequence of using the 45000rpm Faemat retrofit spindle) while feed rate and depth of cut were varied between 2000–5000mm/min and 5–15 μm/pass respectively. Tests were completed to a workpiece depth of at least 300μm. Evaluation included: grinding forces (tangential – F(y) and normal

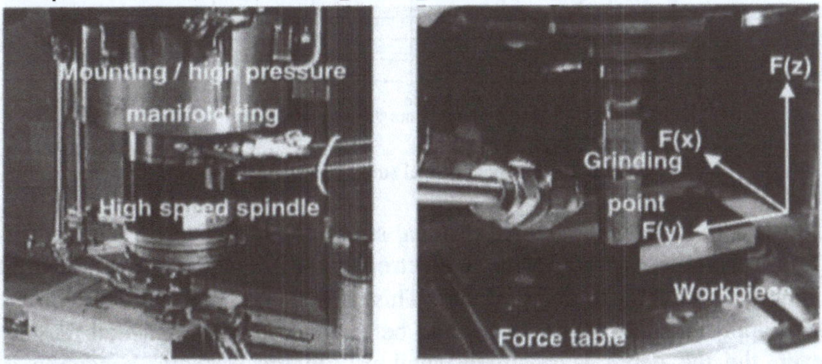

(a) General view (b) Grinding point / workpiece arrangement

Figure 1 – High speed grinding set-up

F(x)), workpiece surface roughness (across the lay) and surface/subsurface residual stress. The subsequent Taguchi experiments involved both Inconel 718 and Udimet 720 with the 60000rpm spindle and an L12 orthogonal array with 8 variable factors each at 2 levels (this arrangement did not allow evaluation of interactions). All wheels used inthese experiments were electroplated with a concentration of ~50% by volume and had a plain profile. The cutting fluid flow rate measured at the nozzles was 20l/min. Evaluation was limited to normal grinding forces and workpiece surface roughness. See Table 1 for variable factors and levels (some of the data have been coded).

	Variable factor		Level	
			Low (1)	High (2)
Workpiece	A	Material	Inconel 718	Udimet 720
Wheel	B	Grit material	CBN	Diamond
	C	Grit size	Low	High
	D	Abrasive toughness	Low friability	High friability
	E	Wheel speed (m/s)	35	45
	F	Depth of cut (μm)	5	15
	G	Workpiece feed rate (mm/min)	1000	4000
Fluid	H	Pressure (bar)	25	100

Table 1 – Taguchi experiment factors and levels

2.3 Results and discussion

In the feasibility tests, surface roughness varied from ~0.4 – 4.5μm Ra. The lowest value was produced using a CBN vitrified wheel with a depth of cut of 15μm and a feed of 5000mm/min. Tangential and normal forces were typically between 20-30N and 60-80N respectively, irrespective of wheel type. Minimum forces were obtained when using an electroplated diamond wheel with a depth of cut of 5μm and a feed rate of 2000mm/min.

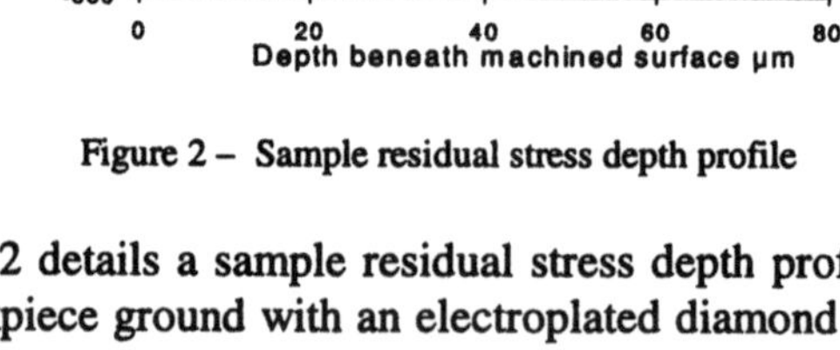

Figure 2 – Sample residual stress depth profile

Figure 2 details a sample residual stress depth profile (in the direction of feed) for a workpiece ground with an electroplated diamond wheel at a feed rate of 4000mm/min and a 15μm depth of cut. This shows a compressive layer down to a depth of approximately 50μm and would be expected to provide good fatigue life, however, the the depth is relatively shallow. Figure 3 shows a sample cross-sectional photomicrograph of a polished and etched workpiece, point ground with an electroplated diamond wheel. No significant thermal damage is evident and the microstructure shows minimal deformation.

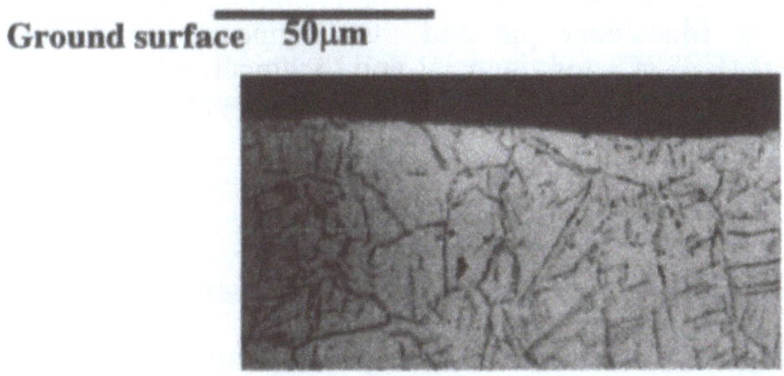

Figure 3 –Sample subsurface microstructure

The main effects plots for the Taguchi experiments are detailed in Figure 4. The steepness of the response lines indicates the significance of the factor in relation to the output measure at the levels tested. The forces generated when grinding Udimet 720 were higher than with Inconel 718, due to the former material's higher strength. The fact that forces increased with speed was unexpected. Diamond grits would be expected to give a lower force than CBN due their higher hardness and retention of edge geometry. The main factor affecting workpiece Ra was grit size where, as expected, larger grits produced higher values. The fact that higher depth of cut and feed rate both resulted in reduced roughness is more difficult to explain. It is possible that this was due to increased stiffness and less vibration throughout the system. All other Ra trends are within the range of scatter (up to ~20%) normally associated with such measurements.

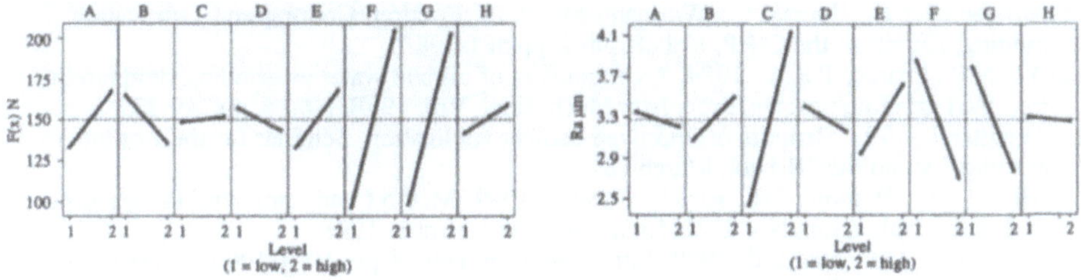

Figure 4 – Main effects plot analysis of grinding force (F(x)) and surface roughness (Ra)

3. Conclusions

- The work demonstrated that the use of mounted grinding points on a high speed prismatic machining centre is a feasible production technique.
- Workpiece surface roughness values were 0.4–4.5µm Ra and observed to be lowest when using a CBN wheel with high feed and high depth of cut
- Workpiece surface microstructural analysis showed that little or no damage had occurred and residual stress measurements indicated a compressive (up to –400MPa) surface layer of 50µm.
- Normal grinding force increased with depth of cut, feed rate and wheel speed

Acknowledgements

We are indebted to Rolls-Royce plc, DeBeers Industrial Diamonds UK Ltd, Unicorn International plc, Matsuura Machinery plc and the Engineering and Physical Sciences Research Council (EPSRC) for financial and technical support. Finally, we would like to acknowledge the help of Dr C.Small, Rolls-Royce plc, for his help with residual stress measurement.

References

[1]Bradley, E.F., 1988, Superalloys – A Technical Guide, ASM, Metals Park, Ohio, ISBN: 0-87170-327-0

[2]Fischer, J.J., Weber, J.H., 1990, Mechanical Alloying spreads its Wings, Advanced Materials and Processes, Vol. 138, Issue 4, pp.21-28

[3]Kovach, J.A., Malkin, S., 1988, Superalloy Grinding Damage, Third International Grinding Conference, SME, Chicago, Illinois, pp.112-117

[4] Tso, P-L., 1995, Study on the grinding of Inconel 718, Journal of Mat. Proc. Tech., Vol.55, pp.421-426

[5] Gardinier C.F., 1988, Physical properties of superabrasives, American Ceramic Society Bulletin, Vol.67, No.6, pp.1006-1009.

[6]Parrott, E., 1983, Development and application of continuous dress creep feed grinding, Proc. Instn. Mech. Eng., Vol. 197B, pp.231-235

[7] Popov, M.A., 1981, Grinding nickel-titanium aluminides, Soviet Engineering Research, Vol.1, No.3, pp.90-92

[8] Kumar K.V., Koskey J., Ishikawa T., 1991, Superabrasive grinding of superalloys, Proc. Superabrasive Conference, Vol.17, pp.17-30

[9] Kumar K.V., 1993, Comparison of monocrystalline and microcrystalline CBN abrasives in grinding, 5th Int. Grinding Conf., Ohio, pp.1-16.

[10] Zhang L.C., Suto T., Noguchi H., Waida T., 1995, A study of creep feed grinding of metallic and ceramic materials, Journal of Mat. Proc, Tech., Vol.48, pp.267-274.

[11] Brinksmeier E., Heinzel C., Wittmann M., 1999, Friction, Cooling and Lubrication in Grinding, Annals of the CIRP, Vol.48, No.2, pp.581-598.

[12] Ye, N.E., Pearce, T.R.A., 1984, A comparison of oil and water as grinding fluids in the creep feed grinding process, Proc. Istn. Mech. Eng., Vol. 198B, No.14, pp.229-237

[13] Webster J., 1999, Benefits of improved coolant application, Seminar for the Institute of Grinding Technology, Bristol, March 16

[14] Hill C.P.R., Watkins J.R., Ray C., Ray S., 1999, Method and apparatus for grinding, UK Patent Application, No. 9827353.5, Filed 14 December 1998

[15] Bridgeport Machines Ltd, 1999, Introducing the new Viper VMC1000 5 axis turbine blade machine, Product literature.

[16] EPSRC GR/M66790 Grant, 1999, Machining of Nickel Based Superalloys Using Mounted Grinding Points Fitted to High Speed Machining Centre.

LASER PROCESSING I

Laser Spot Welding of Electrical Wires

B.S Yilbas and S.J. Hyder
Mechanical Engineering Department,
King Fahd Univeristy of Petroleum & Minerals, Dhahran – 31261, Saudi Arabia.

Abstract: In the present study, laser spot welding of copper wires with different diameters on steel, nickel and titanium are investigated. The laser power intensity of resulting welds is carried out to determine the mechanical strength of the resulting welds. SEM is carried out to examine the metallurgical changes across the weld zone. The study is extended to include a statistical analysis to identify the most affecting parameters on the resultant weld properties. It is found that a successful welding of single core copper wires onto a different base material can be possible. The tensile strength of the resulting welds vary within 20-40 N applied loads provided that the base material having low thermal diffusivity and high absorption depth results in low tensile strength.

1. Introduction

Laser machining finds wide application in industry due to its precise of operation, low cost and rapid processing. The laser machining processes include cutting, drilling, welding, marking, cladding, surface alloying and etc.

The use of laser in welding of metals has a number of advantages over conventional techniques. These include, the weld and the heat-affected zone adjacent to the weld are very narrow, a vacuum is not needed. In laser welding, the transition from a heat conduction controlled process to deep penetration welding is determined by the laser intensity. Above or certain threshold laser intensity, the energy transfer from the laser beam to the workpiece increases rapidly and deep penetration welding occurs. It is generally accepted that in deep penetration welding a keyhole is formed and plays a dominant part in the whole process [1].

The laser spot welding, which depends on the penetration of the fusion isotherm through the full depth of the weld, is considered as conduction limited process [2]. The principal disadvantages of the conduction limited process is that depth to width ratio of the weld is limited to about unity [3]. Moreover, it was suggested that under suitable conditions, laser welding need not to be conduction limited and some specific examples of weld depth to width ratios exceeding four have been reported. The laser spot welding of electrical wires onto different metallic materials is fruitful, since the difficulties associated with the welding and soldering are minimized such as copper wire welding onto titanium.

In the present study, laser spot welding of copper wires with different diameters on steel, nickel, and titanium are investigated. The laser power intensity and focusing conditions are varied in the welding process. The tensile testing of resulting welds are carried out to determine the mechanical strength of the resulting welds. SEM is conducted to examine the metallurgical changes across the weld zone. The study is extended to include a statistical analysis (Taguchi method) to identify the most affecting parameters on the resultant weld properties.

2. Robust Design Using Taguchi Methodology

By the end of 1940's G. Taguchi had developed the Taguchi Method for quality improvement that involved System, Parameter and Tolerance Design [4]. The idea of performing these designs is to minimize the effect of hidden factors known as noise that cause the performance characteristics to deviate from its nominal target value. Taguchi recommended the use of signal-to-noise ratio (SNR) for evaluating the effect of each factor on the objective function. SNR is a measure of the functional performance of the objective function in the presence of noise factors where maximizing it results in minimizing the quality loss. There are three types of SNR namely smaller the better, larger the better and nominal the best. In this study our goal is to minimize the weld defects that are dependent on various parameters hence, smaller the better characteristic of SNR is chosen. This is given by

$$SNR = -10 \log_{10} [\Sigma f_i^2/n]$$

Where f_i are the individual response values and n is the number of experimental runs.

The steps involved in this approach are explained briefly as follows:

1) Select the levels of different parameter such as diameter, metal and power intensity affecting the weld quality depending upon the literature survey and previous practical experience.
2) Identify the Taguchi Orthogonal Array (OA) that fits into the design depending on the number of parameters and thier levels. In this study modified eight unit's OA (L_8) is used.
3) Perform the experiment and obtain the objective function value.
4) Evaluate SNR and determine the best levels associated with the experiment.
5) The effect of each factor on the objective function is obtained by performing the Analysis of Variance (ANOVA) test.

3. Experimental

The experimental setup consists of a Nd:YAG laser, which delivered output energy in the range of 10 – 21 Joules with a nominal pulse length of 1.48 ms. A focusing lens of 51 mm nominal focal length lens was used to focus the laser beam. The lens was mounted on a graduated barrel providing a vertical movement. To alter the laser power intensity on the workpiece surface, the focal position of the lens was changed by turning the barrel.

4. Discussion

Three workpiece materials (Nickel, Titanium and Stainless Steel) and four different wire diameters are considered, which are given in Table 1. In Figure 1 the tensile test results are plotted for laser welded multi core wire having 2.5 mm outer diameter. It is evident that the tensile strength of the welds are not very high when they are compared to those given in Figure 2.

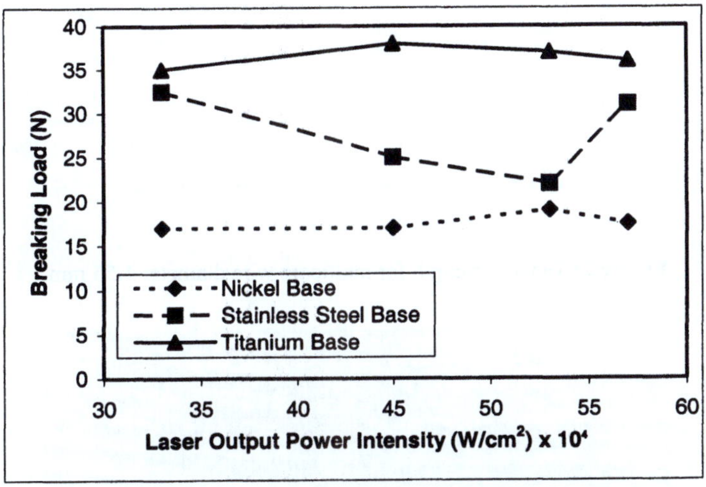

Figure 1: Tensile strength for multi core wire diameter 2.5 mm

This may be explained by either or both of the following circumstances:

1) with multi core wire, there are always air gaps between the wires. This increases convective heat losses during welding.

2) the focal position relative to the wire surface is not constant across the surface, since the free surface of the wire is not flat and consequence of this is that the power intensity varies across the surface, resulting in non-uniform welding sections. In Figure 2 the tensile test results are plotted for 1.55 mm thick laser welded multi core wire. It can be seen that the tensile strength of the welds are relatively higher than those obtained for 2.5 mm thick multi core wire. This may be due to the effect of the thickness of the wire. In this case relatively less energy is required for melting of the wire and this leaves more available energy to vaporise the base material. Therefore, the vapour pressure of the base material increases and a relatively larger crater diameter results. This produces a large area of base material to interface with the core material, However, the effect of base material on tensile strength can be seen in both Figures 1 and 2. The tensile strength is relatively higher for materials having relatively higher thermal diffusivities and lower absorption depths (such as Ni). In addition, the effect of power intensity is obvious, i.e. welding improves for any given material at a particular power intensity. This case is obtained at $0.443*10^6$ W/cm^2 for 1.5 mm thick multicore wire material.

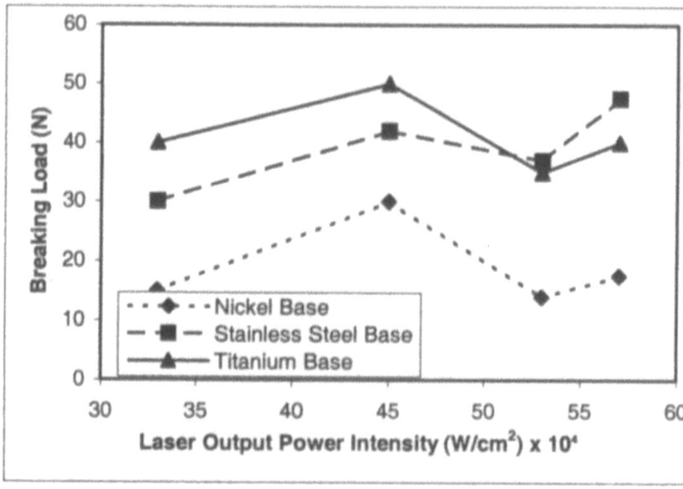

Figure 2: Tensile strength for multicore wire diameter 1.55 mm

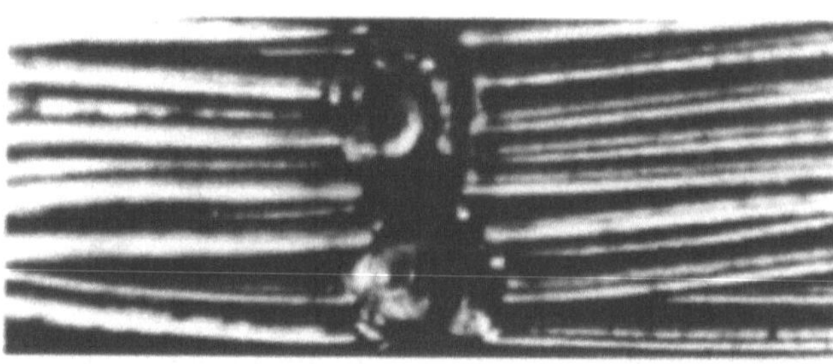

Figure 3: Multicore wire welds after cutting by laser beam

Photographs of multicore wire welds are given in Figure 3. It can be seen that some wires in the surface region are cut by the laser beam. This is especially true for 2.5mm thick wires. However, this reduces the tensile strength of the welds.

Table 1: Percentage effect of each factor along with F-test results

Design Variables	Percent Effect	Variance Ratio (F-Test)
Dia	85.84	24 (Significant at 95% C.L.)
Power	Ns	Ns
Dia x Power	Ns	Ns
Metal	5.54	5.5 (Significant at 90% C.L.)
Dia x Metal	Ns	Ns

Note: Ns denotes that the Factors are not significant and C.L. represents Confidence Level Value

Table 1 gives the F-test results for the parameters used in the equipment. The diameter of the wire has most significant effect on the tensile properties of the resulting weld. In this case, as the wire diameter increases, the tensile strength of the weld reduces as indicated earlier. The effect of power intensity appears to be less significant. This may occur because the power intensity levels used in the equipment are close to each other. In addition, the welding of the wire onto a base metal was achieved for all power intensity levels. The effect of base metal on the welding strength is significant. In this case, the tensile strength of the welds changes as the base substrate changes, i.e. the amount of the wire material in the weld pool changes for different base metal. Moreover, all the interactions of the parameters selected are found to have no significant effect on the resulting weld strength.

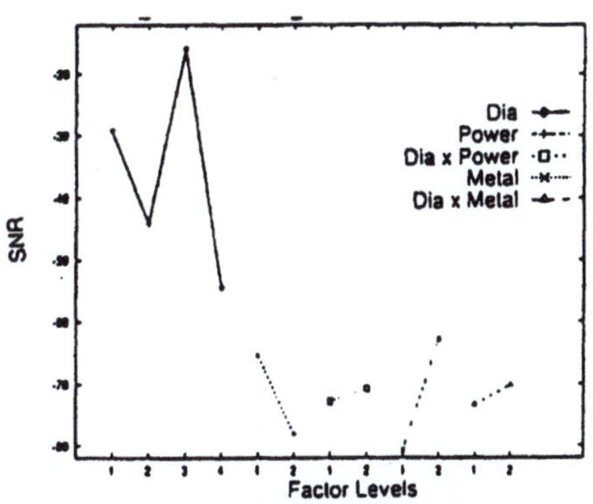

Figure 4: SNR ratios at different levels for all factors

Figure 4 shows the signal-to-noise ratio (SNR) determined at each level of the parameters employed. SNR attains relatively lower values for diameter of the wire as compared to other parameters including levels of laser output power and base material. The low SNR indicates that the influence of wire diameter on the weld strength is considerable. This is especially true for the first and third levels of the wire diameter. Moreover, the effect of other parameters on the weld strength is not considerable, since SNR are low for these parameters.

6. Conclusion

The conclusions derived from the present work can be listed as follows:

1) At some power intensities the mechanical properties of the welds improve.
2) Multicore wires can be welded to any base material giving a sufficient tensile strength. However in the case of 2.5 mm thick multi core wires, the tensile strength of the welds decrease to 20-40 N.
3) A base material having low thermal diffusivity and high absorption depth gives low tensile strength and a large heat affected zone. This may be due to the existence of conduction effects during the welding.

References

1. Dowden J. and K. Papadia, 1996, *The penetration depth in keyhole welding with pseudo-continuous Nd-YAG and CO_2 lasers investigated mathematically*, Appl. Surf. Sin. Vol. 106, pp. 235-239.
2. Yilbas B.S., Sami M. and Al-Farayedhi A., 1998, *Closed-form and numerical solutions to the laser heating process*, Proc. Inst. Mech. Engrs., Part c, J. Mech. Eng. Science, Vol 212, pp. 141-151.
3. Steen W. H., 1991, *Material Processing*, Springer-Verlag, New York.
4. Dehnad K., 1989, *Quality Control, Robust Design and Taguchi Method*, Statistical/Probability Series. Wansworth and Brooks, California, Ist edition.

An Analytical Model for Quality Inspection of Laser Marking/Glazing Clay Tiles

A.A. Peligrad[1], E. Zhou[2], D. Morton[3] and L.Li[4]

[1]Faculty of Technology, Bolton Institute, Deane Road, Bolton BL3 5AB, U.K. Email: aap1res@bolton.ac.uk
[2]Faculty of Technology, Bolton Institute, Email: ez1@bolton.ac.uk
[3]Faculty of Technology, Bolton Institute, Email: D.Morton@bolton.ac.uk
[4]Manufacturing Division, Department of Mechanical Engineering, UMIST, Sackville Street, Manchester M60 1QD, U.K. Email: l.li@umist.ac.uk

Abstract. In this article the quality control of laser marking/engraving of clay tiles using a high power diode laser (HPDL) has been investigated. A one-dimensional analytical model with quasi-stationary situations in an isotropic and inhomogeneous workpiece with a parabolic melt pool geometry being assumed, was successfully developed. The theoretical results were compared with the experimental data. The predicted melt pool depth and the experimental values were in relatively good correspondence for values of the incident laser energy density (>15Wmm^{-1}s$^{-1/2}$), in spite of simplifications introduced during modeling. At the relatively large values of laser energy density, the assumption of parabolic melt pool shape and one-dimensional heat transfer no longer holds true.

1. Introduction

The development of an accurate analysis procedure for many laser applications, including marking/glazing of building materials, is extremely complicated due to the many process parameters involved. Variations in the quality of the marking process may be observed between processing cycles performed with the same laser equipment, and apparently constant operating conditions and material properties. This poor reproducibility arises from the high sensitivity of the laser marking process to small changes in the operating parameters (such as laser power and beam velocity) as well as to process disturbances (such as varying absorptivity).

The parameters that can be measured on-line during laser marking/glazing are the melt pool surface temperature and the melt pool surface area. The development of a suitable model to predict these parameters, can not only eliminate the need to perform experiments within the range of possible operating conditions, but can also provide information outside of this range. This paper presents a one-dimensional analytical model for the melt pool geometry in case of laser engraving clay tiles.

2. Predicted Melt Pool Geometry

The depth of the melt pool during laser marking determines the thickness of the resulting surface layer after solidification (i.e. the glaze thickness), which is related to the quality of the mark. A simple analytical model can be obtained by assuming that the shape of the solid-liquid interface is parabolic and the heat transfer considered as one-dimensional (see Fig. 1). Some similar assumptions have been considered by Römer [1], for laser surface treatment of titanium nitride.

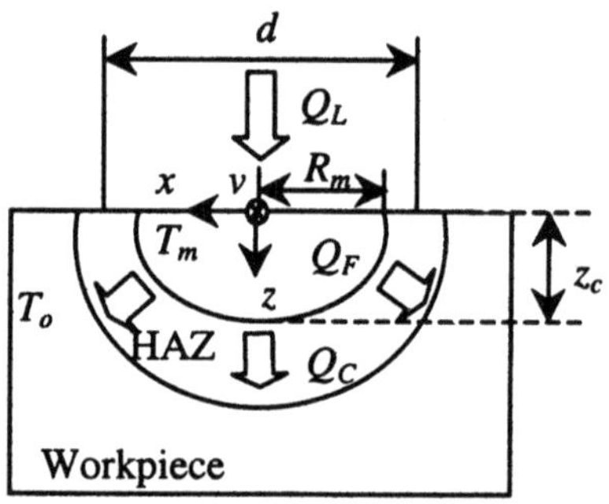

Key

d = beam diameter
R_m = melt pool radius
v = traverse speed
HAZ = heat affected zone
z_c = melt pool depth
T_m = melt pool temperature
T_0 = workpiece temperature
x, z = axis

Figure 1. Melt pool geometry and energy flows in a quasi-stationary situation.

The energy balance within the sample consists of three terms: the absorbed laser energy Q_L, the energy Q_C transported by heat conduction from the liquid-solid interface of the melt pool into the non-moltn material, and the energy Q_F required to create a melt pool. It is also assumed that the heat transfer under the melt pool may be considered as one-dimensional. Then the heat balance equation is:

$$Q_L = Q_C + Q_F \tag{1}$$

Melt pool depth

From the analysis of the physical phenomena, occurring during laser surface glazing and substitution of the energies into the energy balance equation (1), the melt pool depth z_c yields to [1]:

$$z_c = \frac{2A}{\rho c_p C_2 T_m \sqrt{k}} \frac{P_L}{\sqrt{dv}} \tag{2}$$

where $k = K/\bullet c_p$ [m^2s^{-1}] is the thermal diffusitivity, $\bullet$ [$kg{\cdot}m^{-3}$] is the density of the material, c_p [$J{\cdot}kg^{-1}{\cdot}K^{-1}$] is the material thermal capacity, and K [$W{\cdot}m^{-1}{\cdot}K^{-1}$] is the material thermal conductivity, P_L [W] is the laser power, A [%] is the absorptivity of the material, and $C_2 \bullet 4.31$ is a constant.

This expression shows a linear dependence of the melt pool depth with the ratio $P_L/\sqrt{dv}$, which is sometimes referred to as the specific energy [2]. The same dependence was derived by Römer [1] and proved by experimental analysis [3] in the case of laser transformation hardening for metals. No references regarding to this dependence for laser marking/glazing ceramic materials (e.g. tiles) were found. Since, ceramics are composed of at least two elements or often more [4], their crystal structures are generally more complex than those of metals and therefore the melt pool depth model was corrected by a factor (C_3). This factor was empirically deduced ($C_3 = 2.29$).

$$z_c = \frac{1}{C_3}\frac{2A}{\rho c_p C_2 T_m \sqrt{k}}\frac{P_L}{\sqrt{dv}} \qquad (3)$$

Expression (3) has been verified experimentally and described in the next section.

3. Experimental Set-up

The laser used in the study was a surgical HPDL (Diomed Ltd.), emitting at 810nm ± 20nm and operating in continous mode (CW) with rated optical powers ranging from 0-120 W. The laser beam was delivered to the work area through a 3m long, 600μm core diameter optical fibre, the end of which was connected to a 2:1 focusing lens assembly mounted on the z-axis of a 3-axis CNC gantry table (see Fig.2). The samples used were UK standard construction clay tiles and for experimental purpose were cut into squares sized 50 x 50 mm using a diamond edged cutting blade.

A series of lines was marked onto the surface of the tile samples using the output power of 17-51 W-cw (measured at the workplace using a Power Wizard power meter), at rates of 2-8 mm/s and 1.9 mm beam diameter. These output laser power values were chosen after examination of the samples. Lower power would not melt the clay tiles sufficiently, while higher power would vaporise the surface, leading to a bad quality mark. The laser-marked samples were then cut across the mark lines using a sharp diamond edged cutting blade and examined using optical microscopy (Kyowa SD-2PL). A method based on computer software called Visilog 5 was used in measuring accurately the depth and width of the mark from the images recorded and grabbed in the computer.

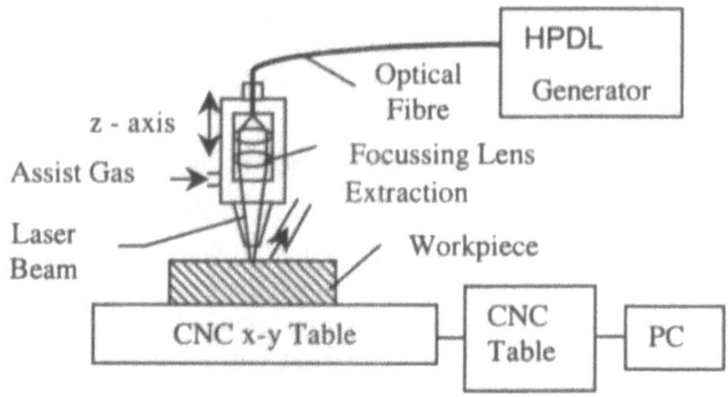

Figure 2. Schematic of the experimental set-up for the HPDL marking/glazing

4. Test Results

A comparison of measured and calculated melt pool depth for marking/glazing of clay tiles is presented in Figure 3. It shows the measured depths as well as the calculated depths, as a function of the ratio $P_L/\sqrt{dv}$. For the convenience the graph displays only the extreme values of the laser power (e.g.17W and 51W for beam diameter $d = 1.9$mm).

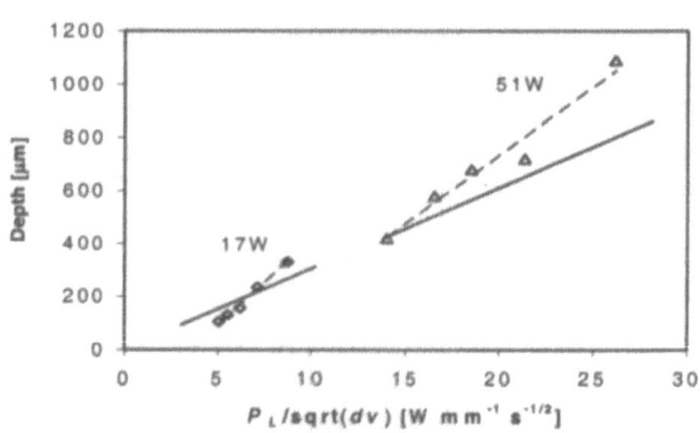

Figure 3. Measured (tick marks) and calculated (solid curve) melt pool depth for clay tiles

5. Discussions

The model predicted HPDL induced melt depths in clay quarry tiles is in close agreement with those obtained experimentally. It was observed, however, that as the incident laser energy density increased over 15Wmm^{-1}s$^{-1/2}$, the calculated and the experimental melt depths are not in correspondence anymore.

The poor correspondence at high values of $P_L/\sqrt{dv}$ can be attributed to the shape of the melt pool, which is not parabolic at these operating parameters, whereas the model assumes a parabolically shape melt pool. Changing the parabolic shape by increasing the laser energy density is due mainly to the fact that the analytical model neglects sideways conduction, assuming all the heat transfer is by conduction downwards (in *z*-direction). The depths of melt measured from the samples were, to varying degrees, different from those predicted. In reality, the heat conduction flows in all directions. For low traverse speed and high laser power, a large amount of sideways conduction energy was neglected and therefore, the errors between the predicted model and the measured depth were increased. It is believed that at depths which are smaller compared to the spot size (1.9mm in this case), the heat transfer is approximately in one dimension only [5]. This can be observed clearly from Fig.3 by comparing the measured and calculated depth.

The inhomogeneous substrate of the clay tile is the second source of errors. Crystal structures of the clay minerals are relatively complicated, because of the large numbers of elements. Laser-induced heating changes the optical properties of materials via their temperature dependence. The change in absorbed laser light intensity, in turn, changes the temperature distribution. This feedback is strong enough, to cause new phenomena in the laser-matter interaction (e.g. thermal runaway or oscillation [6]). As a result, the reflectivity and absorptivity of the laser beam by the material, becomes a multiple dimension complex function.

The mark depth predicted by the analytical model for value $P_L/\sqrt{dv}$ < 15 Wmm^{-1}s$^{-1/2}$ will be reasonable accurate, whereas at the relatively large values of $P_L/\sqrt{dv}$ the assumption of a parabolic meltpool shape and one dimensional heat transfer no longer holds true.

6. Conclusions

In this article the quality control of laser marking/engraving of clay tiles using a HPDL has been investigated. A one-dimensional analytical model with quasi-stationary situations in an isotropic and inhomogeneous workpiece with a parabolic melt pool geometry being assumed, was successfully developed. The theoretical results were compared with experimental data. The predicted melt pool depth and the experimental values were in relatively good correspondence, for values of the incident laser energy density (>15Wmm^{-1}s$^{-1/2}$), in spite of simplifications introduced during modeling. It is believed that this observed increasing discrepancy could be attribute to the inhomogeneous substrate of the clay tiles and to the fact that the model developed neglects sideways conduction energy.

References

[1] Römer G.R.B.E., 1999, *Modeling and Control of Laser Surface Treatment.* PhD Thesis. ISBN 90-365-1304-9. Eiberger, the Netherlands.

[2] Schneider M.F., 1998, *Laser cladding with powder.* PhD Thesis. University of Twente. Enschede, the Netherlands.

[3] Derouet H., Caillibotte G. and Kechemair D., 1994, *Towards Real Time Control of Surface Treatments with Liquid Phase: Molten Depth On-line Estimation.* Laser Material Processing: Industrial and Microelectronics Applications. SPIE 2207. p. 321-329.

[4] Callister W.D. Jr., 1994, *Materials Science and Engineering – An Introduction.* Third Edition. John Wiley & Sons, Inc. U.S.A.

[5] Lawrence J., 1999, *The Characteristics and Feasibility of a Two-stage Ceramic Tile Grout Sealing Process Using a High Power Diode Laser.* PhD. Thesis. UMIST. U.K.

[6] Bäuerle D., 1996, *Laser Processing and Chemistry.* Second Edition. Springer-Verlag Berlin Heidelberg New York.

Selection of Optimal Part Orientation for Rapid Prototyping

Hochan Kim[1], Juho Lee[2] and Seokhee Lee[3]

[1]Pusan Nat'l Univ., Kumjunggu Pusan, Korea, hchakim@hyowon.pusan.ac.kr
[2]Pusan Nat'l Univ., Kumjunggu Pusan, Korea, leejuho@hyowon.pusan.ac.kr
[3]Pusan Nat'l Univ., Kumjunggu Pusan, Korea, sehlee@hyowon.pusan.ac.kr

Abstract: Layered manufacturing is used for various purposes as one technique of Rapid prototyping. Because it builds three dimensional objects by adding materials layer-by-layer, the stair-stepping effects worsen surface roughness and volumetric error. Using build orientation, quality of parts can be improved.

This paper proposes a method to simplify a model with equivalent accuracy which evaluates each parameter efficiently, First, some build directions satisfying given acceptable range of each parameter are selected. This algorithm determines the optimal direction by the parameter's priority. Using this proposed algorithm, the optimal layered manufacturing can be efficiently determined.

1. Introduction

Build direction in layered manufacturing impacts part accuracy, build time, and surface roughness. Generation of a final product differs according to build direction. Each laminated manufacturing system may require a different build orientation due to the system's characteristics, even with the same object. Each build parameter needs to be adjusted because they may conflict with each other.

In order to determine the optimal build direction, the characteristics of each layered manufacturing are analysed and the build parameters are evaluated. Previously proposed algorithms for selecting orientation are complex and time-consuming..

In this research, part accuracy, build time, and system dependant parameters are considered in order to determine the proper build direction. For the evaluation of each parameter, a simplified model with equivalent accuracy is proposed. An algorithm to evaluate parameters accurately, quickly and efficiently is implemented with the proposed model. And an objective function is used to select the specific build direction as optimal build direction.

2. Parameters in Consideration

2.1 Build Time

In the layered manufacturing system, the total build time T and the number of layers n_{layer} is obtained from the following equation.

$$\begin{cases} T = T_{contour} + T_{hatch} + n_{layer} \times T_{delay} \\ n_{layer} = (\max(v_z) - \min(v_z))/t = height \end{cases}$$

$$\therefore T \propto height \text{ of the object with build direction}$$

where t denotes the layer thickness, n_{layer} denotes the number of layers, $T_{contour}$ is the total contouring time, T_{hatch} is the total hatching time for filling the inner area, and T_{delay} is the solidifying time per layer. Generally, $T_{contour}$ and T_{hatch} are related not to the build direction, only to the shape of the object. Thus, build height is the only criterion impacting the build time.

2.2 Quality of Surface

There are two types of volumetric errors. One is positive volumetric error which means there is excessive volume. The other is negative volumetric error which means there is insufficient volume intended for building. Thus the volumetric error for each facet is calculated by the sum of the accumulated errors with no sign. Assuming that the area of the surface is significantly larger than the layer thickness. Volumetric error is obtained by following equation. But when the interval angle between the build direction and normal vector of a facet is less than the specific angle(eg. 5°), it is reasonable that there is shifting error not stair-stepping effect on the surface. Volumetric error divided by the total surface area is the average surface roughness.

$$V_{error} = \begin{cases} \frac{1}{2} \sum_{i=0}^{total} \alpha \cdot \left| A_i \cdot t \cdot \cos \theta_i \right| & (\theta \geq 5) \\ 0 & (\theta < 5) \end{cases}$$

$$L_{error} = V_{error} / A$$

where • denotes the interval angle between the normal vector of the surface and build direction vector, α denotes compensation factor for the laser contour offset, A denotes the area of the surface, and t denotes the layer thickness. When the normal vector of a facet is parallel to build direction, the volumetric error changes along with z-coordinate shifting. Suitable shift is needed for the minimization of the errors. In this research, the total volumetric error for the given orientation can be represented as the sum of the absolute value of stair-stepping error volume and shifting error volume.

2.3 System dependant parameters

There are various laminated manufacturing systems according to methodology and material, and each system has its own characteristics. For the SLA system, supports are necessary but they worsen the supported surface and increase post-processing time. Taking the interval angle between the build direction and the normal vector of a surface to be •, the area requiring supports are calculated by following equation:

$$\theta = \cos^{-1}\frac{\overrightarrow{B}\cdot\overrightarrow{N}}{\left|\overrightarrow{B}\right|\left|\overrightarrow{N}\right|} = \cos^{-1}\overrightarrow{B}\cdot\overrightarrow{N} = \cos^{-1} n_z$$

$$A' = \sum_i \cos(\pi - \theta), \quad \textit{for all facets with } \theta > \theta_{support}$$

where $\overrightarrow{B}$ denotes the unit vector of build direction, $\overrightarrow{N}$ denotes the unit vector of surface normal, A denotes the area of surface and $\bullet_{support}$ is the minimum angle which needs support structure.

In the SLS system, distortion called curling effect occurs frequently at the lowest sliced plane. Heat stress from unbalanced energy dissipation causes this shape distortion when the first sliced plane is larger than a specific area. Therefore the orientation should be decided that the lowest sliced plane takes smaller area than the specified area.

3. Simplified Model with Equivalent Geometric Characteristics

3.1 Simplified model with equivalent build time

The build time difference for each orientation is determined by height. At any orientation, only vertices on convex hull can be at the highest or lowest vertex. Therefore build time difference can be predicted by the rotation of convex hull.

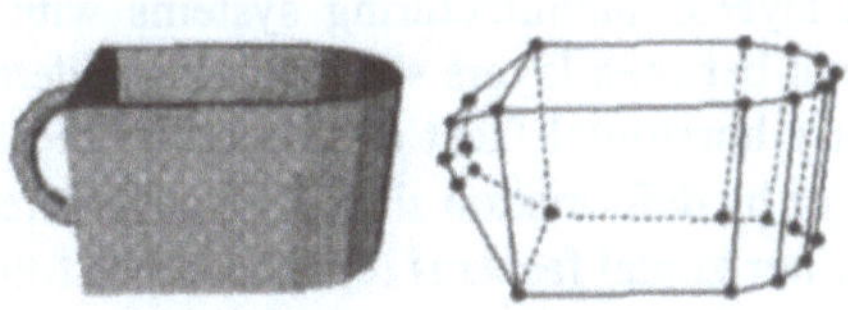

(a) original model (b)Equivalent build time model

Fig .1 Equivalent build time model

3.2 Simplified model with equivalent roughness

Surface roughness and volumetric error are related to the surface and interval angle between the surface normal vector and build direction vector. The facets in the same normal direction are treated as one equivalent facet whose area is the sum of each facet. In Fig. 2, (a) illustrates an original model, (b) illustrates a simplified model grouped by normal, (c) illustrates a simplified model with equivalent

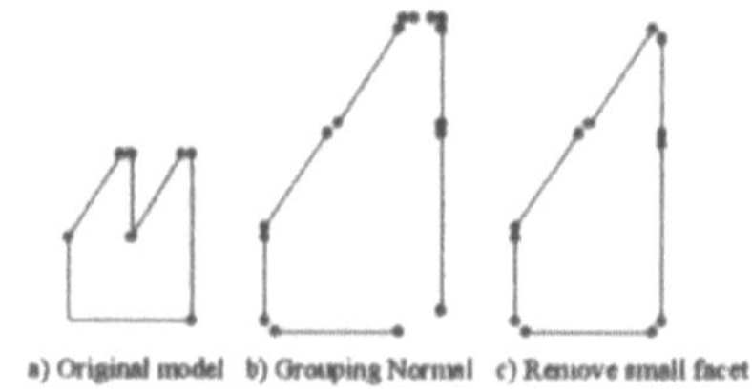

Fig .2 Equivalent roughness model.

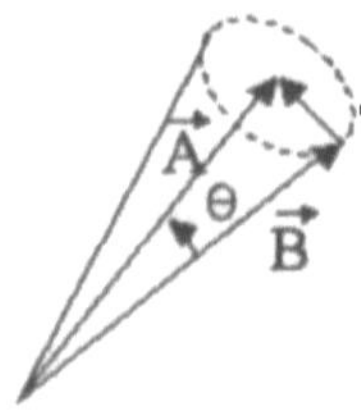

Fig .3 Grouping facets

Fig .4 Roughness view

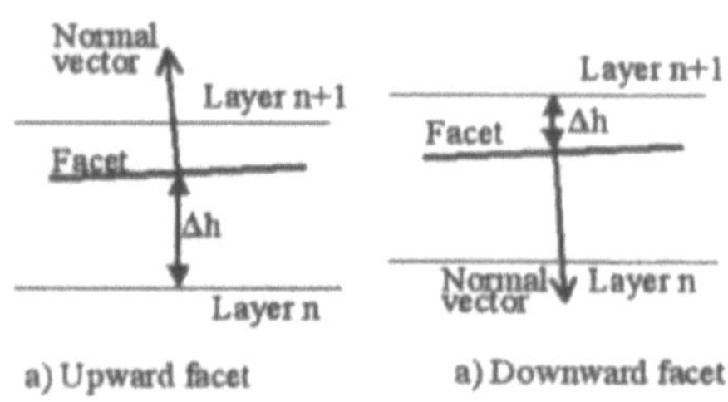

Fig .5 Shift error

accuracy of (b). Fig.3 illustrates the method for constructing equivalent facet of which normal is equal. The relationship between two unit normal vectors of facets $\vec{A}$, $\vec{B}$, and their interval angle is as follows.

$$\left|\vec{A}-\vec{B}\right| = 1 \cdot \theta$$

Equivalent facet models which are grouped by the normal vector reduces the processing time for roughness calculation.

3.3 Simplified model with equivalent accuracy

In layered manufacturing systems with uniform layer thickness, the error of a facet between layers should be considered for volumetric error. Volumetric error for a horizontal facet can be optimised by upward/downward shifting. Accuracy error by deformation should be considered as system characteristics. Shifting error for horizontal facets is represented as follows.

$$V_{shift\ error} = \sum_{i=0}^{n} \Delta h_i \times A_i$$

4. Determination of Optimal Build Direction

Object functions and weight factors were discussed as criteria. It is not sufficient to select the build direction by weight only, because of low accuracy and large build time deviation with each build direction.

$$f_i = \alpha \times \frac{(T_i - T_{min})}{T_{max} - T_{min}} + \beta \times \frac{(A_i - A_{min})}{A_{max} - A_{min}} + \tau \times \frac{(S_i - S_{min})}{S_{max} - S_{min}}$$

$$\begin{pmatrix} f_i : \textit{Object function}, \ T_i : \textit{Build time}, \ A_i : \textit{Accuracy} \\ S_i : \textit{System dependent parameter}, \ i : \textit{build direction} \end{pmatrix}$$

Thus it is reasonable for users to select from among candidate orientations proposed by acceptable limitation, priority, and weight factor for each parameter. The orientation beyond limitation is ignored and the order of the optimal orientation for each parameter is determined by the priority of the parameters. The other candidate build direction is determined by object function with weight factor. The build direction which minimizes the object function is optimal orientation.

5. Example for Evaluation

Fig. 6 shows the wheel feature experimented in this example. Table 1 includes the model specification, the solution from rotating by 1 degree about the x and z axes, and the results using the simplified model proposed in this research. Fig. 7 and Fig. 8 show the plot of the accuracy and build time of the wheel. Fig. 9 and Fig. 10 show the plot of the solution obtained from the simplified model in this research. The simplified model shortens the time to compute optimal orientation and suggests good orientation similar to that of manual computation by experts.

Fig .6 example of wheel

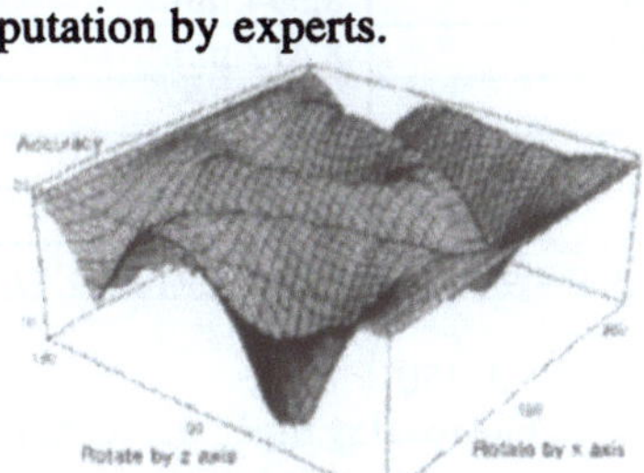

Fig .7 Accuracy surface

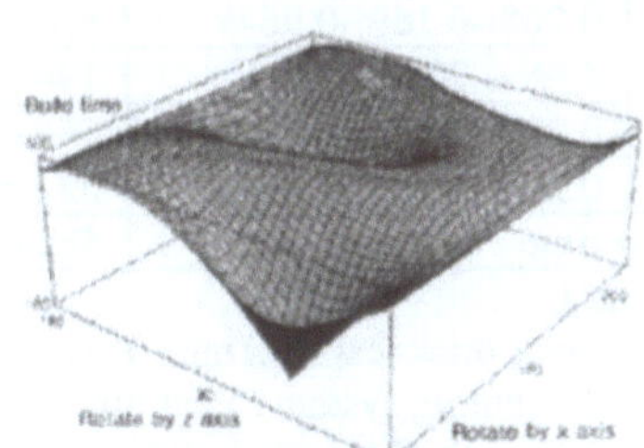

Fig .8 Build time surface

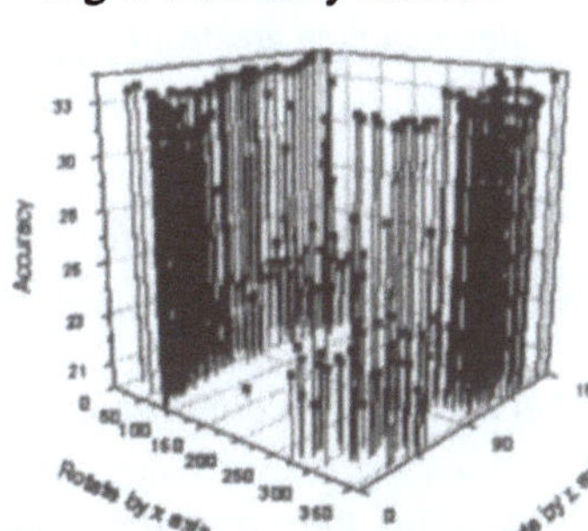

Fig .9 Accuracy of simplified model

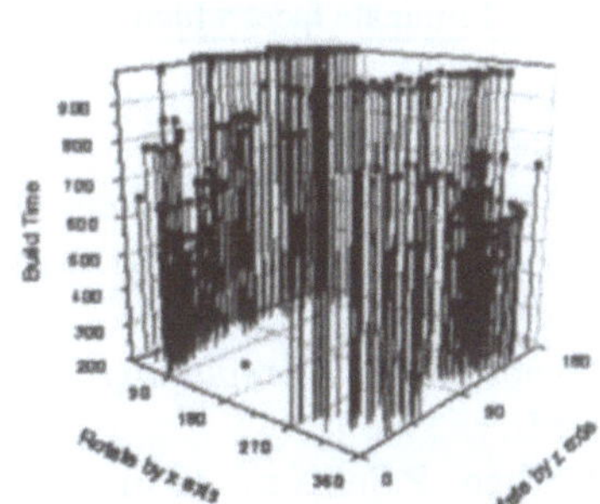

Fig .10 Build time of simplified model

6. Conclusion

In this research, the parameters needed for determination of orientation are classified by build time, part accuracy and system dependant factor. The method of calculating each parameter efficiently by means of simplified models, and proposed algorithm were found to determine the optimal orientation fast and efficiently.

Table 1: Results for sample

Model	Type	File size	Facets	Vertices	Unique vertices	Surface area
Wheel	Binary	28MB	574,542	2,263,626	287,273	40,526mm^2

Item			Rotate by step 1°	Equivalent model
Accuracy	Best	Over cure	388 mm^3	434 mm^3
		Under cure	368 mm^3	314 mm^3
		Avg. accuracy	0.0186	0.0184mm(99%)
		Build direction	0.464i-0.537j+0.704k	0.480i-0.523j+0.705k
	Worst	Over cure	707 mm^3	722 mm^3
		Under cure	694 mm^3	689 mm^3
		Avg. accuracy	0.0346 mm	0.0348mm(100.6%)
		Build direction	-0.125i-0.150j-0.980k	0.726i-0.687j+0.002k
Build Time	Best	Build time	180minute(h=22.0)	185minute(h=23.3)
		Build direction	0.500i-0.500j+0.707k	0.480i-0.523j+0.705k
	Worst	Build time	583minute(h=120.6)	583(h=120.6)
		Build direction	0.469i-0.640j-0.609k	0.526i-0.395j-0.735k
Area Require support	0.499i-0.708j+0.499k		12,874 mm^2	12,172 mm^2
	0.231i+0.649j-0.725k		8,984 mm^2	8,594 mm^2
	0.208i-0.169j-0.964k		8,716 mm^2	8,412 mm^2
Lowest layer's Area	499i-0.708j+0.499k		194 mm^2	162 mm^2
	0.231i+0.649j-0.725k		56 mm^2	38 mm^2
	0.208i-0.169j-0.964k		52 mm^2	37 mm^2
Computing time			5hours 27minutes	44seconds(0.08%)

[Conditions]
Common : Layer thickness : 0.1mm, Target system : SLA 350
Computing system : Pentium III 500MHz, File reading time : 58seconds
Rotation : step method : rotate build direction with x axis and z axis by step 1°
Equivalent model method : Facet grouping angel 4°,
Eliminate facet which has 200 times smaller area then greatest facet

References

[1] P.T.Lan, S.Y.Chou, L.L.Chen, D.Gemmill, 1997, *Determining fabrication orientations for rapid prototyping with stereolithography apparatus*, Computer Aided Design

[2] P.Jacobs, 1996, *Stereolithography and other RP & M technologies*, SME, Dearborn, MI,

[3] C.C.Chen, P.Sullivan, 1996, *Predicting total build time and the resultant cure depth of 3D stereolithography process*, Rapid Prototyping Journal

[4] H.T.Loh, F.Xu, Y.S.Wong, 1998, *Optimal Orientation selection in different rapid prototyping process*, ICRPM

LASER PROCESSING II

The effects of the metallic substrate type during high power diode laser removal of chlorinated rubber coatings

Marc J J Schmidt and Lin Li

Manufacturing Division, Department of Mechanical Engineering, University of Manchester Institute of Science and Technology (UMIST), Sackville Street, Manchester, M60 1QD, United Kingdom, email: m.schmidt@umist.ac.uk

Abstract: The removal of Chlorinated Rubber (CR) from concrete surfaces using a variety of lasers has been well characterised over the past years regarding wavelength effects and processing parameters. However, the effects of different substrate materials on the removal rates and other removal effects have not yet been investigated. In this paper work is presented on investigating the effects of mild steel, stainless steel and aluminium substrates on the efficiency of the removal of CR surface coatings, using High Power Diode Laser (HPDL) systems. The results are interpreted regarding the thermal effects of the substrates on the removal and the combustion processes.

1. Introduction

The laser removal of paint coatings from metal surfaces has been dominated by the use of pulsed laser systems, due to the fact that they are based on photolytic, photoacoustic phenomena and photothermal effects [1,2,3]. The avoidance of using continuous beams (cw) results in the usually desirable controllability of the removal depth, and a reduction in thermal side effects from substrates and coating materials alike. The disadvantage of using these pulsed lasers is the low removal rate, which can only be improved, give stable pulse parameters, by increasing the pulse repetition frequencies. This can be difficult, if the process relies on precise timing of the pulses to deliver the correct photo-acoustic effects.

Therefore, cw laser systems have come back into focus for such applications, despite the complications involved in the thermic dissociation of coatings. Furthermore, the necessity for decontamination and decommissioning (D&D) procedures in a number of industries, has led to the necessity of higher stripping rates, with a reduced requirement for surface finish. This, together with the recent developments in the field of diode laser systems has led to the investigation of a process termed *laser sustained combustion* in which the laser irradiation supports and controls a combustion process which on its own is not self-sustaining. This results in good process control and a moderate accuracy, which is sufficient for the D&D applications described above [4].

One of the main factors in the process is the substrate type on which the coating system is applied. For a mainly thermally based process, substrates with a high thermal conductivity can lead to the loss of energy and ultimately temperature through the conduction process. It was therefore felt necessary to investigate these properties and effects to understand the process, which was well characterised on ceramic substrate types. In this study, the substrate material includes a number of metals, more specifically aluminium, mild steel (EN8, Grade 1040) and stainless steel (Grade 304).

2. Experimental Procedure

The experiments were conducted using a fibre delivered Diomed D120 HPDL system, emitting at 810 nm, through imaging optics capable of focusing to a circular spot size of approximately 1 mm diameter. Figure 1 illustrates the experimental arrangements, including the CNC table system which was used for transverse motion across the target surface. The system has been described in more detail elsewhere [5,6].

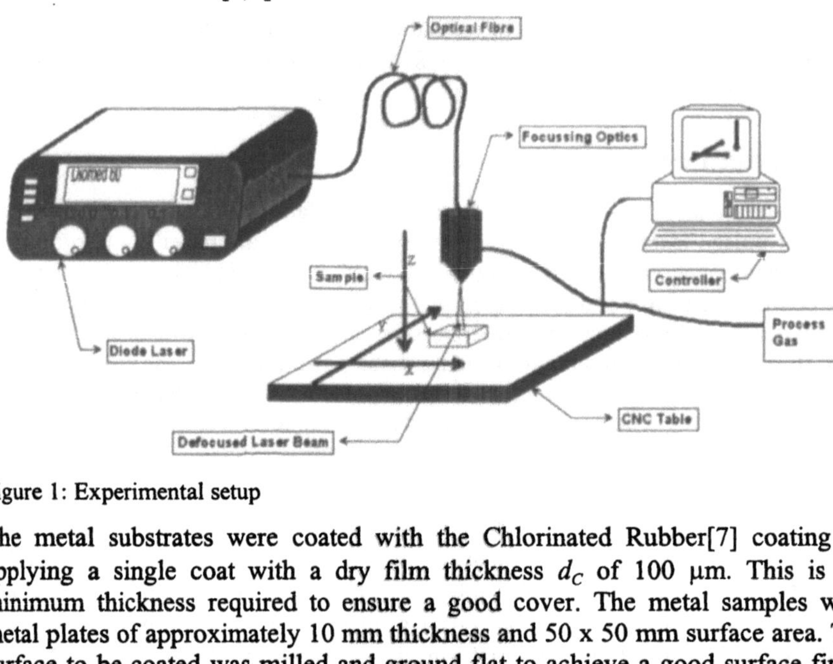

Figure 1: Experimental setup

The metal substrates were coated with the Chlorinated Rubber[7] coating by applying a single coat with a dry film thickness d_C of 100 μm. This is the minimum thickness required to ensure a good cover. The metal samples were metal plates of approximately 10 mm thickness and 50 x 50 mm surface area. The surface to be coated was milled and ground flat to achieve a good surface finish and to avoid surface morphology effects to interfere with the characterisation. The paint material adheres well to the machined metal surface without any special pre-treatment. To ensure the coating was free from solvent, the samples were dried for three days in a ventilated chamber and then kept at the elevated temperature of 40 °C for 24 hrs.

The experiments were conducted with a laser output power of approximately 35 W, resulting in an irradiance of approximately 1.2 kW cm^{-2} and therefore a fluence (energy density) of 60 to 175 J cm^{-2} for scanning speeds from 60 to 20 mm s^{-1}. A typical profile of a laser generated groove on the paint coating, by two passes over the surface, is shown in Figure 2. It is noteworthy, that the actual width of the track as such (defined via the Full Width Half Maximum (FWHM) of the track depth) is actually wider than the width of the two laser tracks of 1 mm each, due to the effects of the flame during the process, resulting in slanted track walls. Both, depth and profile measurements were performed using a "Talysurf" surface profiling system.

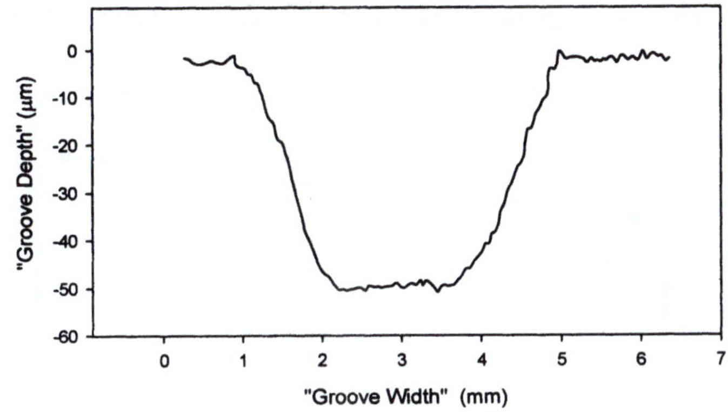

Figure 2: Typical groove profile for two tracks.

3. Results and Discussion

The most obvious aspect is the dependency of the removal track depth on the thermal properties of the substrate, most specifically the thermal diffusivity, which is defined as

$$D_f = \frac{\kappa}{c_p \cdot \rho}$$

Equation 1

where κ represents the thermal conductivity, c_p the specific heat capacity and ρ the density of the material in question. For simplification purposes, the temperature of the combustion area is assumed to be constant. It is possible to assume an equilibrium state and interpret the process using the energy balance. This allows one to avoid the use of the full heat diffusion equations, and simplifies the interpretation of the results.

When plotting the track depth against the inverse of the diffusivity for a specific set of parameters, as shown in Figure 3, it can be seen that the relationship is linear. The least squares fitting equation is shown to be

$$d = d_0 + a \cdot D_f^{-1} = 44.1 + 0.82 \cdot D_f^{-1}$$

Equation 2

The standard error for d_0 is ±0.20 and for a it is ±0.01. The inverse of the diffusivity, D_f^{-1}, can, under these circumstances, be interpreted as a thermal diffusional resistance.

The two constants can be interpreted in the following manner. d intercepting at d_0 is the track depth at "infinite" diffusivity, i.e. a substrate that would immediately distribute the thermal energy it receives throughout its body. The diffusivity values for the substrates used are as follows:

Aluminium: 0.95 $cm^2 s^{-1}$; Grade 1040 steel: 0.07 $cm^2 s^{-1}$; Grade 304 steel: 0.05 $cm^2 s^{-1}$.

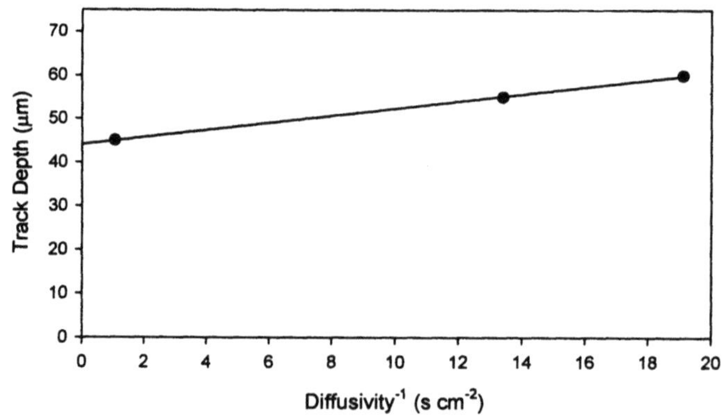

Figure 3: The track depth for measurements performed at 20 mm s^{-1} scanning speeds, plotted over the inverse of the thermal diffusivity.

Even with infinite diffusivity, a removal track would be formed, suggesting that the diffusion of thermal energy through the substrate is only of a limited effect when performing combustive removal. This would suggest that high thermal diffusivity of the substrate material results in a minor hindrance to the process, *provided it is possible to induce combustion* in the coating layer. This is further supported by the fact that the dependency on a, from the inverse of the diffusivity is comparatively small.

If t is defined as the dwell time of the energy input, then the thermal penetration depth for this time period is approximated by [8]:

$$d_T = 2\sqrt{D_f t}$$

Equation 3

For the paint coating involved ($D_f < 0.01$ W cm^{-2} [9]) at the slowest speed of 20 mm s^{-1} the thermal penetration depth is approximately 140 µm. This is of a similar magnitude as the coating thickness itself. Should the penetration depth be significantly higher than the coating thickness, the effects of the substrate material would be much more significant, due to the increased losses of energy into the substrate. However in the circumstances presented this is not the case.

Figure 4 shows the relationship between the removal depth and the speed of the transverse scan. As can be seen from Figure 4b, the relative development of the track depth with speed is very similar for all three material types.

One notable result, however, is the fact that the removal depth appears to be reducing linearly with increasing scanning speed. Were this governed by a mainly thermal process, one would expect the depth to be dependent on the dwell time, i.e. the inverse of the scanning speed (see Equation 3). Also a purely optical process (as e.g. the Beer-Lambert law) would be expected to result in similar relationships.

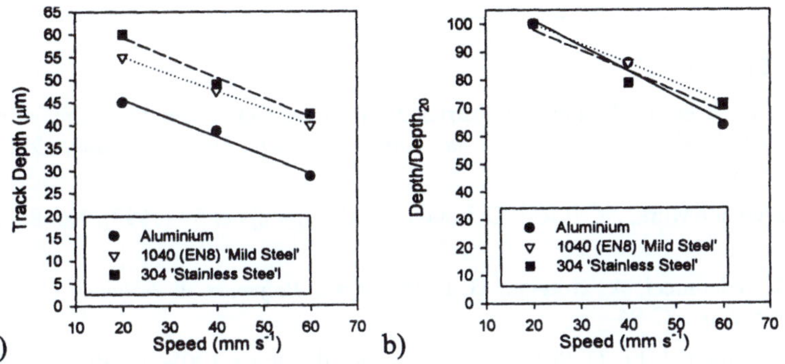

Figure 4: The (a) absolute and (b) relative (to 20 mm s^{-1} scanning speed) track depth dependent on the scanning speed.

This observed linearity must, therefore, result from an effect originating from the combustion process. This suggests that there may be changes in the optical absorption coefficient of the material, as well as primary absorption during the development of the combustion flame. After all, the laser in this case does *not directly* cause the decomposition of the polymer material, unlike in most other laser paint stripping processes, but rather supplies the energy for the continuation of the combustion process [4]. The effect could then be described as the absorption of the laser into the flame and the subsequent change in the combustion temperature. The combustion temperature in turn would have an effect on the removal depth, resulting in the observed phenomena. This would be supported by the different combustion temperatures observed during the variation of the process parameters [5].

4. Conclusions

A number of paint coated metal substrates were used to show the effects of a continuous wave laser removal process under such circumstances. It has been shown that the substrate material has an effect on the removal process. A certain minimal removal track depth can then be achieved, independent of the thermal properties of the substrate material involved, as long as a combustion process can be initiated. Lower thermal diffusivity can result in improved removal depths. This is mainly true for thin coatings ($d_T \leq d_C$).

The removal depth has been shown to be inversely proportional to the thermal diffusivity of the substrate, as well as linearly dependent on the transverse scanning speed.

5. Acknowledgements

The authors would like to thank the Mr A. Sumner for the preparation of the metal substrates and Dr G Kennedy of Diomed Ltd. for the supply and assistance with the diode laser system.

References

[1] G Schweitzer, L Werner; "Industrial 2 kW TEA CO_2 Laser for Paint stripping of Aircraft", *SPIE Vol 2502: Gas Flow and Chemical Lasers*, Friedrichshafen, 1995, pp. 57-62

[2] I GobernadoMitre, AC Prieto, C. Fotakis, et al.; *App Spect* 51 (1997), No. 8, pp. 1125-1129

[3] L.M. Galantucci, A. Gravina, G. Chita, et al.; *Polymers & Polymer Composites* 5 (1997), No 2, pp. 87-94

[4] MJJ Schmidt, L Li, JT Spencer; "Paint removal using a 120W Diode Laser System", presented at the *AMPT Conference 99* (Dublin)

[5] MJJ Schmidt, L Li, JT Spencer; "High power diode laser removal of chlorinated rubber coatings from concrete surfaces", *Opt Las Tech* 31 (1999), 171-180

[6] MJJ Schmidt, L Li, JT Spencer, PH Key; "The effects of wavelength of laser wavelength on laser removal of chlorinated rubber", *App Surf Sci* 139 (1999), 418-423

[7] Tremco Tetrochlor EAX Finish Airless Spray, Prod Ref. 356 000, Originally Produced by Tretol Plc., Supplied by Tremco Ltd., Wigan, UK.

[8] HS Carslaw, JC Jaeger; Conduction of Heat in Solids, 2nd Edition; Clarendon Press, Oxford, 1959

[9] MJJ Schmidt, L Li, JT Spencer; "Removal of a polymer ceramic mixture with a pulsed Nd:YAG system", *App Surf Sci* 154-155 (2000), 53-54

Taper Formation Mechanisms and Taper Control during Laser Percussion Drilling of Ni Alloy

D.K.Y. Low[1], L. Li[1] and P.J. Byrd[2]

[1]Manufacturing Division, Department of Mechanical Engineering,
University of Manchester Institute of Science and Technology (UMIST),
PO Box 88, Sackville Street, Manchester M60 1QD, UK
[2]Manufacturing Technology, Rolls-Royce plc,
PO Box 3, Filton, Bristol BS34 7QE, UK

Abstract: This paper reports a novel technique for controlling hole taper in laser percussion drilling. The method is termed Sequential Pulse Delivery Pattern Control (SPDPC) and it is based on the alteration of the overall laser pulse pattern during a pulse train required to drill a hole. The work has shown that SPDPC is effective in producing through holes with good overall parallelism in Ni alloy (2.65 mm thick), giving an average hole taper of 12 μm for 0.55 mm holes. Experiments were conducted to understand the taper formation mechanisms. The effectiveness of the SPDPC method in reducing hole taper has been identified as being primarily due to the reduction of upward liquid melt ejection which in turn reduced the erosion of hole due to melt ejection, as well as the optimisation of the laser pulse intensity distribution during hole propagation.

1. Introduction

Drilling was the first industrial application of the laser when a ruby laser was employed to drill holes in diamond [1]. Over the past thirty-five years many technological advances have been made in the field of laser drilling. It has now advanced to be a well-accepted hole-making technique, with its current applications ranging from electronic to medical to aerospace industries [2, 3]. Nevertheless, the current level of laser drilling technology is far from maturity. Researchers and manufacturers alike are actively seeking new ways to optimise the process with particular efforts being focussed on achieving better quality holes. An area for improvement is the laser drilling of metals wherein the holes are inherently associated with defects such as spatter [4] and hole taper. The magnitude of hole taper can be a problem in many applications, especially in fluid flow applications in which the resulting flow characteristics are dependent on the hole geometry.

To date, some work has been reported on the characteristics of hole taper during laser drilling. Many investigators reported that the degree of taper may be minimised by the careful selection of the process parameters such as the pulse energy, pulse width, as well as by varying the design of the optical system [5]. In work studying into the effect of beam waist position on the hole formation in some metals, Yilbas and Sami [6] reported that by altering the beam waist positions, converging and diverging hole walls could be produced for negative and positive positions respectively. More recently, Murray and Tyrer [7] optimised the drilling cycle to obtain low taper holes in 8.26 mm thick ZrO_2.

Whereas the previous work generally investigated the reduction of hole taper through the optimisation of the process parameters, this current work presents a technique developed for controlling the hole taper during laser percussion drilling [8, 9]. The method is termed as Sequential Pulse Delivery Pattern Control (SPDPC). SPDPC is a method of altering the overall pulse delivery pattern within a pulse train needed to drill a hole, as opposed to the normal delivery pattern (NDP) traditionally employed in laser percussion drilling where the parameters of successive laser pulses are identical. Laser drilling experiments were conducted for both the SPDPC and NDP methods in Ni alloy sheets. The fundamental effects of the SPDPC method has been investigated in terms of the hole taper formation mechanisms that are primarily associated with the NDP method. This has been carried out by analysing the melt ejection process and the laser intensity distribution during the hole propagation.

2. Methodology

Predominantly, in both research and commercial areas, laser percussion drilling is carried out with the normal delivery pattern (NDP) whereby a series of laser pulses having the same pulse parameters are repeatedly used throughout the pulse train needed to drill a hole. In other words, the total laser energy, E_T, of a NDP pulse train is the product of the energy per pulse, E, and the number of pulses, n, employed. Although the degree of hole taper could be minimised and controlled by optimising the laser parameters (e.g. peak power and number of pulses), its elimination to achieve good overall through hole parallelism can be difficult when drilling with the NDP. As a consequence, a marked difference between the entrance and exit diameters is usually observed, with the entrance diameter being consistently larger than the exit diameter.

SPDPC is essentially a method of varying the parameters of individual laser pulses during laser percussion drilling. A viable method found was to linearly increase the laser pulse energies throughout the pulse train (by increasing the peak power), whilst maintaining the pulse shape of the individual laser pulse used in the NDP (i.e. standard rectangular) as shown in Figure 1.

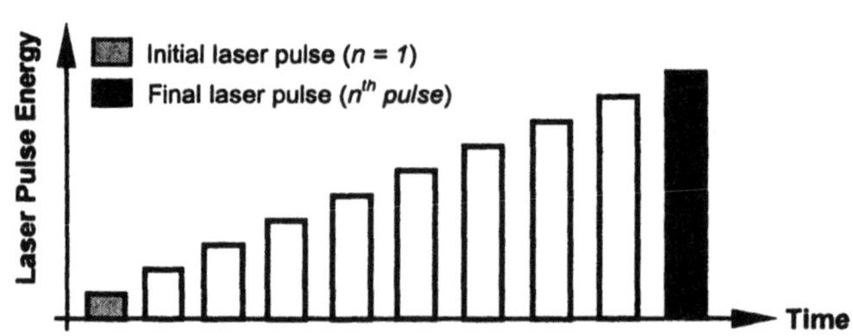

Figure 1: Schematic diagram of a linearly increasing SPDPC method.

With respect to the linearly increasing SPDPC method shown in Figure 1, the rate of energy deposition can be described by the following equation.

$$\frac{\mathrm{d}E}{\mathrm{d}t} = \frac{(E_2 - E_1)\,\mathrm{f}}{(n-1)} \qquad (1)$$

where E_2 and E_1 are the final and initial pulse energies of the pulse train, n is the number of pulses used in the pulse train and f is the pulse frequency. Subsequently, the total energy, E_T, delivered by the entire train can be expressed as:

$$E_{\mathrm{T}} = E_1 + \sum_{i=1}^{n-1} \left[\frac{\mathrm{d}E}{\mathrm{d}t} \left(\frac{1}{\mathrm{f}} \right) i + E_1 \right] \tag{2}$$

In essence, SPDPC is a method of temporal pulse shaping the overall delivery pattern of a pulse train as opposed to the case of temporal pulse shaping within a single laser pulse and repeatedly using it throughout the pulse train. Unlike in laser welding, where the latter case of single pulse shaping has been shown to improve the weld quality [10], this was found not to be the case in terms for hole taper benefits during laser drilling. This result was perhaps expected since laser percussion drilling utilises a pulse train to drill a hole, therefore, temporal pulse shaping should be considered in the light of the entire pulse train employed.

3. Experimental Procedure

A fibre-optic delivered 400 W Nd:YAG laser was used in the experiments to drill through-holes in 2.65 mm thick Ni alloy sheets using both the NDP and SPDPC methods. An O_2 assist gas (3 bar) was employed through a conical nozzle. The laser was focussed on the material surface giving a spot size of approximately 0.5 mm. Standard rectangular laser pulses containing 7 J/pulse and 0.1-7 J/pulse were employed for the NDP and linearly increasing SPDPC methods respectively. For given set of SPDPC and NDP drilling experiments, the laser parameters such as the total laser energy, pulse width and pulse frequency were kept constant for both delivery patterns.

4. Results

Figure 2(a) and (b) compare the cross section geometries of the through-holes drilled with the NDP and linearly increasing SPDPC methods respectively, under the same operating parameters. Since hole taper angle is typically determined by the equation, $\alpha = (d_{ent} - d_{exit})/t$ [11], where d_{ent} is entrance diameter, d_{exit} is the exit diameter and t is the material thickness, then a hole having $d_{ent} \approx d_{exit}$, is generally considered to possess no taper. As can be seen from Figure 2(a), the d_{ent} and d_{exit} of the hole produced by the SPDPC method are relatively equal, whilst the hole produced by the NDP contained a marked difference between the d_{ent} and d_{exit}. In addition, the NDP drilled holes are strongly characterised with an inlet cone, hole wall undulation and some degree of barrelling, whilst the SPDPC drilled hole possessed good overall hole wall parallelism. The SPDPC method gave an average d_{ent} of 555 μm and an average hole taper of 12 μm (i.e. d_{ent} - d_{exit}), whilst the NDP drilled holes gave an average d_{ent} of 688 μm and an average hole taper of 76 μm (15 holes). Indeed, such a reduced hole taper was consistently observed in SPPDC drilling for different sets of comparative experiments.

Figure 2: Typical cross section micrographs of holes obtained with (a) linearly increasing SPDPC and (b) NDP methods using similar laser pulse parameters.

5. Discussion

The finite hole taper as shown in Figure 2(a) is a feature commonly found in holes that were percussion drilled with the NDP [12]. The reduced hole taper in SPDPC drilled holes can be explained by first examining the two main taper formation mechanisms which are predominant during NDP drilling: (a) erosion of hole wall as a result of the upward melt ejection (before beam breakthrough) and (b) the reducing laser pulse intensity with hole propagation.

Firstly, the material removal in laser drilling with intensities in the range of 10^5-10^7 W/cm^2 is dominated by liquid melt ejection, as in the case of this study. As a result of surface vaporisation a recoil pressure is generated in the opposite direction such that the liquid melt is pushed out of the interaction zone (hole bottom) towards hole entrance and subsequently out of the hole [13]. This hot flowing melt can further induce melting of additional solid material on the hole wall during laser drilling, or on side of the weld or cut in the case of laser welding and cutting [14]. As the hole propagates with repeated melt ejection due to the successive laser pulses, more melt will flow past the upper section of the hole as compared to the lower section. As such, the interaction time between the melt and the hole wall was the greatest at the entrance side of the hole and reduced with depth. As a consequence of this variation in hole wall erosion throughout the hole depth, a hole taper can thus be generated.

Secondly, owing to the fact that the laser peak power in the NDP was identical throughout the pulse train, the laser pulse intensity would subsequently be reduced with hole propagation. This is because of the increasing beam diameter with beam divergence along the laser axis and the absorption of any part of the incident beam at the upper section of the hole. In other words, during NDP drilling the hole bottom would be subjected to reducing laser pulse intensities as the hole deepened. Since the amount of material removed generally increases with the incident laser beam intensity, then the reducing laser intensities with hole propagation will therefore result in a smaller d_{exit} as compared to the d_{ent}.

By using the linearly increasing SPDPC, it was found that the fraction of upward melt ejection (before beam breakthrough) was significantly reduced from 80% to 60%. As such, the absolute mass of the upward ejected material was effectively reduced by up to half of that measured for NDP drilling. Using high-speed photography (9000 frames/sec) it was found that after the occurrence of beam breakthrough most of the material was ejected in a downward manner out of the hole exit, showing that hole wall erosion only occurred before bream breakthrough. Therefore, the reduction in upward melt ejection before beam breakthrough in SPDPC drilling will result in a reduction of hole wall erosion and consequently minimising the undesired hole wall enlargement particularly at the upper section of the hole. Also as a result of this reduced upward melt ejection during SPDPC drilling, the spatter deposited on the material surface was reduced by up to 2.5 times in terms of the spatter area.

The laser pulse intensity variation with the experimental hole depth propagation was approximated by taking into account of the beam divergence and the hole wall absorption as shown in Figure 3.

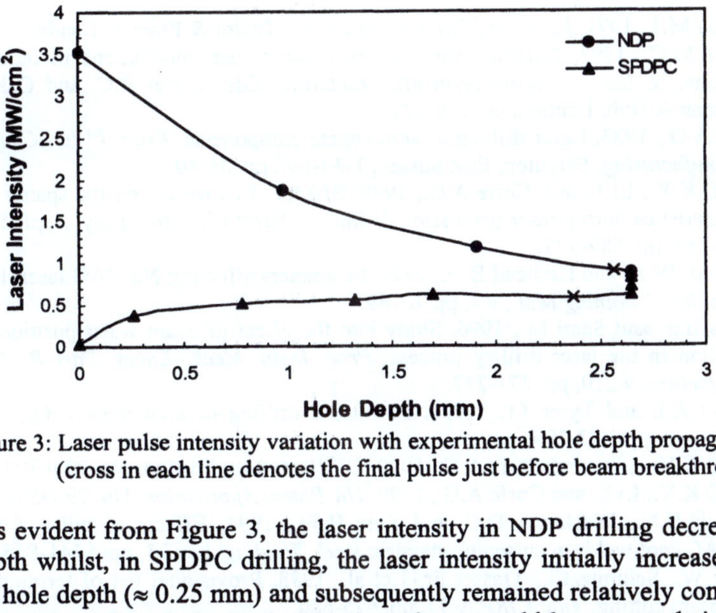

Figure 3: Laser pulse intensity variation with experimental hole depth propagation. (cross in each line denotes the final pulse just before beam breakthrough)

As is evident from Figure 3, the laser intensity in NDP drilling decreased with hole depth whilst, in SPDPC drilling, the laser intensity initially increased up to a shallow hole depth ($\approx$ 0.25 mm) and subsequently remained relatively constant with the propagation of the hole depth. This is perhaps would be expected since during SPDPC drilling the peak power was linearly increased throughout the entire pulse train. As such, the reducing laser pulse intensity with hole propagation that was associated with NDP drilling was prevented when employing the linearly increasing SPDPC method and as a result, an effective flushing (reaming) at the lower section of hole was possible in order to increase the d_{exit}. Consequently, the combination of reduced hole wall erosion and improved laser intensity with hole propagation that were associated with the linearly increasing SPDPC drilling enabled the possibility of producing holes with reduced hole taper during laser percussion drilling.

6. Conclusion

The work has shown that the linearly increasing SPDPC method is capable of drilling low taper holes in Ni alloy sheets with average entrance diameter of 555 μm and hole taper of 12 μm being achieved whilst maintaining good overall parallelism throughout the hole. This represents a significant improvement from the NDP drilled holes, which possessed an average hole taper of 76 μm. The reduction in hole taper during SPDPC drilling has been identified to be principally due to the reduction/prevention of hole taper formation mechanisms that were predominantly associated in NDP drilling. Using the SPDPC method the melt eject fraction has been reduced from 80% to 60% and as a direct result the extent of melt erosion of the hole wall was reduced. As a result of the linearly increasing peak power, the laser pulse intensity distribution in SPDPC drilling was optimised to be relatively constant throughout the hole propagation in order to improve material removal at the lower section of the holes.

References

[1] Beesley M.J., 1971, *Laser and Their Applications*, Taylor & Francis, London.
[2] Gower M.C., 1993, Excimer lasers: current and future applications in industry and medicine, In *Laser Processing in Manufacturing*, Eds: Crafer R.C. and Oakley P.J., Chapman & Hall, London, pp. 189-271.
[3] Corfe A.G., 1983, Laser drilling of aero engine components, *Proc. 1st Int. Conf. Lasers in Manufacturing*, Brighton, East Sussex, 1-3 Nov., pp. 31-40.
[4] Low D.K.Y., Li L. and Corfe A.G., 1999, Effects of assist gas on the spatter physical characteristics during laser percussion drilling in NIMONIC 263 alloy, *Appl. Surf. Sci.*, v 154-155, pp. 689-695.
[5] Hamoudi W.K. and Rasheed B.G., 1995, Parameters affecting Nd:YAG laser drilling of metals, *Int. J. Joining Mat.*, v 7, pp. 63-69.
[6] Yilbas B.S. and Sami M., 1996, Study into the effect of beam waist position on hole formation in the laser drilling process, *Proc. Instn. Mech. Engrs. Part B: J. Engng. Manufacture*, v 210, pp. 271-277.
[7] Murray A.J. and Tyrer J.R., 1998, Low taper drilling of 8.26 thick ZrO_2, *Lasers in Engng.*, v 6, pp. 273-289.
[8] Low D.K.Y., Li L. and Corfe A.G., 1999, *UK Patent Application*, No. 9908709.0.
[9] Low D.K.Y., Li L. and Corfe A.G., 1999, *UK Patent Application*, No. 9909504.4.
[10] Bransch H.N., Weckman D.C. and Kerr H.W., 1994, Effects of pulse shaping on Nd:YAG spot welds in austenitic stainless steel, *Welding J.*, v 73, pp. S141-S151.
[11] Konig W., Schmitz Cl., Trasser Fr.-J et al., 1988, Provisional list of terminology for laser beam cutting, *Ann. CIRP*, v 73, pp. 675-680.
[12] Chen X., Ortiz A.L., Staver P.R. et al., 1997, Improved hole drilling using a high peak power Nd:YAG laser at the second harmonic wavelength. *J. Laser Applications*, v 9, pp. 287-290.
[13] Allmen M.V., 1976, Laser drilling velocity in metals, *J. Appl. Phys.*, v 47, pp. 5460-5463.
[14] Semak V. and Matsunawa A., 1997, The role of recoil pressure in energy balance during laser materials processing, *J. Phys. D: Appl. Phys.*, v 30, pp. 2541-2552.

Planning Model for Active Gas Assisted Laser Cutting

Giovanni Tani[1], Luca Tomesani[2], Andrea Zucchelli[2]

[1] University of Florence, Department of Mechanics and Industrial Technologies, Via S.Marta 3, 50100 Italy - email: tani@.ing.unifi.it
[2] University of Bologna, Department of Mechanical Construction Engineering, Viale Risorgimento 2, 40136 Bologna, Italy - email: luca.tomesani@mail.ing.unibo.it

Abstract: A new theoretical model for kerf geometry evaluation in CW laser cutting operations has been implemented. The model is based on the analytical solution of the energy balance of four terms: laser beam absorbed intensity, oxidized melt film reaction energy, conduction losses, latent heat of melting material. No restrictions are made on kerf geometry. Experimental verification of model predictions are made on mild steel (2-5 mm thickness) and Aluminum alloy 6061 (2-4mm), evaluating the shape of the kerf.

1. Introduction

In laser cutting technology, detailed process planning is always required for the purpose of minimising setup and thus keeping machining costs as low as possible. In fact, process planning activities are subjected to severe limitations due to the inherent complexity of the process, where several physical phenomena are involved: phase change, plasma formation, surface reflection, convection, three-dimensional conduction, all occurring within limited spatial and temporal constraints. A wide range of analytical models has been developed in order to consider specific phenomena, but none of them includes all these variables.

Modest and Abakians [1] have developed a linear steady-state model for cutting with a CW laser of punctiform section; Sheng and Cai [2] have studied the geometry of the cutting front on rectilinear and curved paths when inert gas is used, by considering heat losses on mono-dimensional conduction paths; Vicanek and Simon [3] have analized the interactions between shielding gas and melting material; Ivarson, Powell and Magnusson [4] have studied the oxidation dynamic in laser cutting of mild steel and stainless steel; Kaplan [5] has developed a model for metal cutting with a gaussian laser beam which entails the role of oxidation, by considering a simplified kerf geometry.

Moreover, experimental investigations conducted to verify the model are usually limited to a small set of materials to be cut within a narrow range of beam power.

Accordingly, a mixed approach for parameters optimization is thus frequently adopted [6], with the aim to select the process variables by means of both theoretical models and experimental support.

In an attempt to resolve these problems, a new theoretical model for kerf geometry evaluation in CW laser cutting operations has been implemented. The model is based on analytical solution of the energy balance of four terms: laser beam absorbed intensity, oxidized melt film reaction energy, conduction losses, latent heat. No restrictions are made on kerf geometry. The model takes into consideration the variables related to the material to be cut, the beam source, the shielding gas and process parameters. Experimental investigations on different materials and under different operating conditions have been conducted, and the results have been used to refine the system.

2. Analytical model

2.1 Kerf geometry

The cutting front is divided into areolas dA to each of which is associated the normal direction n along which the thermal conduction differential equation will be integrated. From the value of the depth s for each areola the profile of the cutting front can be determinated. The individual elementary flat surfaces can be identified by the inclination of the angles θ in the x direction (parallel to the beam path) and φ in the y direction (perpendicular to the beam path). The geometrical relationship between the value of the elementary area dA and its projection dxdy on the local reference point is [2]:

$$dA = \sqrt{1+\tan^2 \vartheta + \tan^2 \varphi} \cdot dxdy$$

2.2 Laser beam energy

The accumulated energy in a generic point (x, y) can be evaluated with the integral in time of the laser beam energy, which was assumed to have a TEM_{00} gaussian distribution. In a rectilinear cut, the speed is constant and the increment of time can be transformed to an increment of space by means of dt=du/V , where u is the coordinate in x direction. Thus, the beam energy in a generic point (x,y) can be determined as a function of spatial coordinates as follows:

$$E_{beam}(x,y) = \int_{-\infty}^{x} I(u,y)\frac{du}{V} = \int_{-\infty}^{x} \frac{a(u,y)P}{\pi R^2 V} \exp\left(-\frac{2(u^2+y^2)}{R^2}\right) du$$

where a is the surface absorptivity, dependent on the element inclination, P is the beam power and R is the spot radius, dependent on the depth s due to beam divergence.

2.3Oxidation Energy

In the present model, no evaporative cutting has been considered. Assuming that all the melt material is expulsed from the bottom of the kerf, we can write:

$V \cdot D \cdot w = v_m \cdot s_m \cdot w$. Here, V is the beam speed, D is the sheet thickness, w the kerf width, v_m and s_m the melt film velocity and thickness. Thus, the melt film thickness can thus be expressed as:

$$s_m = V \cdot D / v_m$$

The melting thickness is the sum of the oxidized melt thickness S_{oxide} and the non-oxidized melt thickness S_{No} :

$$s_m = S_{oxide} + S_{No}$$

where S_{oxide} can be computed by the parabolic Tamman law of diffusion, giving a oxidized thickness of:

$$S_{oxide} = (2 \cdot D_e \cdot t_{eff})^{1/2} \leq s_m$$

where D_e is the diffusion constant of the oxigen in the considered metal. The effective time of reaction t_{eff} is the ratio between the sheet thickness and the mean velocity of the melt: $t_{eff} = D/v_m$. In these equations, unknown terms are: v_m, s_m, S_{oxide}, S_{No}. One more equation is needed to solve the system, which can be found by the equilibrium of forces on the melt phase:

$$F_o + F_n + F_t = F_a + F_{st} + F_d + F_m$$

Here, the forces applied on the melt thickness by the shielding gas are balanced by friction forces between melting phase and solid material [5]. The melt surface in question is approximately a half cylinder of diameter w and depth D, with surface $D \cdot w \cdot \pi / 2$. The expressions used for each term are the following:

$F_o = D \cdot w \cdot \pi \cdot p_g / 2$	Gas force [p_g = gas static pressure at the top]
$F_n = D \cdot w \cdot \pi \cdot \rho_g \cdot v_g^2 \cdot \tan(\theta)/2$	Normal dynamic pressure [ρ_g = gas density]
$F_t = (D \cdot \rho_g \cdot \mu_g)^{1/2} \cdot 2 v_g^{3/2} \cdot w \cdot \pi/2$	Tangential dyn. pressure [μ_g = gas viscosity]
$F_a = s_m \cdot w \cdot p_a$	Static pressure at the bottom [P_a=amb. pressure]
$F_{st} = D \cdot w \cdot \pi \cdot \sigma \cdot$	Tangential friction force [σ=surface tension]
$F_d = w \cdot \pi \cdot \rho \cdot v_m^2 \cdot s_m / 6$	Average melt dynamic force [ρ =melt density]
$F_m = D \cdot w \cdot \pi \cdot \mu \cdot v_m / (2 \cdot s_m)$	Friction internal to the melt [μ= melt viscosity]

Concerning evaluation of the physical state of the gas within the kerf, it has been considered that the gas on outlet from the nozzle is in a critical state [5], and it can be considered that, due to adiabatic expansion from the internal pressure p_i, it is possible to evaluate its density ρ^* and speed c^*. After the outlet, only a fraction of the gas goes into the kerf, and its pressure has been found by evaluating the stagnation pressure p_g from the condition taken into consideration: $p_g = \rho^* \cdot v_g^2$, $v_g = c^*$, $\rho_g = \rho^*$.

The phisical state of the gas in the kerf can be considered as constant without significant errors, because pressure variations are considerable only for steep slopes

of the cutting front [3]. Thus, the physical state of the gas at the top of the kerf has been considered as rapresentative the entire cutting front.
Once the values of v_m, s_m, S_{oxide}, S_{No} are known, the oxidation reaction is dependent on the kind of oxide resulting from the cut: above all FeO for mild steel, Al_2O_3 for aluminum [Ivarson]. Thus, the reaction speed is: $N_{oxide}=v_m \cdot w \cdot S_{oxide} \cdot \rho / A_{metal}$ [mol/s] where A_{metal} is the relative atomic mass of the base metal and ρ its density. The expression for the power of reaction is: $P_r=N_{oxide} \cdot H_{react}$, where H_{react} is the heat of oxidation. The term to be introduced in the energy balance is then in the form:

$$E_{oxide}(x,y) = \int_{-\infty}^{x} 2yv_m S_{oxide} \frac{\rho}{A_{metal}} H_{react} \frac{du}{V}$$

2.4 Phase change Energy

The energy for phase change is the energy required to melt a column of material of height s which extends from the upper surface of the sheet to the local position of the element taken into consideration:

$$E_{phase}(x,y)=\rho \cdot (L+c_p(T_{melt}-T_{amb})) \cdot s(x,y)$$

Where L is the latent heat and c_p is the specific heat capacity of the material.

2.5 Conduction Energy

The heat of conduction in the material is governed by the Fourier's equation, the solution of which has been obtained, according to Sheng hypothesis [2], by integrating it on a rectilinear (and mono-dimensional) conduction path orthogonal to the elementary area dA up to the lower surface of the piece. The conduction path δ can be expressed as a function of D, s, θ and φ. The conduction energy can be written as [7]:

$$E_{cond}(x,y) = \int_{-\infty}^{x} -k\left(\frac{\partial T(x,y,n)}{\partial n}\right)_{n=0} \cdot \sqrt{1+\tan^2 \vartheta + \tan^2 \varphi}\, \frac{du}{V}$$

2.6 Energy balance

For each surface element of the cutting front, the energy balance is expressed by:
$E_{beam}\,dxdy + E_{oxide}\,dxdy = E_{phase}\,dxdy + E_{cond}\,dA$, from which the depth of the cutting front can be derived.

3. Experimental

Cutting experiments have been performed with the aim to evaluate the shape of the kerf and relating it to the results of the model. A 3 kW CW CO_2 laser with TEM_{00} spatial mode has been used on low carbon steel (2-5mm thickness) with oxigen

assisting gas and Aluminum alloy 6061 (2-4mm thickness) with air shielding gas. Two pressure levels of the gas and three levels of power were adopted for each cutting condition. Cutting operations were performed at different beam speeds, starting from 0.5 m/min with increments of 0.5 m/min up to the limiting velocity for that condition. For each sequence, the optimal velocity was recorded as that for which the best cutting edge is obtained.
The cuts were interrupted by shutting off the power source, but without stopping the head speed and the shielding gas, so to reveal also the cutting front geometry. The starting point was kept outside the sheet sample. Geometrical observations on the transverse section was made with the back part of the sample still uncut and rigid, in order to avoid errors in the measures. The section plane was taken at a distance from the sheet edge proportional to the sheet thickness, in order to avoid inaccuracies deriving from the border effect.

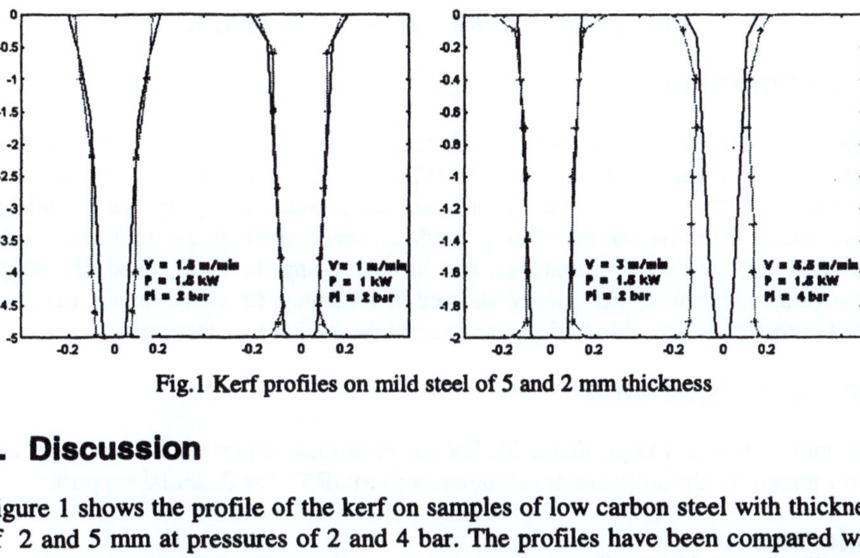

Fig.1 Kerf profiles on mild steel of 5 and 2 mm thickness

4. Discussion

Figure 1 shows the profile of the kerf on samples of low carbon steel with thickness of 2 and 5 mm at pressures of 2 and 4 bar. The profiles have been compared with the values calculated by the model. At low pressure (Fig. 1a-1b-1c) excellent fit with the real profile can be observed, except for the typical flaring on the bottom of the sheet, which the proposed model is not able to provide. This represents a limitation in all conditions in which the kerf presents inverse conicity, as is the case at high pressures and high speeds (Fig.1d). The calculation of the profile is good only in areas with direct conicity. In spite of this, the model very precisely calculates the limit speed, for any value of pressure, speed and power.
On aluminum the solutions calculated by the model are always very accurate, at both high and low pressures, as long as the cutting speed remains below the limit speed (Fig.2a-2b-2c). On the other hand, the proposed model has been shown to decidedly overestimate said speed, so that in the entire field between real limit speed and simulated one, the model provides for the execution of very thin kerfs, which are in reality impossible.

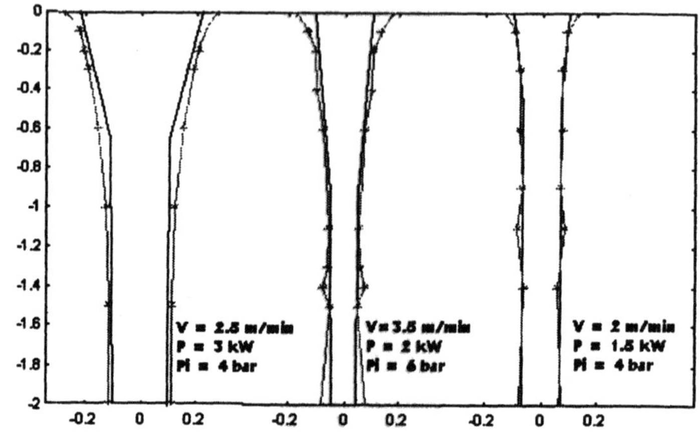

Fig.2 Kerf profiles on aluminum of 2 mm thickness

5. Conclusions

A free-geometry model has been developed for calculating kerfs in laser cutting operations with reactive gas. The model takes into account most of the variables in question, with the exception of the presence of plasma. The proposed model has shown itself effective in calculating kerfs on mild steel at the pressure values commonly utilized for that material, and in calculating the limit speed of cutting. As regards aluminum, the model showed a tendency to overestimate the limit speed, although it was able to describe exactly the kerf before this limit.

Acknowledgements

The authors would like to thank EL.EN for providing experimental facilities, eng. A. Fortunato for the software development and MURST for financial support.

References

[1] Modest M.F., Abakians H., Evaporative cutting of a semi-infinite body with a moving CW lase, *Transaction of the ASME, J. Heat Transfer*, 602-607, 1986

[2] P. Sheng, L. Cai, Analysis of laser evaporative and fusion cutting. In: *Transaction of the ASME, J. of Man. Sci. Eng.*, 118 (2), pp.225-234 (1996)

[3] Vicanek M. and Simon G., Momentum and heat transfer of an inert gas jet to the melt in laser cutting, *J. of Physics D, applied Physics*, 20, pp. 1191-1196, 1987

[4] Ivarson A., Powell J., Magnusson L., The role of oxidation in laser cutting stainless and mild steel, *Journal of laser applications*, 3, n.3, 1991

[5] Kaplan A. F. H., An analytical model for metal cutting with a laser beam, Journal of applied physics, 5, pp.79-90, 1996

[6] Tani G., Tomesani L., "Heat affected zone evaluation in laser drilling processes" Acts of the Int. Conf. AMPT'97, July 1997, Guimares, Portugal, pp.483-488

[7] Tani G., Tomesani L., "Kerf geometry evaluation in laser processing of stainless steel" Acts of the IV AITEM Conf., Brescia, 13-15 September 1999, pp.467-476

MACHINE TOOL ENGINEERING

DESIGN AND EVALUATION OF MACHINE TOOL COMPONENTS USING CEMENTITIOUS COMPOSITES

Md. Bazlul Karim[1], M. Rahman[1] and M. A. Mansur[1]
[1]National University Of Singapore, 10 Kent Ridge Crescent, Singapore 119260.

Abstract: The utilization of micro fibers has produced cementitious composites with significantly improved toughness or ductility. Research presented here is the feasibility study of micro-fiber reinforced cement composite (MFRCC) material for a machine tools structure to replace conventional cast iron material. A series of static and dynamic tests have been conducted to find out the most suitable design mix of MFRCC, which possesses many beneficial properties for fabricating machine tools structures. Further, Latex is used as an additive for enhancing the vibration damping capacity of cement composites. A grinding machine column has been designed and fabricated using the developed cementitious composite and its performance has been evaluated. Test results revealed better dynamic properties for MFRCC column than cast iron column.

1. Introduction

For a considerable amount of time gray cast iron was the material of choice as a machine tools structure [1]. The need for higher material damping in very high and the ultra-precision machining situation, combined with long term dimensional and geometrical stability, to support the ever-rising working speeds made possible by the development of tools and machining processes, the initial cost of fabricating patterns, molds and the generally poor environment of operating foundries, exert the impetus to find alternatives to cast iron. Although various degrees of success have been achieved with granite, ceramics, polymer concrete and cementitious composites material, some problems still exist. Fiber reinforced concrete is increasingly attracting the attention of engineers' world wide because of such favourable performance characteristics as high strength to weight ratio, resistance to corrosion and magnetic neutrality. It is ordinary concrete containing discrete fibers of short length and small diameter. Investigation carried out with the utilization of the large steel fibers (lengths between 20 to 60 mm and diameter 0.3 to 1.0 mm) indicate that a fiber reinforced concrete column with 3% volume fraction of fibers possesses adequate dimensional stability, strength and stiffness to justify its use as a suitable replacement material [2-3]. The incorporation of large steel fibers with cement composite has been shown no significant improvement in tensile strength. The reason often proposed for this incapability of the large fibers to improve the tensile strength is that the tensile failure is initiated at the micro-cracking level followed by progressive coalescence of micro cracks into micro cracks and the large steel fibers are too far apart to arrest, deflect, or blunt these micro cracks in any significant way. It may be hypothesized that very fine fibers (herein called 'micro-fibers'), which would effectively modify the behaviour of cracks at the micro level, may be expected to improve the tensile strength of cement-based materials [4]. A variety of micro-fiber types, including steel, synthetic and natural fibers have been applied to concrete. However, polyvinyl alcohol (PVA) fiber reinforced cement has not been paid much attention by researchers and users. The research reported here is concerned with assessing the feasibility of PVA micro-fiber reinforced cement composite (MFRCC) as a machine tool structure. Effort has been made to study the PVA micro-fiber reinforced cement composite modified by latex, which is used for enhancing the vibration damping capacity of the cement composite [5].

2. Materials

The high-modulus PVA micro fiber (diameter 14 μ, cut length 6mm, Young's modulus 36 Gpa, tensile strength 1.5Gpa, bundle form) was used in this experiment, which has commercial name RMS 182. In application to cement paste, the PVA fiber is highly effective in increasing the flexural strength and toughness characteristics and impact resistance of the material. In the cementitious mixture of this study 5% by weight of cement was substituted with a fine pozzolan (Silica fume) with a relatively small particle size to improve fiber dispersibility of the matrix. When mixed with water, silica fume provides a cohesive mixture that is capable of attracting fiber balls and coating the individual fibers, the fiber-to- matrix interfacial bond strength also improves in the presence of silica fume. To improve the vibration damping properties of the cement composite styrene butadiene latex was used with varying proportion of cement. Silica fume, as well as fibers, tends to decrease the workability of the fresh mix; a superplasticizer Rehobuild 1000 was used to improve workability. 10 different recipes were prepared and tested containing different amount of fiber volume fractions and latex proportions. An omni –mixer was employed for applying varying acceleration to the mix particles and fibers in many different directions. This action forces all inclusions to come into intimate contact with the cementicious paste in a very short time, there by eliminating the possibility of dry fiber balls forming in the mixer. The prepared fresh mixtures were then cast inside molds and vibrate externally on a vibrating table for compacting. All test specimens were cured for 24hr in the mold covered with polyethylene sheeting, followed by 27 days in frost room. The following static tests were performed on mortar specimens (three replicated tests were performed for each ten composition): flexural (ASTM C78-84) on 40 X 40 X 160 mm prismatic specimen; Compression (ASTM C109-86) on 100 mm cubes, Elastic modulus on φ100 X 200 mm cylinders and splitting tensile strength (ASTM C496-96) on φ75 X 150 mm cylinders. Three 40 X 40 X 160 mm mortar specimen were manufactured and tested under free-free boundary condition to find out natural frequency and damping ratios. Three phases investigation process have been conducted according to the parameter studied in each phase. The specimens were manufactured to find out the static (cube compressive strength, flexural strength, splitting cylinder tensile strength and elastic modulus) and dynamic (natural frequency, damping ratios and mode shapes) properties with respect to fiber volume fractions, water-cement ratios and latex proportions by the weight of cement. From the first two phases of investigations 1.5% PVA micro-fiber with the cement, sand, water ratio 1: 1.5: 0.45 mix showed the optimal design mix for machine tools structure fabrication, which require high flexural and tensile strength. By varying the latex proportions (10%, 15% and 20%) by weight of cement with the mix selected from the first two phases it has been found that 20% latex by weight of cement was the most suitable design mix. The comparisons of static and dynamic test results of the mortar specimens are shown in Table 1 and 2.

Table 1: Static test results:

Mix Type	Fiber Vol. fraction Vf (%)	Latex %	Compressive strength	Flexural strength	Splitting tensile strength	Elastic modulus
Plain			51.30	4.71	5.14	32.20
With PVA micro-fiber	1%		53.77	6.58	5.01	26.29
	1.5%		53.94	8.88	8.97	27.03
	2%		51.795	10.42	8.81	26.50
With fiber and latex	1.5%	10%	45.73	7.01	7.90	22.42
		15%	47.12	7.05	8.02	23.27
		20%	48.49	7.90	8.35	24.06

Table 2: Dynamic test results:

Mix Type	Fiber vol. Fraction Vf (%)	Latex %	1st Mode		2nd Mode		3rd Mode	
			Fn	D.R	Fn	D.R	Fn	D.R
Plain			5290	0.11	7370	0.86	14290	0.82
With PVA micro-fiber	1%		5320	0.55	7100	1.25	14590	0.58
	1.5%		5150	0.89	7050	1.37	14820	1.05
	2%		5130	0.42	7140	0.55	14300	1.30
With fiber and latex		10%	5090	0.48	7150	1.14	13820	1.03
		15%	5010	0.83	6840	1.42	13500	1.33
		20%	5060	1.13	6970	1.48	13500	1.83

Note: Fn: Natural frequency, D. R.: Damping Ratio.

3. Manufacturing the micro-fiber reinforced cement composite column

The column of the surface-grinding machine Model ELLIOT 921 produced by ELLIOT was chosen as our reference. The column of that machine was designed and manufactured with the optimum composition of latex modified micro-fiber reinforced cement composite.

3.1 Design consideration

The structural design of machine frame components is fundamentally dependent upon the duties, which they are expected to perform. The basic form of this grinding machine column is a rectangular cross-section. The inside of the column is kept free to accommodate the belt, pulleys which are required to transmit motion from the motor, connected with the lower portion of that column, to the head stock, supported by the upper part of that column. In the grinding process the grinding force is transferred from the grinding wheel to the column and this force may cause bending and tension of the column. In order to ensure adequate strength and stiffness, it is necessary to establish the overall forces acting on the column. The power requirement P (1.5kw in present study) for the grinding operation depends on the tangential force component F_t and the tangential velocity V_t. The usual range of V_t is 2140- 2850 rpm (1680 – 2220 m/min). If the turning power is kept unchanged, the smaller the tangential velocity the larger the tangential force will be. However, V_t is assumed to be 1000 rpm (800 m/min). With these values the tangential and normal components of the grinding force become 112.5N and 225N respectively. Thus it can be observed that the grinding force is very small. Since the specific modulus of the selected mortar is little smaller than that of the cast iron, the mortar column should be designed a little heavier to give the same stiffness.

Fig.1 Photograph of MFRCC column.

3.2 Fabrication of the column

Attention has been paid for simplification of column geometry without sacrificing required strength and stiffness. Casting skeleton consisted of two dove tail guideways and one bottom plate all made with mild steel. Mild steel studs (length 50 mm and dia 5 mm) were designed to provide required anchorage. Due to the incapability of Concrete structure to sustain machining operation metal inserts have been designed to provide boltholes for connecting the column with other parts of the machine. A wooden casting mold has been constructed with exactly same size of the designed column. Selected material micro-fiber reinforced cement composite (the ratio of cement, sand and water is 1:1.5:0.45, 1.5% fiber volume fraction used with 20% latex by weight of cement) was prepared and poured into the mold. During filling vibration was provided for sufficient compaction of the mortar. The column was demolded 24 hours after casting and curing was done under wet gunnysacks for 28 days before testing. Fig 1 shows the manufactured micro-fiber reinforced cement composite column for surface grinding machine.

4. Dynamic test for the machine tool structure

The dynamic characteristics of the columns (cast iron and micro-fiber reinforced cement composites) were measured by giving impulse signals and measuring responses through experimental modal analysis method. The testing apparatus consisted of a dual channel fast fourier transform (FFT) analyser, charge amplifier, an impulse hammer, an accelerometer, a force transducer and a PC using software STAR. The experimental set up for the test has been shown in fig. 2.

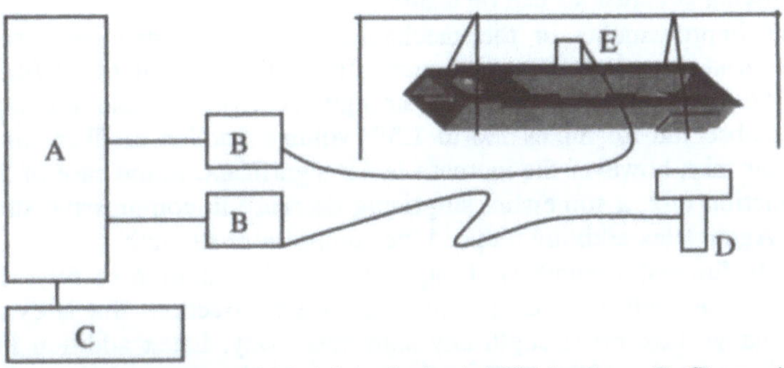

Fig.2. Set up for Experimental Modal Analysis: A: FFT Analyser, B: Charge Amplifier, C: Computer, D: Impact hammer, E: Accelerometer.

Table 3: Dynamic-test results.

Resonance No	Mode shape	Cast Iron Column		Mortar Column	
		Fn(Hz)	D.R	Fn(Hz)	D.R.
1	1st bending	451	0.267	442	1.16
2	1st torsional	525	0.218	520	0.972
3	2nd bending	756	0.170	671	0.68
4	2nd torsional	803	0.179	815	0.532

Comparative dynamic test results are summarised in table 3. Comparing the modal parameters of cast iron and MFRCC machine tool structure, it can be seen that mortar is a better material in terms of dynamic performance. The damping ratio of the MFRCC column is much higher than that of cast iron column. But in case of natural frequencies there is no significant difference between two columns for each mode of vibration. The natural frequencies for the vibration of a continuous beam with a free boundary condition are given by equation [6]

$$f_n = \frac{\lambda_i^2}{2\pi L^2}\sqrt{\frac{EI}{\rho S}} \quad \text{Where, } \lambda_i = (i+0.5)\pi,\ i = 1,2,3....n$$

Where ρ, EI, L and S are the density, bending stiffness, length and cross sectional area of the beam respectively. For cast iron ρ is 7.3 and the ρ of MFRCC used in this study is 2.22, which is about one third of that of cast iron At the same time bending stiffness of the MFRCC column also decrease 2~ 3 times compared with cast iron column. Therefore, the effect of higher density of cast iron and lower bending stiffness of MFRCC column resulted in similar natural frequencies for both of the columns.

5. Conclusions

In this work, the optimal composition of the microfiber reinforced cement composite for the machine tool column was investigated. Based on the results, the column of the grinding machine was fabricated with that material and comparative dynamic performance tests were conducted for both mortar and cast iron column. From the results the following conclusions can be made.

- Significant improvements in the mechanical properties may be observed in cement materials by reinforcing with micro-fibers. The addition of the fiber seems to have little effect on the compressive strength. A slight increase in compressive strength is observed for mixes 1% to 1.5% volume fraction of fiber compare to plane mortar mix; however the increase is not significant. In the case of 2% fiber volume fraction mix, a somewhat surprising decrease in compressive strength is observed. Again latex addition reduced the compressive strength.
- The ultimate flexural strength is strongly affected by addition of fiber. Flexural strength increases with the increase of fiber volume fraction. But latex addition does not change flexural strength any significant way. Latex addition improves the dynamic properties of the mortar mix compare to the plane mix.
- The damping ratios of the MFRCC column have been found to be about four times higher than the cast iron column at each corresponding mode of vibration. As far natural frequencies are concerned, no significant difference exists between two columns at each mode of vibration.

References

[1] S. B. Rao,1999, Journal of Manufacturing Science and Engineering,.
[2] Zoh F., Rahman M. and Mansur M. A., 1994, Int. Conf. on Prod. Engg.,
[3] M. Rahman, M. A. Mansur, Zhou feng,,1995, Materials & Design.
[4] Nemkumer Banthia and Jiakang Sheng, 1996, Cement & Concrete Composites.
[5] Xuli Fu, Xiaohui Li and D. D. L. Chung,, 1998, Journal of material science.
[6] Blevins, Robert D., 1979, Van Nostrand Reinhold Co., New York.

Analysis and Design of a Translational 3 dof PKM for Light Deburring Operations

L. Molinari-Tosatti, G. Bianchi, I. Fassi, R. Maj

ITIA CNR,
Institute of Industrial Technologies and Automation, National Research Council
Viale Lombardia 20/a, 20131, Milano, Italy
molinari@itia.mi.cnr.it

Abstract: This paper presents the conceptual design of a 3 d.o.f. parallel kinematic machine (here and after PKM), to be applied in the shoe manufacturing industry, starting from selection of the most suitable architecture to kineto-static optimization, actuator sizing (considering the effects of gravity and inertial forces), internal loads evaluation, including also the effect of manufacturing and assembly errors. Finally, the complete deburring cell is shown.

Keywords: Translational parallel manipulator, conceptual design, kinetostatic optimization

1. Introduction

Shoe manufacturing is nowadays facing all around the world new challenges in terms of increased market turbulence, of new competition and of other unpredictable factors that are forcing shoe companies to reconsider their position in the market. In order to cope with the new production requirements, a three years research project, which involves a selected group of Italian shoe machinery companies, has been launched by the Italian Ministry for Scientific and Technological Research (MURST). The scope of the project is developing new families of flexible and highly automated machinery for all the main phases of shoe manufacturing, and integrating them in an 'agile' Production System.

At the machinery level, a major research topic is the design of a new generation of robots, conceived as integral functional parts of a rotary machine for the production of directly injected PU soles. Nowadays most of these operations, sometimes dangerous and obnoxious for humans, are manually done.

The need to have low cost flexible machines leads to investigate the possible application of a PKM for multitasking operations, that combines the dexterity of robots with the precision of machine tools. PKMs are claimed to have interesting performances in respect to their serial counterparts, namely: higher payload to weight ratio (the payload is carried by several links in parallel); non cumulative joint error; higher structural rigidity (the load is carried by the parallel links, and in some structures there are only compression-traction mode); modularity (each kinematic chain is composed by the same physical modules), and this cuts down manufacturing costs; location of motors close to the fixed base; simpler solution of the inverse kinematic problem. Conversely, they suffer from singular configurations, low workspace to footprint ratio, a more complicated direct kinematic solution, and more difficult control techniques. The process needs require good performances, in terms of cycle times and speed, dexterity and precision, as listed below.

2. Industrial Requirements & Process Needs

The automation of the injected sole shoe manufacturing process should cover 3 different steps:

1. Deburring of the shoe profile: the tool, generally a mill for wood machining, removes a strip of leather 0.8 mm thick and 8 mm high;
2. Deposition, around the same profile, of a glue curtain
3. Spraying a silicon mixture over the mould for the PVA injection.

The strictest requirements are given by the first two operations, in terms of:

- working volume: 400 x 200 x 100 mm^3
- max. speed: 30 m/min, acceleration: 5 m/s^2
- precision: +/- 0.1 mm
- loads: 45 N deburring force

To execute the contour following in operation 2, 3 translational D.o.f. (degrees of freedom) for positioning and 2 rotations for tool orientation are required, with +/- 180° around the nozzle axis (A axis) and at least +/- 90°around the B axis. These requirements can be hardly achieved using a fully parallel PKM, and trying to cut down costs being competitive with traditional serial robots. Furthermore, the issue of flexibility and future applications drives to the configuration of a hybrid device (a 3 d.o.f. PKM plus a 2 d.o.f. serial wrist for orientation). The detailed structure of the PKM positioning device is described in the next paragraph.

3. PKM Architecture

Only recently researchers have focused on the study of kinematics and dynamics of mechanisms with reduced d.o.f. [1-3]. A class of 3 d.o.f. PKM with only translational movements has been presented in [4] and Di Gregorio, Parenti-Castelli [5,6] gave the minimum set of geometric conditions to provide the platform with a pure translational motion. A schematic view of this so called 3 UPU mechanism (3 identical legs, each composed by an extensible link connected to the base and to the moving platform by universal joints) is shown in Fig. 1.

The displacement analysis is solved by Eq. (1):

$$\mathbf{x} + \mathbf{R}\mathbf{b}_i = \mathbf{a}_i + q_i\mathbf{w}_i \qquad (1)$$

where:

$\mathbf{x}$ defines the position of O_1UVW with respect to OXYZ
$\mathbf{b}_i$ defines the position of the ideal joint B_i center with respect to O_1UVW
$\mathbf{a}_i$ defines the position of the ideal joint A_i center with respect to OXYZ
q_i is the length of the ith leg,
$\mathbf{w}_i$ is the unit vector defining the ith leg with respect to OXYZ

The velocity analysis is obtained differentiating Eq.(1) and projecting both the member on the $\mathbf{w}_i$ direction:

$$\dot{\mathbf{q}} = \mathbf{J}_x\dot{\mathbf{x}} \qquad (2)$$

being $\mathbf{J}_x$ the Jacobian that maps the end effector velocity into the joints' velocity. Under the hypothesis of pure translational motion [5,6], $\mathbf{J}_x$ can be reduced to a 3 by 3 square matrix, whose columns are the components of the unit vectors $\boldsymbol{w}_i$ of the legs.

For each leg, the actuator velocity is always less or equal to the module of the Cartesian velocity (Eq. (2)).

The relation between the Cartesian acceleration and the actuators' acceleration, obtained differentiating Eq. (2) is described by:

$$\ddot{\mathbf{q}} = \mathbf{J}_x \ddot{\mathbf{x}} + \mathbf{A}(\dot{\mathbf{x}} \cdot \dot{\mathbf{x}}) \quad (4)$$

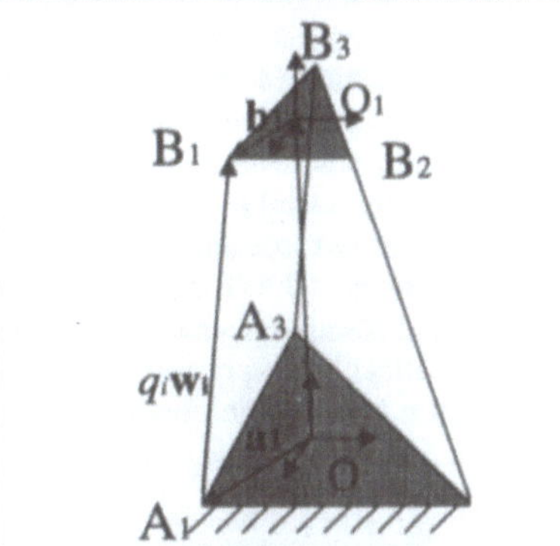

Fig.1 A schematic representation of the manipulator

where:

$$\mathbf{A} = \begin{bmatrix} \dfrac{(sin\alpha_1)^2}{q_1} \\ \dfrac{(sin\alpha_2)^2}{q_2} \\ \dfrac{(sin\alpha_3)^2}{q_3} \end{bmatrix} \qquad \ddot{\mathbf{q}} = \begin{bmatrix} \ddot{q}_1 \\ \ddot{q}_2 \\ \ddot{q}_3 \end{bmatrix} \quad (5)$$

being α_i the angle between the Cartesian velocity vector $\dot{\mathbf{x}}$ and the ith leg unit vector $\mathbf{w}_i$. Eq.(2) and Eq.(4) describe the kinematic behavior of the manipulator. Using the virtual works principle and starting from Eq.(2), the relation between the Cartesian forces, acting on the moving platform, and the joints' forces is defined as:

$$\mathbf{f} = ((\mathbf{J}_x)^t)^{-1} \cdot \mathbf{F} \quad (6)$$

where **F** is the Cartesian forces vector and **f** is the joints' forces vector.
If, for example, we consider $\mathbf{F}=[1\ 0\ 0]^T$ a unit force load applied to the end effector along the x-axis the associated actuators' forces are given by the first column of the matrix $((\mathbf{J}_x)^T)^{-1}$. The maximum of the absolute values of this column vector represent the maximum amplification factor for a force applied along the x axis.
Considering the stiffness of each leg, as a linear spring', with the same spring constant k, the displacement $\mathbf{x}$ in OXYZ, caused by a known load **F**, is described by Eq.(7):

$$\mathbf{F} = k\mathbf{J}_x{}^t \cdot \mathbf{J}_x \cdot \mathbf{x} \quad (7)$$

Assuming that all the geometric conditions illustrated in [5] are satisfied, a displacement of the actuators (i.e. a variation of legs' length) produces a simple translation of the moving platform. It's quite clear that the three rotations are inhibited by the mechanical constraints introduced by the universal joints; with respect to classical Stewart platforms (6 d.o.f. and 6 legs), the three legs are loaded not only by axial forces but also by torsional and bending loads. This implies a more accurate design of the legs, in order to verify the expected life of the employed components to all the above mentioned internal loads.

4. Kinetostatic optimization

Since the direction of loads applied to the center of mass of the mobile parts (both inertial and working loads) is unpredictable, the optimal configuration should have an isotropic behavior, i.e. a well conditioned Jacobian matrix. The obtained PKM is characterized by a manipulability ellipsoid [7,8,9] almost equal to a sphere at least in one configuration within the workspace. The optimization procedure that has been set up was based on well-known performance indexes: maximizing the minimum singular value of the Jacobian matrix and its determinant over all the defined workspace; requiring a condition number of the Jacobian matrix equal to 1 over all the defined workspace.

Considering the definition of Jacobian matrix, the only influencing parameter is the relative orientation of the three legs. When the three legs are mutually orthogonal the Jacobian matrix becomes orthonormal: its three singular values are equal to 1, the condition number is equal to 1 and the manipulability index is equal to 1, and the PKM assumes an isotropic behavior.

To obtain an almost isotropic manipulator overall the workspace, the machine is designed such that the optimum configuration is in the middle of the workspace with all the three legs at the middle of their elongation range.

5. Actuator sizing

Having chosen the optimal architecture, the next steps in machine development are: design of the mechanical components and actuators sizing. For both analyses, the manipulator has been modeled in a multibody package (ADAMS, from M.D.Inc), to include in the analysis not only the kinematic relations, but also the inertial effects and the machine weight. This dynamic analysis is of fundamental importance for high speed machines, where inertial loads are often larger then loads due to the machining process or to machine weight.
In order to support the extensible leg's actuators selection, the required force and velocity are computed over a selected grid of points in the manipulator workspace. The computation has been automated using a custom set of macros developed in the multi-body package, while the most critical loading condition is identified by a mathematical post-processing. The following figure shows the various contributions to the maximum load on the horizontal leg actuator .

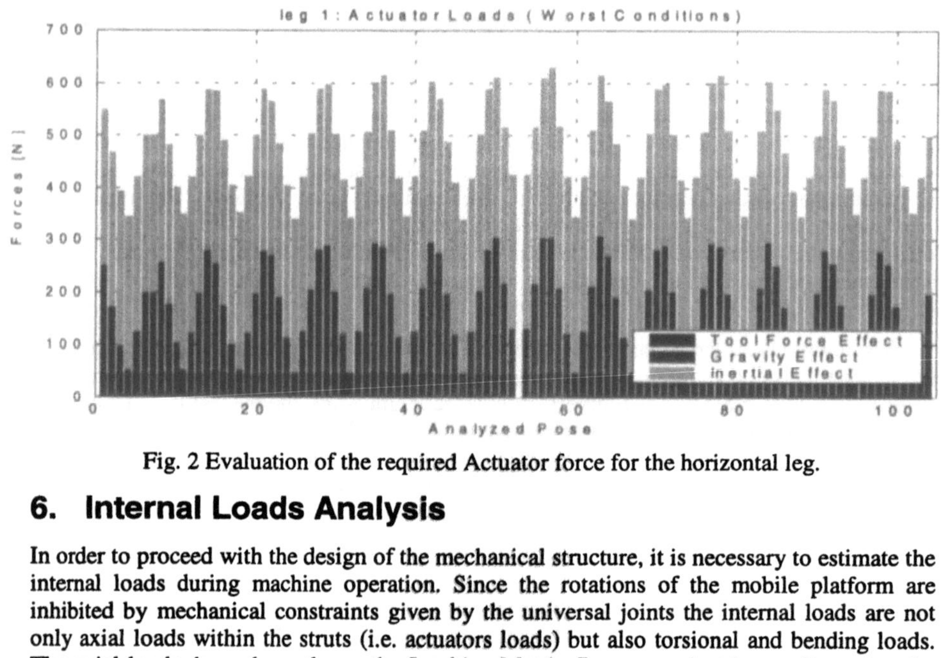

Fig. 2 Evaluation of the required Actuator force for the horizontal leg.

6. Internal Loads Analysis

In order to proceed with the design of the mechanical structure, it is necessary to estimate the internal loads during machine operation. Since the rotations of the mobile platform are inhibited by mechanical constraints given by the universal joints the internal loads are not only axial loads within the struts (i.e. actuators loads) but also torsional and bending loads. The axial loads depend mostly on the Jacobian Matrix $\mathbf{J_x}$: near a singular configuration the required force in at least one actuator tends to infinity, generating therefore high stresses in the mechanical structure. But the actuator's loads don't describe completely the existing stresses in the structure: some components of the external forces and torques applied to the end effector don't produce any actuator reaction (because the machine has less then six degrees of freedom). In our casean applied torque is not balanced by an actuator effort. For this reason, the analysis based on $\mathbf{J_x}$ doesn't provide any indication about the optimal location, referred to the head center point, of the end effector and of the legs ($\mathbf{J_x}$ depends on their orientation only), even if these design parameters are important, because they strongly influence the load on the passive joints (mainly the torsional load on the legs, in this case).
The analysis is performed computing an "Extended Inverse Jacobian (EIJ)":

$$\delta x_{\text{end effector}} = EIJ\ \delta q_{\substack{\text{fictitious}\\ \text{joints}}} \Leftrightarrow f_{\substack{\text{fictitious}\\ \text{joints}}} = EIJ^T\ f_{\text{end effector}} \quad (8)$$

that relates the motion (i.e. velocities) at the end effector with the velocities at different fictitious joints, located where the internal forces have to be computed. The EIJ is obtained from the multi-body package and analyzed by the mathematical post-processor, to identify, for each examined machine pose, the combination of external forces (inertial effects aren't taken into account at the moment) that produces the maximum value of the structural load considered. Considering a machining force of 45 N and a maximum acceleration of 5 m/s^2, the leg torsional loads for the examined PKM have been evaluated: 40 Nm for the horizontal leg, and 70 Nm for the inclined legs. These values have been used to appropriately select the bearings in the universal joints.

7. Influence of manufacturing errors and machine compliance

Basing on the duality between forces and velocities, the same extended jacobian is used to evaluate the end effector displacement generated by a displacement at one or more fictitious joint, that can be provoked in reality by manufacturing errors and/or elastic deformations [10].

Nine modifications of the nominal machine geometry are here considered: for each leg two bending at the mid point and the torsion around the leg axis.

The previous two evaluations can be unified when, having the first drawings of the mechanical components, is possible to produce an estimate of their stiffness: in this case the extended jacobian can be used to compute how localized structural compliances influence the global compliance at the end effector:

$$\begin{cases} \delta x^T_{\substack{\text{end}\\\text{effector}}} \cdot f_{\substack{\text{end}\\\text{effector}}} = \delta q^T_{\substack{\text{fictitious}\\\text{joints}}} \cdot f_{\substack{\text{fictitious}\\\text{joints}}} & \begin{pmatrix}\text{system}\\\text{equilibrium}\end{pmatrix} \\ f_{\substack{\text{fictitious}\\\text{joints}}} = K_{FJ} q_{\substack{\text{fictitious}\\\text{joints}}} & \begin{pmatrix}\text{elastic reaction}\\\text{at fictitious joints}\end{pmatrix} \\ x_{\text{end effector}} \equiv C_{EE} f_{\text{end effector}} & \begin{pmatrix}\text{compliance at}\\\text{the end effector}\end{pmatrix} \end{cases} \Rightarrow C_{EE} = EIJ \cdot K_{FJ}^{-1} \cdot EIJ \qquad (9)$$

Having estimated , after the mechanical design, a minimum torsional stiffness of 26.4 Nm/mrad for the horizontal leg and 28.9 Nm/mrad for the other legs, the software computes the corresponding stiffness at the end-effector (considering as rigid, in this case, the rest of the machine):

minimal translational stiffness: 110 N/mm
minimal rotational stiffness: 24 Nm/mrad

The relatively low translational stiffness is due to the position of the tool tip, 460 mm lower than the axis of the horizontal leg (see figure 3).

8. Conclusions

The conceptual design process of a 3 d.o.f. PKM has been described, starting from the industrial requirements. The suitable architecture has been selected and the kineto static optimization was performed. The actuator sizing has been performed considering also the effect of gravity and inertial forces.

The internal loads have been evaluated and providing indication for machine and components design that were unobtainable from the basic kinematic analysis. Finally the effect of manufacturing and assembly errors has been considered.

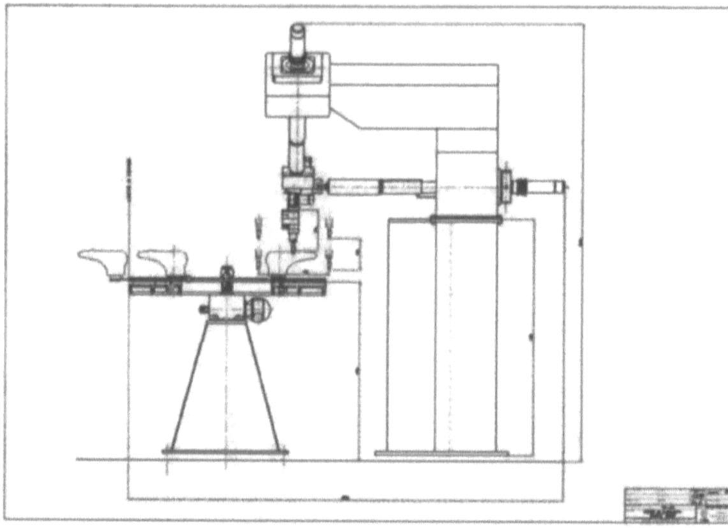

Fig. 3 XY Lay-out of the conceived PKM deburring cell

9. Acknowledgments

The financial support of MURST is gratefully acknowledged. This work has been done within the National Research Project on Innovative Production Systems SPI6: "Fully Automatic and Integrated Shoe Production Systems". The authors acknowledge all the ITIA-PKM team: G. Di Bernardo, M. Leonesio, S. Negri.

References

[1] R. Di Gregorio, V. Parenti-Castelli, 1998, *A Translational 3-DOF Parallel Manipulator*, Lenar• i• and Husty (eds.), Advances in robot Kinematics: Analysis and Control, pp. 49-58

[2] R. Di Gregorio, V. Parenti-Castelli, 1999, *Influence of leg flexibilityon the kinetostatic behaviour of a 3-DOF fully-parallel manipulator*, Proc. of the 10th World Congress on the Theory of Machines and Mechanisms, vol.3, pp. 1091-1098.

[3] C. Gosselin, J. Angeles, 1989, *The optimum kinematic design of a spherical 3 degrees of freedom parallel manipulator*, ASME Journal of Mechanisms, Transmission and Automation in Design, vol.111,n.2 pp. 202-207

[4] F. Pierrot, C. Reynaud, A. Fournier 1990, *DELTA: A Simple and Efficient Parallel Robot.* Robotica 8:105-109

[5] L. W. Tsai, R. Stamper, 1996, *A parallel manipulator with only translational degrees of freedom*, ASME 96 DETC-MECH-1152, Irvine (CA), USA

[6] L. W. Tsai, 1999, *Sistematic Enumeration of Parallel Manipulators*, Parallel Kinematic Machines: Theoretical Aspects and Industrial Requirements, Springer-Verlag, C.R. Boër, L. Molinari-Tosatti, K. S. Smith (Eds), pp. 295-312

[7] J. Angeles, 1997, *The mechanical design of redundant manipulators*, Proceedings of the RAAD97, pp. 15-25

[8] C. Gosselin, 1990, *Stiffness Mapping for Parallel Manipulators*, IEEE Trans. On Robotics and Automation, vol.6, n. 3, pp.377-382

[9] T. Huang, D.J. Whitehouse, J. Wang, 1998, *The Local Dexterity, Optimal Architecture and Design Criteria of Parallel Machine Tools*, Annals of the CIRP, vol. 47/1/98

[10] Patel A.J., Ehmann K.F., 1997, "Volumetric Error Analysis of a Stewart Platform-Based Machine Tool", *Annals of the CIRP*, 46:287-290.

A CAD Tool for the Conceptual Mechanical Design of 3-Axis Machine Tools

István Németh and Hendrik Van Brussel

Katholieke Universiteit Leuven, Department of Mechanical Engineering
Division of Production engineering, Machine design and Automation (PMA)
Celestijnenlaan 300B, 3001 Heverlee, Belgium
Istvan.Nemeth@mech.kuleuven.ac.be

Abstract: The paper presents a CAD tool for the conceptual design of the mechanical structures of 3-axis machine tools. The design tool helps the designer to synthesise thousands of different machine tool configurations, to optimise the geometry of the synthesised structures, and to search for optimal configurations. Five analyses are developed for the evaluation of the generated machine variants. The optimisations of the configurational and of the geometrical parameters are carried out by SQP algorithms and/or by genetic algorithms on a parallel computational platform. Optimisation objectives can be to minimise the total volume, and/or to maximise the static stiffness, and/or to maximise the lowest natural frequency, and/or to minimise the geometrical interferences.

1. Introduction

Presently, in machine tool design the time from concept creation to production is considerably long because of the lack of efficient conceptual design tools that support the designers in the early stages of the design process when the major decisions affecting the commitment of resources are made [1]. At the Katholieke Universiteit Leuven a research project has been launched in order to develop a design tool with which the generation, the evaluation and the optimisation of the mechanical conceptual structure of 3-axis machine tools can be realised. A summary of this design system is presented in the paper.

2. Overview of the Design System

Figure 1 depicts an overview of the design system. Using some configurational parameters the designer can prescribe one or more machine tools of interest. The machines are synthesised (configured from a component library) applying the prescribed configurational parameters and some input data (e.g. workspace dimensions, size of tool). After configuration, with the help of some analysis tools, the machines can be evaluated. Using the analysis tools objective functions can be formulated for the optimisation of the geometry of the machine structures. Then some geometrical parameters, which are chosen by the designer, can be optimised using special optimisation algorithms. A higher level optimisation can also be performed for the optimisation of the configurational parameters (if these parameters describe a large number of possible machine configurations). During this configuration optimisation the machine tools having optimal geometry (after geometry optimisation) are compared. These comparisons are based on the analysis results of the geometry optimisation loops. Finally, the best machine tool configuration can be found using specific optimisation algorithms.

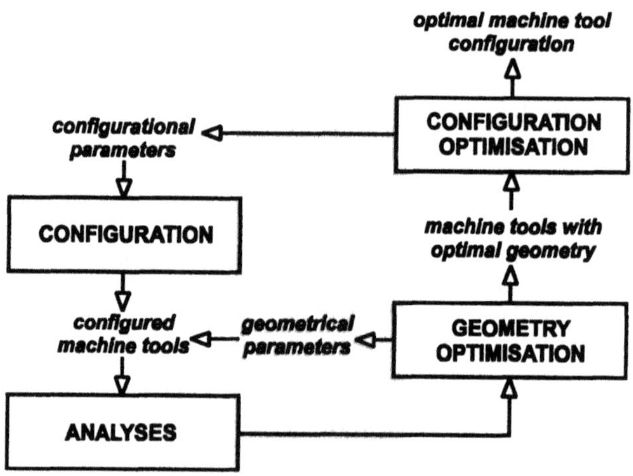

Figure 1: System overview.

3. Conceptual Model and Description of Machine Tools

The conceptual model of the generic 3-axis machine tool structure under study is represented in Figure 2. This model and its description system are adopted from Lipóth [2]. The generic machine structure is described as an open chain of four members: the headstock (HST) with the main spindle, the member next to the headstock (NT_HST), the table (TBL), and the member next to the table (NT_TBL). One of the four members is fixed to the ground. The fixed member is always constructed from two parts, therefore one machine is constructed from five components. (In the figures below dots indicate the two components of the fixed member.) The four members are connected by three orthogonal linear guideways: the headstock guideway (*HST_GW*), the middle guideway (*MID_GW)* and the table guideway (*TBL_GW*). One of these three guideways is vertical. The guideways have a 'long part' and a 'short part' connecting two adjacent members. The direction of the spindle, the direction of the table normal and the directions of the normals of the three guideways are parallel with the direction of one of the guideways.

From this description system 10 configurational parameters are defined as follows:

1. *fixed_memb*: which member is fixed to the ground;

2. *vert_gw*: which guideway is the vertical guideway;
3. *spindle_dir*: direction of the spindle axis;
4. *table_norm*: direction of the normal of the table;
5. *hst_gw_norm*: direction of the normal of the *HST_GW* plane;
6. *mid_gw_norm*: direction of the normal of the *MID_GW* plane;
7. *tbl_gw_norm*: direction of the normal of the TBL_GW plane;
8. *hst_gw_short*: which member has the short part at the *HST_GW;*
9. *mid_gw_short*: which member has the short part at the *MID_GW;*
10. *tbl_gw_short*: which member has the short part at the *TBL_GW.*

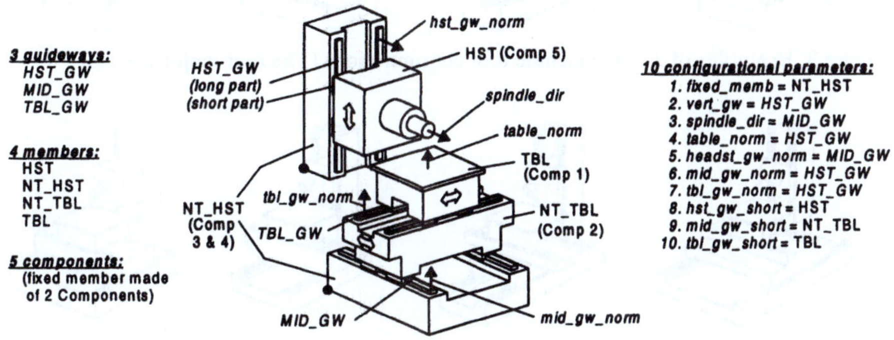

Figure 2: Conceptual model of a 3-axis machine tool.

On the right hand side of Figure 2 the values of the 10 configurational parameters show the description of the particular machine tool depicted on the figure. A more detailed explanation of this description system and of the 10 configurational parameters with several examples was previously published [3].

4. Configuration of the Machine Variants

The linear guideways in the description system have a 'long part' and a 'short part' that move relatively on each other (short-on-long or long-on-short). In the design system a frequently used guideway concept is applied, where the 'long' part is constructed from two rails, and the 'short' part is constructed from four carriages.

The machine variants are configured from a component library where each component has parametric shapes and dimensions. The components and their assemblies are modelled in a commercial CAD system using parametric multi-part solid modelling. Currently there are 52 components in the library. For more information about the design and classification of components consult [3].

The machine tools are grouped in four major classes depending on which member is the fixed member. These four classes can be seen in Figure 3. Each machine tool class has special selection rules that determine the components to be used for configuration. During configuration time the geometrical parameters of the components are set and linked with parameter expressions. The strokes of the linear guideways and the order of magnitude of the main dimensions of the components are derived from the workspace dimensions. After configuring the machines, the values and/or the expressions of the geometrical parameters can be modified by the user (manually) or by the optimisation algorithms (automatically; see Section 6). Figure 4 demonstrates some examples for the modification of the geometrical parameters of the machine tool of Figure 1.

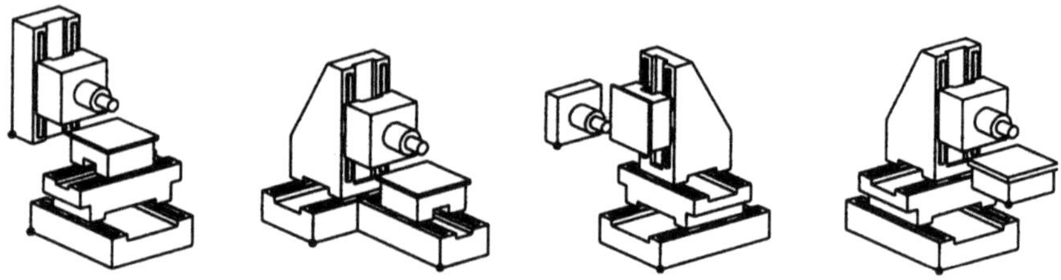

Figure 3: Examples of 3-axis machine tool configurations (The four major classes).

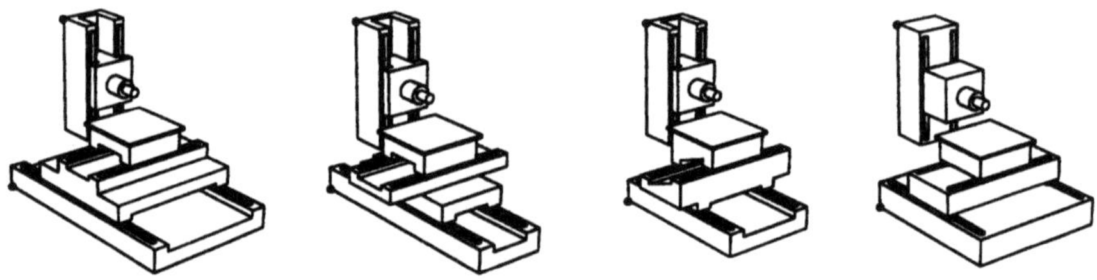

Figure 4 : Modification of the geometrical parameters.

5. Analyses of the generated machine variants

There are five types of analyses developed to evaluate the generated machine tool variants: a geometrical analysis, a volume analysis, a static analysis, a dynamic analysis, and a clearance analysis.

For the static and dynamic analyses, the guideways and the transmission elements are represented by linear springs and the bodies of the components are assumed to be rigid. Every carriage is modelled by two linear springs and the actuator (e.g. lead screw) is also modelled by a linear spring, resulting in 27 springs in the entire mechanical model.

5.1 Geometrical analysis

The total number of possible, unique machine tool configurations, hereinafter referred to as legal design space, is 56,832. This number is obtained by some geometrical reasoning with which the illegal variants can be filtered out from all the possible configurations. There are several types of illegal variants defined. For example one type of illegal variant is a particular machine configuration where the direction vector of the spindle and that of the table normal are parallel and they point in the same direction.

The legal design space is extremely huge, therefore some configurational constraints have been introduced in order to restrict the design variants of interest. A legal variant that satisfies the configurational constraints hereinafter is referred to as feasible variant. Currently, it is possible to use 12 configurational constraints, such as for example the designer would need only either symmetric or asymmetric machines; he or she might not need machines that are mirrored variant of other machines etc. More configurational constraints and some examples for specific sub-spaces of the legal design space (feasible zones) using different configurational constraint settings can be found in [4].

5.2 Volume analysis

It is a calculation of the total volume of the solid bodies of the machine models. In most cases the better the machine tool the smaller its total volume is.

5.3 Static analysis

It is a calculation of the relative displacement between the table (workpiece) and the spindle applying a particular cutting force as a static load at the end of the spindle. For a detailed description of the calculations of the static deformations consult [3]. From the static analysis point of view, the better the machine the stiffer its structure.

5.4 Dynamic analysis

ROBOTRAN, a software tool developed for multibody system analysis [5], is used for generating the symbolic dynamic equations for the configured machine variants. Each guideway is modelled in a way that 3D motion can be allowed between the adjacent components (6 DOF joint). Therefore, each machine tool has $3 \cdot 6 = 18$ degrees of freedom for the dynamic analysis. A program code has been developed to calculate the natural frequencies of the synthesised machine structures. This program numerically solves the symbolic dynamic equations generated by ROBOTRAN. The input (geometry, masses, inertia tensors, etc.) of this numerical process is extracted from the CAD system in which the machine tool models are constructed. For more detailed description of this dynamic analysis and for some examples consult [4]. From the point of view of dynamic analysis the better the machine the higher the lowest (the fundamental) natural frequency is.

5.5 Clearance analysis

Since the components can move on their guideways, they can possibly interfere with each other, or, in case of improper design, they can steadily interfere. During clearance analysis the interference between the components of the assemblies is checked and calculated. The aggregate volume of the interference objects has been chosen as a measure of the clearance of the structure. From the clearance analysis point of view, the better the machine tool the lower this value. For more detailed description of this clearance analysis and for some examples consult [3].

6. Optimisation of the generated machine variants

Genetic algorithms [6] and/or sequential quadratic programming (SQP) [7] are applied for optimisation of the configuration or/and of the geometry, so that the optimal values of the ten configurational parameters and/or the optimal values of the geometrical parameters of the components can be searched for. The optimisation criteria can consist of one or more of the following objectives: (1) to minimise the total volume; (2) to maximise the static stiffness; (3) to maximise the lowest natural frequency; (4) to minimise the geometric interference. The optimisations can be performed on a parallel computational platform [8].

Figure 5 shows an optimal configuration for an optimisation example. The requirements for configuration of this optimisation were set as follows: the fixed member of the required machine tools should be the NT_TBL, the machine should have horizontal spindle and horizontal table facing up, and the mirrored variants should not be taken into account. In this case the number of feasible variants is 114. In Figure 5 the best configuration of these 114 feasible variants can be seen. The right side of Figure 5 shows the machine with optimal geometry, and the left side shows the machine with rough geometry (machine newly generated, having default geometry). Three parameters were chosen for the optimisation of the geometry.

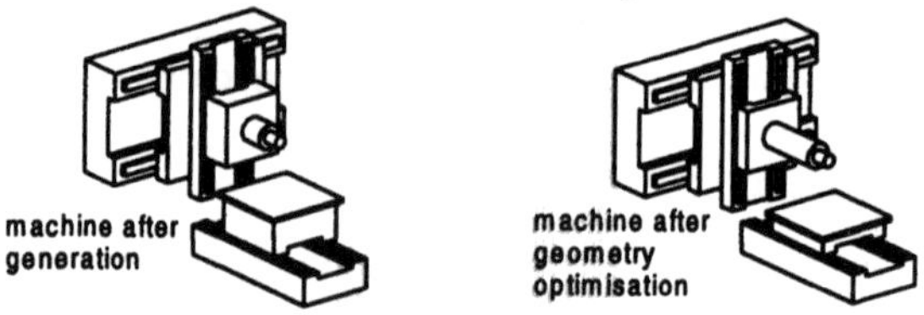

Figure 5 : Optimal configuration for an optimisation example.

References

[1]French, M.J., 1985, *Conceptual design for engineers.* Second Edition, The Design Council London, Springer-Verlag.

[2]Lipóth, A., 1993, *Systematic Creation and Evaluation of the Global Structures of Machining Centres.* Doctoral dissertation (in Hungarian), Technical University of Budapest, Department of Manufacturing Engineering.

[3]Németh, I. and Van Brussel, H., 1998, *Conceptual design of 3-axis machine tools.* Proceedings of the CIRP Seminar, STC Design, "New Tools and Workflows for Product Development", Berlin, Germany, pp. 269-281

[4]Németh, I. and Van Brussel, H., 1999, *Synthesis, analysis and optimisation of 3-axis machine tools.* Proc. of the Computer Aided Conceptual Design '99 (CACD'99), 1999 Lancaster Int. Workshop on Engineering Design, Lancaster, United Kingdom, pp. 49-64

[5]Schiehlen, W. ed., 1989, *ROBOTRAN.* In Multibody Systems Handbook, Springer-Verlag Berlin, pp 246-264

[6]Wall, M., *Galib: A C++ library of Genetic Algorithm Components.* Release 2.4.4, Mechanical Engineering Department, Massachusetts Institute of Technology

[7]Spellucci, P., *DONLP2: an SQP method (written in C),* Department of Mathematics, Technical University Darmstadt, Germany.

[8]*Parallel Virtual Machine (PVM) for networked parallel computing,* Release 3.4.1, Computer Science & Mathematics Division, Oak Ridge National Laboratory, USA.

New Ways to Manufacture Using Rotary Transfer Machine

Enzo Gentili[1], Stefano Moinari[2]

[1]Dip.to di Ingegneria Meccanica, Università di Brescia, Via Branze 38, 25123 Brescia, Italy, gentili@bsing.ing.unibs.it
[2]Dip.to di Meccanica, Politecnico di Milano, Piazza Leonardo da Vinci 32, 20133 Milano, Italy

Abstract: In the past all rotary transfer machines were rigid and dedicated, able to work only big batches (greater than 10000 pieces). Quick obsolescence and vulnerability to failure (serial system) were critical problems. Some important technologies changed rotary transfer machines and gave the opportunity to machine complex pieces on 5-6 faces in a single clamping. Now there are two innovative architectures of transfers which were devised recently. Thus it is really important to study and improve them.

1. New Architecture in Rotary Transfer Machine

Rotary transfer machines were born to machine brass fittings and valves, where usually pieces were simple and it was necessary to work 2 or 3 faces. Later the machining of more and more difficult pieces was possible. A recent investigation, about Italian machine tools shows, that 32% of rotary transfers work more than 3 faces and this is possible with the new generation of rotary transfer machines. New technologies give the possibility to machine complex pieces on 4,5 or 6 faces in a single clamping and it is possible to machine pieces with RTMs that yesterday were impossible [1,2].

Some important technologies have changed rotary transfer machines: CNC reduced set-up time and improved precision. The possibility to machine a piece on 4,5 or 6 faces in a single clamping is possible by means of rotary jaws and rotary clamping devices.

Now we can look at two new generations of RTMs:

- Transfers that execute all the machining in parallel (as traditional transfers [4]). These machines are installed with a fixed operating unit, and 3,4 axis operating units. They have similar machining centres. This kind of architecture gives high flexibility and productivity. In Italy some recent transfers can machine more than five hundred different pieces with a low set-up time (less than one hour). Other transfers allow to obtain high productivity, up to 10000 pcs/h, doubling the productivity. The most important constraint in this kind of machining is the dimension of the parts (less than 200x200x200mm), and the longer machining time that the piece needs. A recent investigation, about Italian machine tools shows that 58% of RTMs machine brass. There is correlation between brass and short time machining. This is the good workability of the brass [11,14].

- A modular transfer with 3-8 operating units, 3 axis CNC, performs sequential machining. Each unit has a revolver head with 4-8 spindles. The chief characteristic is the good performance in reconversion flexibility, but productivity is lower than in the precedent architecture. Today a modular transfer can machine infinite shapes and configurations in a 300mm cube. The chief constraints of modular RTMs are productivity and cost (more than 1.000.000$) [12].

Differences between the two classes are :

- The architecture and geometry (shape, size).
- The dimension of the batches is big in first class, medium and big in modular RTMs.
- The design of the working schedule with modular transfer is more difficult than that of the first class of rotary transfer.
- The difficulties in the change over and to lead the production is more problematic with modular transfers.
- The range of pieces worked. While the first class allows to machine one or the same family of pieces, the modular rotary transfer gives the opportunity to machine infinite shapes and configurations of pieces. In the first class group technology is really important. In fact the possibility to join the different pieces in one or few similar families gives the opportunity to sum the single productivity volume therefore obtaining a high productivity volume. It is necessary to have high productivity volume to justify the buying of a RTM [1].
- The ways to lead the production: in a traditional transfer all the machining is executed in parallel, while modular transfer machines execute both parallel and serial machining. This is the great difference between the two classes, and this is the cause of the worse cycle time in a modular transfer.

2. Parallel and Series Machining

It is necessary to talk about the meaning of in parallel and serial working in a RTM and the meaning of «worse cycle time» in a modular rotary transfer.
In a RTM we can have four levels of parallelism:

- The first level permits to use more stations in a RTM (See Fig.1 in this document). There are 3 kinds of stations: charge/discharge stations, auxiliary stations, and working stations.

A RTM usually has 1-2 charge and discharge stations: in some RTMs there is just one station for charge and discharge, in other ones there are 2 different stations (Sc = 1-2). The working stations can vary from 2 to 15 (Sw = 2-15).

Usually RTMs do not have auxiliary stations (Sa). This kind of station is necessary only in particular situations. During the machining, the number of the works in progress is the same as the number of stations (WIP = S = Sc+Sw+Sa) and in every working station there is a different working stage.

- The second level of parallelism consists of utilising more units in every station. Usually traditional transfers utilise up to 3 operating units per station. In particular situations it is possible to have 4 operating units per station. Modular transfers utilise one or two operating units per station: the limit is given by the dimensions of revolver heads (See Fig. 1 in this document.)

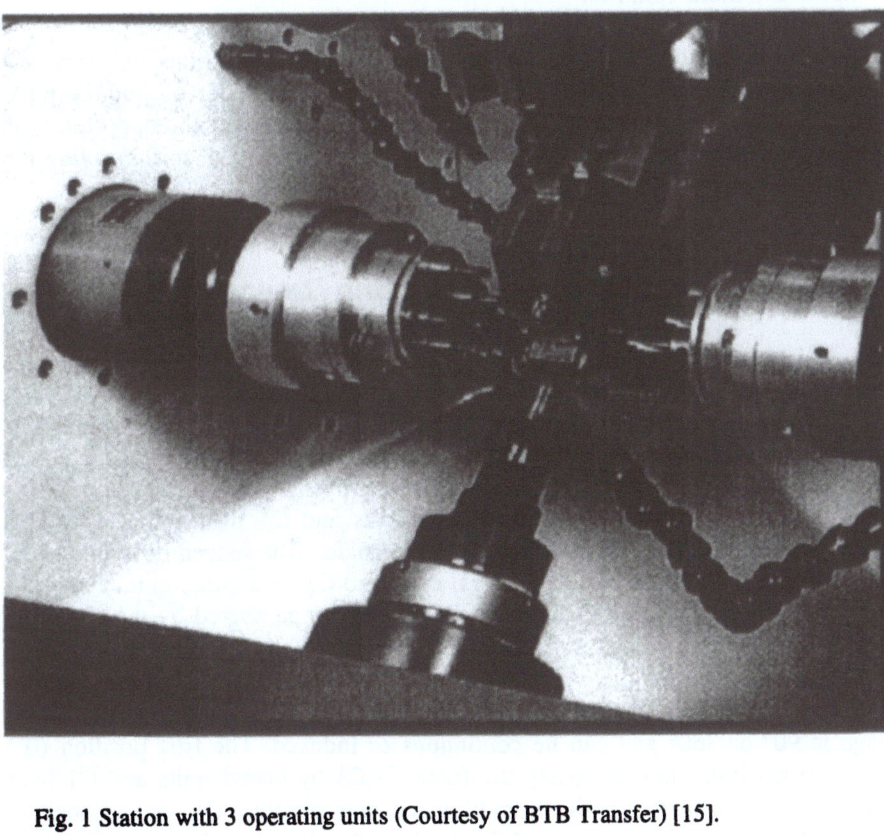

Fig. 1 Station with 3 operating units (Courtesy of BTB Transfer) [15].

- The third level of parallelism consists of utilising multi-spindle heads on the operating units. In this way it is possible to utilise an operating unit to impart a spinning motion to many spindles: many parallel workings can be made simultaneously by a unit.
- The fourth level utilises special tools called forming or profile tools. These tools permit to execute more different and coaxial machining operations, such

as drilling, milling and static turning. The important constraint is the rate of the spindle that involves different cutting speeds with different diameters of machining: for example if a profile tool executes simultaneously three coaxial holes: Φ5, Φ15, Φ25, the smallest one will have the lowest cutting speed and the largest one will have the highest cutting speed.

For every level the limits of the two classes are observed:

- First level: in the first class it is possible to have more than 16 stations whereas in the modular RTMs there are just 8 (the half).
- Second level: in the first class it is possible to have 1-4 units per stations (usually 3) whereas in the modular RTMs there are just 1-2.
- Third level: multi-spindle heads are often utilised in the first class, but seldom in the modular RTMs because of the encumbrance of the system of revolver heads and multi-spindles.
- Fourth level: in both these cases profile tools are used.

The capability of a traditional rotary transfer machine to carry out simultaneously more machining operations is better than the one of modular RTM (and this is the biggest drawback), but modular transfers offer the opportunity to change tools using revolver heads (up to 8 tools per operating unit) carrying out serial machining.

3. Rotary Jaws and Rotary Clamping Devices

The possibility to machine pieces on 4,5 or 6 faces in a single clamping is possible by means of rotary jaws and rotary clamping devices.

Rotary jaws rotate the pieces. The axis is radial to the rotary table. The rotary motion is realised by hydraulic device, and using CNC. The range is 90° or 360° and can be continuous or indexed. The first position (0°) allows to machine simultaneously the faces C2,C3 by lateral units and C1 by a radial unit: this situation is like the one of a traditional 3-way transfer. The second position (+90°) allows to machine faces C4,C5 by lateral units and C1 by a radial unit. There is a problem because the faces C4 and C5 are partially encumbered by the clamping device, so it is impossible to machine all the faces.

Rotary clamping devices allow the pieces to rotate. The axis is tangent to the rotary table. The rotary motion consists of a hydraulic device and is CNC. The range is 90° or 360° and can be continuous or indexed. The first position (0°) allows to machine simultaneously the faces C2,C3 by lateral units and C1 by a radial unit. This situation is like a traditional 3-way transfer. The second position (+90°) allows to machine faces C1,C6 by lateral units and C2 by a radial unit. These devices allow to machine completely four faces.

Rotary clamping devices and rotary jaws are bigger than traditional jaws, making it necessary to have a longer lateral area of the table (See Fig. 2 in this document.) So to machine pieces on 4 or 5 faces a bigger table is necessary than for 3 faces [3].

Fig. 2 Rotary clamping devices (Courtesy of Buffoli transfer S.p.a.) [14].

4. Recent Studies and Achievements

The biggest achievement of the last few years of research is the possibility to machine (with a RTM) complex pieces of brass, aluminium, steel, iron (e.g., brass body of hydro-cleaners and complex pieces for the automotive sector).

For each situation there is the best kind of architecture and the right level of flexibility.

Flexibility is getting more and more important in rotary transfer machines. The most important dimensions of flexibility are:

- Mix flexibility, measured by change-over time / saturation time;
- Product flexibility, measured by time and costs to reach the markets, or measured by the number of kinds of articles that the RTM can machine;
- Reconversion flexibility, measured by time and costs of reconversion;
- Volume flexibility measured by the maximum output obtainable.

A recent investigation about Italian machine tools [9-15] shows as a mix, product and reconversion flexibility are important for Italian use of RTMs, while volume flexibility is not so important. Usually RTM machines work all day and they are never turned off. The only way to obtain the volume flexibility is to work a different number of shifts: two for low production, three for high production.

Now it is possible to obtain the necessary mix flexibility (given by low changeover time) also in a mechanic RTM (without numerical control), using the possibilities given by the presetting.

The actual challenges [5,6,7,8] are:

- to obtain high capability and to have high productivity: to machine pieces in a single clamping eliminating accumulative errors due to re-clamping,

- to obtain high efficiency and reduction of the encumbrance of the production system: to integrate several technologies in transfer machines, to increase automation (now transfer machines load and slot pieces by a robot with artificial vision), to use control in process, to use unmanned production with a machine surveillance system and tool control,
- in other words, to give customers, not simply a machine tool, but a backup service to help them resolve problems they might meet.

The actual study about Group Technology will give the possibility of producing different kinds of pieces simultaneously. This important result is useful for example when it is necessary to produce a piece made in two parts: the RTM carries out the machining of two different parts (with a sequence A-B-A-B.) and another transfer assembles the parts: just two machines for the production.

The opportunity to machine mixed batches means a better scheduling and processing of the orders.

References

[1] Koepfer C.,1993 *A new spin in rotary transfer machine.* Modern Machine Shop.
[2] Koepfer C., 1998, *Why not change. What you're doing?,* Modern Machine Shop.
[3] Molinari S., Gentili E., 1999, *New technology in rotary transfer machine.* Proceedings IV AITEM Conference.
[4] Consoli G., 1983, *Transfer di lavorazione*, Nuove Edizioni S. Rocco. (In Italian)
[5] Buffoli M., 1997, *Il transfer da barra,* RM. (In Italian)
[6] Agnesi F., 1996, *Tecnologia di qualità,* TM. (In Italian)
[7] Morelli R.,1998, *Il settore dei transfer,* TM. (In Italian)
[8] Costa P., 1998, *Italia in pole position,* TM. (In Italian)
[9] Giuliani *http://www.giulianico.com*
[10] Gnutti transfer *http://www.gnutti.com*
[11] Goss Deleeuw-Porta *http://www.goss-deleeuw.com*
[12] Riello macchine utensili *http://www.riellomu.it*
[13] Sinico S.p.a. *http://www.sinico.com*
[14] Buffoli Transfer S.p.a. *http:/www.buffoli.com*
[15] BTB Transfer *http://www.btb.com*

Linear Motor based High Speed CNC milling Machine

Seamus Gordon[1], Michael T. Hillery[2]

[1]University of Limerick, Limerick, Ireland. Email: seamus.gordon@ul.ie
[2]University of Limerick, Limerick, Ireland. Email: michael.hillery@ul.ie

Abstract: The traditional approach to constructing CNC machine tools has been to use rotary drive motors and ballscrews to achieve table motion. Linear Motors offer several advantages over this approach, including low inertia, better performance, increased accuracy and reduced complexity. At the University of Limerick a low cost, high speed routing machine was built using linear motors in order to conduct machining tests on a composite material. The machine consisted of a gantry type X/Y motion system on which a high speed electrospindle was mounted. Both were interfaced with a standalone CNC controller for which a Microsoft Windows based interface was written in C++ . The machine has the usual functionality of a CNC router and is capable of table speeds of over 25 m/min, with a spindle speed of 30,000 rev/min. The machine is at present being used to carry out tests on tool wear as part of an ongoing research project in the University.

1. Introduction

The term High Speed Machining has been used for a number of years to describe end milling with small diameter tools at high rotational speeds. High spindle speeds are usually partnered with high table feed rates. A high speed milling machine can typically provide spindle rotation of 25,000 rev/min or greater and can perform contour machining at feedrates up to 25 m/min. The advantages of High Speed Machining include increased material removal rates, reduced tool wear, improved surface finish and the ability to machine hard materials [1] . High speed machine tools are expensive because they need to be more rigid in construction then conventional CNC machines, and need higher than normal specification CNC controls to perform contour machining found in Mould and Die cavity work at high speeds. The traditional approach in building machine tools has been to use rotary motors and drive screws to achieve motion. Limiting factors in the performance of this type of machine are limited maximum rotation speed, inertia, windup, backlash and hysterisis[1]. Linear motors offer significant advantages over traditional drive systems. Velocities in excess of 6m/sec and accelerations of 200m/sec^2 are possible. As there are no moving parts except the motor block, repeatability and accuracy are also improved [1][2]. In addition, the relatively low cost of linear drive technology made it attractive as a basis for developing the High Speed Milling (HSM) test rig.

2. Machine Build

This consisted of the following tasks that were carried out in the following order.

1. Specification of a high speed electrospindle

2. Calculation of payload and specification of a linear motor system.
3. Construction of the machine chassis
4. Interfacing of the electrospindle with the linear motor CNC control.
5. Provision of a PC based user interface for the system.

2.1 Electrospindle

Spindles are available which cover a wide range of rotational speeds and power ratings. The spindle chosen was a Faemat 110 CU liquid cooled electrospindle capable of 500 to 30,000 rev/min controlled by a frequency converter. The system incorporated a PLC which allowed spindle rotation and speed to be governed by means of 20v digital and 0-5v analogue inputs. The overall weight of the spindle was 29kg resulting in a total payload of 35kg including ancilliary equipment. The HSK (ISO/DIN 12164-1,ISO/DIN 12164-2) toolholding system has gained popularity in HSM in recent years. This system uses a hollow shank and a taper of 1:10. It has half the mass of a conventional toolholder and provides improved repeatability and clamping pressure at high speeds. The spindle chosen was equipped with a HSK 50 Form B.

2.2 Motor specification

The X/Y motion system consisted of three LinearDrives LD3806 Powered Motor Modules in a gantry arrangement. The X axis was 2m in length and was driven by a dual parallel motor pair set 1m apart. The Y (transverse) axis was then mounted on top of these. This in turn carried the Z axis slideways and electrospindle. The Z axis was controlled by a conventional rotary servo motor and leadscrew. The motors were powered by three LinearDrives LDA 310 3200W brushless servo Amplifiers. Linear encoders were incorporated in the Y and primary and Secondary X motors. The stated thrust and accelerations for the motors exceeded the requirements of the system [2].

2.2.2 CNC Control

The control used was a Baldor Nextmove BX 3-Axis servo motion controller which was supplied with the linear motor system. The control was a standalone unit that communicated with a PC in terminal mode via an RS-232 serial link. The control was capable of storing and executing entire programs as well as being controlled by individual typed commands. In addition, the control had eight opto-isolated digital inputs and outputs as well as two analogue outputs which could be used for interfacing with other devices. The programming language used was MINT, which is a structured form of BASIC designed for motion control applications [3]. In addition to the normal BASIC syntax, it has additional motion specific keywords built in. These allow control of motor position, speed and synchronization of axes. Table 1 shows a fragment of a typical MINT program. MINT can also be used to raise and lower individual digital outputs as well as test inputs. For example, the command `OUT.1=1 : PAUSE IN.3` would raise output 1 and then wait until input 3 was raised. This facilitated interfacing the control with the spindle.

2.3 Interfacing the control and spindle.

Table 1 Sample MINT code

```
COMMSMODE._tmRS232=1
ACTIVEINLEVEL =0xF8FF
ERRORIN[0,1,2]=8,9,10
REPEAT
HM[0,1]=0.00,0.00
PAUSE ID [0,1]
UNTIL VEL[0,1]=0
SPEED[0,1]=100,100
MOVEA [0,1]= 300.00,0.00 :GO
```

The electrospindle system provided 20v outputs corresponding to the following functions: **Spindle switched to CNC, Spindle OFF, Spindle at target speed.** The inputs provided were: - **Start spindle** and **Stop spindle**. In addition one 0-5v analogue input governed spindle speed. Once connected to the controller, it was possible to write MINT subroutines to start and stop the spindle. These subroutines tested whether the spindle was in automatic mode before setting the speed and starting motion. The analogue output was governed by a Digital to Analogue Converter which gave a maximum resolution of 8192 increments which resulted in spindle speed increments of 3.6 rev/min.

2.4 User Interface

It was considered desirable to provide a user interface for the machine which would, as far as possible, provide the same functions as found on a conventional machine tool control. It was decided to incorporate the following features in the interface.

1. Support for ISO 'G-Code' programs with an interactive program edit screen.
2. Machine Homing, Manual Jog and Datum Setting.
3. Graphical Simulation of program.
4. Positional readout of machine axes.

2.4.1 Programming language

Source Code in C++ for an Application Program Interface (API) was available which provided support for communication with the control[4]. The source code consisted of a class hierarchy which allowed an instance of a class CNextMoveBX to be declared. The member functions of the CNextMoveBX object performed file transfer and other functions between the control and a host computer. It was decided to create a Microsoft Windows based graphical user interface for the system. The programming environment used was Borland C++ Builder 3, which allowed the visual development of the program interface while integrating with the existing source code and any additional code written. The interface is shown in Figure 1. The main features of the program are now discussed.

2.4.2 Software Features

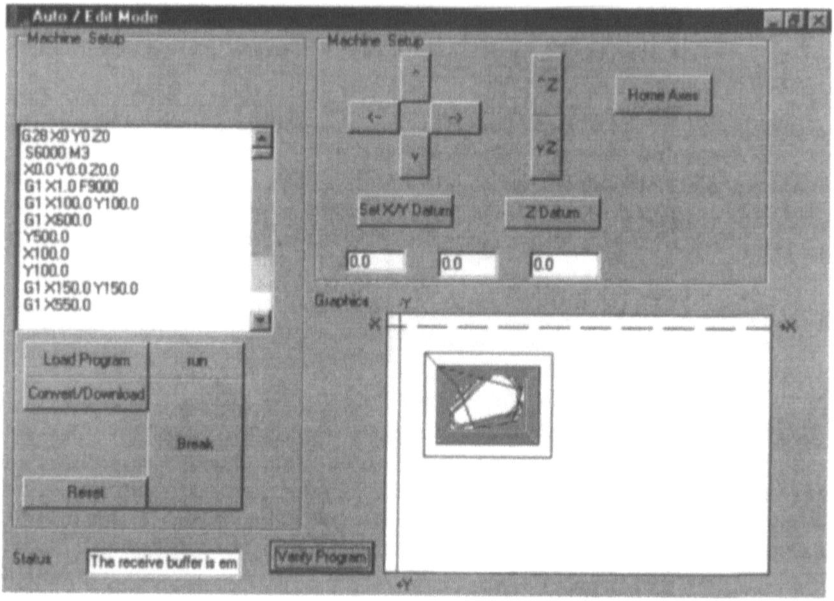

Figure 1

When the RUN function was executed, a check was made on whether the contents of the program text window had changed. If so, the program text was passed to a translation routine which interpreted the code and then wrote an equivalent MINT program which was in turn written to an ASCII file on disk. This file was then

<u>Table 2 Switch construct</u>

```
switch(ch){
     case 'G' : fscanf(infile,"%d", &gcode);
                 switch(gcode){
                 case 0: movemode = RAPID;
                         break;
                 case 1: movemode= LINE;
                         break;  // And so on for other codes
                }
                break;
     case 'X' :fscanf(infile,"%f", &xpos);
                break; // and so on for other labels
     case '\n' : nextmove(); // End of block -
                              //process and output move.
                break;
     default : ; // Unrecognised character - ignore
     }
```

downloaded to the control and executed using the API described above. The translation function was based on a **switch** construct to which the program text was passed, one character at a time. A simplified version is shown in Table 2. The switch worked on the principle that if the character was recognised i.e. G, X, Y, M etc., then the corresponding parameter value was then read from the file and stored. When the newline (\n) character was encountered, the function **nextmove()** then processed the information gathered in the preceding block of code. Any characters not recognised by the switch were ignored, thus making the program more robust. The usual G-Codes for linear and circular interpolation, Machine homing and Tool datum setting were supported.

Graphical program display was achieved in a similar manner to the program translation, except that the output was a series of calls to graphics methods instead of text output.

3. Discussion

The finished machine is shown in Figure 2. It is currently being used to carry out High Speed Machining tests on a composite material. The machine has successfully been run at spindle speeds of 30,000 rev/min combined with 40,000 mm/min table feed. A potential problem with the system is the attraction between ferrous particles and the magnetic thrust rods of the motors which are exposed. Protective gaiters are available from the manufacturers in addition to a laminating process which provides a protective coating to the rods. The user interface performed satisfactorily. In particular the G-Code translation function proved useful and allowed programs to be generated for the system using standard CAD/CAM software. At this time tests on the repeatability and positional accuracy have not yet been carried out. However, results so far have been encouraging in this regard.

Figure 2

4. Acknowledgements

This project was in part funded by Enterprise Ireland Ltd and QTEK Manufacturing Ltd.

References

[1] Weidner,C., and D. Quickel, 1999, *High-Speed Machining with Linear Motors*, Manufacturing Engineering, Vol 122 No.3, pp80-87.

[2] Internet, *http://www.lineardrives.com*

[3] Baldor Ltd., 1997, *Mint for NextMove programmers Manual*, Issue3.1, Baldor Optimised Control Ltd.

[4] Baldor Ltd., 1997, *Mint Interface Library*, Version 4.2, Baldor Optimised Control Ltd, 1997.

COMPUTER-AIDED TOLERANCING

A Knowledge-Based System for Tolerance Allocation in Reverse Engineering

G.J. Kaisarlis[1], S.C. Diplaris[2] and M.M. Sfantsikopoulos[3]

[1]Department of Mechanical Engineering, National Technical University of Athens, Heroon Politechniou 9, 157 73, Zografou (Athens), Greece, gkaiss@central.ntua.gr, [2]stdiplar@central.ntua.gr, [3]1msfntua@central.ntua.gr

Abstract: Tolerance assignment in mechanical engineering is a typical procedure that is mainly based on experiment and human expertise and is critical for product quality, performance and for the manufacturing cost. In reverse engineering the allocation of tolerances has still a significant impact in both manufacturing cost and performance of the part. This task is even more delicate in case that the remodeled component is to be used as a spare part expected to fit and well perform in an existing assembly. The paper presents the development of a knowledge-based computer aided system for the designation of the tolerances for parts that work in an assembly and are reconstructed through reverse engineering techniques. The proposed system can be fully integrated in a PC environment and its effectiveness is illustrated in a case study.

1. Introduction

Tolerances have a direct impact on the quality, performance, manufacturability and cost of a part or assembly. A part cannot be produced exactly to the dimensions specified in the blueprint. Therefore a tolerance, as the maximum amount of variation allowed, is given to the production shop in achieving a given dimension. Clearance is the practical result of mating two parts, with the most typical example of a shaft in a hole. If two parts can be interchanged in an assembly without altering the function or quality of the assembly they are deemed to be interchangeable and can be considered to be the same part. The interests and principles of interchangeable manufacturing have been incorporated in the series of ANSI and ISO standards that concern tolerances, limits and fits. Still, many factors such as the length of engagement, bearing load, speed, lubrication, operating temperatures, surface texture and materials must be taken into account in tolerance allocation and fit selection for particular applications. Mainly based on human expertise, experiment and industry available databases and guidelines, proper tolerance assignment is one of the least well -understood engineering tasks.

In reverse engineering a CAD model is constructed by digitizing a physical part, for which drawings are not accessible or do not exist, and then using it to manufacture the component. Though conceptually simple, reverse engineering encompasses a great variety of tasks so much so that it has proved difficult to carry out. Tolerance allocation, having a significant impact in both manufacturing cost and assemblability of the reverse engineered component, is considered as one of the most delicate and complicated of such tasks. This task becomes even more delicate in the case that the remodeled component is to be used as a spare part expected to be interchangeable, to fit and perform without altering the function or quality of an existing assembly. In that case the objective is not the designation of the most appropriate tolerances and type of fit as in the design of a new similar assembly, but the closest possible match to the originally assigned ones.

The purpose of this paper is to present the development of a knowledge-based computer aided system for the designation of the tolerance and type of fit of reverse engineered mating parts that are expected to perform in an existing assembly. In the following sections the conceptual framework of the tolerance allocation problem in reverse engineering is presented and the expert system approach is justified. The overview of the proposed system is focused on the knowledge base construction. Its use is demonstrated in a case study.

2. Tolerance Allocation in Reverse Engineering

The general concept of tolerance design can be found in several books [1], [2]. Extensive research in the area of dimensioning and tolerancing from both product design and process planning points of view has been carried out with the advancement of computers since the 1970's. Ngoi and Ong [3] presented a comprehensive review of the state-of-the-art dimensioning and tolerancing techniques published over the last 30 years. However, most of the published work is focused on tolerance optimization using tolerance - cost optimization models. An

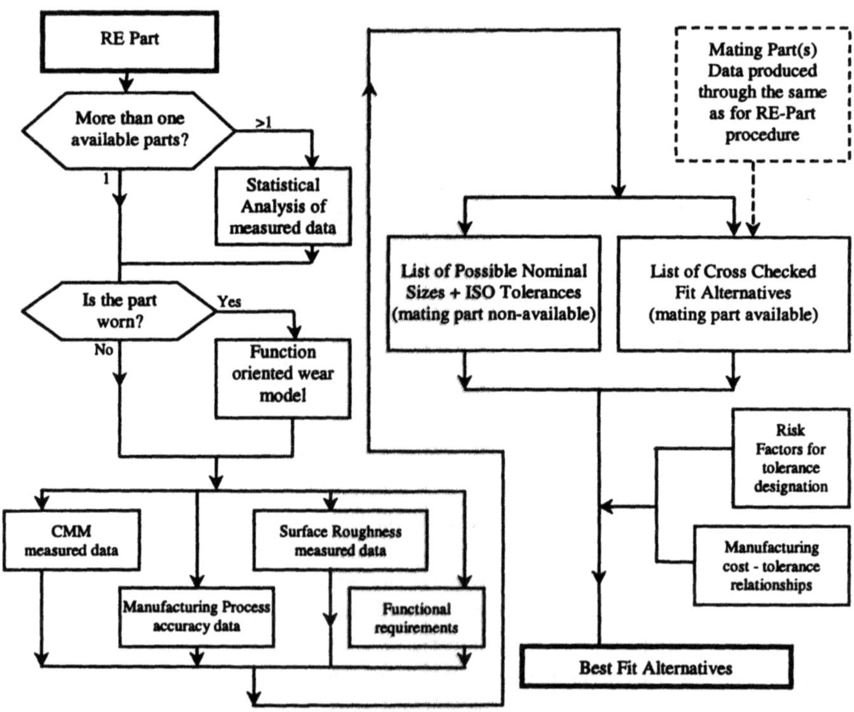

Fig. 1: Scheme of the system framework

early example of a knowledge-based system for the assignment of dimensions and tolerances according to the ISO standards for limits and fits is given in the work of Manivannan et al. [4]. This rule-oriented system for computer aided tolerancing specifies the ISO fit for the manufacture of rotational mating components that is best suited for the given manufacturing conditions. Janakiram et al. [5] developed an expert system for the selection of the optimum manufacturing process and tolerances based on the accumulated processing cost and the assembly functional requirements. As in [4] the method is only applicable to two cylindrical mating parts. Panchanal et al. [6] propose a computer aided tolerance assignment procedure that automatically assigns tolerances in CAD drawings. The procedure envelops three major stages: feature extraction, feature inferancing and rule based reasoning for the final tolerance allocation. Tolerance determination issues in reverse engineering are briefly addressed by Ingle [7]. An appropriate tolerance allocation and fit determination is very difficult to achieve when the CMM measured data are the only ones available. Especially in cases where a few or just one of the original reverse engineered components have been digitized, long term use has led to considerable wear or one of the mating parts to be reconstructed is missing, the problem gets rather sophisticated. So far in existing literature a knowledge-based system for the tolerance allocation in reverse engineering was not possible to be identified.

3. Knowledge-Base System

Knowledge-based systems (KBS) have been built to facilitate and automate tasks in many areas that require significant human expertise [8]. As in many problems in engineering, solutions generated by KBS are not deterministic, may provide more than one alternative and are strongly dependent on the expertise of the problem solver. Hence, tolerancing is the kind of problem that is amenable to be solved via this type of systems. In Fig.1 the factors that influence the problem of tolerance allocation in reverse engineering and the developed expert system approach are presented. TORE (Tolerance for Reverse Engineering) is a knowledge-based system for the establishment of the possible alternatives – i.e. tolerance range and type of fit - of a reverse engineered component that is expected to work in an existing assembly. A typical reverse engineering application that such a system can be used is, for instance, the production of parts for machines for which spares are not available. The proposed system has been developed within the frame of a commercially available KBS shell KAPPA-PC [9]. This tool runs on a PC under windows, is reasonably user-friendly and supports both object-oriented and rule-based programming. Tasks include object representation of the domain, knowledge base and user interface construction.

At its present state TORE is focused on reverse engineered parts with negligible or no wear at all that have to fit in an existing assembly for which the mating parts are available. Because of the ISO, ANSI and BS standards [10]-[14] implemented in the knowledge base, the system is applicable in metric cylindrical parts with nominal sizes 1-500mm. As it is pointed out in [12] the terms hole or shaft can be nevertheless taken to mean the space contained by or containing two parallel faces or tangent planes of any part, such as the width of a slot or the thickness of a key. It is also assumed that in the original design an *ISO standard clearance fit* was assigned of either hole or shaft basis system.

Functional requirements and production method used for the component manufacture in conjunction with CMM and surface roughness measurement data are the input introduced into the system through a user-friendly interface in an interactive way. An overview of the TORE architecture is presented in Fig. 2. The boxes included in the dashed line knowledge base area represent "knowledge islands" that are implemented to the knowledge base either as rule sets or functions and are manipulated by the inference engine. Based on the input data Tolerance Ranges 1,2 and 3 are obtained [15] for both shaft and hole in terms of IT grades. The Final Tolerance Range is also expressed as a set of IT grades and is the intersection

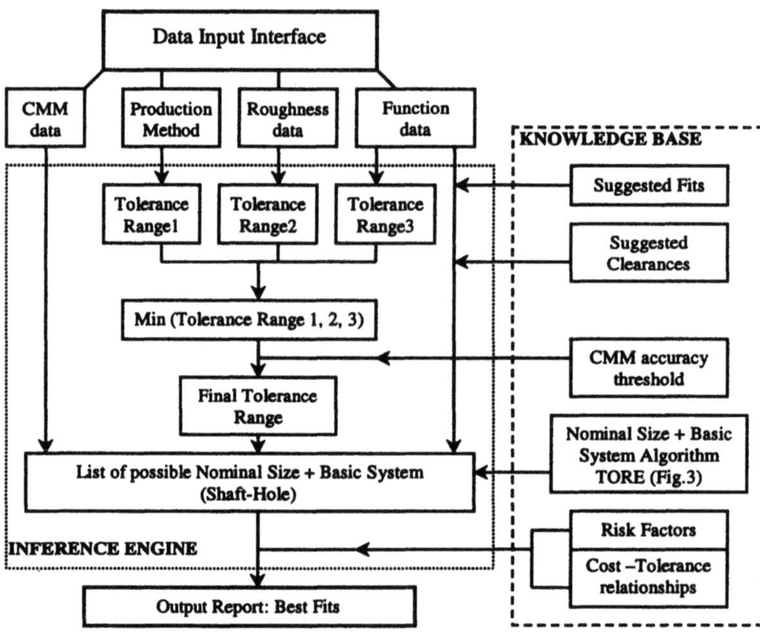

Fig. 2: Overview of the TORE KBS function structure

of the above ranges, taking into account that the minimum reasonable tolerance range should be at least greater than five times the CMM accuracy threshold [16]. Suggested fits, CMM data and the Final Tolerance Ranges for both shaft and hole are used as input in TORE algorithm, (Fig. 3). The algorithm produces a list of possible nominal sizes and Basis Systems (Shaft-Hole) by performing an extensive search for the establishment of all possible combinations of nominal sizes, fundamental deviations and IT grades that satisfy the input data. Fig. 3 shows only the part of the algorithm that refers to hole basis system (TORE-Hole). A similar algorithm, TORE-Shaft, runs following that one searching for shaft basis system fits. The best fits suggested in the KBS output report are finally derived through the additional consideration of the cost – tolerance relationships and risk factors.

4. Case study

To illustrate the effectiveness of the proposed system an application of a reverse engineered valve guide of an internal combustion engine is presented. Table 1 gives the details of the input to the system, denoted in bold, and the final results obtained.

5. Conclusions

This paper presents TORE, a computer aided knowledge-based system integrated in PC environment for the tolerance allocation and best I.S.O. fit selection of reverse

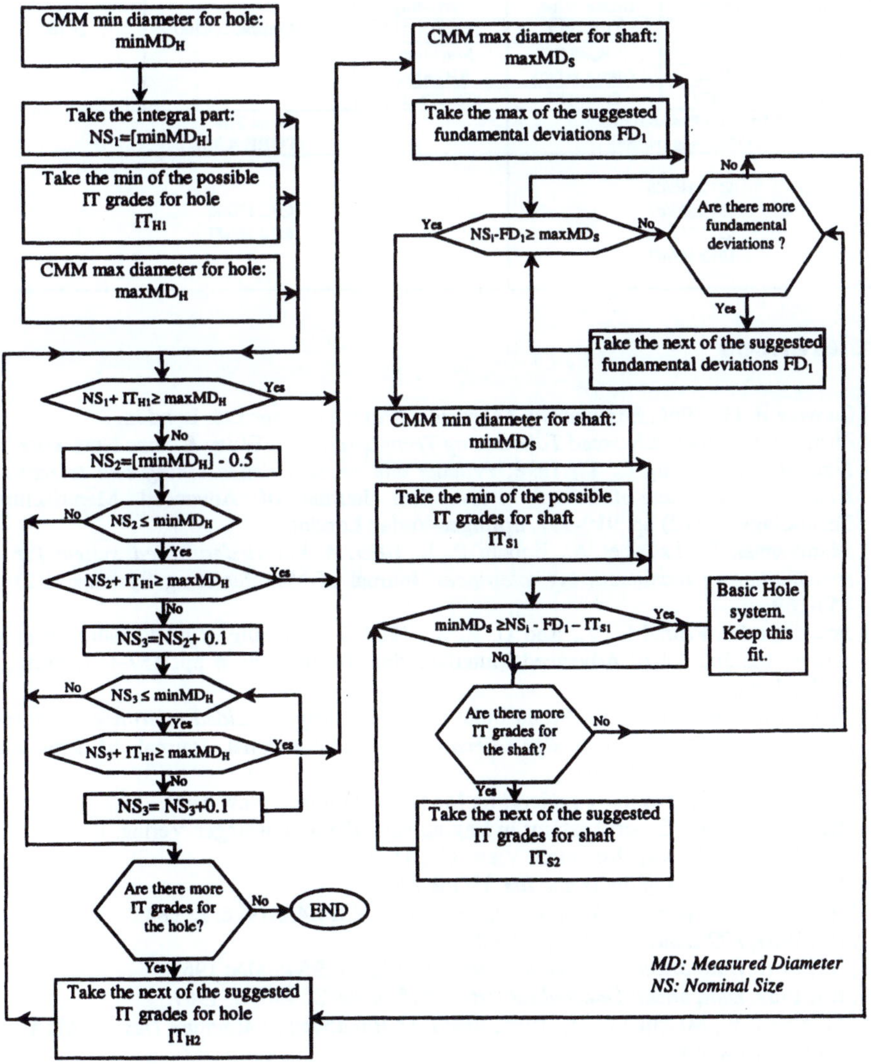

Fig. 3: TORE (Hole) Algorithm overview (Hole Basis System search)

engineered parts that work in existing assemblies. TORE, the KBS algorithm for the establishment of possible nominal sizes and ISO fits is introduced. The flexibility of the system makes the enrichment of its knowledge base and the integration of new rule sets easily achievable. Future work is oriented towards the system enhancement with the integration of function oriented wear models, statistical and cost-tolerance analysis techniques and the introduction of probability risk factors. An enhanced version of TORE implementing the above issues is currently under development.

Table 1: TORE KBS application example input and output report

<table>
<tr><th></th><th>CMM data(• m)</th><th>Roughness Data (• m)</th><th>Production Methods</th><th>Function Requirement (Clearance) (• m)</th><th>CMM Accuracy Threshold</th></tr>
<tr><td>Hole1</td><td>12.010</td><td rowspan="2">R_a=1
(mean value)</td><td rowspan="2">Machining
/Grinding</td><td rowspan="6">40<Clearance<100</td><td rowspan="6">2• m</td></tr>
<tr><td>Hole2</td><td>12.006</td></tr>
<tr><td>Hole3</td><td>12.004</td><td>IT5 – IT7</td><td>IT5–IT6–IT7</td></tr>
<tr><td>Shaft1</td><td>11.948</td><td rowspan="2">R_a=0.8
(mean value)</td><td rowspan="2">Machining
/Grinding</td></tr>
<tr><td>Shaft2</td><td>11942</td></tr>
<tr><td>Shaft3</td><td>11.939</td><td>IT5 – IT7</td><td>IT5–IT6–IT7</td></tr>
<tr><td colspan="3">List of all Possible Fits
(TORE Algorithm)</td><td colspan="3">Best Fits
(TORE KBS)</td></tr>
<tr><td colspan="3">Ø12 H6/d6
Ø12 H6/d7
Ø12 H7/d6
Ø12 H7/d7</td><td colspan="3">Ø12 H7/d6
Ø12 H7/d7</td></tr>
</table>

References

[1] Conway H. G., 1966, *Engineering Tolerances*, Pitman & Sons Ltd. London.

[2] Zhang H. C., 1997, *Advanced Tolerancing Techniques*, John Wiley & Sons New York.

[3] Ngoi B. K. A., Ong C. T., 1998, *Product and process dimensioning and tolerancing techniques. A state-of-the-art review,* Int. Journal of Advanced Manufacturing Technology, 14(12) pp. 910-917, Springer-Verlag London.

[4] Manivannan S., Lehtihet A., Egbelu P. J., 1989, *A Knowledge based system for the specification of manufacturing tolerances*, Journal of Manufacturing Systems, 8(2) pp. 153-160.

[5] Janakiram D., Psarad L. V., Rao U. R. K., 1989, *Tolerancing of parts using an expert system*, Int. Journal of Advanced Manufacturing Technology, 4 pp. 157-167, Springer-Verlag London.

[6] Panchal K., Shivakumar R., Simin Pulat P., 1992, *Computer aided tolerance assignment procedure (CATAP) for design dimensioning*, International Journal of Production Research, 30(3) pp. 599-610.

[7] Ingle K. A., 1994, *Reverse engineering*, Mc Graw- Hill Inc., New York.

[8] Pham D. T., 1988, Expert Systems in Engineering, IFS and Springer-Verlag.

[9] KAPPA-PC, ItelliCorp, Mountain View, CA, USA

[10] ISO, 1962, *System of limits and fits*, ISO/R 286

[11] ANSI, 1978, *Preferred Metric Limits and Fits*, ANSI B4.2-1978, R1984.

[12] BS, 1969, *ISO Limits and Fits*, BS 4500:1969.

[13] BS, 1969, *Data Sheet: Selected ISO fits – hole-basis*, BS4500A: 1969.

[14] BS, 1969, *Data Sheet: Selected ISO fits –shaft-basis*, BS4500B: 1969.

[15] Shigley J. E., Mischke C. R., 1996, *Standard Handbook of Machine Design*, Mc Graw-Hill Inc., New York.

[16] Pfeifer T., 1992, *Koordinaten – Messtechnik fur die Qualitatssicherung*, VDI-Verlag GmbH

Comparative Evaluation of Composite Position Tolerance specifications for patterns of holes

S. Ranade[1], E.A. Lehtihet [2], T.M. Cavalier[3]

[1] i2 Technology, Dallas, Tx, Srikant_Ranade@i2.com

[2] Penn State University, University Park, Pa, lvo@psu.edu

[3] Penn State University, University Park, Pa, tmc7@psu.edu

Abstract : The ASME Y14.5M-1994 Standard provides three distinct composite position tolerance constraints for patterns of features. A simulation is carried out to explore differences in the producibility of these three geometric position tolerance constraints. The influence of the loss of degrees of freedom asociated with the FRTZF of the new variants of composite position tolerance specification.

1. Introduction

The ASME Y14.5M-1994 [1] provides a comprehensive class of geometric tolerance constraints which can exercise a variety of control actions on the position characteristics of a hole or pattern of holes. Of particular interest in this paper is the class of constraints referred to as "composite position tolerances" which exercise a dual control action on the position of a pattern of holes by constraining both the pattern as a group, and the hole-to-hole relationship within the pattern. The 1994 revision of the ASME Y14.5M-1994 [1] Standard has expanded composite position tolerance control to include three broad classes of control. The practical application and implementation of this class of position tolerance controls in design and manufacturing is somewhat hampered by a distinct lack of information relative to the ease or difficulty involved in the production of these tolerance constraints and the sensitivity of producibility to various parameters used to specify these constraints. The technological aspects associated with the realization of holes by machining have been studied by several researchers [2-3] and machining experiments have been used to characterize size and position variability associated with machined holes [4-6]. The metrology of patterns of holes subject to the original composite position tolerance specification has been the subject of several publications [7-9] which present graphical and algorithmic methods to verify compliance with composite position tolerance specification. This paper uses simulation to provide a comparative evaluation of the producibility of the three composite position tolerance controls embodied in ASME Y14.5M-1994 and the sensitivity of producibility to material condition modifiers and the number of holes in the pattern.

2. Simulation of Producibility

Figure 1 shows a part with three patterns of holes controlled by composite position tolerance constraints. The circular pattern of holes is controlled by the original version of composite position tolerance control frame, which will henceforth be referred to as Variant 1. The linear and rectangular patterns, however, are controlled by the new additions to the Standard, which will be referred to respectively as Variant 2 (shown applied to the linear pattern) and Variant 3 (shown applied to the rectangular pattern).

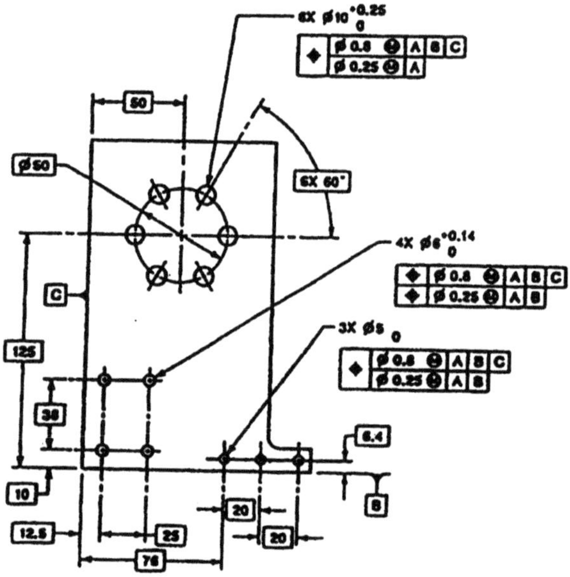

Figure 1: Three variants of composite position tolerance

Given a pattern of holes and a composite position tolerance specification, the process starts by sampling an arbitrary error model for size and position error data for each hole in the pattern. The accuracy and precision of manufactured features are dependent on a large number of physical causes associated with the machine tool, tool, fixture, work material, and work environment. The use of an arbitrary size and position error model is merely an artificial device to generate scattered hole centers and sizes to drive the simulation and carry out a comparative evaluation of the producibility of Variants 1-3. Hole center deviations in the x and y directions relative to true position are sampled from two identically and independently distributed normal density functions with means equal to 0 and standard deviations σ_{tool}. Hole sizes are also sampled from a normal density function. The choice of normal density functions to describe the position and size errors follows available experimental data [5-7] which suggest that non-repeatable

adequately approximated by such a distribution. The generated hole size and position sample is passed on to the next block which checks if size and composite position tolerance constraints imposed by the specification are satisfied. Upon completion of the check, a pattern of holes is either accepted or rejected, and yield of the simulated operation is updated. The next sample of holes is generated and the process is repeated until steady state statistics are obtained. Compliance with size and composite position tolerance specification is checked as follows.Consider a pattern of n-holes such as the circular pattern of Figure 1, controlled by size and composite position tolerance Variant 1. Let

ϕ_0 : The MMC hole size (ie. the smallest hole size)

t_ϕ : The hole size tolerance

ϕ_i : A measured hole size for the *i*th hole in the pattern

$(a_{i,1}^{(0)}, b_{i,1}^{(0)})$: (x,y) position deviations of top hole *i*

$(a_{i,2}^{(0)}, b_{i,2}^{(0)})$: (x,y) position deviations of bottom hole *i*

t_p : The specified pattern tolerance zone

t_f : The specified feature relating tolerance zone

$t_{p,i}$: The effective pattern tolerance zone for hole *i*

s_i: The effective feature relating tolerance zone for hole *i*

(p_i, q_i) : The nominal position of hole center *i*

Hole size tolerance is satisfied if $\phi_0 \le \phi_i \le \phi_0 + t_\phi$, $\quad i = 1,2,\dots,n$

Pattern locating requirement is satisfied if

$$(a_{i,j}^{(0)})^2 + (b_{i,j}^{(0)})^2 \le (t_{p,i}/2)^2 \quad , \quad i = 1,2,\dots,n \quad j = 1,2$$

$$\text{where} \quad t_{p,i} = \begin{cases} t_p + \phi_i - \phi_0 & \text{, under MMC} \\ t_p & \text{, under RFS} \end{cases}$$

The feature relating requirement is verified by taking the part through a succession of small displacements consisting of a translation (x, y) and rotation (θ) until each actual hole center is contained within the FRTZF. At the *k*th iteration, the new deviations $(a_{i,j}^{(k+1)}, b_{i,j}^{(k+1)})$ of top and bottom hole centers are given by :

$$\begin{pmatrix} a_{i,j}^{(k+1)} \\ b_{i,j}^{(k+1)} \end{pmatrix} = \begin{pmatrix} x \\ y \end{pmatrix} + \begin{pmatrix} \cos\theta & -\sin\theta \\ \sin\theta & \cos\theta \end{pmatrix} \begin{pmatrix} p_i + a_{i,j}^{(k)} \\ q_i + b_{i,j}^{(k)} \end{pmatrix} - \begin{pmatrix} p_i \\ q_i \end{pmatrix}$$

The feature relating requirement is satisfied if we can find a triplet (x, y, θ) such that

$$(a_{i,j}^{(k+1)})^2 + (b_{i,j}^{(k+1)})^2 \le s_i^2 \quad , \qquad i = 1,2,\ldots,n, \quad j = 1,2$$

$$\text{where} \quad s_i = \begin{cases} (t_f + \phi_i - \phi_0)/2 & \text{, under MMC} \\ t_f/2 & \text{, under RFS} \end{cases}$$

This latter inequality can be rewritten as a general constraint $F_{i,j}(x, y, \theta)$ given by

$$F_{i,j}(x,y,\theta) = [(p_i + a_{i,j}^{(k)})\cos\theta - (q_i + b_{i,j}^{(k)})\sin\theta + x - p_i]^2$$
$$+ [(p_i + a_{i,j}^{(k)})\sin\theta + (q_i + b_{i,j}^{(k)})\cos\theta + y - q_i]^2 - s_i^2 \le 0$$
$$i = 1,2,\ldots,n \qquad j = 1,2 \qquad k \in \{0,1,2,\ldots\ldots\}$$

Each constraint $F_{i,j}(x,y,\theta) \le 0$ defines a region in X, Y, Θ space. The feature relating requirement is satisfied if the intersection of such regions for i=1,2,...n is not null. This problem is conveniently solved as an optimization problem. The strategy is to minimize (over x, y, θ) the maximum surplus associated with any constraint as follows :

Minimize $f = z$ subject to ,

$$g_{i,j}(u) = g_{i,j}(x,y,\theta,z) = F_{i,j}(x,y,\theta) - z \le 0 \qquad i = 1,2,\ldots,n \quad j = 1,2$$

Failure to find a feasible triplet (x, y, θ) of small displacements means that the feature relating requirement is violated. A pattern of holes is acceptable if size, pattern locating and feature relating constraints are all satisfied.

The analysis for Variants 2 and 3 follows a similar approach and yields constraints corrected with the supression of appropriate degrees of fredom associated with the FRTZF. The Generalized Reduced Gradient method is used to solve the resulting non linear optimization problems.

3. Simulation Results

The numerical values selected for the parameters of the simulation were chosen to emphasize FRTZF violations. Hole size variability σ_ϕ was selected from a range of experimentally observed values. The ratio of hole size design tolerance t_ϕ to hole size variability σ_ϕ was deliberately set at $t_\phi / \sigma_\phi = 12$ to minimize part rejection due to out of bound holes. Pattern tolerance t_p was made sufficiently large in relation to hole position error to also minimize part rejections due to pattern violations since this component can be computed analytically. Finally, the ratio of feature tolerance t_f to hole position variability σ_{tool} was deliberately varied to observe variations in yield across a useful range. With t_f=0.15 mm, σ_{tool} was set at the following levels: 0.025, 0.028, 0.032, 0.0375, 0.0425 mm, and these levels were selected from representative manufacturing process capability values reported in the literature.

Figure 2 shows a plot of simulation results for a 32 hole rectangular pattern with all three variants specified with the MMC (Maximum Material Condition) modifier present for both hole-to-hole and pattern position tolerance. As anticipated, when manufacturing hole positioning error σ_{tool} increases, the ratio $t_f / R_{tool} = t_f / 6\sigma_{tool}$ decreases, and the yield of all three variants decreases as both graphs show increased percentage defects. The significant differences in yield among the three variants offer a quantitative assessment of the effect of loss of degrees of freedom associated with the movement of the FRTZF within the boundaries of the PLTZF.

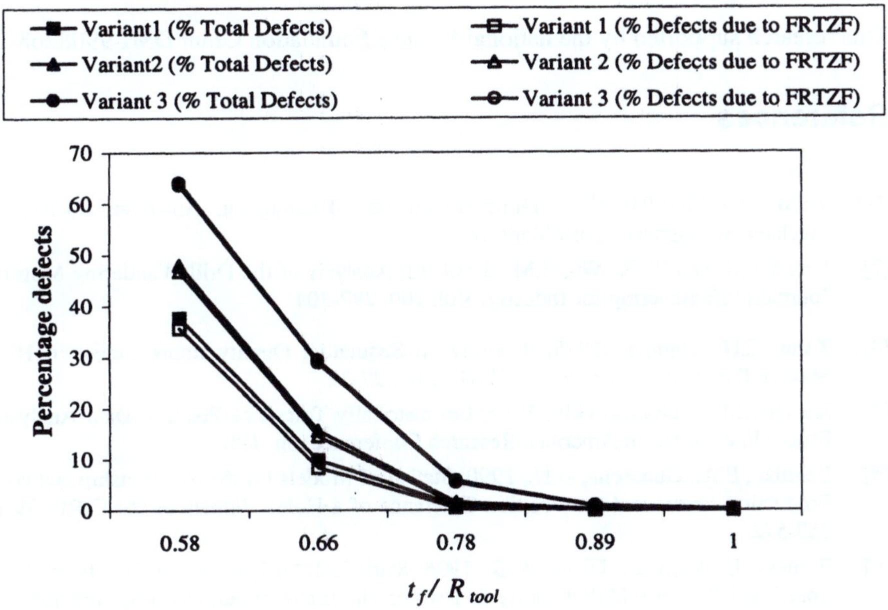

Figure 2: Producibility of Variants 1-3 simulated on a 32 hole rectangular pattern with all tolerances quoted at Maximum Material Condition.

The difference in yield between Variants 1 and 2 is directly related to the loss of the rotational degree of freedom of the FRTZF of Variant 2, whereas the difference in yield between Variants 2 and 3 is directly related to the loss of a translational and rotational degrees of freedom inherent in Variant 3. Yield differences are also seen to increase as hole positioning errors become worse, since the possibility of finding a feasible position of the FRTZF becomes remote.

4. Conclusions

A simulation of the producibility of composite position tolerancing variants was carried out to show that the loss of degrees of freedom associated with the FRTZF of Variants 2 and 3 have a significant impact on the yield of a hypothetical production operation. Additional simulations carried out in this study show the significant advantage conferred by the use of the Maximum Material Condition Modifier and the large influence of the number of holes in the pattern on the yield of the operation.

Acknowledgment

This research supported by the national Science Foundation Grant DMI-9908268

References

[1] ASME Y14.5-1994, 1995, Dimensioning and Tolerancing, American Society of Mechanical Engineers, New York, NY.

[2] Lee, S.J., Eman, K.F., Wu, S.M., 1989, An Analysis of the Drill Wandering Motion, Journal of Engineering for Industry, Vol. 109: 297-304.

[3] Kahn, C.H., Ham, I., 1975, A Study on Sequential Quality Improvement in Hole Making Processes, Annals of the CIRP, 24/1: 27-32.

[4] Heaphy, M.S., Gruska, G.F., 1983, Geometrically Tolerance Position Data Analysis, Proceedings of North American Research Conference, pp. 1-5.

[5] Lehtihet, E.A., Gunasena, U.N., 1990, Statistical Models for the Relationship Between Production Errors and the Position Tolerance of a Hole, Annals of the CIRP, 39/1: 569-572.

[6] Furness, J., Wu, C.L., Ulsoy, A.G., 1996, Statistical Analysis of the Effects of Feed, Speed and Wear on Hole Quality in Drilling, Journal of Manufacturing Science and Engineering, Vol. 118:367-375.

[7] O'Leary, J.R., 1981, A Computer Simulation of True Position Pattern Gauge, ASME Technical Paper No. 81-DE-9: 1-9.

[8] Lehtihet, E.A., Gunasena, N.U., 1991, On the Composite Position Tolerance for Patterns of Holes, Annals of the CIRP, 40/1: 495-498.

[9] McCann, M.J., 1992, An Algorithm for Locating Feature Patterns for the Verification of Composite Position Tolerancing, Transactions of the ASME, Vol. 114: 626-632.

Tolerancing for Enhanced Quality and Optimum Cost

S. C. Diplaris, M. M. Sfantsikopoulos

Department of Mechanical Engineering, National Technical University of Athens, Heroon Polytechniou 9, 157 73, Zografou (Athens), Greece, stdiplar@central.ntua.gr, 1msfntua@central.ntua.gr

Abstract: The two component clearance (or interference) cylindrical fits predominate in mechanical engineering. Industrial data, experimental research and / or theoretical analysis establish for these fits upper and lower clearance limits. In accordance with the Taguchi concept any other clearance than an ideal one will contribute to lower product quality. Closer dimensional tolerances that are required in order to pursue nominal clearances impose, nevertheless, higher manufacturing costs. The paper presents a new methodology for Cost – and – Quality optimum tolerancing for both the assembly fit dimensions that takes into consideration not only the relevant manufacturing parameters and the reject costs but also enhanced quality requirements as those represented by the Taguchi approach.

1. Introduction

Most mechanical products consist of assemblies of individual parts. There are generally three fundamental categories of fits: cylindrical, longitudinal, multiple hole fits. The two component - bore and shaft - cylindrical fit is the one that predominates in mechanical engineering. Depending on operational requirements, a particular assembly may have a clearance or an interference cylindrical fit. A clearance (or interference), on the other hand, as the result of the combination of two nominal dimensions apparently has also a tolerance zone in the same way as each of the contributing dimensions have their own tolerance zones. Industrial experience, experimental research and theoretical analysis establish for clearance fits (e.g. hydrodynamic journal bearing) upper and lower tolerance limits, for the bore and the shaft diameters. Taking into account that dimensional tolerances follow a more-or-less normal distribution [1], it is to be acknowledged that a dimensional clearance will also possess a normal distribution. The ideal clearance value is consequently represented by the mean value of the respective normal curve. This mean value of clearance, is the clearance that will provide for the *nominal* functional characteristics of the considered assembly (e.g. ideal lubrication of the bearing). In accordance with the Taguchi approach [2], any other clearance value - although lying within prescribed and functionally permissible limits- will contribute to a lower product quality and to a related with it customer cost. Closer dimensional tolerances that are required in order to pursue nominal clearances impose, nevertheless, higher production costs. An optimum cost- and – quality tolerance assignment for both the assembly fit dimensions is therefore always needed. A methodology that faces this problem through the use of the Cost – Tolerance function and the Taguchi Loss Function concept is presented in the following sections in conjunction with the discussion of a relevant case-study.

2. Theoretical Analysis

2.1 Total cost and quality function

The nominal value X_0 of the clearance of a cylindrical clearance fit in terms of the nominal values of the contributing dimensions (diameters) D_0 for the bore and d_0 for the shaft is in general equal to,

$$X_0 = D_0 - d_0 \tag{1}$$

If $\pm t$, $\pm t_D$, $\pm t_d$ are the corresponding to X_0, D_0, d_0 dimensional tolerances, it may then be written,

$$t = t_D + t_d, \quad X = X_0 \pm t = (D_0 - d_0) \pm (t_D + t_d) \tag{2}$$

where X is the actual clearance value. Equations (2) correspond to the worst case approach. For a statistical approach these equations are replaced by,

$$X = X_0 \pm t \quad , t = (t_D^2 + t_d^2)^{1/2} \tag{3}$$

The cost for machining both diameters D_0, d_0 to the t_D, t_d tolerances is equal to,

$$C_M = C_{RF} [\, C_D(t_D) + C_d(t_d)] \tag{4}$$

where $C_D(t_D)$, $C_d(t_d)$ are the respective to bore and shaft relative cost-tolerance functions, [3, 4, 5], and C_{RF} the reference machining cost (i.e. the cost for rough machining without any specific accuracy requirements, taken the same for bore and shaft). To this cost the Taguchi's quality cost has now be added, if a cost and quality optimum assembly is to be produced. The latter is, [2],

$$C_q = A\frac{t^2}{t_0^2}, \qquad t \leq t_0 \tag{5}$$

where A is the rejection (or rework) cost of an out-of-tolerance assembly, t is the actual and t_0 the designated tolerance of the considered clearance. From equations (4) and (5) the *total accuracy* cost of the bore and shaft assembly may now be formulated, as,

$$C_T = w_m.C_m + w_q.C_q = w_m C_{RF} [C_D(t_D) + C_d(t_d)] + w_q.A.t^2/t^2_0 =$$

$$= C_{RF} \{ w_m [C_D(t_D) + C_d(t_d)] + w_q(A/C_{RF}).t^2/t_0^2\}$$

or

$$C_T^R = \frac{C_T}{C_{RF}} = w_m\,[C_D(t_D) + C_d(t_d)] + w_q(A/C_{RF}).t^2/t_0^2\,,\; w_m + w_q = 1 \tag{6}$$

where C_T^R is the *relative total accuracy* cost and w_m , w_q are weighting factors which create a suitable each time balance between the machining and quality costs depending each time on the requirements of the particular application. The assignment of the machining tolerances for the bore and the shaft which will result to a quality and cost optimum tolerance zone for the *assembly clearance* can be consequently obtained through the minimisation of the total relative cost function as this is given by the above equation (6).

2.2 Estimating the cost A

An unsuccessful clearance fit can be caused either because of an out-of-tolerance shaft or because an out-of-tolerance bore or, perhaps, because both shaft and bore are out of their specified tolerance zones. This last case, as most improbable, for the sake of this analysis will not be taken here into account. Let now for n failed fits, n_1 shafts and n_2 bores are responsible, i.e. $n = n_1 + n_2$. Due to their normal distribution, 50% of the above n_1 , n_2 quantities will obviously be overdimensioned and 50% of the same will be underdimensioned. Overdimensioned shafts and underdimensioned bores can be considered, at least in principle, as repairable, whereas underdimensioned shafts and overdimensioned bores constitute a total loss.
Rejected underdimensioned shafts and overdimensioned bores will cost,

$$C_d^{RJ} = C_{RF}\, C_d\,(t_d) + C_{db}, \qquad C_D^{RJ} = C_{RF}\, C_D\,(t_D) + C_{Db} \tag{7}$$

where C_{db}, C_{Db} are the costs of the blanks.
Overdimensioned shafts and underdimensioned bores, on the other hand, it is reasonable to be assumed that will be reworked with costs C_d^{RW} and C_D^{RW} which will certainly be much lower than the costs of their initial machining and equal to,

$$C_d^{RW} = C_{dm}(t_d) \text{ and } C_D^{RW} = C_{Dn}(t_D) \tag{8}$$

where $C_{dm}(t_d)$, $C_{Dn}(t_D)$ are the relative machining costs of the final finishing (m^{th} and n^{th}) operations.
The sum of the above costs C_d^{RJ} , C_D^{RJ} , C_d^{RW} , C_D^{RW} for the n_1 shafts and n_2 bores, constitutes the total manufacturing cost that is due to the out-of-tolerance assemblies, i.e.,

$$C_{Total} = n_1.(C_d^{RJ} + C_d^{RW})/2 + n_2.(C_D^{RJ} + C_D^{RW})/2 \tag{9}$$

The cost A per one failed fit will therefore be equal to,

$$A = C_{Total}/n = n_1.(C_d^{RJ} + C_d^{RW})/2n + n_2\,.(C_D^{RJ} + C_D^{RW})/2n \tag{10}$$

Equation (10) may be rewritten in the form,

$$\frac{A}{C_{RF}} = \frac{1}{2}\{k_1[(1+c_1)C_d(t_d)+C_{dm}(t_d)]+k_2[(1+c_2)C_D(t_D)+C_{Dn}(t_D)]\} \tag{11}$$

where

$$c_1 = \frac{C_{db}}{C_{RF}C_d(t_d)}, \quad c_2 = \frac{C_{Db}}{C_{RF}C_D(t_D)}, \quad k_1 = \frac{n_1}{n}, \quad k_2 = \frac{n_2}{n}$$

c_1, c_2 are the relative (with reference to machining cost) material costs and k_1, k_2 are reject percentages.

2.3 Optimum tolerance assignment

Introduction of equation (11) to equation (6) results to the following two relationships for the relative cost function of the cylindrical fit,

<u>Worst Case Approach</u>

$$C_T^R = w_m[C_D(t_D)+C_d(t_d)]+$$
$$+\frac{w_q}{2}\{k_1[(1+c_1)C_d(t_d)+C_{dm}(t_d)]+k_2[(1+c_2)C_D(t_D)+C_{Dn}(t_D)]\}$$
$$(t_D+t_d)^2/t_0^2 \rightarrow \min \tag{12}$$

under the constrains

$$t_d + t_D \leq t_0 \tag{13}$$

and

$$t_d \geq k_d.T_d/2, \quad t_D \geq k_D.T_D/2 \tag{14}$$

where k_d, k_D are the machining process capability indexes ($k_d, k_D \geq 2{,}0$) [6] , T_d , T_D the machining accuracies of the available production facilities.

<u>Statistical Approach</u>

$$C_T^R = w_m[C_D(t_D)+C_d(t_d)]+$$
$$+\frac{w_q}{2}\{k_1[(1+c_1)C_d(t_d)+C_{dm}(t_d)]+k_2[(1+c_2)C_D(t_D)+C_{Dn}(t_D)]\}$$
$$(t_d^2+t_D^2)/(t_{d0}^2+t_{D0}^2) \rightarrow \min \tag{15}$$

under the constrain,

$$t_d^2 + t_D^2 \leq t_0^2 \tag{16}$$

and the constrains (14) above.

The minimization of the cost functions (12) or (15) through the use of a suitable algorithm can in this way furnish the required bore and shaft machining tolerances.

3. Case Study and Discussion

For reasons of comparison the application example of [7, 5] is studied under the presented here approach. For the cylinder-piston assembly of a Diesel engine [D_0 = 50.856 mm, d_0 = 50.800 mm, $X_0 \pm t_0$ = 0.056mm ± 25 μm] the cost-tolerance functions are, [5] :

a. From blanks to finished parts

$$C_d(t_d) = 23{,}33 + \frac{74}{t_d^{0,9}} \qquad C_D(t_D) = 64{,}5 + \frac{65}{t_D^{0,7}}$$

b. For the finishing operations only

$$C_{dm}(t_d) = 10{,}80 + \frac{42{,}50}{t_d^{0,9}} \qquad C_{Dm}(t_D) = 14{,}68 + \frac{330}{t_D^{2,15}}$$

and the problem constraints, [7],

$t_d + t_D \le 25$ μm, (or $t_d^2 + t_D^2 \le 25^2$ μm^2), $t_d \ge 5$ μm, $t_D \ge 8$ μm.

For $k_1 = k_2 = 0.5$, $c_1 \approx c_2 \approx 0$ (negligible material cost) – 0.25 – 1 – 5 – 10 (high material cost) the contribution of the Taguchi quality input to the tolerance optimisation process is shown in Fig. 1 (worst case) and Fig. 2 (statistical). In these figures C_{m0} is the machining cost of accuracy as obtained in [5] i.e. without taking into account the Taguchi Quality Loss Function concept. C_{q0} , on the other hand, is the calculated for this case Quality Loss. As it can be seen, for not very expensive materials ($c_1 = c_2 \le 0{,}25$) significant quality improvements start to be observed for $W_q \ge 0.15$, whereas the required additional machining cost does not exceed a maximum increase of + 13%. Quality improvement starts to be observed for expensive materials ($c_1 = c_2 \ge 5$) for much lower W_q values. In this case the maximum machining cost increase is practically the same + 13%.

The tolerance zones of the *conventional* method, as they were obtained in [5] were respectively t_D(cylinder) = 13 μm, t_d(piston) = 12 μm. Here, depending on W_q value, they vary 8 μm ≤ t_D ≤ 13 μm, 5 μm ≤ t_d ≤ 12 μm.

4. Conclusions

A new methodology for cost – and – quality optimum tolerancing was presented in the paper that it is based on:

(i) The use of conventional cost – tolerance functions and the Taguchi Loss Function concept

(ii) A systematic evaluation of the reject costs

(iii) The introduction of weighting factors for the contribution of the quality and manufacturing cost issues to the final industrial result.

The analysis of this methodology in conjunction with the discussion of the case study have demonstrated that high product quality can in this way be achieved (through the calculated tolerances) without, on the other hand, any considerable increase of the machining cost. The use of the methodology has encouraged its further development and adaptation for process – planning and CAM applications.

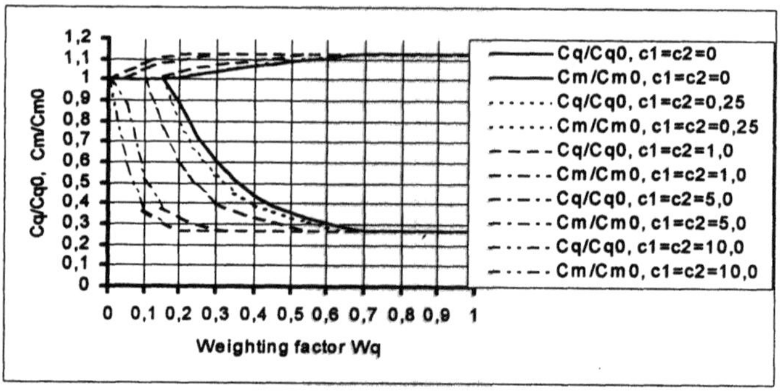

Fig. 1 Worst case

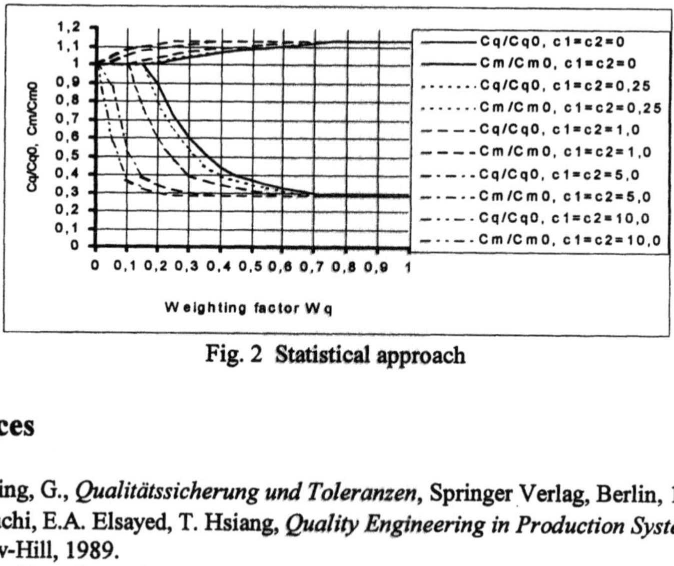

Fig. 2 Statistical approach

References

[1] Kirschling, G., *Qualitätssicherung und Toleranzen*, Springer Verlag, Berlin, 1988.

[2] G. Taguchi, E.A. Elsayed, T. Hsiang, *Quality Engineering in Production Systems*, McGraw-Hill, 1989.

[3] M. .M. Sfantsikopoulos, "A Cost- Tolerance analytical approach for Design and Manufacturing", Int. J. Adv. Manuf. Technology (1990), 5, p.p. 126 - 134.

[4] S.H. Yeo, B.K.A. Ngoi and H. Chen, "A Cost - Tolerance Model for process Sequence Optimisation", Int. J. Adv. Manuf. Technology (1996), 12, pp. 423 - 431.

[5] S.C.Diplaris, M.M. Sfantsikopoulos " Cost - Tolerance Function: A new approach for cost optimum machining accuracy", Int. J. Adv. Manuf. Technology (2000) 16: 32-38

[6] H.-C. Zhang, *Advanced Tolerancing Techniques*, John Wiley & Sons, Inc., New York, 1997

[7] C. Zhang, H.P. Wang, J.K. Li, "Simultaneous optimization of Design and Manufacturing - Tolerance with Process (Machine) Selection", Annals of the CIRP Vol. 41/1/1992.

MIX
Papier aus verantwortungsvollen Quellen
Paper from responsible sources
FSC® C105338

If you have any concerns about our products,
you can contact us on
ProductSafety@springernature.com

In case Publisher is established outside the EU,
the EU authorized representative is:
Springer Nature Customer Service Center GmbH
Europaplatz 3, 69115 Heidelberg, Germany

Printed by Libri Plureos GmbH
in Hamburg, Germany